D1338102

**Springer Proceedings in Mathematics & Statistics**

Volume 289

**Springer Proceedings in Mathematics & Statistics**

This book series features volumes composed of selected contributions from workshops and conferences in all areas of current research in mathematics and statistics, including operation research and optimization. In addition to an overall evaluation of the interest, scientific quality, and timeliness of each proposal at the hands of the publisher, individual contributions are all refereed to the high quality standards of leading journals in the field. Thus, this series provides the research community with well-edited, authoritative reports on developments in the most exciting areas of mathematical and statistical research today.

More information about this series at http://www.springer.com/series/10533

Samuel N. Cohen · István Gyöngy ·
Gonçalo dos Reis · David Siska ·
Łukasz Szpruch
Editors

# Frontiers in Stochastic Analysis - BSDEs, SPDEs and their Applications

Edinburgh, July 2017
Selected, Revised and Extended Contributions

*Editors*
Samuel N. Cohen
Mathematical Institute
University of Oxford
Oxford, UK

István Gyöngy
School of Mathematics
University of Edinburgh
Edinburgh, UK

Gonçalo dos Reis
School of Mathematics
University of Edinburgh
Edinburgh, UK

David Siska
School of Mathematics
University of Edinburgh
Edinburgh, UK

Łukasz Szpruch
School of Mathematics
University of Edinburgh
Edinburgh, UK

ISSN 2194-1009 ISSN 2194-1017 (electronic)
Springer Proceedings in Mathematics & Statistics
ISBN 978-3-030-22284-0 ISBN 978-3-030-22285-7 (eBook)
https://doi.org/10.1007/978-3-030-22285-7

Mathematics Subject Classification (2010): 60H15, 60H30, 60G55, 60A99, 91B16, 91B28



This Springer imprint is published by the registered company Springer Nature Switzerland AG
The registered company address is: Gewerbestrasse 11, 6330 Cham, Switzerland

International Workshop on
BSDEs, SPDEs and their Applications
Edinburgh, 3-7 July 2017

# Preface

It was our pleasure to be among the organizers of the *International workshop on BSDEs, SPDEs, and applications*, held at the University of Edinburgh in July 2017. The workshop brought together more than 200 active researchers in probability theory, for over 150 research talks, in addition to poster presentations and networking events. The meeting also included the *8th World BSDE symposium*.

The papers in this volume give a taste of those areas presented at the meeting, covering a range of actively researched areas. We hope that they act as a stimulus for further research in this exciting subfield of probability theory. We now summarize the key themes of each of the papers in the volume:

The first paper, by Dirk Becherer, Martin Büttner, and Klebert Kentia, considers the monotone stability approach to BSDEs with jumps. This is an approach to studying basic questions of existence and uniqueness of solutions to backward SDEs, by leveraging the result of the "comparison theorem" for these equations. This is made more difficult than in the standard case due to the presence of jumps, which imply that additional requirements on the generator of the BSDE must be imposed. This paper uses this result to provide existence results without a standard Lipschitz continuity condition and then further explores how these equations appear in some applied problems in mathematical finance.

The second paper, by Mireille Bossy and Jean-François Jabir, studies McKean stochastic differential equations, in particular, a framework where the dynamics of a process $Y$ depend on the (conditional) distribution of $Y$ given a related process $X$. The well-posedness of this equation is proven, under appropriate continuity and regularity assumptions.

The third paper, by Philippe Briand and Adrien Richou, studies the uniqueness of solutions to BSDEs with drivers which may grow quadratically, without an assumption of convexity. If the driver and terminal value are assumed to be bounded, the uniqueness of solutions to these equations is well known; however in the unbounded case, the study of these equations is significantly more difficult. This paper studies the case where the terminal value is unbounded and is determined by the path of a forward SDE.

The fourth paper, by Antonella Calzolari and Barbara Torti, studies the question of martingale representation, when a filtration is enlarged by additional information. In particular, a model is studied in which information arrives from two sources—a Brownian motion and the occurrence of a random time. In this setting, they show that while the Brownian motion and the martingale associated with the random time have the predictable representation property in each of their filtrations, the combination of these two sources can introduce the necessity for a third martingale in a representing set (alternatively, the *multiplicity* of the joint filtration may be three).

The fifth paper, by Samuel N. Cohen and Martin Tegnér, considers the pricing of European options in a setting with estimation uncertainty. The paper considers estimating the parameters in a Heston stochastic volatility model for stock prices, along with their statistical uncertainties. It then explores, if the future dynamics of the price are only constrained to lie within the estimated bounds on the parameters, how to find the range of possible prices for a financial option. This is done by means of numerical solutions of BSDEs.

The sixth paper, by Gonçalo dos Reis and Greig Smith, studies a class of transport PDEs which have a representation from a stochastic perspective, in terms of branching processes with regime switching. This is then used to study the convergence of Monte Carlo approximations to these equations, and a comparison with alternative Laplacian–perturbation methods is given.

The seventh paper, by Nicole El Karoui, Caroline Hillairet, and Mohamed Mrad, gives a method of constructing an aggregate consistent utility from a collection of heterogeneous agents. Working in a setting of a financial market and assuming no arbitrage, they consider the marginal utilities of each agent and their corresponding investment preferences, and from these, construct a utility function which gives the same aggregate preferences. This is then applied to studying the yield curve in bond markets.

The eighth paper, by Monique Jeanblanc and Dongli Wu, returns to the theme of enlargement of filtrations, in this case studying how BSDEs vary when additional information is given. This question is then extended to the related question of how an optimal control (in particular the choice of an optimal investment in a financial market) would change under an increase in the information available.

The final paper, by Mauro Rosestolato, focusses on path-dependent stochastic differential equations in Hilbert spaces. This paper in particular focuses on the continuity and (Gâteaux) differentiability of the solution to such an equation with respect to the initial value given and with respect to perturbations of the other coefficients.

Oxford, UK Samuel N. Cohen
Edinburgh, UK István Gyöngy
Edinburgh, UK Gonçalo dos Reis
Edinburgh, UK David Siska
Edinburgh, UK Łukasz Szpruch

# Contents

**On the Monotone Stability Approach to BSDEs with Jumps: Extensions, Concrete Criteria and Examples** . . . . . 1
Dirk Becherer, Martin Büttner and Klebert Kentia

**On the Wellposedness of Some McKean Models with Moderated or Singular Diffusion Coefficient** . . . . . 43
Mireille Bossy and Jean-François Jabir

**On the Uniqueness of Solutions to Quadratic BSDEs with Non-convex Generators** . . . . . 89
Philippe Briand and Adrien Richou

**An Example of Martingale Representation in Progressive Enlargement by an Accessible Random Time** . . . . . 109
Antonella Calzolari and Barbara Torti

**European Option Pricing with Stochastic Volatility Models Under Parameter Uncertainty** . . . . . 123
Samuel N. Cohen and Martin Tegnér

**Construction of an Aggregate Consistent Utility, Without Pareto Optimality. Application to Long-Term Yield Curve Modeling** . . . . . 169
Nicole El Karoui, Caroline Hillairet and Mohamed Mrad

**BSDEs and Enlargement of Filtration** . . . . . 201
Monique Jeanblanc and Dongli Wu

**An Unbiased Itô Type Stochastic Representation for Transport PDEs: A Toy Example** . . . . . 221
Gonçalo dos Reis and Greig Smith

**Path-Dependent SDEs in Hilbert Spaces** . . . . . 261
Mauro Rosestolato

# On the Monotone Stability Approach to BSDEs with Jumps: Extensions, Concrete Criteria and Examples

**Dirk Becherer, Martin Büttner and Klebert Kentia**

**Abstract** We show a concise extension of the monotone stability approach to backward stochastic differential equations (BSDEs) that are jointly driven by a Brownian motion and a random measure of jumps, which could be of infinite activity with a non-deterministic and time-inhomogeneous compensator. The BSDE generator function can be non-convex and needs not satisfy global Lipschitz conditions in the jump integrand. We contribute concrete sufficient criteria, that are easy to verify, for results on existence and uniqueness of bounded solutions to BSDEs with jumps, and on comparison and a-priori $L^{\infty}$-bounds. Several examples and counter examples are discussed to shed light on the scope and applicability of different assumptions, and we provide an overview of major applications in finance and optimal control.

**Keywords** Backward stochastic differential equations · Random measures · Monotone stability · Lévy processes · Step processes · Utility maximization · Entropic risk measure · Good deal valuation bounds

**MSC2010** 60G57 · 60H20 · 93E20 · 60G51 · 91G80

---

D. Becherer (✉)
Institut für Mathematik, Humboldt-Universität zu Berlin,
Unter den Linden 6, 10099 Berlin, Germany
e-mail: becherer@math.hu-berlin.de

M. Büttner
Hochschule für Technik und Wirtschaft,
Treskowallee 8, 10313 Berlin, Germany
e-mail: buettnm@htw-berlin.de

K. Kentia
Institut für Mathematik, Goethe-Universität Frankfurt,
60054 Frankfurt am Main, Germany
e-mail: kentia@aims.ac.za



S. N. Cohen et al. (eds.), *Frontiers in Stochastic Analysis - BSDEs, SPDEs and their Applications*, Springer Proceedings in Mathematics & Statistics 289,
https://doi.org/10.1007/978-3-030-22285-7_1

# 1 Introduction

We study bounded solutions $(Y, Z, U)$ to backward stochastic differential equations with jumps

$$Y_t = \xi + \int_t^T f_s(Y_{s-}, Z_s, U_s)\,\mathrm{d}s - \int_t^T Z_s\,\mathrm{d}B_s - \int_t^T\int_E U_s(e)\,\widetilde{\mu}(\mathrm{d}s, \mathrm{d}e)\,,$$

which are jointly driven by a Brownian motion $B$ and a compensated random measure $\widetilde{\mu} = \mu - \nu^{\mathbb{P}}$ of some integer-valued random measure $\mu$ on a probability space $(\Omega, \mathcal{F}, \mathbb{P})$. This is an extension of the classical BSDE theory on Wiener space towards BSDEs which involve jumps (JBSDEs), that are driven by the compensated random measure $\widetilde{\mu}$, and do evolve on non-Brownian filtrations. Such JBSDEs do involve an additional stochastic integral with respect to the compensated jump measure $\widetilde{\mu}$ whose integrand $U$, differently from $Z$, typically takes values in an infinite dimensional function space instead of an Euclidean space.

Comparison theorems for BSDEs with jumps require more delicate technical conditions than in the Brownian case, see [4, 15, 54]. The starting point for our article will be a slight generalization of the seminal $(\mathbf{A}_\gamma)$-condition for comparison due to [54]. Our first contribution are extensions of comparison, existence and uniqueness results for bounded solutions of JBSDEs to the case of infinite jump activity for a family (2.6) of generators, that do not need to be Lipschitz in the $U$-argument. This shows how the monotone stability approach to BSDEs with jumps, pioneered by [44] for one particular generator, permits for a concise proof in a setting, that may be of particular appeal in a pure jump case without a Brownian motion, see Corollary 4.12. While the strong approximation step for this approach is usually laborious, we present a compact proof with a $\mathcal{S}^1$-closedness argument and more generality of the generator in the $U$-argument for infinite activity of jumps. To be useful towards applications, our second contribution are sufficient concrete criteria for comparison and wellposedness that are comparably easy to verify in actual examples, because they are formulated in terms of concrete properties for generator functions $f$ from a given family (2.6) w.r.t. to basically Euclidean arguments, instead of assuming inequalities to hold for rather abstract random processes or fields. This is the main thrust for the sufficient conditions of the comparison results in Sect. 3 (see Theorem 3.9 and Proposition 3.11, compared to Proposition 3.1 or the result by [54] and respective enhancements [38, 52, 57]) and of the wellposedness Theorem 4.13 (in comparison to Theorem 4.11, whose conditions are more general but more abstract). A third contribution are the many examples and applications which illustrate the scope and applicability of our results and of the, often technical, assumptions that are needed for JBSDE results in the literature. Indeed, the range of the imposed combinations of several technical assumptions is often not immediately clear. We believe that more discussion of examples and counter examples may help to shed light on the scope and the differences of some assumptions prevailing in the literature, and might also caution against possible pitfalls.

The approach in this paper can be described in more detail as follows: The comparison results will provide basic a-priori estimates on the $L^\infty$-norm for the $Y$-component of the JBSDE solution. This step enables a quick intermediate result on existence and uniqueness for JBSDEs with finite jump activity. To advance from here to infinite activity, we approximate the generator $f$ by a monotone sequence of generators for which solutions do exist, extending the monotone stability approach from [37] and (for a particular JBSDE) [44]. For the present paper, the compensator $\nu(\omega, \mathrm{d}t, \mathrm{d}e)$ of $\mu(\omega, \mathrm{d}t, \mathrm{d}e)$ can be stochastic and does not need to be a product measure like $\lambda(\mathrm{d}e) \otimes \mathrm{d}t$, as it would be natural e.g. in a Lévy-process setting, but it is allowed to be inhomogeneous in that it can vary predictably with $(\omega, t)$. In this sense, $\nu$ is only assumed to be absolutely continuous to some reference product measure $\lambda \otimes \mathrm{d}t$ with $\lambda$ being $\sigma$-finite, see Eq. (2.1). Such appears useful, but requires some care in the specification of generator properties in Sect. 2. For the filtration we assume that $\widetilde{\mu}$ jointly with $B$ (or alone) satisfies the property of weak predictable representation for martingales, see (2.2). As explained in Example 2.1, such setup permits for a range of stochastic dependencies between $B$ and $\widetilde{\mu}$, which appear useful for modeling of applications, and encompasses many interesting driving noises for jumps in BSDEs; This includes Lévy processes, Poisson random measures, marked point processes, (semi-)Markov chains or much more general step processes, connecting to a wide range of literature, e.g. [3, 14, 15, 17, 25–27].

The literature on BSDE started with the classical study [50] of square integrable solutions to BSDEs driven solely by Brownian motion $B$ under global Lipschitz assumptions. One important extension concerns generators $f$ which are non-Lipschitz but have quadratic growth in $Z$, for which [37] derived bounded solutions by pioneering a monotone stability approach, and [56] by a fixed point approach. Square integrable solutions under global Lipschitz conditions for BSDEs with jumps from a Poisson random measures are first studied by [4, 55]. There is a lot of development in JBSDE theory recently. See for instance [2, 22, 38, 39, 49] for results under global Lipschitz conditions on the generator with respect to on $(Z, U)$. In the context of non-Lipschitz generators that are quadratic (also in $Z$, with exponential growth in $U$), JBSDEs have been studied to our knowledge at first by [44] using a monotone stability approach for a specific generator that is related to exponential utility, by [23] using a quadratic-exponential semimartingale approach from [6], and by [40] or [35] with again different approaches, relying on duality methods or, respectively, the fixed-point idea of [56] for quadratic BSDEs. For extensive surveys of the active literature with more references, let us refer to [38, 57], who contribute results on $L^p$-solutions for generators, being monotone in the $Y$-component, that are very general in many aspects. Their assumptions on the filtrations or generator's dependence on $(Y, Z)$ are for instance more general than ours. But the present paper still contributes on other aspects, noted above. For instance, [57] assumes finite activity of jumps and a Lipschitz continuity in $U$. More relations to some other related literature are being explained in many examples throughout the paper, see e.g. in Sect. 5. Moreover, it is fair to say that results in the JBSDE literature often involve combinations of many technical assumptions; To understand the scope, applicability and differences of those assumptions, it appears helpful to discuss concrete examples and applications.

The paper is organized as follows. Section 2 introduces the setting and mathematical background. In Sects. 3, 4, we prove comparison results and show existence as well as uniqueness for bounded solutions to JBSDEs, both for finite and infinite activity of jumps. Last but not least, Sect. 5 surveys key applications of JBSDEs in finance. We discuss several examples to shed light on the scope of the results and of the underlying technical assumptions, and discuss connections to the literature.

## 2 Preliminaries

This section presents the technical framework, sets notations and discusses key conditions. First we recall essential facts on stochastic integration w.r.t. random measures and on bounded solutions for Backward SDEs which are driven jointly by Brownian motions and a compensated random measure. For notions from stochastic analysis not explained here we refer to [28, 31].

Inequalities between measurable functions are understood almost everywhere w.r.t. an appropriate reference measure, typically $\mathbb{P}$ or $\mathbb{P} \otimes \mathrm{d}t$. Let $T < \infty$ be a finite time horizon and $(\Omega, \mathcal{F}, (\mathcal{F}_t)_{0 \le t \le T}, \mathbb{P})$ a filtered probability space with a filtration $\mathbb{F} = (\mathcal{F}_t)_{0 \le t \le T}$ satisfying the usual conditions of right continuity and completeness, assuming $\mathcal{F}_T = \mathcal{F}$ and $\mathcal{F}_0$ being trivial (under $\mathbb{P}$); Thus we can and do take all semimartingales to have right continuous paths with left limits, so-called càdlàg paths. Expectations (under $\mathbb{P}$) are denoted by $\mathbb{E} = \mathbb{E}_{\mathbb{P}}$. We will denote by $\mathbf{A}^T$ the transpose of a matrix $\mathbf{A}$ and simply write $\boldsymbol{x}\boldsymbol{y} := \boldsymbol{x}^T \boldsymbol{y}$ for the scalar product for two vectors $\boldsymbol{x}, \boldsymbol{y}$ of same dimensionality. Let $H$ be a separable Hilbert space and denote by $\mathcal{B}(E)$ the Borel $\sigma$-field of $E := H \backslash \{0\}$, e.g. $H = \mathbb{R}^l, l \in \mathbb{N}$, or $H = \ell^2 \subset \mathbb{R}^{\mathbb{N}}$. Then $(E, \mathcal{B}(E))$ is a standard Borel space. In addition, let $B$ be a $d$-dimensional Brownian motion. Stochastic integrals of a vector valued predictable process $Z$ w.r.t. a semimartingale $X$, e.g. $X = B$, of the same dimensionality are scalar valued semimartingales starting at zero and denoted by $\int_{(0,t]} Z \mathrm{d}X = \int_0^t Z \mathrm{d}X = Z \bullet X_t$ for $t \in [0, T]$. The *predictable* $\sigma$-field on $\Omega \times [0, T]$ (w.r.t. $(\mathcal{F}_t)_{0 \le t \le T}$) is denoted by $\mathcal{P}$ and $\widetilde{\mathcal{P}} := \mathcal{P} \otimes \mathcal{B}(E)$ is the respective $\sigma$-field on $\widetilde{\Omega} := \Omega \times [0, T] \times E$.

Let $\mu$ be an integer-valued random measure with compensator $\nu = \nu^{\mathbb{P}}$ (under $\mathbb{P}$) which is taken to be absolutely continuous to $\lambda \otimes \mathrm{d}t$ for a $\sigma$-finite measure $\lambda$ on $(E, \mathcal{B}(E))$ satisfying $\int_E 1 \wedge |e|^2 \lambda(\mathrm{d}e) < \infty$ with some $\widetilde{\mathcal{P}}$-measurable, bounded and non-negative density $\zeta$, such that

$$\nu(\mathrm{d}t, \mathrm{d}e) = \zeta(t, e)\, \lambda(\mathrm{d}e)\, \mathrm{d}t = \zeta_t\, \mathrm{d}\lambda\, \mathrm{d}t, \tag{2.1}$$

with $0 \le \zeta(t, e) \le c_\nu$ $\mathbb{P} \otimes \lambda \otimes \mathrm{d}t$-a.e. for some constant $c_\nu > 0$. Note that $L^2(\lambda)$ and $L^2(\zeta_t \mathrm{d}\lambda)$ are separable Hilbert spaces since $\lambda$ (and $\lambda_t := \zeta_t\, \mathrm{d}\lambda$) is $\sigma$-finite and $\mathcal{B}(E)$ is finitely generated. Since the density $\zeta$ can vary with $(\omega, t)$, the compensator

$\nu$ can be time-inhomogeneous and stochastic. Such permits for a richer dependence structure for $(B, \widetilde{\mu})$; For instance, the intensity and distribution of jump heights could vary according to some diffusion process. Yet, it also brings a few technical complications, e.g. function-valued integrand processes $U$ from $\mathcal{L}^2(\widetilde{\mu})$ (as defined below) for the JBSDE need not take values in one given $L^2$-space (for a.e. $(\omega, t)$), like e.g. $L^2(\lambda)$ if $\zeta \equiv 1$, and the specifications of the domain and of the measurability for the generator functions should take account of such.

For stochastic integration w.r.t. $\widetilde{\mu}$ and $B$ we define sets of $\mathbb{R}$-valued processes

$$\mathcal{S}^p := \mathcal{S}^p(\mathbb{P}) := \Big\{Y \text{ càdlàg} \,:\, |Y|_p := \Big\| \sup_{0\le t\le T} |Y_t| \Big\|_{L^p(\mathbb{P})} < \infty\Big\} \quad \text{for } p \in [1, \infty],$$

$$\mathcal{L}^2(\widetilde{\mu}) := \Big\{U \ \widetilde{\mathcal{P}}\text{-measurable} \,:\, \|U\|^2_{\mathcal{L}^2(\widetilde{\mu})} := \mathbb{E}\Big(\int_0^T\int_E |U_s(e)|^2\, \nu(\mathrm{d}s, \mathrm{d}e)\Big) < \infty\Big\},$$

and the set of $\mathbb{R}^d$-valued processes

$$\mathcal{L}^2(B) := \Big\{\theta \ \mathcal{P}\text{-measurable} \,:\, \|\theta\|^2_{\mathcal{L}^2(B)} := \mathbb{E}\Big(\int_0^T \|\theta_s\|^2\, \mathrm{d}s\Big) < \infty\Big\},$$

where $\widetilde{\mu} = \widetilde{\mu}^{\mathbb{P}} = \mu - \nu$ denotes the compensated measure of $\mu$ (under $\mathbb{P}$). Recall that for any predictable function $U$, $\mathbb{E}(|U| * \mu_T) = \mathbb{E}(|U| * \nu_T)$ by the definition of a compensator. If $(|U|^2 * \mu)^{1/2}$ is locally integrable, then $U$ is integrable w.r.t. $\widetilde{\mu}$, and $U * \widetilde{\mu}$ is defined as the purely discontinuous local martingale with jump process $\big(\int_E U_t(e)\, \mu(\{t\}, \mathrm{d}e)\big)_t$ by [31, Definition II.1.27] noting that $\nu$ is absolutely continuous to $\lambda \otimes \mathrm{d}t$. For $Z \in \mathcal{L}^2(B)$ and $U \in \mathcal{L}^2(\widetilde{\mu})$ we recall that $Z \bullet B$ and $U * \widetilde{\mu} = (U * \widetilde{\mu}_t)_{0\le t\le T}$ with $U * \widetilde{\mu}_t = \int_0^t \int_E U_s(e)\, \widetilde{\mu}(\mathrm{d}s, \mathrm{d}e)$ are square integrable martingales by [31, Theorem II.1.33]. For $Z, Z' \in \mathcal{L}^2(B)$ and $U, U' \in \mathcal{L}^2(\widetilde{\mu})$ we have for the predictable quadratic covariations that $\langle U * \widetilde{\mu}, U' * \widetilde{\mu}\rangle_t = \int_0^t\int_E U_s(e) U'_s(e)\, \nu(\mathrm{d}s, \mathrm{d}e)$ by [31, Theorem II.1.33], $\langle \int Z\, \mathrm{d}B, \int Z'\, \mathrm{d}B\rangle_t = \int_0^t Z_s^T Z'_s\, \mathrm{d}s$ and $\langle \int Z\, \mathrm{d}B, U * \widetilde{\mu}\rangle_t = 0$ by [31, Theorem I.4.2].

We denote the space of square integrable martingales by $\mathcal{M}^2$ and its norm by $\|\cdot\|_{\mathcal{M}^2}$ with $\|M\|_{\mathcal{M}^2} = \mathbb{E}(M_T^2)^{1/2}$. We recall [28, Theorem 10.9.4] that the subspace of BMO($\mathbb{P}$)-martingales BMO($\mathbb{P}$) contains any square integrable martingale $M$ with uniformly bounded jumps and bounded conditional expectations for increments of the quadratic variation process:

$$\sup_{0\le t\le T} \Big\|\mathbb{E}\big((M_T - M_t)^2 \,|\, \mathcal{F}_t\big)\Big\|_{L^\infty(\mathbb{P})} = \sup_{0\le t\le T} \big\|\mathbb{E}\big(\langle M\rangle_T - \langle M\rangle_t \,|\, \mathcal{F}_t\big)\big\|_{L^\infty(\mathbb{P})} \le \text{const} < \infty.$$

We will assume that the continuous martingale $B$ and the compensated measure $\widetilde{\mu}$ of an integer-valued random measure $\mu$ (or $\widetilde{\mu}$ alone, see Example 2.1 and Corollary 4.12 with trivial $B = 0$) jointly have the weak predictable representation property (weak PRP) w.r.t. the filtration $(\mathcal{F}_t)_{0\le t\le T}$, in the sense that every square integrable martingale $M$ has a (unique) representation, i.e.

$$\text{for all } M \in \mathcal{M}^2 \text{ there exists } Z, U \text{ such that } M = M_0 + \int Z \,\mathrm{d}B + U * \widetilde{\mu}, \quad (2.2)$$

with (unique) $Z \in \mathcal{L}^2(B)$ and $U \in \mathcal{L}^2(\widetilde{\mu})$. Let us note that in the literature [31, III.§4c] or [28, XIII.§2] the weak representation property is defined as a decomposition like (2.2) for any local martingale $M$ with integrands $Z, U$ being integrable in the sense of local martingales. Such clearly implies our formulation above. Indeed, for a (locally) square integrable martingale $M$ in such a decomposition both integrands must be at least locally square integrable and $\langle M\rangle = \int |Z|^2 \,\mathrm{d}t + |U|^2 * \nu$ by strong orthogonality of the stochastic integrals. Then $E[\langle M\rangle_T] < \infty$ implies that $Z, U$ are in the respective $\mathcal{L}^2$-spaces. We exemplify how (2.2) connects with a wide literature.

*Example 2.1* The weak predictable representation property (2.2) holds in the cases below. Cases 1.–4. are well known from classical theory [28], see [7, Example 2.1] for details.

1. Let $X$ be a Lévy process with $X_0 = 0$ and predictable characteristics $(\alpha, \beta, \nu)$ (under $\mathbb{P}$). Then the continuous martingale part $X^c$ (rescaled to a Brownian motion if $\beta \neq 0$, or being trivial if $\beta = 0$) and the compensated jump measure $\widetilde{\mu}^X = \mu^X - \nu$ of $X$ have the weak PRP w.r.t. the usual filtration $\mathbb{F}^X$ generated by $X$. An example for a Lévy process of infinite activity is the Gamma process. One can add that weak PRP even holds in the sense of Theorem III.4.34 from [31] for the more general class of PII-processes with independent increments. This class encompasses the more familiar Lévy processes without requiring time-homogeneity or stochastic continuity.
2. Assume that $B$ and $\widetilde{\mu}$ satisfy (2.2) under $\mathbb{P}$. Let $\mathbb{P}'$ be an equivalent probability measure with density process $Z$. Then the Brownian motion $B' := B - \int (Z_-)^{-1}\,\mathrm{d}\langle Z, B\rangle$ and $\widetilde{\mu}' := \mu - \nu^{\mathbb{P}'}$ have the weak PRP (2.2) also w.r.t. $\mathbb{P}'$ under the same filtration. This offers plenty of scope to construct examples where $W$ and $\widetilde{\mu}$ are not independent, based on other examples.
3. Let $B$ be a Brownian motion independent of a step process $X$ (in the sense of [28, Chap. 11]). Then $B$ and $\widetilde{\mu}$, the compensated measure of the jump measure $\mu^X$ of $X$, have the weak PRP w.r.t. the usual filtration generated by $X$ and $B$. An example for a step process is a multivariate (non-explosive) point process, as appearing in [17].
4. A (semi-)Markov chain $X$, possibly time-inhomogeneous, is a step process. Thus weak PRP (2.2) holds for a filtration generated by a Brownian motion and an independent Markov chain, relating later results to literature [3, 15, 16] on BSDEs driven by compensated random measures of the respective pure-jump (semi-)Markov processes. Markov chains $X$ on countable state spaces can be chosen [15] to take values in the set of unit vectors $\{e_i : i \in \mathbb{N}\}$ of the sequence space $\ell^2 \subset \mathbb{R}^{\mathbb{N}}$, with jumps $\Delta X$ taking values $e_i - e_j$, $i, j \in \mathbb{N}$.

5. The pure jump martingale $U * \widetilde{\mu}$ (for $U \in \mathcal{L}^2(\widetilde{\mu})$) may be written as a series of mutually orthogonal martingales. More precisely, assume that the compensator coincides with the product measure $\lambda \otimes dt$, i.e. $\zeta = 1$. Let $(u^n)_{n\in\mathbb{N}}$ be an orthonormal basis (ONB) of the separable Hilbert space $L^2(\lambda)$ with scalar product $\langle u, v\rangle := \int_E u(e)v(e)\,\lambda(de)$. Let $U_t = \sum_{n\in\mathbb{N}}\langle U_t, u^n\rangle u^n$ be the basis expansion of $U_t$ for $U \in \mathcal{L}^2(\widetilde{\mu})$, $t \le T$. Then it holds (in $\mathcal{M}^2$)

$$U * \widetilde{\mu} = \sum_{n\in\mathbb{N}} \int_0^\cdot \langle U_t, u^n\rangle \int_E u^n(e)\,\widetilde{\mu}(dt, de) =: \sum_{n\in\mathbb{N}} \int_0^\cdot \alpha_t^n \, dL_t^n = \sum_{n\in\mathbb{N}} \alpha^n \bullet L^n, \tag{2.3}$$

for $\alpha_t^n := \langle U_t, u^n\rangle$ and $L^n := u^n * \widetilde{\mu}$. Indeed, setting $F_t^n := \sum_{k=1}^n \langle U_t, u^k\rangle u^k = \sum_{k=1}^n \alpha_t^k u^k$ one sees that $\| \sum_{k=1}^\infty |\alpha^k|^2 \|_{L^1(\mathbb{P}\otimes dt)} \le \|U\|^2_{\mathcal{L}^2(\widetilde{\mu})} < \infty$. By dominated convergence one obtains as $n \to \infty$

$$\|F^n - U\|^2_{\mathcal{L}^2(\widetilde{\mu})} = \mathbb{E}\Big(\int_0^T\int_E |F_t^n(e) - U_t(e)|^2 \lambda(de)\,dt\Big) = \mathbb{E}\Big(\int_0^T \sum_{k=n+1}^\infty |\alpha_t^k|^2\,dt\Big) \to 0.$$

Isometry implies that the stochastic integrals $F^n * \widetilde{\mu}$ converge to $U * \widetilde{\mu}$ in $\mathcal{M}^2$, proving (2.3).
In particular, we see how the PRP (2.2) w.r.t. a random measure can be rewritten as series of ordinary stochastic integrals w.r.t. scalar-valued strongly orthogonal martingales $L^n$, which are in fact Lévy processes with deterministic characteristics $(0, 0, \int u^n(e)\,\lambda(de))$. In this sense, the general condition (2.2) links well with results on PRP and BSDEs for Lévy processes in [46, 47] who study a specific Teugels martingale basis consisting of compensated power jump processes for Lévy processes which satisfy exponential moment conditions. For a systematic analysis of related PRP results, comprising general Lévy processes, see [20, 21].
6. Previous arguments could extend to the general case with $\zeta \not\equiv 1$ in (2.1). To this end, suppose $U^n$ to be in $\mathcal{L}^2(\widetilde{\mu})$ such that for all $t \le T$ the sequence $(U_t^n)_{n\in\mathbb{N}}$ is ONB of $L^2(\lambda_t)$ for $d\lambda_t = \zeta_t d\lambda$ with scalar product $\langle u, v\rangle_t := \int_E u(e)v(e)\,\zeta(t, e)\,\lambda(de)$. Analogously to case 5. above, with $\alpha_t^n := \langle U_t, U_t^n\rangle_t$ and $L^n := U^n * \widetilde{\mu}$ one gets equalities of martingales (in $\mathcal{M}^2$)

$$U * \widetilde{\mu} = \sum_{n\in\mathbb{N}} \int_0^\cdot \langle U_t, U_t^n\rangle_t \int_E U_t^n(e)\,\widetilde{\mu}(dt, de) =: \sum_{n\in\mathbb{N}} \alpha^n \bullet L^n.$$

To proceed, we now define a solution of the Backward SDE with jumps to be a triple $(Y, Z, U)$ of processes in the space $\mathcal{S}^p \times \mathcal{L}^2(B) \times \mathcal{L}^2(\widetilde{\mu})$ for a suitable $p \in (1, \infty]$ that satisfies

$$Y_t = \xi + \int_t^T f_s(Y_{s-}, Z_s, U_s)\,ds - \int_t^T Z_s\,dB_s - \int_t^T\int_E U_s(e)\,\widetilde{\mu}(ds, de), \quad 0 \le t \le T, \tag{2.4}$$

for given data $(\xi, f)$, consisting of a $\mathcal{F}_T$-measurable random variable $\xi$ and a generator function $f_t(y, z, u) = f(\omega, t, y, z, u)$. The values $p$ will be specified below in the respective results, although a particular focus will be on bounded BSDE solutions (i.e. $p = \infty$). Because we permit $\nu$ to be time-inhomogeneous with a bounded but possibly non-constant density $\zeta$ in (2.1), it does not hold in general that $U_t$ takes values a.e. in one space $L^2(\lambda)$ for $U \in \mathcal{L}^2(\widetilde{\mu})$. This requires some extra consideration about the domain of definition and measurability of $f$, as the generator function $f$ needs to be defined for $u$-arguments from a suitable domain, which cannot be some fixed $L^2$-space in general (and needs to be larger than $L^2(\lambda)$), as integrability of $u = U_t(\omega, \cdot)$ over $e \in E$ may vary with $(\omega, t)$. On suitable larger domains, one typically may have to admit for $f$ to attain non-finite values. To this end, let us denote by $L^0(\mathcal{B}(E), \lambda)$ the space of all $\mathcal{B}(E)$-measurable functions with the topology of convergence in measure and define

$$|u - u'|_t := \Big( \int_E |u(e) - u'(e)|^2 \, \zeta(t, e) \, \lambda(\mathrm{d}e) \Big)^{\frac{1}{2}}, \tag{2.5}$$

for functions $u, u'$ in $L^0(\mathcal{B}(E), \lambda)$. Terminal conditions $\xi$ for BSDE considered in this paper will be taken to be square integrable $\xi \in L^2(\mathcal{F}_T)$ and often even as bounded $\xi \in L^\infty(\mathcal{F}_T)$. Generator functions $f : \Omega \times [0, T] \times \mathbb{R} \times \mathbb{R}^d \times L^0(\mathcal{B}(E), \lambda) \to \overline{\mathbb{R}}$ are always taken to be $\mathcal{P} \otimes \mathcal{B}(\mathbb{R}^{d+1}) \otimes \mathcal{B}(L^0(\mathcal{B}(E), \lambda))$-measurable. Main Theorems 3.9 and 4.13 are derived for families of generators having the form

$$f_t(y, z, u) := \widehat{f}_t(y, z) + \int_A g_t(y, z, u(e), e) \zeta(t, e) \lambda(de) \quad \text{(where finitely defined)} \tag{2.6}$$

and $f_t(y, z, u) := \infty$ elsewhere, or more specially (for a $g$-component not depending on $y, z$)

$$f_t(y, z, u) := \widehat{f}_t(y, z) + \int_A g_t(u(e), e) \, \zeta(t, e) \, \lambda(de) \quad \text{(where finitely defined)} \tag{2.7}$$

and $f_t(y, z, u) := \infty$ elsewhere, for a $\mathcal{B}(E)$-measurable set $A$ and component functions $\widehat{f}$, $g$ where $\widehat{f} : \Omega \times [0, T] \times \mathbb{R}^{1+d} \to \mathbb{R}$ is $\mathcal{P} \otimes \mathcal{B}(\mathbb{R}^{d+1})$-measurable and $g : \Omega \times [0, T] \times \mathbb{R}^{1+d} \times \mathbb{R} \times E \to \mathbb{R}$ is $\mathcal{P} \otimes \mathcal{B}(\mathbb{R}^{d+2}) \otimes \mathcal{B}(E)$-measurable. Clearly statements for generators of the form (2.6) are also true for those of the (more particular) form (2.7). (In)finite activity relates to generators with $\lambda(A) < \infty$ (respectively $\lambda(A) = \infty$). A simple but useful technical Lemma clarifies how we can (and always will) choose a bounded representative for $U$ in a BSDE solution $(Y, Z, U)$ with bounded $Y$.

**Lemma 2.2** *Let $(Y, Z, U) \in \mathcal{S}^\infty \times \mathcal{L}^2(B) \times \mathcal{L}^2(\widetilde{\mu})$ be a solution of some JBSDE (2.4) with data $(\xi, f)$. Then there exists a representative $U'$ of $U$, bounded pointwise by $2|Y|_\infty$, such that $U' = U$ in $\mathcal{L}^2(\widetilde{\mu})$ and $\mathbb{P} \otimes \mathrm{d}t$-a-e., and $(Y, Z, U')$ solves the BSDE $(\xi, f)$.*

*Proof* We reproduce a brief argument sufficient to our general setting, similarly to e.g. [44, Corollary 1] or [7, proof of Theorem 3.5]. Use that $\mu(\omega, dt, de) = \sum_{s\geq 0} \mathbb{1}_D(\omega, s)\, \delta_{(s,\beta_s(\omega))}(dt, de)$ for an optional $E$-valued process $\beta$ and a thin set $D$, since $\mu$ is an integer-valued random measure [31, II.§1b]. Clearly the jump $\Delta Y_t(\omega) = (Y_t - Y_{t-})(\omega) = \int_E U_t(\omega, e)\, \mu(\omega; \{t\}, de)$ is equal to $\mathbb{1}_D(\omega, t) U_t(\omega, \beta_t(\omega))$ and bounded by $2|Y|_\infty$. For $U'_t(\omega, e) := U_t(\omega, e)\mathbb{1}_D(\omega, t)\mathbb{1}_{\{\beta_t\}}(e)$, we have $U_t(\omega, \beta_t(\omega)) = U'_t(\omega, \beta_t(\omega))$ on $D$, and $\sum_{s\geq 0} \mathbb{1}_D(\omega, s)|U_s - U'_s|^2(\omega, \beta_s(\omega)) = 0$ implies $E[|U - U'|^2 * \nu_T] = E[|U - U'|^2 * \mu_T] = 0$. Since $U = U'$ in $\mathcal{L}^2(\widetilde{\mu})$ and $U_t = U'_t$ in $L^0(\mathcal{B}(E), \lambda)$, the BSDE is solved by $(Y, Z, U')$. □

Under these conditions, we can and will take $U$ to be bounded by twice the norm of $Y$; Defining $|U|_\infty := \text{ess sup}_{(\omega,t,e)} |U_t(e)|$ for $U \in \mathcal{L}^2(\widetilde{\mu})$ yields $|U|_\infty \leq 2|Y|_\infty$ for any bounded BSDE solution $(Y, Z, U)$. The next lemma notes that the stochastic integrals of bounded JBSDE solutions are BMO-martingales when some truncated generator function is bounded from above (below) by $+(-)\langle M\rangle$ for a BMO-martingale $M$; Moreover, their BMO-norms depend only on $|Y|_\infty$, the BMO-norm of $M$ and the horizon $T$. See [36, Lemma 1.3] for details of the proof, and note that BMO-properties of integrals of (bounded) BSDEs are of course a well-studied topic, cf. [42] and references therein.

**Lemma 2.3** *Let $(Y, Z, U) \in \mathcal{S}^\infty \times \mathcal{L}^2(B) \times \mathcal{L}^2(\widetilde{\mu})$ be a bounded solution to the BSDE $(\xi, f)$. Assume there is $M \in \mathrm{BMO}(\mathbb{P})$ such that $\int_t^T f_s(Y_{s-}, Z_s, U_s)\, ds \leq \langle M\rangle_T - \langle M\rangle_t$ or $-\int_t^T f_s(Y_{s-}, Z_s, U_s)\, ds \leq \langle M\rangle_T - \langle M\rangle_t$. Then $\int Z\, dB$ and $U * \widetilde{\mu}$ are BMO-martingales and their BMO-norms (resp. $L^2$-norms) are bounded by a constant depending on $|Y|_\infty$ and $\|M\|_{\mathrm{BMO}(\mathbb{P})}$ (resp. on $|Y|_\infty$, $\|M\|_{\mathcal{M}^2}$).*

## 3 Comparison Theorems and A-Priori-Estimates

The stage for the main comparison Theorem 3.9 and the a-priori-$L^\infty$-estimate of Proposition 3.11 in this section is set by the next proposition. Its line of proof follows the seminal Theorem 2.5 by [54], with slight generalizations that are needed in the sequel. Just some details for the change of measure argument are elaborated a bit differently, measurable dependencies of the random field $\gamma$ are specified in more detail, and less is assumed on the generators. Instead of imposing specific conditions on the generators which imply existence of solutions, we only insist that we have solutions and impose a generalized $(\mathbf{A}_\gamma)$-condition as explained in Example 3.8.

**Proposition 3.1** *Let $(Y^i, Z^i, U^i) \in \mathcal{S}^\infty \times \mathcal{L}^2(B) \times \mathcal{L}^2(\widetilde{\mu})$ be solutions to the BSDE (2.4) for data $(\xi_i, f_i)$, $i = 1, 2$. Assume that $f_2$ is Lipschitz continuous w.r.t. $y$ and $z$. Let $\gamma : \Omega \times [0, T] \times \mathbb{R}^{d+3} \times E \to [-1, \infty)$ with $(\omega, t, y, z, u, u', e) \mapsto \gamma_t^{y,z,u,u'}(e)$*

*be a $\mathcal{P} \otimes \mathcal{B}(\mathbb{R}^{d+3}) \otimes \mathcal{B}(E)$-measurable function such that for $\overline{\gamma} := \gamma^{Y^2_-, Z^2, U^1, U^2}$ it holds*

$$f_2(t, Y^2_{t-}, Z^2_t, U^1_t) - f_2(t, Y^2_{t-}, Z^2_t, U^2_t) \leq \int_E \overline{\gamma}_t(e)\,(U^1_t(e) - U^2_t(e))\,\zeta(t,e)\,\lambda(\mathrm{d}e), \ \mathbb{P} \otimes \mathrm{d}t\text{-a.e.} \tag{3.1}$$

*and the stochastic exponential $\mathcal{E}(\int \beta\,\mathrm{d}B + \overline{\gamma} * \widetilde{\mu})$ is a martingale for $\beta$ from (3.2). Then a comparison result holds, that means that the inequalities $\xi_1 \leq \xi_2$ and $f_1(t, Y^1_{t-}, Z^1_t, U^1_t) \leq f_2(t, Y^1_{t-}, Z^1_t, U^1_t)$, $\mathbb{P} \otimes \mathrm{d}t$-a.e., together imply $Y^1_t \leq Y^2_t$ for all $t \leq T$.*

In results like the above, in [54] and further enhancements [38, 52, 57], the key assumption needed for comparison is the existence of an abstract random field $\gamma$ such that inequalities are satisfied between processes. In contrast, the subsequent results of this section offer sufficient criteria for comparison that can be verified more easily by checking concrete dependencies w.r.t. to basically Euclidean arguments for generator functions $f$ of the type (2.6). See also [24] for a simpler version in a setting with a jump measure of Lévy-type on $E = \mathbb{R}^1 \setminus \{0\}$ and $\zeta \equiv 1$.

*Proof* We define $\widehat{\xi} := \xi_1 - \xi_2$, $\widehat{Y} := Y^1 - Y^2$, $\widehat{Z} := Z^1 - Z^2$ and $\widehat{U} := U^1 - U^2$. The processes

$$\begin{aligned}
\alpha_s &:= \mathbb{1}_{\{Y^1_{s-} \neq Y^2_{s-}\}} \frac{f_2(s, Y^1_{s-}, Z^1_s, U^1_s) - f_2(s, Y^2_{s-}, Z^1_s, U^1_s)}{(Y^1_{s-} - Y^2_{s-})}, \\
\beta_s &:= \mathbb{1}_{\{Z^1_s \neq Z^2_s\}} \frac{f_2(s, Y^2_{s-}, Z^1_s, U^1_s) - f_2(s, Y^2_{s-}, Z^2_s, U^1_s)}{\|Z^1_s - Z^2_s\|^2}(Z^1_s - Z^2_s)
\end{aligned} \tag{3.2}$$

and $R_t := \exp(\int_0^t \alpha_s\,\mathrm{d}s)$ are bounded due to the Lipschitz assumption on $f_2$. As in [54], applying Itô's formula to $R\widehat{Y}$ between $\tau \wedge t$ and $\tau \wedge T$ for some stopping times $\tau$ yields

$$\begin{aligned}
(R\widehat{Y})_{\tau\wedge t} = (R\widehat{Y})_{\tau\wedge T} &+ \int_{\tau\wedge t}^{\tau\wedge T} R_s\big(f_1(s, Y^1_{s-}, Z^1_s, U^1_s) - f_2(s, Y^2_{s-}, Z^2_s, U^2_s)\big)\,\mathrm{d}s \\
&- \int_{\tau\wedge t}^{\tau\wedge T} R_s \widehat{Z}_s\,\mathrm{d}B_s - \int_{\tau\wedge t}^{\tau\wedge T}\int_E R_s \widehat{U}_s(e)\,\widetilde{\mu}(\mathrm{d}s, \mathrm{d}e) - \int_{\tau\wedge t}^{\tau\wedge T} R_s \alpha_s \widehat{Y}_{s-}\,\mathrm{d}s.
\end{aligned}$$

Set $M := \int R\widehat{Z}\,\mathrm{d}B + (R\widehat{U}) * \widetilde{\mu}$ and $N := \int \beta\,\mathrm{d}B + \overline{\gamma} * \widetilde{\mu}$. Then $\mathrm{d}\mathbb{Q} := \mathcal{E}(N)_T \mathrm{d}\mathbb{P}$ defines an absolutely continuous probability by the martingale property of the stochastic exponential $\mathcal{E}(N) \geq 0$; cf. [28, Lemma 9.40]. By Girsanov $L := M - \langle M, N\rangle$ is a local $\mathbb{Q}$-martingale, and the inequality

$$f_1(s, Y^1_{s-}, Z^1_s, U^1_s) - f_2(s, Y^2_{s-}, Z^2_s, U^2_s) \leq \alpha_s \widehat{Y}_{s-} + \beta_s \widehat{Z}_s + \int_E \overline{\gamma}_s(e)\widehat{U}_s(e)\,\zeta_s(e)\,\lambda(\mathrm{d}e)\ \mathbb{P} \otimes \mathrm{d}s\text{-a.e.}$$

$$\text{implies} \quad (R\widehat{Y})_{\tau\wedge t} \leq (R\widehat{Y})_{\tau\wedge T} - (L^\tau_T - L^\tau_t). \tag{3.3}$$

Localizing $L$ along a sequence of stopping times $\tau_n \uparrow \infty$ and taking conditional expectations, we obtain $\mathbb{E}_{\mathbb{Q}}\big((R\widehat{Y})_{t\wedge\tau^n} \,\big|\, \mathcal{F}_t\big) \le \mathbb{E}_{\mathbb{Q}}\big((R\widehat{Y})_{\tau^n\wedge T} \,\big|\, \mathcal{F}_t\big)$ for each $n \in \mathbb{N}$. Dominated convergence yields the estimate $R_t\widehat{Y}_t \le \mathbb{E}_{\mathbb{Q}}\big(R_T\widehat{\xi}\,\big|\,\mathcal{F}_t\big) \le 0$ and thus $Y_t^1 \le Y_t^2$. □

*Remark 3.2* 1. Switching roles of $f_1$ and $f_2$, one gets that if $f_1$ is Lipschitz in $y$,$z$ and satisfies (3.1) instead of $f_2$, then $\xi_1 \le \xi_2$ and $f_1(t, Y_{t-}^2, Z_t^2, U_t^2) \le f_2(t, Y_{t-}^2, Z_t^2, U_t^2)$ imply $Y_t^1 \le Y_t^2$.

2. The result of Proposition 3.1 remains valid (with a similar proof) if one requires that the $Y$-components of JBSDE solutions to compare are in $\mathcal{S}^2$ instead of $\mathcal{S}^\infty$, and the stochastic exponential $\mathcal{E}(\beta \bullet B + \overline{\gamma} * \widetilde{\mu})$ is in $\mathcal{S}^2$. However, as it is stated, Proposition 3.1 is exactly what we will need to apply in the sequel to derive, e.g., Proposition 4.3 and Theorem 4.13.

*Example 3.3* Sufficient conditions for $\mathcal{E}(\overline{\gamma} * \widetilde{\mu})$ to be a martingale are, for instance,

1. $\Delta(\overline{\gamma} * \widetilde{\mu}) > -1$ and $\mathbb{E}\big(\exp(\langle\overline{\gamma} * \widetilde{\mu}\rangle_T)\big) = \mathbb{E}\big(\exp\big(\int_0^T \int_E |\overline{\gamma}_s(e)|^2\,\nu(\mathrm{d}s, \mathrm{d}e)\big)\big) < \infty$; see [51, Theorem 9]. This holds i.p. if $\int_E |\overline{\gamma}_s(e)|^2\,\zeta(s, e)\,\lambda(\mathrm{d}e) < const. < \infty$ $\mathbb{P}\otimes \mathrm{d}s$-a.e. and $\overline{\gamma} > -1$.
2. $\Delta(\overline{\gamma} * \widetilde{\mu}) \ge -1 + \delta$ for $\delta > 0$ and $\overline{\gamma} * \widetilde{\mu}$ is a BMO($\mathbb{P}$)-martingale due to Kazamaki [33].
3. $\Delta(\overline{\gamma} * \widetilde{\mu}) \ge -1$ and $\overline{\gamma} * \widetilde{\mu}$ is a uniformly integrable martingale and $\mathbb{E}\big(\exp(\langle\overline{\gamma} * \widetilde{\mu}\rangle_T)\big) < \infty$; see [41, Theorem I.8]. Such a condition is satisfied when $\overline{\gamma}$ is bounded and $|\overline{\gamma}| \le \psi$, $\mathbb{P}\otimes \mathrm{d}t \otimes \lambda$-a.e. for a function $\psi \in L^2(\lambda)$ and $\zeta \equiv 1$. The latter is what is required for instance in the comparison Theorem 4.2 of [52].

Note that under above conditions, also the stochastic exponential $\mathcal{E}(\int \beta \mathrm{d}B + \overline{\gamma} * \widetilde{\mu})$ for $\beta$ bounded and predictable is a martingale, as it is easily seen by Novikov's criterion.

Let us also refer to [14, Sects. 19 and A.9] for related so-called balance conditions on generators for JBSDE comparison by change of measure arguments.

In the statement of Proposition 3.1, the dependence of the process $\overline{\gamma}$ on the BSDE solutions is not needed for the proof as the same result holds if $\overline{\gamma}$ is just a predictable process such that the estimate on the generator $f_2$ and the martingale property (3.1) hold. The further functional dependence is needed for the sequel, as required in the following

**Definition 3.4** We say that an $\overline{\mathbb{R}}$-valued generator function $f$ satisfies condition $(\mathbf{A}_\gamma)$ if there is a $\mathcal{P}\otimes\mathcal{B}(\mathbb{R}^{d+3})\otimes\mathcal{B}(E)$-measurable function $\gamma : \Omega\times[0,T]\times\mathbb{R}^{d+3}\times E \to (-1,\infty)$ given by $(\omega, t, y, z, u, u', e) \mapsto \gamma_t^{y,z,u,u'}(e)$ such that for all $(Y, Z, U, U') \in \mathcal{S}^\infty \times \mathcal{L}^2(B) \times (\mathcal{L}^2(\widetilde{\mu}))^2$ with $|U|_\infty < \infty$, $|U'|_\infty < \infty$ it holds for $\overline{\gamma} := \gamma^{Y_-,Z,U,U'}$

$$f_t(Y_{t-}, Z_t, U_t) - f_t(Y_{t-}, Z_t, U_t') \le \int_E \overline{\gamma}_t(e)(U_t(e) - U_t'(e))\zeta(t, e)\lambda(\mathrm{d}e),\ \mathbb{P}\otimes \mathrm{d}t\text{-a.e.}$$

and $\mathcal{E}(\int \beta \mathrm{d}B + \overline{\gamma} * \widetilde{\mu})$ is a martingale for every bounded and predictable $\beta$.

(3.4)

We will say that $f$ satisfies condition $(\mathbf{A}'_\gamma)$ if the above holds for all bounded $U$ and $U'$ with additionally $U * \widetilde{\mu}$ and $U' * \widetilde{\mu}$ in $\mathrm{BMO}(\mathbb{P})$.

Clearly, existence and applicability of a suitable comparison result for solutions to JBSDEs implies their uniqueness. In other words, if there exists a bounded solution for a generator being Lipschitz w.r.t. $y$ and $z$ which satisfies $(\mathbf{A}_\gamma)$ or $(\mathbf{A}'_\gamma)$, we obtain that such a solution is unique.

*Example 3.5* The natural candidate for $\gamma$ for generators $f$ of the form (2.6) is given by

$$\gamma_s^{y,z,u,u'}(e) = \frac{g_s(y,z,u,e) - g_s(y,z,u',e)}{u-u'} \mathbb{1}_A(e)\, \mathbb{1}_{\{u\neq u'\}}, \tag{3.5}$$

which is $\mathcal{P} \otimes \mathcal{B}(\mathbb{R}^{d+3}) \otimes \mathcal{B}(E)$-measurable since $g$ is. Assuming absolute continuity of $g$ in $u$, we can express $\gamma_s^{y,z,u,u'}(e) = \int_0^1 \frac{\partial}{\partial u} g_s(y,z,tu+(1-t)u',e)\,\mathrm{d}t\, \mathbb{1}_A(e)$, by noting that

$$(u-u')\int_0^1 \frac{\partial}{\partial u} g_s(y,z,tu+(1-t)u',e)\,\mathrm{d}t\, \mathbb{1}_A(e) = \int_0^1 \frac{\partial}{\partial t}\left[(g_s(y,z,tu+(1-t)u',e))\right]\mathrm{d}t\, \mathbb{1}_A(e).$$

For generators of type (2.7) the $\gamma$ simply is

$$\gamma_s^{y,z,u,u'}(e) = \int_0^1 \frac{\partial}{\partial u} g_s(tu+(1-t)u',e)\mathrm{d}t\, \mathbb{1}_A(e).$$

**Definition 3.6** We say that a generator $f$ satisfies condition $(\mathbf{A}_{\mathbf{fin}})$ or $(\mathbf{A}_{\mathbf{infi}})$ (on a set $D$) if

1. $(\mathbf{A}_{\mathbf{fin}})$: $f$ is of the form (2.6) with $\lambda(A) < \infty$, is Lipschitz continuous w.r.t. $y$ and $z$ uniformly in $(t,\omega,u)$, and the map $u \mapsto g(t,y,z,u,e)$ is absolutely continuous (in $u$) for all $(\omega,t,y,z,e)$ (in $D \subseteq \Omega \times [0,T] \times \mathbb{R} \times \mathbb{R}^d \times E$), i.e. $g(t,y,z,u,e) = g(0) + \int_0^u g'(t,y,z,x,e)\mathrm{d}x$, with density function $g'$ being strictly greater than $-1$ (on $D$) and locally bounded (in u) from above, uniformly in $(\omega,t,y,z,e)$.
2. $(\mathbf{A}_{\mathbf{infi}})$: $f$ is of the form (2.7), is Lipschitz continuous w.r.t. $y$ and $z$ uniformly in $(t,\omega,u)$, and the map $u \mapsto g_t(u,e)$ is absolutely continuous (in $u$) for all $(\omega,t,e)$ (in $D$), i.e. $g(t,u,e) = g(0) + \int_0^u g'(t,x,e)\mathrm{d}x$, with density function $g'$ being such that for all $c \in (0,\infty)$ there exists $K(c) \in \mathbb{R}$ and $\delta(c) \in (0,1)$ with $-1+\delta(c) \leq g'(x)$ and $|g'(x)| \leq K(c)|x|$ for all $x$ with $|x| \leq c$.

*Remark 3.7* Note that under condition $(\mathbf{A}_{\mathbf{infi}})$ the density function $g'$ is necessarily locally bounded, in particular with $|g'(x)| \leq K(c)c =: \bar{K}(c) < \infty$ for all $x \in [-c,c]$. Observe that the conditions are not requiring the function $g$ to be convex and moreover refrain from requiring it to be continuously differentiable in $u$. Both can be helpful in application examplres, see Sect. 5.1.2.

*Example 3.8* Sufficient conditions for condition ($\mathbf{A}_\gamma$) and ($\mathbf{A}'_\gamma$) are

1. $\gamma$ is a $\mathcal{P} \otimes \mathcal{B}(\mathbb{R}^{d+3}) \otimes \mathcal{B}(E)$-measurable function satisfying the inequality in (3.4) and
$$C_1(1 \wedge |e|) \le \gamma_t^{y,z,u,u'}(e) \le C_2(1 \wedge |e|)$$
on $E = \mathbb{R}^l \setminus \{0\}$ $(l \in \mathbb{N})$, for some $C_1 \in (-1, 0]$ and $C_2 > 0$. In this case $\exp(\langle \int \beta \mathrm{d}B + \overline{\gamma} * \widetilde{\mu} \rangle_T)$ is clearly bounded and the jumps of $\int \beta \mathrm{d}B + \overline{\gamma} * \widetilde{\mu}$ are bigger than $-1$. Hence $\mathcal{E}\left(\int \beta \mathrm{d}B + \overline{\gamma} * \widetilde{\mu}\right)$ is a positive martingale [51, Theorem 9]. Thus Definition 3.4 generalizes the original ($\mathbf{A}_\gamma$)-condition introduced by [54] for Poisson random measures.
2. ($\mathbf{A_{fin}}$) is sufficient for ($\mathbf{A}_\gamma$). This follows from Example 3.3, (3.5) and $\lambda(A) < \infty$.
3. ($\mathbf{A_{infi}}$) is sufficient for ($\mathbf{A}'_\gamma$). To see this, let $u, u'$ be bounded by $c$ and $\gamma$ be the natural candidate in Example 3.5. Then $|\gamma_s^{y,z,u,u'}(e)| \le \int_{u'}^{u} |g'(x)| \mathrm{d}x / (u - u') \le K(c)(|u| + |u'|)$. Hence $\int \beta \mathrm{d}B + \overline{\gamma} * \widetilde{\mu}$ is a BMO-martingale by the BMO-property of $U * \widetilde{\mu}$ and $U' * \widetilde{\mu}$ with some lower bound $-1 + \delta$ for its jumps. And $\mathcal{E}(\int \beta \mathrm{d}B + \overline{\gamma} * \widetilde{\mu})$ is a martingale by part 2 of Example 3.3.
4. Condition ($\mathbf{A_{fin}}$) above is satisfied if, e.g., $f$ is of the form (2.6) with $\lambda(A) < \infty$, is Lipschitz continuous w.r.t. $y$ and $z$, and the map $u \mapsto g(t, y, z, u, e)$ is continuously differentiable for all $(\omega, t, y, z, e)$ (in $D$) such that the derivative is strictly greater than $-1$ (on $D \subseteq \Omega \times [0, T] \times \mathbb{R} \times \mathbb{R}^d \times E$) and locally bounded (in $u$) from above, uniformly in $(\omega, t, y, z, e)$.
5. Condition ($\mathbf{A_{infi}}$) is valid if for instance $f$ is of the form (2.7), is Lipschitz continuous w.r.t. $y$ and $z$, and the map $u \mapsto g_t(u, e)$ is twice continuously differentiable for all $(\omega, t, e)$ with the derivatives being locally bounded uniformly in $(\omega, t, e)$, the first derivative being (locally) bounded away from $-1$ with a lower bound $-1 + \delta$ for some $\delta > 0$, and $\frac{\partial g}{\partial u}(t, 0, e) \equiv 0$.

As an application of the above, we can now provide simple conditions for comparison in terms of concrete properties of the generator function, which are easier to verify than the more general but abstract conditions on the existence of a suitable function $\gamma$ as in Proposition 3.1 or the general conditions by [15]. Note that no convexity is required in the $z$ or $u$ argument of the generator. The result will be applied later to prove existence and uniqueness of JBSDE solutions.

**Theorem 3.9** (Comparison Theorem) *A comparison result between bounded BSDE solutions in the sense of Proposition 3.1 holds true in each of the following cases:*

1. (finite activity) *$f_2$ satisfies ($\mathbf{A_{fin}}$).*
2. (infinite activity) *$f_2$ satisfies ($\mathbf{A_{infi}}$) and $U^1 * \widetilde{\mu}$ and $U^2 * \widetilde{\mu}$ are* BMO($\mathbb{P}$)-*martingales for the corresponding JBSDE solutions $(Y^1, Z^1, U^1)$ and $(Y^2, Z^2, U^2)$.*

*Proof* This follows directly from Proposition 3.1 and Example 3.8, noting that representation (3.5) in connection with condition ($\mathbf{A_{fin}}$) resp. ($\mathbf{A_{infi}}$) meets the sufficient conditions in Example 3.3. □

Unlike classical a-priori estimates that offer some $L^2$-norm estimates for the BSDE solution in terms of the data, the next result gives a simple $L^\infty$-estimate for the $Y$-component of the solution. Such will be useful for the derivation of BSDE solution bounds and for truncation arguments.

**Proposition 3.10** *Let $(Y, Z, U) \in \mathcal{S}^\infty \times \mathcal{L}^2(B) \times \mathcal{L}^2(\widetilde{\mu})$ be a solution to the BSDE $(\xi, f)$ with $\xi \in L^\infty(\mathcal{F}_T)$, $f$ be Lipschitz continuous w.r.t. $(y, z)$ with Lipschitz constant $K_f^{y,z}$ and satisfying $(\mathbf{A}_\gamma)$ with $f_\cdot(0, 0, 0)$ bounded.*
*Then $|Y_t| \le \exp\big(K_f^{y,z}(T-t)\big)\big(|\xi|_\infty + (T-t)|f_\cdot(0,0,0)|_\infty\big)$ for $t \le T$.*

*Proof* Set $(Y^1, Z^1, U^1) = (Y, Z, U)$, $(\xi^1, f^1) = (\xi, f)$, $(Y^2, Z^2, U^2) = (0, 0, 0)$ and $(\xi^2, f^2) = (0, f)$. Then following the proof of Proposition 3.1, Eq. (3.3) becomes

$$(RY)_{\tau\wedge t} \le (RY)_{\tau\wedge T} + \int_{\tau\wedge t}^{\tau\wedge T} R_s f_s(0,0,0)\,\mathrm{d}s - (L_T^\tau - L_t^\tau), \quad t \in [0, T],$$

for all stopping times $\tau$ where $L := M - \langle M, N\rangle$ is in $\mathcal{M}_{\mathrm{loc}}(\mathbb{Q})$, $M := \int RZ\,\mathrm{d}B + (RU) * \widetilde{\mu}$ is in $\mathcal{M}^2$, $N := \int \beta\,\mathrm{d}B + \overline{\gamma} * \widetilde{\mu}$ with $\overline{\gamma} := \gamma^{0,0,U,0}$ and the probability measure $\mathbb{Q} \approx \mathbb{P}$ is given by $\mathrm{d}\mathbb{Q} := \mathcal{E}(N)_T \mathrm{d}\mathbb{P}$. Localizing $L$ along some sequence $\tau^n \uparrow \infty$ of stopping times yields

$$\mathbb{E}_{\mathbb{Q}}\big((RY)_{\tau^n\wedge t} \,\big|\, \mathcal{F}_t\big) \le \mathbb{E}_{\mathbb{Q}}\Big((RY)_{\tau^n\wedge T} + \int_{\tau\wedge t}^{\tau\wedge T} R_s f_s(0,0,0)\,\mathrm{d}s \,\Big|\, \mathcal{F}_t\Big).$$

By dominated convergence, we conclude that $\mathbb{P}$-a.e

$$Y_t \le \mathbb{E}_{\mathbb{Q}}\Big(\frac{R_T}{R_t}\xi + \int_t^T \frac{R_s}{R_t} f_s(0,0,0)\,\mathrm{d}s \,\Big|\, \mathcal{F}_t\Big) \le \mathrm{e}^{K_f^{y,z}(T-t)}\big(|\xi|_\infty + (T-t)|f_\cdot(0,0,0)|_\infty\big).$$

Analogously, if we define $\overline{N} := \int \beta\,\mathrm{d}B + \overline{\overline{\gamma}} * \widetilde{\mu}$ with $\overline{\overline{\gamma}} := \gamma^{0,0,0,U}$, and $\overline{\mathbb{Q}}$ equivalent to $\mathbb{P}$ via $\mathrm{d}\overline{\mathbb{Q}} := \mathcal{E}(\overline{N})_T \mathrm{d}\mathbb{P}$, we deduce that $\overline{L} := M - \langle M, \overline{N}\rangle$ is in $\mathcal{M}_{\mathrm{loc}}(\overline{\mathbb{Q}})$ and

$$(RY)_{\tau\wedge t} \ge (RY)_{\tau\wedge T} + \int_{\tau\wedge t}^{\tau\wedge T} R_s f_s(0,0,0)\,\mathrm{d}s - (\overline{L}_T^\tau - \overline{L}_t^\tau), \quad t \in [0, T],$$

for all stopping times $\tau$. This yields the required lower bound. □

Again, we can specify explicit conditions on the generator function that are sufficient to ensure the more abstract assumptions of the previous result.

**Proposition 3.11** *Let $(Y, Z, U) \in \mathcal{S}^\infty \times \mathcal{L}^2(B) \times \mathcal{L}^2(\widetilde{\mu})$ be a solution to the BSDE $(\xi, f)$ with $\xi$ in $L^\infty(\mathcal{F}_T)$, $f$ being Lipschitz continuous w.r.t. $(y, z)$ with Lipschitz constant $K_f^{y,z}$ such that $f_\cdot(0, 0, 0)$ is bounded. Assume that one of the following conditions holds:*

1. (finite activity) $f$ *satisfies* ($\boldsymbol{A_{fin}}$).
2. (infinite activity) $f$ *satisfies* ($\boldsymbol{A_{infi}}$) *and* $U * \widetilde{\mu}$ *is a* BMO($\mathbb{P}$)*-martingale.*

*Then* $|Y_t| \leq \exp\big(K_f^{y,z}(T-t)\big)\big(|\xi|_\infty + (T-t)|f_s(0,0,0)|_\infty\big)$ *holds for all* $t \leq T$, *in particular* $|Y|_\infty \leq \exp\big(K_f^{y,z}T\big)\big(|\xi|_\infty + T|f_s(0,0,0)|_\infty\big)$.

*Proof* This follows directly from Proposition 3.10 and Example 3.8, since $f$ satisfies condition ($\mathbf{A}_\gamma$) (resp. ($\mathbf{A}'_\gamma$)) using Eq. (3.5). □

In the last part of this section we apply our comparison theorem for more concrete generators. To this end, we consider a generator $f$ being truncated at bounds $a < b$ (depending on time only) as

$$\widetilde{f}_t(y,z,u) := f_t\big(\kappa(t,y),\ z,\ \kappa(t,y+u) - \kappa(t,y)\big), \tag{3.6}$$

with $\kappa(t,y) := \big(a(t) \vee y\big) \wedge b(t)$. Next, we show that if a generator satisfies ($\mathbf{A}_\gamma$) within the truncation bounds, then the truncated generator satisfies ($\mathbf{A}_\gamma$) everywhere.

**Lemma 3.12** *Let $f$ satisfy (3.4) for $Y$, $U$ such that*

$$a(t) \leq Y_{t-}, Y_{t-} + U_t(e), Y_{t-} + U'_t(e) \leq b(t), \quad t \in [0,T]$$

*and let $\gamma$ satisfy one of the conditions of Example 3.3 for the martingale property of $\mathcal{E}(\overline{\gamma} * \widetilde{\mu})$. Then $\widetilde{f}$ satisfies (3.4). Especially, if $f$ satisfies ($\boldsymbol{A_{fin}}$) on the set where $a(t) \leq y, y+u \leq b(t)$ then $\widetilde{f}$ is Lipschitz in $(y,z)$, locally Lipschitz in $u$ and satisfies ($\boldsymbol{A}_\gamma$).*

*Proof* Using monotonicity of $x \mapsto \kappa(t,x)$, we get that

$$\widetilde{f}_t(Y_{t-}, Z_t, U_t) - \widetilde{f}_t(Y_{t-}, Z_t, U'_t)$$

equals

$$\begin{aligned}
&f_t\big(\kappa(t,Y_{t-}), Z_t, \kappa(t,Y_{t-}+U_t) - \kappa(t,Y_{t-})\big) - f_t\big(\kappa(t,Y_{t-}), Z_t, \kappa(t,Y_{t-}+U'_t) - \kappa(t,Y_{t-})\big)\\
&\quad \leq \int_E \overline{\gamma}_t(e)\big(\kappa(t,Y_{t-}+U_t(e)) - \kappa(t,Y_{t-}+U'_t(e))\big)\,\zeta(t,e)\,\lambda(\mathrm{d}e)\\
&\quad \leq \int_E \overline{\gamma}_t(e)\big(\mathbb{1}_{\{\overline{\gamma}\geq 0, U\geq U'\}} + \mathbb{1}_{\{\overline{\gamma}<0, U<U'\}}\big)\big(U_t(e) - U'_t(e)\big)\,\zeta(t,e)\,\lambda(\mathrm{d}e).
\end{aligned}$$

Setting $\overline{\gamma^*} := \overline{\gamma}\big(\mathbb{1}_{\{\overline{\gamma}\geq 0, U\geq U'\}} + \mathbb{1}_{\{\overline{\gamma}<0, U<U'\}}\big)$ we see that the stochastic exponential $\mathcal{E}\big(\int \beta \mathrm{d}B + \overline{\gamma^*} * \widetilde{\mu}\big)$ is a martingale for all bounded and predictable processes $\beta$ and $\widetilde{f}$ satisfies (3.4). The latter claim easily follows from the fact that if $f$ satisfies ($\mathbf{A_{fin}}$) on $a(t) \leq y, y+u \leq b(t)$ then $f$ satisfies (3.4) on $a(t) \leq Y_{t-}, Y_{t-} + U_t(e), Y_{t-} + U'_t(e) \leq b(t)$ using Example 3.8. The Lipschitz properties of $\widetilde{f}$ follow from the fact that $\kappa$ is a contraction and $f$ is Lipschitz within the truncation bounds. □

Concrete $L^\infty$-bounds for bounded solutions to BSDE $(\xi, f)$ with suitable $\widehat{f}$-part are provided by

**Proposition 3.13** *Let $f$ be a generator of the form (2.6) with $\left|\widehat{f}_t(y,z)\right| \le K_1 + K_2|y|$ for some $K_1, K_2 \ge 0$, $g_t(y,z,0,e) \equiv 0$ and $\xi \in L^\infty(\mathcal{F}_T)$ with $c_1 \le \xi \le c_2$ for some $c_1, c_2 \in \mathbb{R}$. Assume that there are solutions $a$ and $b$ to the ODEs $y'(t) = K_1 + K_2|y(t)|$, $y(T) = c_1$ and $y'(t) = -(K_1 + K_2|y(t)|)$, $y(T) = c_2$ respectively, such that $a \le b$ on $[0,T]$. If the truncated generator $\widetilde{f}$ in (3.6) satisfies $(\mathbf{A}_\gamma)$ and is Lipschitz in $(y,z)$, then any solution $(\widetilde{Y}, \widetilde{Z}, \widetilde{U}) \in \mathcal{S}^\infty \times \mathcal{L}^2(B) \times \mathcal{L}^2(\widetilde{\mu})$ to the JBSDE $(\xi, \widetilde{f})$ also solves the JBSDE $(\xi, f)$ and satisfies $a(t) \le \widetilde{Y}_t \le b(t)$, $t \in [0,T]$.*

*Proof* We set $Y_t := \kappa(t, \widetilde{Y}_t)$, $Z_t := \widetilde{Z}_t$, $U_t(e) := \kappa(t, \widetilde{Y}_{t-} + \widetilde{U}_t(e)) - \kappa(t, \widetilde{Y}_{t-})$ and

$$f_t^i(y,z,u) := \widehat{f}_t^i\big(\kappa(t,y),z\big) + \int_E g_t\big(\kappa(t,y), z, \kappa(t,y+u) - \kappa(t,y), e\big)\,\zeta(t,e)\,\lambda(\mathrm{d}e)$$

with $\widehat{f}_t^1(y,z) := -(K_1 + K_2|y|)$, $\widehat{f}_t^2(y,z) := \widehat{f}_t(y,z)$ and $\widehat{f}_t^3(y,z) := K_1 + K_2|y|$. By the assumptions on the ODEs, we have that $(a(t), 0, 0)$ solves the BSDE $(c_1, f^1)$ and $(b(t), 0, 0)$ solves the BSDE $(c_2, f^3)$. Taking into account that $\widetilde{f}^1 \le \widetilde{f}^2 \le \widetilde{f}^3$, $c_1 \le \xi \le c_2$ and $\widetilde{f}^2$ satisfies $(\mathbf{A}_\gamma)$, comparison theorem Proposition 3.1 yields $a(t) \le \widetilde{Y}_t \le b(t)$. Hence, $Y$ and $\widetilde{Y}$ are indistinguishable, $U = \widetilde{U}$ in $\mathcal{L}^2(\widetilde{\mu})$ and $(\widetilde{Y}, \widetilde{Z}, \widetilde{U})$ solves the BSDE $(\xi, f)$. □

In the next section, we apply these results to two situations: Using Corollary 4.4, we give an alternative proof of Theorem 3.5 of [7] via a comparison principle instead of an argument with stopping times. Moreover, the estimates in Corollary 4.6 are applied to solve the power utility maximization problem via a JBSDE approach in Sect. 5.2.

## 4 Existence and Uniqueness of Bounded Solutions

This section studies BSDE with jumps by the monotone stability approach. Building on (straightforward) results for finite activity, the infinite activity case is treated by monotone approximations.

### 4.1 *The Case of Finite Activity*

**Definition 4.1** A generator function $f$ satisfies condition $(\mathbf{B}_\gamma)$, if it is Lipschitz continuous in $(y,z)$, locally Lipschitz continuous in $u$ (in the sense that $u \mapsto f_t(y,z,-c \vee u \wedge c)$ is Lipschitz continuous for any $c \in (0,\infty)$), $f_\cdot(0,0,0)$ is bounded, and $f$ satisfies condition $(\mathbf{A}_\gamma)$.

The next result readily leads to Proposition 4.3, for $A$ in (2.6) with $\lambda(A) < \infty$.

**Proposition 4.2** *Let $\xi \in L^\infty(\mathcal{F}_T)$ and $f$ satisfies $(\boldsymbol{B}_\gamma)$. Then there exists a unique solution $(Y, Z, U)$ in $\mathcal{S}^\infty \times \mathcal{L}^2(B) \times \mathcal{L}^2(\widetilde{\mu})$ to the BSDE $(\xi, f)$. Moreover for all $t \in [0, T]$, $|Y_t|$ is bounded by* $\exp\big(K_f^{y,z}(T-t)\big)\big(|\xi|_\infty + (T-t)|f_\cdot(0,0,0)|_\infty\big)$.

*Proof* Consider the Lipschitz generator $f_t^c(y,z,u) := f_t\big(y,z,(u \vee (-c)) \wedge c\big)$ with $c > 0$ and Lipschitz constant $K_{f^c}$. By classical fixed point arguments and a-priori estimates (cf. e.g. [7, Propositions 3.2, 3.3]) there is a unique solution $(Y^c, Z^c, U^c) \in \mathcal{S}^2 \times \mathcal{L}^2(B) \times \mathcal{L}^2(\widetilde{\mu})$ to the BSDE $(\xi, f^c)$; it satisfies

$$|Y_t^c| \le C\mathbb{E}\Big(|\xi|^2 + \int_t^T |f_s^c(0,0,0)|^2\,\mathrm{d}s \,\Big|\, \mathcal{F}_t\Big) \le C\big(|\xi|_\infty^2 + T|f_\cdot(0,0,0)|_\infty^2\big) < \infty,$$

for some constant $C = C(T, K_{f^c})$. Now Proposition 3.10 implies that $|Y_t^c|$ is dominated by $\exp\big(K_f^{y,z}(T-t)\big)\big(|\xi|_\infty + (T-t)|f_\cdot(0,0,0)|_\infty\big)$ for all $c > 0$. Choosing $c \ge 2\exp\big(K_f^{y,z}T\big)\big(|\xi|_\infty + T|f_\cdot(0,0,0)|_\infty\big)$ we get that $(Y^c, Z^c, U^c)$ with $Y^c \in \mathcal{S}^\infty$ solves the BSDE $(\xi, f)$ since $U^c$ is bounded by $c$. Uniqueness follows by comparison. □

This leads to a preliminary result on bounded solutions if jumps are of finite activity.

**Proposition 4.3** *Let $\xi \in L^\infty(\mathcal{F}_T)$ and let $f$ satisfy $(\boldsymbol{A_{fin}})$ (recall Definition 3.6) with $f_\cdot(0,0,0)$ bounded. Then there exists a unique solution $(Y, Z, U)$ in $\mathcal{S}^\infty \times \mathcal{L}^2(B) \times \mathcal{L}^2(\widetilde{\mu})$ to the BSDE $(\xi, f)$. Moreover for all $t \in [0, T]$, $|Y_t|$ is bounded by* $\exp\big(K_f^{y,z}(T-t)\big)\big(|\xi|_\infty + (T-t)|f_\cdot(0,0,0)|_\infty\big)$.

*Proof* Noting that local Lipschitz continuity in $u$ follows from the absolute continuity of $g$ in $u$ with locally bounded density function, the claim follows from Propositions 3.11 and 4.2. □

**Corollary 4.4** *Let $\xi \in L^\infty(\mathcal{F}_T)$ and let $f$ be a generator satisfying $(\boldsymbol{A_{fin}})$, with $g_t(y,z,0,e) \equiv 0$ and $|\widehat{f}_t(y,z)| \le K_1 + K_2|y|$ for some $K_1, K_2 \ge 0$. Set*

$$b(t) = \begin{cases} (|\xi|_\infty + \frac{K_1}{K_2})\exp(K_2(T-t)) - \frac{K_1}{K_2}, & K_2 \neq 0 \\ |\xi|_\infty + K_1(T-t), & K_2 = 0. \end{cases}$$

*Then there exists a unique solution $(Y, Z, U) \in \mathcal{S}^\infty \times \mathcal{L}^2(B) \times \mathcal{L}^2(\widetilde{\mu})$ to the BSDE $(\xi, f)$ and moreover $|Y_t| \le b_t$ for $t \in [0, T]$. Finally $\int Z\,\mathrm{d}B$ and $U * \widetilde{\mu}$ are* BMO($\mathbb{P}$)-*martingales.*

*Proof* By Lemma 3.12 and Proposition 4.3, there is a unique solution $(Y, Z, U)$ in the space $\mathcal{S}^\infty \times \mathcal{L}^2(B) \times \mathcal{L}^2(\widetilde{\mu})$ to the BSDE $(\xi, \widetilde{f})$. By Proposition 3.13, it also solves the BSDE $(\xi, f)$ and $-b(t) \le Y_t \le b(t)$, $\forall t \in [0, T]$. Uniqueness follows from the fact that one can apply the comparison Theorem 3.9 for generators satisfying $(\mathbf{A_{fin}})$. The BMO property follows from Lemma 2.3. □

*Remark 4.5* Corollary 4.4 is similar to Theorem 3.5 in [7], but its proof is different: It relies on previous comparison results for JBSDEs instead of stopping arguments. The stochastic integrals of the BSDE solution are BMO-martingales under the assumptions for Lemma 2.3, which hold e.g. under the conditions for [7, Theorem 3.6]

**Corollary 4.6** *Let $\xi \in L^\infty(\mathcal{F}_T)$ with $\xi \geq C$ for some constant $C > 0$, $K \geq 0$ and set $a(t) := C\exp(-K(T-t))$ and $b(t) = |\xi|_\infty \exp(K(T-t))$, $\forall t \in [0,T]$. Assume $f$ satisfies ($\mathbf{A_{fin}}$) for $c \leq y, y+u \leq d$ for all $c, d \in \mathbb{R}$ with $0 < c < d$, and that $|\widehat{f}_t(y,z)| \leq K|y|$ and $g_t(y,z,0,e) = 0$. Then there exists a unique solution $(Y,Z,U) \in \mathcal{S}^\infty \times \mathcal{L}^2(B) \times \mathcal{L}^2(\widetilde{\mu})$ to the BSDE $(\xi, f)$ with $Y \geq \epsilon$ for some $\epsilon > 0$. Moreover, it holds $a(t) \leq Y_t \leq b(t)$ and $\int Z\,\mathrm{d}B$ and $U * \widetilde{\mu}$ are* BMO($\mathbb{P}$)*-martingales.*

*Proof* This can be shown with a similar argument for the uniqueness as above: Let $(Y', Z', U')$ be another solution to the BSDE $(\xi, f)$ with $Y' \geq \epsilon$ for some $\epsilon > 0$. Then $f$ satisfies ($\mathbf{A_{fin}}$) for $a(t) \wedge \epsilon \leq y,\ y+u \leq b(t) \vee |Y'|_\infty$; hence the solutions coincide by comparison. □

*Example 4.7* As a special case of Corollary 4.6 to be applied in Sect. 5.2, setting $K := \big(\gamma|\varphi|^2_\infty\big)/\big(2(1-\gamma)^2\big)$ for some $\gamma \in (0,1)$ and some predictable and bounded process $\varphi$ we define

$$
\begin{aligned}
f_t(y,z,u) &:= \widehat{f}_t(y,z) + \int_E g_t(y,u,e)\,\zeta(t,e)\,\lambda(\mathrm{d}e) \\
&:= \frac{\gamma}{2(1-\gamma)^2}|\varphi_t|^2 y + \int_E \left(\frac{1}{1-\gamma}((u(e)+y)^{1-\gamma}y^\gamma - y) - u(e)\right)\zeta(t,e)\,\lambda(\mathrm{d}e).
\end{aligned}
$$

From $\frac{\partial g}{\partial y}(t,y,u,e) = \left(\frac{u+y}{y}\right)^{1-\gamma} + \frac{\gamma}{1-\gamma}\left(\frac{u+y}{y}\right)^{-\gamma} - \frac{1}{1-\gamma}$, we see that $f$ is Lipschitz in $y$ within the truncation bounds. Moreover, $g$ is continuously differentiable with bounded derivatives and we have $\frac{\partial g}{\partial u}(t,y,u,e) = \left(\frac{u+y}{y}\right)^{-\gamma} - 1 > -1$, for $c \leq y, y+u \leq d$.

## 4.2 The Case of Infinite Activity

For linear generators of the form

$f_t(y,z,u) := \alpha^0_t + \alpha_t y + \beta_t z + \int_E \gamma_t(e)u(e)\,\zeta(t,e)\,\lambda(\mathrm{d}e)$, with predictable coefficients $\alpha^0$, $\alpha$, $\beta$ and $\gamma$, JBSDE solutions can be represented by an adjoint process. In our context of bounded solutions, one needs rather weak conditions on the adjoint process. This will be used later on in Sect. 5. The idea of proof is standard, cf. [36, Lemma 1.23] for details.

**Lemma 4.8** *Let $f$ be a linear generator of the form above and let $\xi$ be in $L^\infty(\mathcal{F}_T)$.*

1. *Assume that $(Y,Z,U) \in \mathcal{S}^\infty \times \mathcal{L}^2(B) \times \mathcal{L}^2(\widetilde{\mu})$ solves the BSDE $(\xi, f)$. Suppose that the adjoint process $(\Gamma^t_s)_{s\in[t,T]} := (\exp(\int_t^s \alpha_u\,\mathrm{d}u)\mathcal{E}(\int \beta\mathrm{d}B + \gamma * \widetilde{\mu})^s_t)_{s\in[t,T]}$*

*is in* $\mathcal{S}^1$ *for any* $t \le T$ *and* $\alpha^0$ *is bounded. Then* $Y$ *is represented as* $Y_t = \mathbb{E}\big[\Gamma_T^t \xi + \int_t^T \Gamma_s^t \alpha_s^0 \,\mathrm{d}s|\mathcal{F}_t\big]$.

2. *Let* $\alpha^0$, $\alpha$, $\beta$ *and* $\widetilde{\gamma}_t := \int_E |\gamma_t(e)|^2 \zeta(t, e)\lambda(\mathrm{d}e)$, $t \in [0, T]$, *be bounded and* $\gamma \ge -1$. *Then there is a unique solution in* $\mathcal{S}^\infty \times \mathcal{L}^2(B) \times \mathcal{L}^2(\widetilde{\mu})$ *to the BSDE* $(\xi, f)$ *and Part 1. applies.*

Our aim is to prove existence and uniqueness beyond Proposition 4.3 for infinite activity of jumps, that means $\lambda(A)$ may be infinite in (2.6). To show Theorems 4.11 and 4.13, we use a monotone stability approach of [37]: By approximating a generator $f$ of the form (2.7) (with $A$ such that $\lambda(A) = \infty$) by a sequence $(f^n)_{n\in\mathbb{N}}$ of the form (2.7) (with $A_n$ such that $\lambda(A_n) < \infty$) for which solutions' existence is guaranteed, one gets that the limit of these solutions exist and it solves the BSDE with the original data. As in [37], the monotone approximation approach is perceived as being not easy in execution, a main problem usually being to prove strong convergence of the stochastic integral parts for the BSDE. By Proposition 4.9 convergence works for small terminal condition $\xi$. That is why we can not apply this Proposition directly to data $(\xi, f^n)_{n\in\mathbb{N}}$. Instead we sum (converging) solutions for small $1/N$-fractions of the desired terminal condition. This is inspired by the iterative ansatz from [45] for a particular generator. For our generator family, we adapt and elaborate proofs, using e.g. a $\mathcal{S}^1$-closeness argument for the proof of the strong approximation step. Compared to [45], the analysis for our general family of JBSDEs adds clarity and structural insight into what is really needed. It extends the scope of the BSDE stability approach [37, 45], in particular with regards to non-Lipschitz dependencies in the jump-integrand, while the proof shows comparable ease for the (usually laborious) strong approximation step in the setup under consideration. Differently to e.g. [23, 45, 57], no exponential transforms or convolutions are needed here, as our generators are "quadratic" in $U$ but not in $Z$. Despite similarities at first sight, a closer look reveals that Theorem 4.11 is different from [35, Theorem 5.4], both in the method of proof and in scope: They prove existence for small terminal conditions by following the fixed point approach by [56], whereas we show stability for small terminal conditions (Proposition 4.9) and apply a different pasting procedure, approximating not only terminal data but also generators. Here wellposedness of the approximating JBSDEs is obtained directly from classical theory by using comparison and estimates from Sect. 3, which enable us to argue within uniform a-priori bounds for the approximating sequence. Examples in Sect. 5 demonstrate that also the scope of our results is different.

In more detail, the task for the next Theorem 4.11 is to construct generators $(f^{k,n})_{1\le k\le N, n\in\mathbb{N}}$ and solutions $(Y^{k,n}, Z^{k,n}, U^{k,n})$ to the BSDEs with data $(\xi/N, f^{k,n})$ for $N$ large enough such that $(Y^{k,n}, Z^{k,n}, U^{k,n})$ converges if $n \to \infty$ and $(Y^n, Z^n, U^n) := \sum_{k=1}^N (Y^{k,n}, Z^{k,n}, U^{k,n})$ solves the BSDE $(\xi, f^n)$. In this case $(Y^n, Z^n, U^n)$ converges and its limit is a solution candidate for the BSDE $(\xi, f)$. For this program, we next show a stability result for JBSDE.

**Proposition 4.9** *Let* $(\xi^n) \subset L^\infty(\mathcal{F}_T)$ *with* $\xi^n \to \xi$ *in* $L^2(\mathcal{F}_T)$ *and* $(f^n)_{n\in\mathbb{N}}$ *be a sequence of generators with* $f^n(0, 0, 0) = 0$, $\forall n$, *having property* $(B_{\gamma^n})$ *such that* $K_f^{y,z} := \sup_{n\in\mathbb{N}} K_{f^n}^{y,z} < \infty$. *Denote by* $(Y^n, Z^n, U^n) \in \mathcal{S}^\infty \times \mathcal{L}^2(B) \times \mathcal{L}^2(\widetilde{\mu})$

*the solution to the BSDE* $(\xi, f^n)$ *with* $Y^n$ *bounded by* $|\xi|_\infty \exp(K^{y,z}_{f^n}T)$ *and set* $\tilde{c} := |\xi|_\infty \exp(K^{y,z}_f T)$. *Assume that* $Y^n$ *converges pointwise,* $(Z^n, U^n) \to (Z, U)$ *converges weakly in* $\mathcal{L}^2(B) \times \mathcal{L}^2(\widetilde{\mu})$ *and* $|f^n_t(0, 0, u)| \le \widehat{K}|u|_t^2 + \widehat{L}_t$ *for all* $n$ *and* $u$ *with* $|u| \le 2\tilde{c}$, $\widehat{K} \in \mathbb{R}_+$ *and* $\widehat{L} \in L^1(\mathbb{P} \otimes \mathrm{d}t)$. *Then* $(Z^n, U^n)$ *converges to* $(Z, U)$ *strongly in* $\mathcal{L}^2(B) \times \mathcal{L}^2(\widetilde{\mu})$, *if* $|\xi|_\infty \equiv \tilde{c}\exp(-K^{y,z}_f T) \le \exp(-K^{y,z}_f T)/(80 \max\{K^{y,z}_f, \widehat{K}\})$.

*Proof* We note that $(Y^n, Z^n, U^n)$ is uniquely defined by Proposition 4.2. To prove strong convergence of $(Z^n)_{n\in\mathbb{N}}$ and $(U^n)_{n\in\mathbb{N}}$ we consider $\delta Y := Y^n - Y^m$, $\delta Z := Z^n - Z^m$, $\delta U := U^n - U^m$ and apply Itô's formula for general semimartingales to $(\delta Y)^2$ to obtain

$$
\begin{aligned}
(\delta Y_0)^2 = (\delta Y_T)^2 &+ \int_0^T 2\delta Y_{s-}(f^n_s(Y^n_{s-}, Z^n_s, U^n_s) - f^m_s(Y^m_{s-}, Z^m_s, U^m_s))\mathrm{d}s \\
&- \int_0^T \|\delta Z_s\|^2 \mathrm{d}s - 2\int_0^T \delta Y_{s-}\delta Z_s \,\mathrm{d}B_s - \int_0^T\int_E (\delta Y_{s-} + \delta U_s(e))^2 - (\delta Y_{s-})^2 \,\widetilde{\mu}(\mathrm{d}s, \mathrm{d}e) \\
&- \int_0^T\int_E (\delta Y_{s-} + \delta U_s(e))^2 - (\delta Y_{s-})^2 - 2\delta Y_{s-}\delta U_s(e)\,\nu(\mathrm{d}s, \mathrm{d}e).
\end{aligned}
$$

Noting that the stochastic integrals are martingales one concludes that

$$
\begin{aligned}
&\mathbb{E}\Big(\int_0^T 2\delta Y_{s-}(f^n_s(Y^n_{s-}, Z^n_s, U^n_s) - f^m_s(Y^m_{s-}, Z^m_s, U^m_s))\,\mathrm{d}s\Big) \\
&= \mathbb{E}\Big(\int_0^T\int_E \delta U_s(e)^2\,\nu(\mathrm{d}s, \mathrm{d}e)\Big) + \mathbb{E}\Big(\int_0^T \|\delta Z_s\|^2\,\mathrm{d}s\Big) - \mathbb{E}\big((\delta Y_T)^2\big) + \mathbb{E}\big((\delta Y_0)^2\big).
\end{aligned} \tag{4.1}
$$

Using the inequalities $a \le a^2 + 1/4$, $(a+b)^2 \le 2(a^2+b^2)$, $(a+b+c)^2 \le 3(a^2+b^2+c^2)$, the Lipschitz property of $f^n$ in $y$ and $z$ and the estimate for $f^n_t(0, 0, u)$, we have

$$
\begin{aligned}
&|f^n_s(Y^n_{s-}, Z^n_s, U^n_s) - f^m_s(Y^m_{s-}, Z^m_s, U^m_s)| \\
&\le K^{y,z}_{f^n}(|Y^n_{s-}| + \|Z^n_s\|) + K^{y,z}_{f^m}(|Y^m_{s-}| + \|Z^m_s\|) + \widehat{K}|U^n_s|_s^2 + \widehat{L}_s + \widehat{K}|U^m_s|_s^2 + \widehat{L}_s \\
&\le K_1 + 2\widehat{L}_s + K_2(\|\delta Z_s\|^2 + \|Z^n_s - Z_s\|^2 + \|Z_s\|^2 + |\delta U_s|_s^2 + |U^n_s - U_s|_s^2 + |U_s|_s^2),
\end{aligned} \tag{4.2}
$$

where $K_1 := K^{y,z}_f(2\tilde{c} + 1/2) \in \mathbb{R}$, $K_2 := 5\max\{K^{y,z}_f, \widehat{K}\}$ and $|\cdot|_t$ is defined in (2.5). Combing inequalities (4.1) and (4.2) yields

$$
\begin{aligned}
\mathbb{E}\Big(\int_0^T \|\delta Z_s\|^2 + |\delta U_s|_s^2\,\mathrm{d}s\Big) \le 2\mathbb{E}\Big(\int_0^T &|\delta Y_{s-}|(K_1 + 2\widehat{L}_s + K_2(\|\delta Z_s\|^2 + \|Z^n_s - Z_s\|^2 + \|Z_s\|^2 \\
&+ |\delta U_s|_s^2 + |U^n_s - U_s|_s^2 + |U_s|_s^2))\,\mathrm{d}s\Big) + \mathbb{E}\big((\xi^n - \xi^m)^2\big).
\end{aligned}
$$

Let us recall that the predictable projection of $Y$, denoted by $Y^{\mathrm{p}}$, is defined as the unique predictable process $X$ such that $X_\tau = \mathbb{E}(Y_\tau|\mathcal{F}_{\tau-})$ on $\{\tau < \infty\}$ for all predictable times $\tau$. For $Y^n$ it holds $(Y^n)^{\mathrm{p}} = Y^n_-$. This follows from [31, Proposition I.2.35.] using that $Y^n$ is càdlàg, adapted and quasi-left-continuous, as $\Delta Y_\tau = \Delta U * \widetilde{\mu}_\tau = 0$ on $\{\tau < \infty\}$ holds for all predictable times $\tau$ thanks to the absolute continuity of the compensator $\nu$. Noting that $1 - 2K_2|\delta Y_{s-}| \ge 1 - 4K_2\tilde{c} \ge 3/4$ and setting $Y := \lim_{n\to\infty} Y^n$ we deduce by the weak convergence of $(Z^n)_{n\in\mathbb{N}}$ and

$(U^n)_{n\in\mathbb{N}}$, $Y^n_- = (Y^n)^{\mathrm{p}} \uparrow (Y)^{\mathrm{p}}$ as $n\to\infty$ and by Lebesgue's dominated convergence theorem

$$
\begin{aligned}
&\frac{3}{4}\mathbb{E}\Big(\int_0^T \|Z^n_s - Z_s\|^2 + |U^n_s - U_s|_s^2\,\mathrm{d}s\Big)\\
&\le \frac{3}{4}\liminf_{m\to\infty}\mathbb{E}\Big(\int_0^T \|Z^n_s - Z^m_s\|^2 + |U^n_s - U^m_s|_s^2\,\mathrm{d}s\Big)\\
&\le \liminf_{m\to\infty} 2\mathbb{E}\Big(\int_0^T |\delta Y_{s-}|(K_1 + 2\widehat{L}_s + K_2(\|Z^n_s - Z_s\|^2 + \|Z_s\|^2 + |U^n_s - U_s|_s^2 + |U_s|_s^2))\,\mathrm{d}s\Big)\\
&\qquad + \mathbb{E}((\xi^m - \xi^n)^2)\\
&= 2\mathbb{E}\Big(\int_0^T |Y^n_{s-} - (Y_s)^{\mathrm{p}}|(K_1 + 2\widehat{L}_s + K_2(\|Z^n_s - Z_s\|^2 + \|Z_s\|^2 + |U^n_s - U_s|_s^2 + |U_s|_s^2))\,\mathrm{d}s\Big)\\
&\quad + \mathbb{E}((\xi - \xi^n)^2).
\end{aligned}
$$

Noting $3/4 - 2K_2|Y^n_{s-} - (Y_s)^{\mathrm{p}}| \ge 3/4 - 4K_2\tilde{c} \ge 1/2$, one obtains with dominated convergence

$$
\begin{aligned}
&\frac{1}{2}\limsup_{n\to\infty}\mathbb{E}\Big(\int_0^T \|Z^n_s - Z_s\|^2 + |U^n_s - U_s|_s^2\,\mathrm{d}s\Big)\\
&\qquad \le \limsup_{n\to\infty} 2\mathbb{E}\Big(\int_0^T |Y^n_{s-} - (Y_s)^{\mathrm{p}}|(K_1 + 2\widehat{L}_s + \|Z_s\|^2 + |U_s|_s^2)\,\mathrm{d}s\Big) + \mathbb{E}((\xi^n - \xi)^2) = 0.
\end{aligned}
$$

□

We will need the following result which is a slight variation of [37, Lemma 2.5].

**Lemma 4.10** *Let $(Z^n)_{n\in\mathbb{N}}$ be convergent in $\mathcal{L}^2(B)$ and $(U^n)_{n\in\mathbb{N}}$ convergent in $\mathcal{L}^2(\widetilde{\mu})$. Then there exists a subsequence $(n_k)_{k\in\mathbb{N}}$ such that*

$$\sup_{n_k}\|Z^{n_k}\| \in L^2(\mathbb{P}\otimes \mathrm{d}t) \text{ and } \sup_{n_k}|U^{n_k}_t|_t \in L^2(\mathbb{P}\otimes \mathrm{d}t).$$

*Proof* The result for $(Z^n)_{n\in\mathbb{N}}$ is from [37] and the argument for $(U^n)_{n\in\mathbb{N}}$ is analogous. □

**Theorem 4.11** [Monotone stability, infinite activity] *Let $\xi \in L^\infty(\mathcal{F}_T)$ and let $(f^n)_n$ be a sequence of generators satisfying condition $(B_{\gamma^n})$ with $K^{y,z}_f := \sup_{n\in\mathbb{N}} K^{y,z}_{f^n} < \infty$. Assume that*

1. *there is $(\widehat{Y}, \widehat{Z}, \widehat{U})$ in $\mathcal{S}^\infty \times \mathcal{L}^2(B) \times \mathcal{L}^2(\widetilde{\mu})$ with $\widehat{U}$ bounded and $f^n_t(\widehat{Y}_{t-}, \widehat{Z}_t, \widehat{U}_t) \equiv 0$ for all $n$,*
2. *for all $u \in L^0(\mathcal{B}(E), \lambda)$ with $|u| \le |\widehat{U}|_\infty + 2|\xi|_\infty \exp(K^{y,z}_f T)$ there exists $\widehat{K} \in \mathbb{R}_+$ and a process $\widehat{L} \in L^1(\mathbb{P}\otimes \mathrm{d}t)$ such that $|f^n_t(0,0,u)| \le \widehat{K}|u|_t^2 + \widehat{L}_t$ for each $n \in \mathbb{N}$,*
3. *the sequence $(f^n)_{n\in\mathbb{N}}$ converges pointwise and monotonically to a generator $f$,*

4. *there is a* BMO($\mathbb{P}$)*-martingale* $M$ *such that for all truncated generators* $f_t^{n,\hat{c}}(y,z,u) := f_t^n\big((y \vee (-\hat{c})) \wedge \hat{c}, z, (u \vee (-2\hat{c})) \wedge (2\hat{c})\big)$ *with* $\hat{c} := |\widehat{Y}|_\infty + (|\widehat{U}|_\infty/2) + \exp(K_f^{y,z}T)|\xi|_\infty$ *holds* $\int_t^T f_s^{n,\hat{c}}(Y_{s-}, Z_s, U_s)\,\mathrm{d}s \le \langle M\rangle_T - \langle M\rangle_t$ *or* $-\int_t^T f_s^{n,\hat{c}}(Y_{s-}, Z_s, U_s)\,\mathrm{d}s \le \langle M\rangle_T - \langle M\rangle_t$ *for all* $n \in \mathbb{N}$, $(Y,Z,U) \in \mathcal{S}^\infty \times \mathcal{L}^2(B) \times \mathcal{L}^2(\widetilde{\mu})$, *and*
5. *for all* $(Y,Z,U) \in \mathcal{S}^\infty \times \mathcal{L}^2(B) \times \mathcal{L}^2(\widetilde{\mu})$ *and* $(U^n)_{n\in\mathbb{N}} \in \mathcal{L}^2(\widetilde{\mu})$ *with* $U^n \to U$ *in* $L^2(\widetilde{\mu})$ *it holds* $f^n(Y_-, Z, U^n) \longrightarrow f(Y_-, Z, U)$ *in* $L^1(\mathbb{P}\otimes \mathrm{d}t)$.

*Then*

(i) *there exists a solution* $(Y,Z,U) \in \mathcal{S}^\infty \times \mathcal{L}^2(B) \times \mathcal{L}^2(\widetilde{\mu})$ *for the BSDE* $(\xi, f)$, *with* $\int Z\,\mathrm{d}B$ *and* $U * \widetilde{\mu}$ *being* BMO($\mathbb{P}$)*-martingales, and*
(ii) *this solution is unique if additionally* $f$ *satisfies condition* $(\mathbf{A}'_\gamma)$.

*Proof* Let us first outline the overall program of the proof. We want to construct generators $(f^{k,n})_{1\le k\le N, n\in\mathbb{N}}$ and solutions $(Y^{k,n}, Z^{k,n}, U^{k,n})$ to the BSDEs $(\xi/N, f^{k,n})$ for $N$ sufficiently large (to employ Proposition 4.9 and get that $((Y^{k,n}, Z^{k,n}, U^{k,n}))_{n\in\mathbb{N}}$ converges and $(Y^n, Z^n, U^n) := \sum_{k=1}^N (Y^{k,n}, Z^{k,n}, U^{k,n})$ solves the BSDE $(\xi, f^n)$). We show that if for some $k < N$ and all $1 \le l \le k$ and $n \in \mathbb{N}$ we have already constructed generators $(f^{l,n})_{1\le l\le k, n\in\mathbb{N}}$ such that there exist solutions $((Y^{l,n}, Z^{l,n}, U^{l,n}))_{n\in\mathbb{N}}$ to the BSDEs $(\xi/N, f^{l,n})$ converging for $n\to\infty$, with $|Y^{l,n}|_\infty \le \exp(K_f^{y,z}T)|\xi|_\infty/N =: \tilde{c}$, then for $\overline{Y}^{k,n} := \widehat{Y} + \sum_{l=1}^k Y^{l,n}$ with $\overline{Z}^{k,n}$ and $\overline{U}^{k,n}$ defined analogously and

$$f_t^{k+1,n}(y,z,u) := f_t^n\big(y + \overline{Y}_{t-}^{k,n}, z + \overline{Z}_t^{k,n}, u + \overline{U}_t^{k,n}\big) - f_t^n\big(\overline{Y}_{t-}^{k,n}, \overline{Z}_t^{k,n}, \overline{U}_t^{k,n}\big) \tag{4.3}$$

there are solutions $(Y^{k+1,n}, Z^{k+1,n}, U^{k+1,n}) \in \mathcal{S}^\infty \times \mathcal{L}^2(B) \times \mathcal{L}^2(\widetilde{\mu})$ to the BSDEs $(\xi/N, f^{k+1,n})$, converging (in $n$) and satisfying $|Y^{k+1,n}|_\infty \le \tilde{c}$. Starting initially with the triple $(Y^{0,n}, Z^{0,n}, U^{0,n})$ defined by $(Y^{0,n}, Z^{0,n}, U^{0,n}) := (\widehat{Y}, \widehat{Z}, \widehat{U})$, formula (4.3) gives an inductive construction of the generators $f^{k,n}$ and triples $(Y^{k,n}, Z^{k,n}, U^{k,n}) \in \mathcal{S}^\infty \times \mathcal{L}^2(B) \times \mathcal{L}^2(\widetilde{\mu})$ solving the BSDE $(\xi/N, f^{k,n})$ and converging for $n\to\infty$ with $|Y^{k,n}|_\infty \le \tilde{c}$ for each $n \in N$ and $1 \le k \le N$.

Note that $f^{k+1,n}$ is Lipschitz continuous in $y$ and $z$ with Lipschitz constant $K_{f^n}^{y,z}$, locally Lipschitz in $u$ and satisfies condition $(A_{\gamma^{k+1,n}})$ with

$$\gamma_s^{k+1,n}(y,z,u,u',e) := \gamma_s^n\big(y + \overline{Y}_{s-}^{k,n}, z + \overline{Z}_s^{k,n}, u + \overline{U}_s^{k,n}(e), u' + \overline{U}_s^{k,n}(e), e\big)$$

and $f_t^{k+1,n}(0,0,0) \equiv 0$. Hence by the existence and uniqueness result for the finite activity case (see Proposition 4.2), there exists a unique solution $(Y^{k+1,n}, Z^{k+1,n}, U^{k+1,n})$ to the BSDE $(\xi/N, f^{k+1,n})$ such that $Y^{k+1,n}$ is bounded by $\tilde{c}$.
To apply Proposition 4.9, we have to check that the sequence $(Y^{k+1,n})_{n\in\mathbb{N}}$ converges pointwise, that $(Z^{k+1,n}, U^{k+1,n})_{n\in\mathbb{N}}$ converges weakly in $\mathcal{L}^2(B) \times \mathcal{L}^2(\widetilde{\mu})$

and that $f^{k+1,n}(0,0,u)$ can be locally bounded by an affine function in $|u|^2$. Having telescoping sums in (4.3) implies that $(\overline{Y}^{l,n}, \overline{Z}^{l,n}, \overline{U}^{l,n})$ solves the BSDE $(\widehat{Y}_T + l\xi/N, f^n)$. By the comparison result of Proposition 3.1, the sequences $(\overline{Y}^{k,n})_{n\in\mathbb{N}}$ and $(\overline{Y}^{k+1,n})_{n\in\mathbb{N}}$ are monotonic (and bounded) in $n$ so that finite limits $\lim_{n\to\infty} Y^{k+1,n} = \lim_{n\to\infty} \overline{Y}^{k+1,n} - \lim_{n\to\infty} \overline{Y}^{k,n}$ exists, $\mathbb{P}\otimes \mathrm{d}t$-a.e. By Lemma 2.3, the sequences $(\overline{Z}^{k,n}, \overline{U}^{k,n})_{n\in\mathbb{N}}$ and $(\overline{Z}^{k+1,n}, \overline{U}^{k+1,n})_{n\in\mathbb{N}}$ are bounded in $\mathcal{L}^2(B)\times\mathcal{L}^2(\widetilde{\mu})$; hence $(Z^{k+1,n}, U^{k+1,n})$ is weakly convergent in $\mathcal{L}^2(B)\times\mathcal{L}^2(\widetilde{\mu})$ along a subsequence which we still index by $n$ for simplicity. Due to the Lipschitz continuity of $f^n$ and Assumption 2, we get for all $|u|\le 2\tilde{c}$ that

$$\begin{aligned}\left|f_t^{k+1,n}(0,0,u)\right| &\le \left|f_t^n\big(\overline{Y}_{t-}^{k,n}, \overline{Z}_t^{k,n}, u+\overline{U}_t^{k,n}\big) - f_t^n\big(\overline{Y}_{t-}^{k,n}, \overline{Z}_t^{k,n}, \overline{U}_t^{k,n}\big)\right| \\ &\le 2K_{f^n}^{y,z}\big(|\overline{Y}_{t-}^{k,n}| + \|\overline{Z}_t^{k,n}\|\big) + \widehat{K}\big(|u+\overline{U}_t^{k,n}|_t^2 + |\overline{U}_t^{k,n}|_t^2\big) + 2\widehat{L}_t \\ &\le 2\widehat{K}|u|_t^2 + \widetilde{L}_t,\end{aligned}$$

where $\widetilde{L}_t = 2K_f^{y,z}(\hat{c} + \sup_{n\in\mathbb{N}}\|\overline{Z}_t^{k,n}\|^2 + 1/4) + 3\widehat{K}\sup_{n\in\mathbb{N}}|\overline{U}_t^{k,n}|_t^2 + 2\widehat{L}_t$. Here we used that by induction hypothesis $(\overline{Z}^{k,n}, \overline{U}^{k,n})_n$ is convergent so that $\sup_{n\in\mathbb{N}}(\|\overline{Z}_t^{k,n}\|^2 + |\overline{U}_t^{k,n}|_t^2)$ is $\mathbb{P}\otimes \mathrm{d}t$-integrable by Lemma 4.10 along a subsequence which again for simplicity we still index by $n$. This implies that $\widetilde{L}\in L^1(\mathbb{P}\otimes\mathrm{d}t)$, and therefore by Proposition 4.9, the sequence $(Z^n, U^n) := (\overline{Z}^{N,n}, \overline{U}^{N,n})$ converges in $\mathcal{L}^2(B)\times\mathcal{L}^2(\widetilde{\mu})$ to some $(Z,U)$ in $\mathcal{L}^2(B)\times\mathcal{L}^2(\widetilde{\mu})$ while $(Y^n) := (\overline{Y}^{N,n})$ converges to some $Y$. Hence, $f^n(Y_-^n, Z^n, U^n) - f^n(Y_-, Z, U^n)$ converges to 0 in $L^1(\mathbb{P}\otimes\mathrm{d}t)$ and by Assumption 5. we have $f^n(Y_-^n, Z^n, U^n) \to f(Y_-, Z, U)$ in $L^1(\mathbb{P}\otimes\mathrm{d}t)$. The stochastic integrals $(Z^n - Z^m)\bullet B$ and $(U^n - U^m)*\widetilde{\mu}$ belong to $\mathcal{S}^2\subset\mathcal{S}^1$ by Doob's inequality, with $\mathcal{S}^1$-norms being bounded by a multiple of $\|Z^n - Z^m\|_{\mathcal{L}^2(B)}$ and $\|U^n - U^m\|_{\mathcal{L}^2(\widetilde{\mu})}$ respectively. Since $|Y^n - Y^m|_{\mathcal{S}^1}$ is dominated by

$$\|f^n(Y_-^n, Z^n, U^n) - f^m(Y_-^m, Z^m, U^m)\|_{L^1(\mathbb{P}\otimes\mathrm{d}t)} + C(\|Z^n - Z^m\|_{\mathcal{L}^2(B)} + \|U^n - U^m\|_{\mathcal{L}^2(\widetilde{\mu})})$$

for some constant $C>0$ with the bound tending to 0 as $n,m\to 0$, we can take $Y$ in $\mathcal{S}^1$ due to completeness of $\mathcal{S}^1$; see [18, VII. 3, 64].
Finally, $(Y,Z,U)$ solves the BSDE $(\xi, f)$ since the approximating solutions $(Y^n, Z^n, U^n)_{n\in\mathbb{N}}$ of the BSDE $(\xi, f^n)_{n\in\mathbb{N}}$ converge to some $(Y,Z,U)\in\mathcal{S}^\infty\times\mathcal{L}^2(B)\times\mathcal{L}^2(\widetilde{\mu})$ and $f^n(Y_-^n, Z^n, U^n)$ tends to $f(Y_-, Z, U)$ in $L^1(\mathbb{P}\otimes\mathrm{d}t)$. Hence, we have $\int_0^t f_s^n(Y_{s-}^n, Z_s^n, U_s^n)\mathrm{d}s \to \int_0^t f_s(Y_{s-}, Z_s, U_s)\mathrm{d}s$, $\int_0^t Z_s^n\mathrm{d}B_s \to \int_0^t Z_s\mathrm{d}B_s$ and $U^n * \widetilde{\mu}_t \to U*\widetilde{\mu}_t$ $\mathbb{P}$-a.s. (along a subsequence) for all $0\le t\le T$. □

The next corollary to Theorem 4.11 provides conditions under which the $Z$-component of the JBSDE solution vanishes. Such is useful for applications in a pure-jump context (see e.g. Sect. 5.1.2 or [17]) with weak PRP by $\widetilde{\mu}$ alone (cf. Example 2.1, Parts 1, 3, 4.), without a Brownian motion. Clearly an independent Brownian motion can always be added by enlarging the probability space, but it is then natural to ask for a JBSDE solution with trivial $Z$-component, adapted to the original filtration. Instead of re-doing the entire argument leading to Theorem 4.11 but now for JBSDEs solely driven by a random measure $\widetilde{\mu}$ with generators without a $z$-argument, the next result gives a direct argument to this end. An example where the corollary is applied is given in Sect. 5.1.2.

**Corollary 4.12** *Let $\mu = \mu^X$ be the random measure associated to a pure-jump process $X$, such that the compensated random measure $\widetilde{\mu}$ alone has the weak PRP (see (2.2)) with respect to the usual filtration $\mathbb{F}^X$ of $X$. Let $B$ be a $d$-dimensional Brownian motion independent of $X$. With respect to $\mathbb{F} := \mathbb{F}^{B,X}$, let $f, (f^n)_n, \xi$ satisfy the assumptions of Theorem 4.11 with $\widehat{Z} = 0$ and $f$ satisfying $(\mathbf{A}'_\gamma)$. Let $\xi$ be in $L^\infty(\mathcal{F}_T^X)$ and $f, f^n$ be $\mathcal{P}(\mathbb{F}^X) \otimes \mathcal{B}(\mathbb{R}^{d+1}) \otimes \mathcal{B}(L^0(\mathcal{B}(E)))$-measurable. Then the JBSDE $(\xi, f)$ admits a unique solution $(Y, Z, U)$ in $\mathcal{S}^\infty \times \mathcal{L}^2(B) \times \mathcal{L}^2(\widetilde{\mu})$, and we have that $Y$ is $\mathbb{F}^X$-adapted, $Z = 0$, and $U$ can be taken as measurable with respect to $\widetilde{\mathcal{P}}(\mathbb{F}^X)$.*

*Proof* Let $B'$ be a (1-dimensional) Brownian motion independent of $(B, X)$. Then $\bar{B} := (B, B')$ is a $(d+1)$-dimensional Brownian motion independent of $X$. Let $\mathbb{F}' := \mathbb{F}^{B',X}$ and $\bar{\mathbb{F}} := \mathbb{F}^{\bar{B},X}$ denote the usual filtrations of $(B', X)$ and $(\bar{B}, X)$. As in Example 2.1, $(B, \widetilde{\mu})$, $(B', \widetilde{\mu})$ and $(\bar{B}, \widetilde{\mu})$ each admits the weak PRP w.r.t. $\mathbb{F}$, $\mathbb{F}'$ and $\bar{\mathbb{F}}$ respectively. Now consider the generator function $\widetilde{f}$ that does not depend on $z$ and is defined by $\widetilde{f}_t(y, u) := f_t(y, 0, u)$. Because $\widehat{Z} = 0$, the conditions for Theorem 4.11 are met by $\widetilde{f}^n := f^n(\cdot, 0, \cdot)$. In addition, $\widetilde{f}$ satisfies condition $(\mathbf{A}'_\gamma)$ since $f$ does. Since $\xi$ is $\mathcal{F}_T^X$-measurable and $\widetilde{f}$ is $\mathcal{P}(\mathbb{F}^X) \otimes \mathcal{B}(\mathbb{R}) \otimes \mathcal{B}(L^0(\mathcal{B}(E)))$-measurable, then by Theorem 4.11 the JBSDE $(\xi, \widetilde{f})$ simultaneously admits unique solutions $(Y, Z, U)$, $(Y', Z', U')$ and $(\bar{Y}, \bar{Z}, \bar{U})$ in the respective $\mathcal{S}^\infty \times \mathcal{L}^2(\cdot) \times \mathcal{L}^2(\widetilde{\mu})$-spaces for each of the filtrations $\mathbb{F}$, $\mathbb{F}'$ and $\bar{\mathbb{F}}$. Noting that both $\mathbb{F}$ and $\mathbb{F}'$ are sub-filtrations of $\bar{\mathbb{F}}$, we get by uniqueness of $(\bar{Y}, \bar{Z}, \bar{U})$ that $Z \bullet B = Z' \bullet B' = \bar{Z} \bullet \bar{B}$ and that $Y$ is $\mathbb{F}^X$-adapted. The former implies $Z = Z' = 0$ by the strong orthogonality of $B$ and $B'$. The claim follows, by noting that the JBSDE gives the (unique) canonical decomposition of the special semimartingale $Y$ and using weak predictable martingale representation in $\mathbb{F}^X$. □

A natural ansatz to approximate an $f$ of the form (2.7) with $\lambda(A) = \infty$ is by taking

$$f_t^n(y, z, u) := \widehat{f}_t(y, z) + \int_{A^n} g_t(u(e), e)\, \zeta(t, e)\, \lambda(\mathrm{d}e), \tag{4.4}$$

for an increasing sequence $(A_n)_{n \in \mathbb{N}} \uparrow A$ of measurable sets with $\lambda(A_n) < \infty$ (as $\lambda$ is $\sigma$-finite).

**Theorem 4.13** [Wellposedness, infinite activity of jumps] *Let the generator $f$ of the JBSDE be of the form (2.7) and let $\xi$ be in $L^\infty(\mathcal{F}_T)$. Let $\widehat{f}$ be Lipschitz continuous with respect to $(y, z)$ uniformly in $(\omega, t, u)$, and let $u \mapsto g(t, u, e)$ be absolutely continuous in $u$, for all $(\omega, t, e)$, with its density function $g'(t, u, e)$ being strictly greater than $-1$ and locally bounded (in $u$) from above.*

*Assume that*

1. *there exists* $(\widehat{Y}, \widehat{Z}, \widehat{U}) \in \mathcal{S}^\infty \times \mathcal{L}^2(B) \times \mathcal{L}^2(\widetilde{\mu})$ *with* $|\widehat{U}|_\infty < \infty$, $\widehat{f}_t(\widehat{Y}_t, \widehat{Z}_t) \equiv 0$, $g_t(\widehat{U}_t(e), e) \equiv 0$,
2. *the function* $g$ *is locally bounded in* $|u|^2$ *uniformly in* $(\omega, t, e)$*, i.e. locally in* $u$ *(for any bounded neighborhood* $N$ *of* 0*) there exists a* $K > 0$ *such that* $|g_t(u, e)| \leq K|u|^2$ *(for all* $u \in N$*),*
3. *and there exists* $D : \mathbb{R} \mapsto \mathbb{R}$ *continuous such that either* $g \geq 0$ *and* $\widehat{f}_t(y, z) \geq D(y)$ *for* $|y| \leq \hat{c} := |\widehat{Y}|_\infty + (|\widehat{U}|_\infty/2) + |\xi|_\infty \exp(K^{y,z}_{\widehat{f}} T)$*, or* $g \leq 0$ *and* $\widehat{f}_t(y, z) \leq D(y)$ *for* $|y| \leq \hat{c}$*.*

*Then*

*(i) there exists a solution* $(Y, Z, U) \in \mathcal{S}^\infty \times \mathcal{L}^2(B) \times \mathcal{L}^2(\widetilde{\mu})$ *to the JBSDE and for each solution triple the stochastic integrals* $\int Z\,\mathrm{d}B$ *and* $U * \widetilde{\mu}$ *are* BMO-*martingales, and*

*(ii) this solution is unique if moreover the function* $g$ *satisfies condition* $(\mathbf{A_{infi}})$*.*

*Finally, the same statements hold if condition* 1*. is replaced by assuming that* $f$ *is not depending on* $y$*, i.e.* $f_t(y, z, u) = f_t(z, u)$*, and that* $\widehat{f}$ *is bounded.*

*Proof* We check that the assumptions of Theorem 4.11 are satisfied. Clearly conditions 1. and 2. are sufficient for assumptions 1. and 2. in Theorem 4.11. The $f^n$ given by (4.4) satisfy conditions $(B_{\gamma^n})$ (cf. Example 3.8 and note $\lambda(A_n) < \infty$) and the sequence $(f^n)$ is either monotone increasing or monotone decreasing, depending on the sign of $g$. For the next assumption 4, $f^{n,\hat{c}}$ is bounded from above (or resp. below) by $\sup_{|y|\leq\hat{c}} D(y)$ (respectively $\inf_{|y|\leq\hat{c}} D(y)$). To show that also condition 5. of Theorem 4.11 holds, we prove that $g_t(U^n_t(e), e)\,\mathbb{1}_{A_n}(e)$ converge to $g_t(U_t(e), e)$ in $L^1(\mathbb{P} \otimes \nu)$ as $n \to \infty$ for $U^n \to U$ in $\mathcal{L}^2(\widetilde{\mu})$, recalling (2.1). We set $B_n := \big(g_t(U^n_t(e), e) - g_t(U_t(e), e)\big)\,\mathbb{1}_{A_n}(e)$ and $C_n := g_t(U_t(e), e)\,\mathbb{1}_{A^c_n}(e)$. Both sequences $(B_n)_{n\in\mathbb{N}}$ and $(C_n)_{n\in\mathbb{N}}$ converge to 0 $\mathbb{P} \otimes \nu$-a.e. since $U^n \to U$ in $L^2(\mathbb{P} \otimes \nu)$, $g$ is locally Lipschitz in $u$ and $A^c_n \downarrow \emptyset$. Moreover, they are bounded by integrable random variables. In particular, $B_n$ is bounded by $\widehat{K}\big(\sup_{n\in\mathbb{N}} |U^n_t(e)|^2 + |U_t(e)|^2\big)$ for some $\widehat{K} > 0$ which is integrable along a subsequence due to Lemma 4.10. Hence applying the dominated convergence theorem yields the desired result.
In the alternative case without the Assumption 1, existence is still guaranteed. Indeed, let $f_t(y, z, u) = f_t(z, u)$ and $\widehat{f}$ be bounded. Denoting $\widetilde{f}_t(z, u) := f_t(z, u) - f_t(0, 0)$ and $\widetilde{\xi} := \xi + \int_0^T f_t(0, 0)\,\mathrm{d}t$, there exists a unique solution $(\widetilde{Y}, Z, U)$ in $\mathcal{S}^\infty \times \mathcal{L}^2(B) \times \mathcal{L}^2(\widetilde{\mu})$ to the BSDE $(\widetilde{\xi}, \widetilde{f})$ with $\int Z\,\mathrm{d}B$ and $U * \widetilde{\mu}$ being BMO-martingales by the first version of this theorem and noting that $g_t(0, e) \equiv 0$ and $f_t(0, 0) = \widehat{f}_t(0)$ is bounded. Taking $Y_t := \widetilde{Y}_t - \int_0^t \overline{f}_s(0, 0)\,\mathrm{d}s$, we obtain that $(Y, Z, U)$ solves the BSDE with the data $(\xi, f)$. If moreover the function $f$ satisfies $(\mathbf{A_{infi}})$, then $f$ satisfies $(\mathbf{A'_\gamma})$ (cf. Example 3.8) and hence uniqueness follows from applicability of the comparison argument in Proposition 3.1. □

*Example 4.14* A function $g$ is locally bounded in $|u|^2$ in the sense of condition 2. in Theorem 4.13 if, for instance, $u \mapsto g_t(u, e)$ is twice differentiable for any $(\omega, t, e)$, with the second derivative in $u$ being locally bounded uniformly in $(\omega, t, e)$, and $g_t(0, e) \equiv g_t'(0, e) \equiv 0$ vanishing.

*Example 4.15* An example for a generator that satisfies the assumption of Theorem 4.13 but has super-exponential growth is $f$ of the form (2.7) with $\widehat{f} \equiv 0$ and $g_t(u) = \exp(|u^+|^2) - 1$. Here exists, in general, no $\gamma \in (0, \infty)$ such that $-\frac{1}{\gamma}(e^{-u} + u - 1) \le g_t(u) \le \frac{1}{\gamma}(e^u - u - 1)$ holds for all $u$ and $t$. Thus, the example appears not to satisfy exponential growth assumptions as formulated, e.g., in [1] [Assumption (H), Theorem 1], [23] [2.condition, Definition 5.6] or [35] [Assumption 3.1].

Note that convexity is not required for our theorems on comparison, existence and uniqueness for JBSDEs. Many relevant applications are convex in nature but not all, see examples in Sect. 5.1.2.

# 5 Examples and Applications: Optimal Control in Finance

Results for JBSDEs in the literature commonly rely on combinations of several, often quite technical, assumptions. But their scope can be difficult to judge at first sight without examples, and to verify them may be not easy. This section discusses key applications that JBDEs have found in mathematical finance, and it illustrates by concrete examples the applicability and the scope of the theory from previous sections. The examples do also help to shed some light on connections and differences to related literature. Counter examples might caution against potential pitfalls.

The applications in Sect. 5.1 are about exponential utility maximization, possibly with an additive liability or non-convex constraints. This problem is closely related to the entropic risk measure and to (exponential) utility indifference valuation; It has indeed been a standard motivation for much of the (quadratic, non-Lipschitz) JBSDE theory, cf. [5, 7, 8, 35, 40, 44]. A result on existence of a solution for the specific JBSDE of this application has been presented in [45], being more general in some aspects (jump-diffusion stock price) but less so in others (multiple assets, time-inhomogeneous $\mu$). Section 5.2 shows how a change of coordinates can transform a JBSDE, which arises from an optimal control problem for power utility maximization but appears to be out of scope at first, into a JBSDE for which theory of Sect. 4 can be applied to derive optimal controls and fully characterize the solution to the control problem by JBSDE solutions, like in [29, 30], by using martingale optimality principles. To our best knowledge, the considered power-utility problem with jumps and a multiplicative liability is solved for the first time in this spirit. Finally, Sect. 5.3 derives JBSDE solutions for the no-good-deal valuation problem in incomplete markets, which is posed over a multiplicatively stable sub-family of arbitrage-free pricing measures. Also here, where the (non-linear) JBSDE generator is even Lipschitz, the slight generalization of Proposition 3.1 to the classical comparison result by [54] is useful; Indeed, the process $\gamma$ in (5.17) is such that the martingale condition (3.1) for Proposition 3.1 can be readily verified, while the same appears not clear for condition $(\mathbf{A}_\gamma)$ in [54, Theorem 2.3].

Sections 5.1.1, 5.2 and 5.3 consider models for a financial market within the framework of Sect. 2, consisting of one savings account with zero interest rate (for simplicity) and $k$ risky assets ($k \leq d$), whose discounted prices evolve according to the stochastic differential equation

$$\mathrm{d}S_t = \operatorname{diag}(S_t^i)_{1\leq i\leq k}\sigma_t(\varphi_t \mathrm{d}t + \mathrm{d}B_t) =: \operatorname{diag}(S_t)\mathrm{d}R_t, \quad t \in [0, T], \tag{5.1}$$

with $S_0 \in (0, \infty)^k$, where the market price of risk $\varphi$ is a predictable $\mathbb{R}^d$-valued process, with $\varphi_t \in \operatorname{Im} \sigma_t^T = (\operatorname{Ker} \sigma_t)^\perp$ for all $t \leq T$, and $\sigma$ is a predictable $\mathbb{R}^{k\times d}$-valued process such that $\sigma$ is of full rank $k$ (i.e. $\det(\sigma_t\sigma_t^T) \neq 0$ $\mathbb{P} \otimes \mathrm{d}t$-a.e.) and integrable w.r.t. $\widehat{B} := B + \int_0^\cdot \varphi_t \,\mathrm{d}t$. We take the market price of risk $\varphi$ to be bounded $\mathbb{P} \otimes \mathrm{d}t$-a.e. The market is free of arbitrage in the sense that the set $\mathcal{M}^{\mathrm{e}}$ of equivalent local martingale measures for $S$ is non-empty. In particular, $\mathcal{M}^{\mathrm{e}}$ contains the minimal martingale measure

$$\mathrm{d}\widehat{\mathbb{P}} := \mathcal{E}(-\varphi \bullet B)_T \,\mathrm{d}\mathbb{P} = \exp\Big( - \varphi \bullet B_T - \frac{1}{2}\int_0^T |\varphi_t|^2 \,\mathrm{d}t\Big)\mathrm{d}\mathbb{P}, \tag{5.2}$$

under which $\widehat{B}$ is a Brownian motion and $S$ is a local martingale by Girsanov's theorem. Clearly, the market (5.1) is incomplete in general (even if $k = d$ and $\sigma$ is invertible, when the random measure is not trivial, filtration then being non-Brownian), cf. Example 2.1.

## 5.1 *Exponential Utility Maximization*

For a market with stock prices as in (5.1), consider the expected utility maximization problem

$$v_t(x) = \operatorname*{ess\,sup}_{\theta\in\Theta} \mathbb{E}\big(u\big(X_T^{\theta,t,x} - \xi\big)|\mathcal{F}_t\big), \quad t \leq T,\ x \in \mathbb{R}, \tag{5.3}$$

for the exponential utility function $u(x) := -\exp(-\alpha x)$ with absolute risk aversion parameter $\alpha > 0$, with some additive liability $\xi$ and for wealth processes $X^{\theta,t,x}$ of admissible trading strategies $\theta$ as defined below. We are going to show, how the value process $v$ and optimal trading strategy $\theta^*$ for the problem (5.3) can be fully described by JBSDE solutions for two distinct problem cases.

### 5.1.1 Case with Continuous Price Processes of Risky Assets

The set of available trading strategies $\Theta$ consists of all $\mathbb{R}^d$-valued, predictable, S-integrable processes $\theta$ for which the following two conditions are satisfied: $\mathbb{E}(\int_0^T |\theta_t|^2\,\mathrm{d}t)$ is finite, and the family $\{\exp(-\alpha\int_0^\tau \theta_t \mathrm{d}\widehat{B}_t) \mid \tau \text{ stopping time}, \tau \le T\}$ of random variables is uniformly integrable under $\mathbb{P}$. Starting from initial capital $x \in \mathbb{R}$ at some time $t \le T$, the wealth process corresponding to investment strategy $\theta \in \Theta$ is given by $X_s^\theta = X_s^{\theta,t,x} = x + \int_t^s \theta_u\,\mathrm{d}\widehat{B}_u$, $s \in [t,T]$.

For this subsection, we assume $k = d$ (so $f$ will not be quadratic in $z$). Let $(Y,Z,U)$ in $\mathcal{S}^\infty_{\widehat{\mathbb{P}}} \times \mathcal{L}^2_{\widehat{\mathbb{P}}}(\widehat{B}) \times \mathcal{L}^2_{\widehat{\mathbb{P}}}(\widetilde{\mu})$ be the unique solution to the BSDE $Y_t = \xi + \int_t^T f_s(Y_{s-}, Z_s, U_s)\,\mathrm{d}s - \int_t^T Z_s\,\mathrm{d}\widehat{B}_s - \int_t^T\int_E U_s(e)\,\widetilde{\mu}(\mathrm{d}s,\mathrm{d}e)$ under the minimal local martingale measure $\widehat{\mathbb{P}}$ for the generator

$$f_t(y,z,u) := -\frac{|\varphi_t|^2}{2\alpha} + \int_E \frac{\exp(\alpha u(e)) - \alpha u(e) - 1}{\alpha}\,\zeta(t,e)\,\lambda(\mathrm{d}e) \tag{5.4}$$

which does exist by Theorem 4.13. Under $\mathbb{P}$ the BSDE is of the form

$$Y_t = \xi + \int_t^T f_s(Y_{s-}, Z_s, U_s) - \varphi_s Z_s\,\mathrm{d}s - \int_t^T Z_s\,\mathrm{d}B_s - \int_t^T\int_E U_s(e)\,\widetilde{\mu}(\mathrm{d}s,\mathrm{d}e).$$

To prove optimality by a martingale principle one constructs, cf. [30], a family of processes $(V^\theta)_{\theta\in\Theta}$ such that three conditions are satisfied: (i) $V_t^\theta = V_t$ is a fixed $\mathcal{F}_t$-measurable bounded random variable invariant over $\theta \in \Theta$, (ii) $V_T^\theta = -\exp(-\alpha(X_T^\theta - \xi)) = -\exp\big(-\alpha\big(x + \int_t^T \theta_s \mathrm{d}\widehat{B}_s - \xi\big)\big)$, and (iii) $V^\theta$ is a supermartingale for all $\theta \in \Theta$ and there exists a $\theta^* \in \Theta$ such that $V_s^{\theta^*}$ ($s \in [t,T]$) is a $\mathbb{P}$-martingale. Then $\theta^*$ is the optimal strategy and $(V_s^{\theta^*})_{s\in[t,T]}$ is the value process of the control problem (5.3). Indeed, $\mathbb{E}\big(V_T^\theta \mid \mathcal{F}_t\big) \le V_t^\theta = V_t^{\theta^*} = \mathbb{E}\big(V_T^{\theta^*} \mid \mathcal{F}_t\big)$ for each $\theta \in \Theta$ implies $v_t(x) = \operatorname{ess\,sup}_{\theta\in\Theta}\mathbb{E}\big(V_T^\theta \mid \mathcal{F}_t\big) = V_t^{\theta^*}$. An ansatz $V^\theta = u(X^\theta - Y)$ yields

$$V_s^\theta = V_t^\theta \exp\Big(\frac{\alpha^2}{2}\int_t^s \Big|\theta_r - Z_r - \frac{\varphi_r}{\alpha}\Big|^2\mathrm{d}r\Big)\,\mathcal{E}(M)_t^s \quad \text{for all } s \in [t,T], \quad \text{with}$$

$$M_t = -\alpha\int_0^t \theta_r - Z_r\,\mathrm{d}\widehat{B}_r + \int_0^t\int_E \exp(\alpha U_r(e) - 1)\,\widetilde{\mu}(\mathrm{d}r,\mathrm{d}e) \quad \text{and} \quad \mathcal{E}(M)_t^s := \frac{\mathcal{E}(M)_s}{\mathcal{E}(M)_t}.$$

Therefore, $V^\theta$ is a supermartingale for all $\theta \in \Theta$ and a martingale for $\theta^* = Z + \varphi/\alpha$ due to the fact that $\mathcal{E}(M)$ is a (local) martingale of the form

$$\mathcal{E}(M)_s = \exp\Big(-\frac{\alpha^2}{2}\int_0^s |\theta_u - Z_u - \varphi_u/\alpha|^2\,\mathrm{d}u\Big)\exp\Big(-\alpha\Big(Y_0 + \int_0^s \theta_u\,\mathrm{d}\widehat{B}_u - Y_s\Big)\Big).$$

Using the boundedness of $Y$, one readily obtains by arguments like in [30, 45] that $\mathcal{E}(M)$ is uniformly integrable and hence a martingale (see e.g. Eq. (4.19) in [7]). This yields

*Example 5.1* Let $k=d$ and $\lambda(E)\leq\infty$. Let $(Y, Z, U) \in \mathcal{S}^{\infty}_{\widehat{\mathbb{P}}} \times \mathcal{L}^2_{\widehat{\mathbb{P}}}(\widehat{B})\times\mathcal{L}^2_{\widehat{\mathbb{P}}}(\widetilde{\mu})$ be the unique solution to the BSDE $(\xi, f)$ under $\widehat{\mathbb{P}}$ for generator $f$ from (5.4). Then the strategy $\theta^*=Z+\varphi/\alpha$ is optimal for the control problem (5.3) and achieves at any time $t\leq T$ the maximal expected exponential utility $v_t(x)=-\exp(-\alpha(x-Y_t))=V_t^{\theta^*}$.

The exponential utility maximization problem is closely linked to the popular entropic convex risk measure, to which we will further relate in Example 5.3. Moreover, the solution to the utility maximization problem is intimately linked to the indifference valuation (also known as reservation price or compensating variation in economics) for a contingent claim $\xi$ in incomplete markets under exponential utility preferences, see [8]. Indeed, denoting by $Y^{\xi}=Y$ the solution to the JBSDE from Example 5.1 for terminal data $\xi$, one can show that $Y^{\xi}-Y^0$ yields the utility indifference valuation process, see [7, 43].

### 5.1.2 Case with Discontinuous Risky Asset Price Processes

We further illustrate the extend to which results by [44, 45], who has pioneered the stability approach to BSDE with jumps specifically for exponential utility, fit into our framework and demonstrate by concrete examples some notable differences in scope in relation to complementary approaches. To this end, let us consider the same utility problem but now in a financial market with pure-jump asset price processes, possibly of infinite activity (as e.g. in the CGMY model of [12]), and with constraints on trading strategies. We note that a pure-jump setting appears as a natural setup for our JBSDE results, which admit for generators that are (roughly said) 'quadratic' in the $u$-argument but not in $z$-argument, differently from, e.g., [1, 35, 40, 45, 57].

Let $\mu=\mu^L$ be the random measure associated to a pure-jump Lévy process $L$ with Lévy measure $\lambda(de)$, on $E=\mathbb{R}^1\setminus\{0\}$. Let $\mathbb{F}=\mathbb{F}^L$ be the usual filtration generated by $L$. The compensated random measure $\widetilde{\mu}=\widetilde{\mu}^L:=\mu^L-\nu$, with $\nu(dt, de)=\lambda(de)dt$ of $L$ alone has the weak PRP w.r.t. the filtration $\mathbb{F}$ (see Example 2.1). Note that $\mu$ could be of infinite activity, i.e. $\lambda(E)\leq\infty$, for instance for $L$ being a Gamma process. In contrast to the setup of Sect. 5.1.1, we consider now a financial market whose single risky asset prices evolves in a non-continuous fashion, being given by a pure-jump process

$$dS_t=S_{t-}\Big(\beta_t dt+\int_E\psi_t(e)\widetilde{\mu}(dt, de)\Big)\quad \text{for } t\in[0, T],\ \text{with } S_0\in(0,\infty),$$

where $\beta$ is predictable and bounded, and $\psi>-1$ is $\widetilde{\mathcal{P}}$-measurable, in $L^2(\mathbb{P}\otimes\lambda\otimes dt)\cap L^{\infty}(\mathbb{P}\otimes\lambda\otimes dt)$ and satisfies $\int_E|\psi_t(e)|^2\lambda(de)\leq\text{const.}$ $\mathbb{P}\otimes dt$-a.e. The set $\Theta$ of admissible trading strategies consists of all $\mathbb{R}$-valued predictable $S$-integrable processes $\theta\in L^2(\mathbb{P}\otimes dt)$, such that $\theta_t(\omega)\in C$ for all $(t,\omega)$, for a fixed compact set $C \subset \mathbb{R}$ of trading constraint containing 0. Interpreting trading strategies $\theta$ as amount of wealth invested into the risky asset yields wealth process $X^{\theta,t,x}$ from initial capital $x$ at time $t$ as

$$X_s^{\theta,t,x} = X_t^{\theta,t,x} + \int_t^s \theta_u \frac{\mathrm{d}S_u}{S_{u-}} = x + \int_t^s \theta_u \Big(\beta_u \mathrm{d}u + \int_E \psi_u(e)\widetilde{\mu}(\mathrm{d}u, \mathrm{d}e)\Big), \quad s \geq t.$$

Because of the compactness of $C$ and the fact that $\psi \in L^2(\mathbb{P} \otimes \lambda \otimes \mathrm{d}t) \cap L^\infty(\mathbb{P} \otimes \lambda \otimes \mathrm{d}t)$, admissible strategies are bounded and for all $\theta \in \Theta$ one can verify that $\{\exp(-\alpha X_\tau^\theta) | \tau \text{ an } \mathbb{F}\text{-stopping time}\}$ is uniformly integrable; arguments being like in [45, Lemma 1]. Consider the JBSDE

$$-\mathrm{d}Y_t = f(t, U_t)\mathrm{d}t - \int_E U_t(e)\widetilde{\mu}(\mathrm{d}t, \mathrm{d}e), \quad Y_T = \xi, \tag{5.5}$$

with terminal condition $\xi \in L^\infty(\mathcal{F}_T)$ and generator $f$ defined pointwise by

$$f(t, u) := \inf_{\theta \in C} \Big( -\theta\beta_t + \int_E g_\alpha\big(u(e) - \theta\psi_t(e)\big)\lambda(\mathrm{d}e)\Big), \quad t \in [0, T], \tag{5.6}$$

for the function $g_\alpha : \mathbb{R} \to \mathbb{R}$ with $g_\alpha(u) := (\mathrm{e}^{\alpha u} - \alpha u - 1)/\alpha$. We have the following

**Proposition 5.2** *Let $(Y, U) \in \mathcal{S}^\infty \times \mathcal{L}^2(\widetilde{\mu})$ be the unique solution to the JBSDE (5.5). Then the strategy $\theta^*$ such that $\theta_t^*$ achieves the infimum in (5.6) for $f(t, U_t)$ is optimal for the control problem (5.3) and achieves at any $t \in [0, T]$ the maximal expected exponential utility $v_t(x) = -\exp(-\alpha(x - Y_t)) = V_t^{\theta^*}$.*

*Proof* Using the martingale optimality principle one obtains, like in the cited literature and analogously to Sect. 5.1.1, that if $(Y, U) \in \mathcal{S}^\infty \times \mathcal{L}^2(\widetilde{\mu})$ is a solution to the JBSDE (5.5) then the solution to the utility maximization problem (5.3) is indeed given by $v_t(x) = u(x - Y_t)$ (recall that $u$ denotes the exponential utility function) with the strategy $\theta^*$ where $\theta_t^*(\omega)$ achieves the infimum $f(\omega, t, U_t(\omega))$ in (5.6) for all $(\omega, t)$ being optimal (it exists by measurable selection [53]). To complete the derivation of this example, it thus just remains to show that the JBSDE (5.5) indeed admits a unique solution, with trivial $Z$-component $Z = 0$. This is shown by applying Theorem 4.11 and Corollary 4.12 since $\xi \in L^\infty(\mathcal{F}_T^L)$ and the generator $f$ does not have a $z$-argument and is $\mathbb{F}^L$-predictable in $(t, \omega)$. It is straightforward, albeit somewhat tedious, to verify that the conditions 1–5 and $(B_{\gamma^n})$, $n \in \mathbb{N}$, for Theorem 4.11 are indeed satisfied for the sequence of $\mathbb{F}^L$-predictable generators functions

$$f^n(t, u) := \inf_{\theta \in C} \Big( -\theta\beta_t + \int_{A_n} g_\alpha\big(u(e) - \theta\psi_t(e)\big)\lambda(\mathrm{d}e)\Big),$$

where $(A_n)_n$ is a sequence of measurable sets with $A_n \uparrow E$ and $\lambda(A_n) < \infty$ for all $n \in \mathbb{N}$, typically $A_n = (-\infty, -1/n] \cup [1/n, +\infty)$. Let us refer to [36, Example 1.32] for details of this verification, but explain here how to proceed further with the proof.

By the first claim of Theorem 4.11 (together with Corollary 4.12) one then gets existence of a solution $(Y, U) \in \mathcal{S}^\infty \times \mathcal{L}^2(\widetilde{\mu})$ to the JBSDE (5.5), such that $U * \widetilde{\mu}$ is a BMO-martingale. To obtain uniqueness by applying the second claim, we need to check that $f$ satisfies condition $(\mathbf{A}'_\gamma)$: To this end, we define $\gamma_t^{u,u'}(e) := \sup_{\theta\in C} \gamma_t^{\theta,u,u'}(e)\mathbb{1}_{\{u\geq u'\}} + \inf_{\theta\in C} \gamma_t^{\theta,u,u'}(e)\mathbb{1}_{\{u<u'\}}$, for $\gamma_t^{\theta,u,u'}(e) := \int_0^1 g'_\alpha\big(l(u-\theta\psi_t(e)) + (1-l)(u'-\theta\psi_t(e))\big)\mathrm{d}l$. Then (by Examples 3.5 and 3.8.) we get $f(t, U_t) - f(t, U'_t) \leq \int_E \gamma_t^{U,U'}(e)\big(U_t(e) - U'_t(e)\big)\lambda(de)$ for all $U, U'$ with $|U|_\infty < \infty, |U'|_\infty<\infty$. Now let $u, u'$ be bounded by $c > 0$; Since $g'_\alpha(0) = 0$, applying the mean-value theorem to $g'_\alpha$ in the expression of $\gamma^{\theta,u,u'}$ gives $\big|\gamma_t^{\theta,u,u'}(e)\big| \leq \sup_{|x|\leq\tilde{c}}|g''_\alpha(x)|\big(|u| + |u'| + |\theta||\psi_t(e)|\big)$ for all $\theta \in C$, where $\tilde{c} := c + \|\psi\|_\infty \mathrm{diam}(C)$. This implies (for $c = |U|_\infty \vee |U'|_\infty < \infty$)

$$\sup_{\theta\in C}\big|\gamma_t^{\theta,U,U'}(e)\big| \leq \sup_{|x|\leq\tilde{c}}|g''_\alpha(x)|\Big(|U_t(e)| + |U'_t(e)| + \mathrm{diam}(C)|\psi_t(e)|\Big).$$

Since $|\inf_\theta \gamma^\theta| \leq \sup_\theta|\gamma^\theta|$, $|\sup_\theta \gamma^\theta| \leq \sup_\theta|\gamma^\theta|$ and $\psi * \widetilde{\mu}$ is a BMO-martingale (as $\psi$ is bounded and $\int_E|\psi_t(e)|^2\lambda(de) \leq \text{const.}$, $\mathbb{P}\otimes dt$-a.e. by assumption), then $\gamma^{U,U'} * \widetilde{\mu}$ is a BMO-martingale if $U * \widetilde{\mu}$ and $U' * \widetilde{\mu}$ are, thanks to $|U|_\infty < \infty$ and $|U'|_\infty < \infty$. Hence $f$ satisfies $(\mathbf{A}'_\gamma)$. □

*Example 5.3* (Entropic convex risk measure) Let us consider the special case $\beta = \psi \equiv 0$ and $S \equiv 1$, i.e. the exponential utility problem *without* trading opportunities in a risky asset. One gets the important and well known example of the (dynamic) entropic risk measure $Y_t = (1/\alpha)\log\mathbb{E}(\exp(\alpha\xi)|\mathcal{F}_t)$ whose JBSDE description can be identified directly by exponential transformation. In the setup of the present subsection, this JBSDE is covered by the application study of [45] and also by our comparison and wellposedness theorems, without any further conditions on the pure-jump Lévy process. In contrast, let us demonstrate that the same scope is not already offered by the seminal comparison Theorem 2.5 of [54] because her key condition $(\mathbf{A}_\gamma)$ is not satisfied, which is also supposed for results in [35, as Assumption 6.1 for wellposedness in Theorem 6.3(i) and for comparison in Proposition 6.4] (and is further used for applications in [34]): Indeed for $\alpha := 1$, just consider a compound Poisson process $L$ (being of finite activity) with uniformly distributed jump heights, taking $\lambda(dx) := \mathbb{1}_{(0,1]}dx$ $(x \in E = \mathbb{R}\setminus\{0\})$. Clearly, the generator $f(u) = \int_E \exp(u(x)) - u(x) - 1\,\lambda(dx) =: \int_E g(u(x))\lambda(dx)$ is not Lipschitz in $u \in L^2(\lambda)$. With $u^\pm(x) := \frac{1}{2}(\pm x^{-3/2} + nx)\mathbb{1}_{(1/n,1]}(x)$ in $L^2(\lambda)$ for $n \in \mathbb{N}$, we get $\int_0^1(e^{u^+} - e^{u^-} - u^+ + u^-)d\lambda \to \infty$ for $n \to \infty$ while $\int_0^1(u^+ - u^-)(x)\cdot(1\wedge|x|)dx \leq \int_0^1 x^{-1/2}dx < \infty$ for all $n$, noting that $e^{u^+} - e^{u^-} - u^+ + u^- \geq nx^{-1/2}\mathbb{1}_{(1/n,1]}$. Thus, there cannot be constants $c_1 \in (-1, 0]$, $c_2 < \infty$, such that $f(u) - f(v) \leq \int_E(u - v)(x)\gamma^{u,v}(x)\,\lambda(dx)$ for all $u, v$, with $c_1(1\wedge|x|) \leq \gamma^{u,v}(x) \leq c_2(1\wedge|x|)$; This shows that condition $(\mathbf{A}_\gamma)$ in [54] or

Assumption 6.1 in [35], are not satisfied here. Indeed, the seminal $(\mathbf{A}_\gamma)$ condition from [54] implies Lipschitz continuity of the generator in $u$.

Similarly, a related condition on the jump measure has been stated in the assumptions of the main theorem in [1, Theorem 1, see their inequality (4)]; Their article further assumes (like also [57], for instance) finite jump activity in that $\lambda(E) < \infty$ holds (in our notation, they write $\eta$ for $\lambda$), what is true in the present example (but is not required for [54], or our previous sections). But in this example there cannot exist constants $c_1 \in (-1, 0]$, $c_2 < \infty$ such that $f(u) - f(u') \le \int_E (u - u')(x)\gamma^{u,u'}(x)\,\lambda(dx)$ would hold for all $u, u' \in L^2(\lambda)$, with suitable functions $c_1 \le \gamma^{u,u'}(x) \le c_2$. Indeed, the latter would imply (using Cauchy-Schwarz inequality) that $f$ is Lipschitz in $u$ on $L^2(\lambda)$, what is not true. Hence, the assumptions for Theorem 1 in [1] appear not satisfied for the entropic risk example, and its conditions (with inequality (4) assumed for all $u, u'$) would imply that the JBSDE generator has to be Lipschitz continuous in $u \in L^2(\lambda)$. Note that the aforementioned problem could be resolved if, e.g., an additional $L^\infty$-bound is available for the $u$-argument, for instance by an exogenuous a-priori $L^\infty$-estimate on the $U$-component for bounded JBSDE solutions (like from Sect. 3). We note that [1] offer results for unbounded JBSDE solutions (see also [29] for exponential utility in the continuous case without jumps).

*Example 5.4* We continue with the previous entropic risk example, but now take a standard Poisson process instead, i.e. $\lambda(de) = \delta_{\{1\}}(de)$ as Dirac point measure at the fixed jump height 1. Then $L^2(\lambda)$ is isomorphic to $\mathbb{R}$, and $f(u) = g(u) = \exp(u) - u - 1$ for $u \in \mathbb{R}$. Obviously $g'$ and $g''$ are of exponential growth (in $u$) and cannot be bounded globally in $u \in \mathbb{R}$ by an affine function or by constants; hence Assumption 5.1(iii) for [35, Theorem 5.4] cannot be satisfied. Moreover, also Assumption 4.3(iii) for [35, Theorems 4.3, 5.4 and 6.3(ii)], noting their Lemma 5.4, appears clearly violated since (taking $u' = 0$) there exist no $\psi, c \in \mathbb{R}$ such that $|g(u) - \psi u| \le c|u|^2$ for all $u \in \mathbb{R}$.

Since the function $(u, \theta) \mapsto f^\theta(\cdot, u) := -\theta\beta_\cdot + \int_E g_\alpha\big(u(e) - \theta\psi_\cdot(e)\big)\lambda(de)$ is convex, the generator constructed as $f = \inf_{\theta\in C} f^\theta(\cdot, u)$ (cf. 5.6) would be convex in $u$ if the constraint set $C$ were assumed to be convex; But for non-convex trading constraints $C$ the generator can be non-convex in general. Similar constructions of generators are typical in this application context, see e.g. [40, Eq. (15)]. Some results on JBSDE in the literature use convexity of the generator function but that can be restrictive for applications. For results in the present paper, convexity is not being assumed. Next, we give a concrete application example where $f$ in (5.6) for the (primal) control problem is indeed non-convex in $u$. The example shows, how non-convex constraints can lead to JBSDE generators which are non-convex in $u$. To this end, let us consider simple trading constraints that are non-convex by taking $C := \{\theta^0, \dots, \theta^m\} \subset \mathbb{R}$ as a finite set including the zero $\theta^0 := 0$. Here $f$ of JBSDE (5.5) becomes

$$f(t, u) = \inf_{k\in\{0,\dots,m\}} \Big( -\theta^k\beta_t + \int_E g_\alpha\big(u(e) - \theta^k\psi_t(e)\big)\lambda(de)\Big).$$

*Example 5.5* (An application where the generator is not convex and not continuously differentiable) Continuing with the above generator $f$, now let us take the particular case where $\lambda(de) = \delta_{\{1\}}(de)$, i.e. $L$ is a standard Poisson process with constant jump height 1, and let $\alpha = 1$ and $\beta = 0$. Observing that $L^2(\lambda)$ in the case of this simple example is isomorphic to $\mathbb{R}$, we see that

$$f(t, u) = \min_{k \in \{0, \dots, m\}} \left( e^{(u - \theta^k \psi_t)} - (u - \theta^k \psi_t) - 1 \right), \tag{5.7}$$

is clearly non-convex in $u \in \mathbb{R}$, unless $\psi \equiv 0$ or $C = \{0\}$. Also, we observe that $u \mapsto f(t, u)$ is not continuously differentiable in $u \in \mathbb{R}$ for this application. But the function $f$ in (5.7) is still absolutely continuous in $u$ with density function being strictly greater than $-1$, locally bounded in u from above and locally of linear growth. Because this $f$ satisfies condition ($\mathbf{A_{infi}}$), existence and uniqueness for the corresponding JBSDE (5.5) solution can be obtained by Theorem 4.13.

Similarly as before, one can check that in this example the assumptions of [54, Theorem 2.5] and [35, Theorems 5.4, 6.3(i)–(ii), Proposition 6.4] for comparison and wellposedness of JBSDE are not satisfied; The example clearly shows how non-convex constraints can indeed lead to a non-convex $f$, which does not satisfy the conditions for the JBSDE results of [40, Theorem A28, Corollary A29, Proposition A30, being used further in the proofs for Theorems 4.3, 4.5, all involving convexity assumption "(c)"].

We proceed next with examples beyond exponential utility, that was the topic in [44, 45].

## 5.2 *Power Utility Maximization*

Again for the market with stock price dynamics (5.1), we consider the utility maximization problem

$$v_t(x) = \operatorname*{ess\,sup}_{\theta \in \Theta} \mathbb{E}\big(u(X_T^{\theta,t,x})\xi \,\big|\, \mathcal{F}_t\big) = \frac{1}{\gamma} \operatorname*{ess\,sup}_{\theta \in \Theta} \mathbb{E}\big(u(X_T^{\theta,t,x}\xi') \,\big|\, \mathcal{F}_t\big), \quad t \le T,\ x > 0, \tag{5.8}$$

for power utility $u(x) = x^\gamma/\gamma$ with relative risk aversion $1 - \gamma > 0$ for $\gamma \in (0, 1)$, with multiplicative liability $\xi$ (alternatively, $\xi' := (\gamma\xi)^{1/\gamma}$ can be interpreted as an unknown future tax rate). The wealth process of strategy $\theta$ (denoting fraction of wealth invested) is $X_s^\theta = X_s^{\theta,t,x} = x + \int_t^s X_u^\theta \theta_u \,\mathrm{d}\widehat{B}_u = x\mathcal{E}(\int \theta \mathrm{d}\widehat{B})_t^s$ for $s \in [t, T]$, for $\theta \in \Theta$, with the set $\Theta$ of strategies given by all $\mathbb{R}^d$-valued, predictable, $S$-integrable processes such that $\theta \bullet B$ is a BMO($\mathbb{P}$)-martingale, cf. [28].

**Proposition 5.6** *Let $k = d$. Assume that there is a solution $(Y, Z, U) \in \mathcal{S}^\infty \times \mathcal{L}^2(B) \times \mathcal{L}^2(\widetilde{\mu})$ to the BSDE $(\xi, f)$ with $f_t(y, z, u) := (\gamma/(2-2\gamma))\, y\, |\varphi_t + y/z|^2$ and $\int Z\, \mathrm{d}B \in \mathrm{BMO}(\mathbb{P})$ and where $\xi$ is in $L^\infty(\mathcal{F}_T)$ with $\xi \geq c$ for some $c > 0$. Then $Y \geq c$ holds and $V^\theta := u(X^\theta)Y$ is a supermartingale for all $\theta$ in $\Theta$ and $V^{\theta^*}$ is a martingale for $\theta^* := (1-\gamma)^{-1}(\varphi + Z/Y_-) \in \Theta$.*

*Proof* Clearly, $V^\theta$ is adapted. Kazamaki's criterion $\mathcal{E}(\int_0^\cdot \gamma\theta_u \mathrm{d}\widehat{B}_u)$ is an $r$-integrable martingale for some $r > 1$. Hence $\sup_{t\leq s\leq T} \mathcal{E}(\int_0^\cdot \gamma\theta_u \mathrm{d}\widehat{B}_u)_t^s$ is integrable by Doob's inequality. By

$$\mathcal{E}(\theta \bullet \widehat{B})^\gamma = \mathcal{E}(\gamma\theta \bullet \widehat{B}) \exp\Big(-\frac{1}{2}\gamma(1-\gamma)\int_0^\cdot |\theta_u|^2\, \mathrm{d}u\Big) \leq \mathcal{E}(\gamma\theta \bullet \widehat{B}),$$

we conclude that $V^\theta$ is dominated by $\sup_{t\leq s\leq T} U(X_s^\theta)|Y|_\infty \in L^1(\mathbb{P})$. By Itô's formula, $\mathrm{d}V_s^\theta$ equals a local martingale plus the finite variation part

$$u(X_s^\theta)\bigg(-f_s(Y_{s-}, Z_s, U_s) + \gamma\Big(Y_{s-}\Big(\theta_s\varphi_s + \frac{1}{2}(\gamma-1)|\theta_s|^2\Big) + \theta_s Z_s\Big)\bigg)\, \mathrm{d}s\,.$$

The latter part is decreasing for all $\theta \in \Theta$ and vanishes at zero for $\theta = \theta^*$. So $V^\theta$ is a local (super)martingale. Uniform integrability of $V^\theta$ yields the (super)martingale property. By the classical martingale optimality principle of optimal control follows that $v_t(x) = \operatorname{ess\,sup}_{\theta\in\Theta} \mathbb{E}(u(X_T^\theta \xi^{1/\gamma})|\mathcal{F}_t)$ equals $V_t^{\theta^*} = \gamma^{-1}x^\gamma Y_t$, and evaluating at $\theta \equiv 0$ yields $\gamma^{-1}x^\gamma \mathbb{E}(\xi|\mathcal{F}_t) \leq \gamma^{-1}x^\gamma Y_t$ and hence $Y \geq c$. Note that $\theta^*$ is in $\Theta$ since $\varphi$ is bounded, $Y$ is bounded away from 0 and $Z$ is an BMO integrand. □

Let $(Y, Z, U)$ be a solution to the BSDE $(\xi, f)$ with the above data. Since a suitable solution theory for quadratic BSDEs with jumps is not available, we transform coordinates by letting

$$\widetilde{Y}_t := Y_t^{\frac{1}{1-\gamma}}, \quad \widetilde{Z}_t := \frac{1}{1-\gamma} Y_{t-}^{\frac{\gamma}{1-\gamma}} Z_t \quad \text{and} \quad \widetilde{U}_t := (Y_{t-} + U_t)^{\frac{1}{1-\gamma}} - Y_{t-}^{\frac{1}{1-\gamma}}, \tag{5.9}$$

such that $(\widetilde{Y}, \widetilde{Z}, \widetilde{U})$ solves the BSDE for data $(\widetilde{\xi}, \widetilde{f})$ with $\widetilde{\xi} = \xi^{1/(1-\gamma)}$ and $\widetilde{f}_t(y, z, u)$ given by

$$\frac{\gamma|\varphi_t|^2}{2(1-\gamma)^2} y + \frac{\gamma}{1-\gamma}\varphi_t z + \int_E \Big(\frac{1}{1-\gamma}\Big((u(e)+y)^{1-\gamma} y^\gamma - y\Big) - u(e)\Big)\, \zeta(t, e)\, \lambda(\mathrm{d}e).$$

Looking at the proof of Lemma 2.2, we may assume that $U + Y_-$ coincides pointwise with $Y_-$ or $Y$ so that the above transformation is well-defined due to $Y \geq c$. In fact, (5.9) gives a bijection between solutions with positive Y-components to the BSDEs $(\xi, f)$ and $(\widetilde{\xi}, \widetilde{f})$ in $\mathcal{S}^\infty \times \mathcal{L}^2(B) \times \mathcal{L}^2(\widetilde{\mu})$.

Next, we show the existence of a JBSDE solution for data $(\xi, f)$ with $\xi \geq c$ for some $c > 0$. Under the probability measure $\mathrm{d}\widetilde{\mathbb{P}} := \mathcal{E}\big(\gamma(1-\gamma)^{-1}\varphi \bullet B\big)_T\, \mathrm{d}\mathbb{P}$ the process $\widetilde{B} = B - \int_0^{\cdot} \gamma(1-\gamma)^{-1}\varphi_t \,\mathrm{d}t$ is a Brownian motion and the JBSDE

$$\widetilde{Y}_t = \widetilde{\xi} + \int_t^T \widetilde{f}_s(\widetilde{Y}_{s-}, \widetilde{Z}_s, \widetilde{U}_s)\,\mathrm{d}s - \int_t^T \widetilde{Z}_s \,\mathrm{d}B_s - \int_t^T \int_E \widetilde{U}_s(e)\,\widetilde{\mu}(\mathrm{d}s, \mathrm{d}e)$$

under $\mathbb{P}$ is of the following form under $\widetilde{\mathbb{P}}$

$$\widetilde{Y}_t = \widetilde{\xi} + \int_t^T \Big(\widetilde{f}_s(\widetilde{Y}_{s-}, \widetilde{Z}_s, \widetilde{U}_s) - \frac{\gamma\varphi_s}{1-\gamma}\widetilde{Z}_s\Big)\,\mathrm{d}s - \int_t^T \widetilde{Z}_s \,\mathrm{d}\widetilde{B}_s - \int_t^T \int_E \widetilde{U}_s(e)\,\widetilde{\mu}(\mathrm{d}s, \mathrm{d}e), \tag{5.10}$$

noting that $\nu$ is the compensator of $\mu$ under $\mathbb{P}$ and $\widetilde{\mathbb{P}}$ as well. In fact, we have

**Lemma 5.7** *Assume $\lambda(E) < \infty$. Then $(\widetilde{Y}, \widetilde{Z}, \widetilde{U}) \in \mathcal{S}^\infty \times \mathcal{L}^2(B) \times \mathcal{L}^2(\widetilde{\mu})$ solves the BSDE $(\widetilde{\xi}, \widetilde{f})$ such that $\int \widetilde{Z}\,\mathrm{d}B$ is in* BMO($\mathbb{P}$) *if and only if $(\widetilde{Y}, \widetilde{Z}, \widetilde{U}) \in \mathcal{S}^\infty_{\widetilde{\mathbb{P}}} \times \mathcal{L}^2_{\widetilde{\mathbb{P}}}(\widetilde{B}) \times \mathcal{L}^2_{\widetilde{\mathbb{P}}}(\widetilde{\mu})$ solves the BSDE $\big(\widetilde{\xi}, \widetilde{f}(y,z,u) - \gamma(1-\gamma)^{-1}\varphi z\big)$ such that $\int \widetilde{Z}\,\mathrm{d}\widetilde{B}$ is in* BMO($\widetilde{\mathbb{P}}$).

*Proof* Equivalence of $\mathbb{P}$ and $\widetilde{\mathbb{P}}$ imply that $\widetilde{Y} \in \mathcal{S}^\infty$ if and only if $\widetilde{Y} \in \mathcal{S}^\infty_{\widetilde{\mathbb{P}}}$. Given $\lambda(E) < \infty$, $\widetilde{U} \in \mathcal{L}^2(\widetilde{\mu})$ holds if and only if $\widetilde{U} \in \mathcal{L}^2_{\widetilde{\mathbb{P}}}(\widetilde{\mu})$ due to the boundedness of $\widetilde{U}$. By [32, Theorem 3.6], the restriction of the Girsanov transform $\Phi : \mathcal{M}^{\mathrm{loc},0}_{\mathrm{c}}(\mathbb{P}) \longrightarrow \mathcal{M}^{\mathrm{loc},0}_{\mathrm{c}}(\widetilde{\mathbb{P}})$, with $M \mapsto M - \langle M, \int_0^{\cdot} \frac{\gamma\varphi}{1-\gamma}\,\mathrm{d}B_s\rangle$, onto BMO($\mathbb{P}$) yields a bijection between BMO($\mathbb{P}$)-martingales and BMO($\widetilde{\mathbb{P}}$)-martingales. Thus, $\int \widetilde{Z}\,\mathrm{d}B$ is in BMO($\mathbb{P}$) if and only if $\int \widetilde{Z}\,\mathrm{d}\widetilde{B}$ is in BMO($\widetilde{\mathbb{P}}$) for $Z = (1-\gamma)\widetilde{Y}^{\gamma}_{-}\widetilde{Z}$ since $\Phi\big(\int \widetilde{Z}\,\mathrm{d}B\big) = \int \widetilde{Z}\,\mathrm{d}B - \int \gamma(1-\gamma)^{-1}\varphi\widetilde{Z}_s\,\mathrm{d}s = \int \widetilde{Z}\,\mathrm{d}\widetilde{B}$. In particular, $\widetilde{Z} \in \mathcal{L}^2(B)$ iff $\widetilde{Z} \in \mathcal{L}^2_{\widetilde{\mathbb{P}}}(\widetilde{B})$. □

To proceed further, let us note at first that under an equivalent change of measure between $\mathbb{P}$ and $\widetilde{\mathbb{P}}$, the weak predictable representations property for $(B, \widetilde{\mu})$ under $\mathbb{P}$ is equivalent to the respective property of $(\widetilde{B}, \widetilde{\mu})$ under $\widetilde{\mathbb{P}}$ for the same filtration, see [28, Theorem 13.22] and recall Example 2.1, Part 2. According to Corollary 4.6, hence there exists a unique solution $\big(\widetilde{Y}, \widetilde{Z}, \widetilde{U}\big) \in \mathcal{S}^\infty_{\widetilde{\mathbb{P}}} \times \mathcal{L}^2_{\widetilde{\mathbb{P}}}(B) \times \mathcal{L}^2_{\widetilde{\mathbb{P}}}(\widetilde{\mu})$ with positive $Y$-component to the BSDE (5.10) with

$$c^{\frac{1}{1-\gamma}} \exp\Big(-\frac{\gamma|\varphi|^2_\infty}{2(1-\gamma)^2}(T-t)\Big) \le \widetilde{Y}_t \le |\xi|_\infty \exp\Big(\frac{\gamma|\varphi|^2_\infty}{2(1-\gamma)^2}(T-t)\Big)$$

such that $\int \widetilde{Z}\,\mathrm{d}\widetilde{B}$ and $\widetilde{U} * \widetilde{\mu}^{\widetilde{\mathbb{P}}}$ are BMO($\widetilde{\mathbb{P}}$)-martingales. By Lemma 5.7 and the statement of Proposition 5.6 that every bounded solution to the BSDE $(\xi, f)$ is bounded from below away from zero in $Y \ge c > 0$, there is a unique solution $(Y, Z, U)$ in $\mathcal{S}^\infty \times \mathcal{L}^2(B) \times \mathcal{L}^2(\widetilde{\mu})$ with $\int Z\,\mathrm{d}B \in$ BMO($\mathbb{P}$) and it is given by the coordinate transform (5.9). We note that $Y$ (resp. $\widetilde{Y}$) can be interpreted as (dual) opportunity process, see [48, Sect. 4]. Overall, we obtain the next theorem.

**Theorem 5.8** *Assume $\lambda(E) < \infty$ and $d = k$. Let*

$$f_s(y, z, u) = \gamma(2 - 2\gamma)^{-1}\, y\, \left|\varphi_s + z/y\right|^2$$

*and let $\xi \in L^\infty(\mathcal{F}_T)$ with $\xi \geq c$ for some $c > 0$. Then there exists a unique solution $(Y, Z, U) \in \mathcal{S}^\infty \times \mathcal{L}^2(B) \times \mathcal{L}^2(\widetilde{\mu})$ with $\int Z\,\mathrm{d}B \in \mathrm{BMO}(\mathbb{P})$ to the BSDE $(\xi, f)$. Then the strategy $\theta^*_s = (1-\gamma)^{-1}\Big(\varphi_s + Z_s/Y_{s-}\Big)$ is optimal for the control problem (5.8), achieving $v_t(x) = \gamma^{-1}x^\gamma Y_t = V_t^{\theta^*}$.*

## 5.3 Valuation by Good-Deal Bounds

In incomplete financial markets without arbitrage, there exist infinitely many pricing measures and the bounds imposed on valuation solely by the principle of no-arbitrage are typically far too wide for applications in practice. Good-deal bounds [13] have been introduced in the finance literature to obtain tighter bounds, by ruling out not only arbitrage but also trading opportunities with overly attractive reward-for-risk ratios, so-called good deals. See [9, 10] for extensive references and applications under model ambiguity. In [11, 13] good deals have been defined in terms of too favorable instantaneous Sharpe ratios (rate of excess return per unit rate of volatility) for continuous diffusion processes. This has been generalized to a jump-diffusion setup by [11], who describe good-deal bounds as solutions of nonlinear partial-integro differential equations by using (formal) HJB methods. We complement their work here by a rigorous, possibly non-Markovian, description by JBSDEs. See [19] for a study of a case where the measure $\lambda$ has finite support.

In our setting, the following description of the set $\mathcal{M}^{\mathrm{e}}$ of martingale measures is routine.

**Proposition 5.9** *$\mathcal{M}^{\mathrm{e}}$ consists of those measures $\mathbb{Q} \approx \mathbb{P}$ such that $\mathrm{d}\mathbb{Q}/\mathrm{d}\mathbb{P} = \mathcal{E}\left(\beta \bullet B + \gamma * \widetilde{\mu}\right)$, where $\gamma > -1$ is a $\widetilde{\mathcal{P}}$-measurable and $\widetilde{\mu}$-integral function, and $\beta$ is a predictable process with $\int_0^T |\beta_s|^2\,\mathrm{d}s < \infty$, satisfying $\beta = -\varphi + \eta$, such that $\eta \in \mathrm{Ker}\,\sigma$, $\mathbb{P} \otimes \mathrm{d}t$-a.e.*

We will refer to the tuple $(\gamma, \beta)$ for such a density $\mathrm{d}\mathbb{Q}/\mathrm{d}\mathbb{P}$ as the Girsanov kernel of $\mathbb{Q}$ relative to $\mathbb{P}$. Clearly, our market is incomplete in general as there exists infinitely many measures in $\mathcal{M}^{\mathrm{e}}$ if $\widetilde{\mu}$ is non-trivial or $k < d$. Björk and Slinko employed an extended Hansen-Jagannathan inequality [11, see Sect. 2] to bound the instantaneous Sharpe ratio by imposing a bound on market prices of risk. More precisely, Theorem 2.3 of [11] showed that the instantaneous Sharpe ratio $SR_t$ in any extension of the market by additional derivative assets (i.e. by any local $\mathbb{Q}$-martingales) satisfies $|SR_t| \leq \|(\gamma_t, \beta_t)\|_{L^2(\lambda_t)\times\mathbb{R}^d}$ at any time $t$, with a (sharp) upper bound in terms of an $L^2$-norm for Girsanov kernels $(\gamma, \beta)$ of pricing measures in $\mathcal{M}^{\mathrm{e}}$, with $\lambda_t(\omega)(\mathrm{d}e) := \zeta_t(\omega, e)\lambda(\mathrm{d}e)$. As no-good-deal restriction they therefore impose a bound on the kernels of pricing measures

$$\|(\gamma_t, \beta_t)\|^2_{L^2(\lambda_t)\times\mathbb{R}^d} = \|\gamma_t\|^2_{L^2(\lambda_t)} + |\beta_t|^2_{\mathbb{R}^d} \le K^2, \quad t \le T, \tag{5.11}$$

by some given constant $K > 0$. To complement the analysis of the problem posed by [11], we are going to describe the dynamic good deal bounds rigorously by JBSDEs in a more general, possibly non-Markovian, setting with no-good-deal restriction like in (5.11) but, more generally, we allow $K = (K_t)$ to be a positive predictable bounded process instead of a constant.

To this end, for $K$ as above, let the correspondence (set-valued) process $C$ be given by

$$C_t := \left\{(\gamma, \eta) \in L^2(\lambda_t) \times \mathbb{R}^d \,\middle|\, \gamma > -1,\ \eta \in \operatorname{Ker}\sigma_t,\ \text{and}\ \|\gamma\|^2_{L^2(\lambda_t)} + |\eta|^2_{\mathbb{R}^d} + |\varphi_t|^2_{\mathbb{R}^d} \le K_t^2\right\}. \tag{5.12}$$

We will write $(\gamma, \eta) \in C$ to denote that $\eta$ is a predictable process and $\gamma$ is a $\widetilde{\mathcal{P}}$-measurable process with $(\gamma_t(\omega), \eta_t(\omega)) \in C_t(\omega)$ holding for all $(t, \omega) \in [0, T] \times \Omega$. For $(\gamma, \eta) \in C$, we know (cf. Example 3.3) that $\mathcal{E}\big((-\varphi + \eta)\bullet B + \gamma * \widetilde{\mu}\big) > 0$ is a martingale that defines a density process of a probability measure $\mathbb{Q}^{\gamma,\eta}$ which is equivalent to $\mathbb{P}$. The set of such probability measures

$$\mathcal{Q}^{\mathrm{ngd}} := \{\mathbb{Q}^{\gamma,\eta} \mid (\gamma, \eta) \in C\} \subseteq \mathcal{M}^{\mathrm{e}}, \tag{5.13}$$

defines our set of no-good-deal measures. Beyond boundedness of $\varphi$, assume that $|\varphi_t|_{\mathbb{R}^d} + \epsilon < K_t$ holds for for some $\epsilon > 0$ for all $t \le T$. Then, in particular, the minimal martingale measure $\widehat{\mathbb{P}} = \mathbb{Q}^{\widehat{\gamma},\widehat{\eta}}$ is in $\mathcal{Q}^{\mathrm{ngd}} \neq \emptyset$, with $(\widehat{\gamma}, \widehat{\eta}) \equiv (0, 0) \in C$. For contingent claims $X \in L^\infty(\mathbb{P})$, the processes

$$\pi_t^u(X) := \operatorname*{ess\,sup}_{\mathbb{Q}\in\mathcal{Q}^{\mathrm{ngd}}} \mathbb{E}_{\mathbb{Q}}(X|\mathcal{F}_t) \quad \text{and} \quad \pi_t^l(X) := \operatorname*{ess\,inf}_{\mathbb{Q}\in\mathcal{Q}^{\mathrm{ngd}}} \mathbb{E}_{\mathbb{Q}}(X|\mathcal{F}_t), \quad t \le T,$$

define the upper and lower good-deal bounds. Noting $\pi^l_.(X) = -\pi^u(-X)$, we focus on $\pi^u(-X)$. One can check that the good-deal bound process satisfies good dynamic properties, e.g. time-consistency and recursiveness (cf. e.g. [10, Lemma 1]). By applying the comparison result of Proposition 3.1, we are going to obtain $\pi^u(X)$ as the value process $Y$ of a BSDE with terminal condition $X \in L^\infty(\mathbb{P})$. Denoting by $\Pi_t(\cdot)$ and $\Pi_t^\perp(\cdot)$ the orthogonal projections on $\operatorname{Im}\sigma_t^T$ and $\operatorname{Ker}\sigma_t$, we have the following lemma (see [36, Lemmas 2.14, 2.22] for details).

**Lemma 5.10** *For $Z \in \mathcal{L}^2(B)$ and $U \in \mathcal{L}^2(\widetilde{\mu})$ there exists $\bar{\eta} = \bar{\eta}(Z, U)$ predictable and $\bar{\gamma} = \bar{\gamma}(Z, U)$ $\widetilde{\mathcal{P}}$-measurable such that for $\mathbb{P} \otimes \mathrm{d}t$-almost all $(\omega, t) \in \Omega \times [0, T]$ holds*

$$\bar{\eta}_t \Pi_t^\perp(Z_t) + \int_E U_t(e)\bar{\gamma}_t(e)\zeta_t(e)\lambda(\mathrm{d}e) = \max_{(\gamma,\eta)\in\bar{C}} \Big(\eta_t \Pi_t^\perp(Z_t) + \int_E U_t(e)\gamma_t(e)\zeta_t(e)\lambda(\mathrm{d}e)\Big), \tag{5.14}$$

*where*

$$\bar{C}_t = \left\{(\gamma, \eta) \in L^2(\lambda_t) \times \mathbb{R}^d \,\middle|\, \gamma \geq -1,\ \eta \in \operatorname{Ker} \sigma_t,\ \|\gamma\|^2_{L^2(\lambda_t)} + |\eta|^2_{\mathbb{R}^d} \leq K_t^2 - |\varphi_t|^2_{\mathbb{R}^d}\right\}$$

*is the closure of $C_t$ in $L^2(\lambda_t) \times \mathbb{R}^d$ for any $t \leq T$.*

To $(\bar{\gamma}, \bar{\eta}) \in \bar{C}$ of Lemma 5.10, we associate a probability measure $\bar{\mathbb{Q}} \ll \mathbb{P}$ defined via $\mathrm{d}\bar{\mathbb{Q}} = \mathcal{E}\left((-\varphi + \bar{\eta}) \bullet B + \bar{\gamma} * \widetilde{\mu}\right) \mathrm{d}\mathbb{P}$, which may not be equivalent to $\mathbb{P}$ as $\bar{\gamma}$ may be $-1$ on a non-negligible set. While $\bar{\mathbb{Q}}$ might not be in $\mathcal{Q}^{\mathrm{ngd}}$ it belongs to the $L^1(\mathbb{P})$-closure of $\mathcal{Q}^{\mathrm{ngd}}$ in general, as shown in

**Lemma 5.11** *For $Z \in \mathcal{L}^2(B)$ and $U \in \mathcal{L}^2(\widetilde{\mu})$, let $(\bar{\gamma}, \bar{\eta})$ be as in Lemma 5.10. Define the measures $\bar{\mathbb{Q}} \ll \mathbb{P}$ via $\mathrm{d}\bar{\mathbb{Q}} = \mathcal{E}\big((-\varphi + \bar{\eta}) \bullet B + \bar{\gamma} * \widetilde{\mu}\big) \mathrm{d}\mathbb{P}$ and $\mathbb{Q}^n := (1/n)\widehat{\mathbb{P}} + (1 - 1/n)\bar{\mathbb{Q}}$ for $n \in \mathbb{N}$. Then the densities $\mathrm{d}\mathbb{Q}^n/\mathrm{d}\mathbb{P}$ of the sequence $(\mathbb{Q}^n)_{n \in \mathbb{N}}$ in $\mathcal{Q}^{ngd}$ converge to the one of $\bar{\mathbb{Q}}$ in $L^1(\mathbb{P})$ for $n \to \infty$. Consequently, $\pi_t^u(X) \geq \mathbb{E}_{\bar{\mathbb{Q}}}(X|\mathcal{F}_t)$ holds for all $t \leq T$.*

*Proof* Let $n \in \mathbb{N}$. Clearly $\mathbb{Q}^n \approx \mathbb{P}$. Moreover $\mathrm{d}\mathbb{Q}^n/\mathrm{d}\mathbb{P} = Z^n := (1/n)\widehat{Z} + (1 - 1/n)\bar{Z}$ with $\widehat{Z} := \mathrm{d}\widehat{\mathbb{Q}}/\mathrm{d}\mathbb{P} = \mathcal{E}(-\varphi \bullet B)$ and $\bar{Z} := \mathrm{d}\bar{\mathbb{Q}}/\mathrm{d}\mathbb{P}$. Itô formula then yields $Z^n = \mathcal{E}\big((-\varphi + \eta^n) \bullet B + \gamma^n * \widetilde{\mu}\big)$ for $\eta^n = \alpha\bar{\eta}$ being predictable and $\gamma^n = \alpha\bar{\gamma}$ is $\widetilde{\mathcal{P}}$-measurable with $\alpha = (1 - 1/n)(\bar{Z}/Z^n) \in [0, 1)$ thanks to $\widehat{Z} > 0$. Therefore $\eta^n \in \operatorname{Ker} \sigma$ and $\gamma^n > -1$ due to $\bar{\gamma} \geq -1$. Hence $(\eta^n, \gamma^n) \in C$ and so $\mathbb{Q}^n = \mathbb{Q}^{\gamma^n, \eta^n}$ is in $\mathcal{Q}^{\mathrm{ngd}}$. Convergence of $\mathbb{Q}^n$ to $\bar{\mathbb{Q}}$ in $L^1(\mathbb{P})$ as $n \to \infty$ is straightforward by definition of $\mathbb{Q}^n$ and this implies $\pi_t^u(X) \geq \mathbb{E}_{\bar{\mathbb{Q}}}(X|\mathcal{F}_t)$ for all $t \leq T$. □

The dynamic good-deal bound $\pi^u(X)$ of $X \in L^\infty(\mathbb{P})$ is given by the solution to the JBSDE

$$-\,\mathrm{d}Y_t = \Big((-\varphi_t + \bar{\eta}_t)Z_t + \int_E U_t(e)\bar{\gamma}_t(e)\zeta_t(e)\lambda(\mathrm{d}e)\Big)\mathrm{d}t - Z_t \mathrm{d}B_t - \int_E U_t(e)\widetilde{\mu}(\mathrm{d}t, \mathrm{d}e), \quad t \in [0, T], \tag{5.15}$$

for terminal condition $Y_T = X$, with $\bar{\gamma} = \bar{\gamma}(Z, U)$, $\bar{\eta} = \bar{\eta}(Z, U)$ given by Lemma 5.10, according to

**Theorem 5.12** *For $X \in L^\infty(\mathbb{P})$, the JBSDE above with $(\bar{\gamma}, \bar{\eta})$ from (5.14) has a unique solution $(Y, Z, U)$ in $\mathcal{S}^\infty \times \mathcal{L}^2(B) \times \mathcal{L}^2(\widetilde{\mu})$. Moreover there exists $\bar{\mathbb{Q}} \ll \mathbb{P}$ in the $L^1$-closure of $\mathcal{Q}^{ngd}$ (cf. Lemma 5.11), with density $\mathrm{d}\bar{\mathbb{Q}}/\mathrm{d}\mathbb{P} = \mathcal{E}\left((-\varphi + \bar{\eta}) \bullet B + \bar{\gamma} * \widetilde{\mu}\right)$ such that the good-deal bound satisfies*

$$\pi_t^u(X) = \mathop{ess\ sup}_{\mathbb{Q} \in \mathcal{Q}^{ngd}} \mathbb{E}_{\mathbb{Q}}(X|\mathcal{F}_t) = Y_t = \mathbb{E}_{\bar{\mathbb{Q}}}(X|\mathcal{F}_t) \quad \text{for } t \leq T. \tag{5.16}$$

*Proof* Consider the family of BSDE generator functions defined for $(z, u) \in \mathbb{R}^d \times L^2(\zeta.\mathrm{d}\lambda)$ by $f^{(\gamma,\eta)}(\cdot, z, u) := (-\varphi_\cdot + \eta_\cdot)z + \int_E u(e)\gamma_\cdot(e)\zeta_\cdot(e)\lambda(de)$ and $f^{(\gamma,\eta)}(\cdot, z, u) := 0$ elsewhere, for $(\gamma, \eta) \in \bar{C}$, where coefficients $(\gamma_t(\omega), -\varphi_t(\omega) + \eta_t(\omega))$ of $f^{(\gamma,\eta)}$ are bounded in $L^2(\lambda_t(\omega)) \times \mathbb{R}^d$ by $K_f := \|K\|_\infty \in (0, \infty)$ for all $(\gamma, \eta)$ and $(t, \omega)$. By Lemma 5.10, a classical generator function $f$ for the JBSDE (5.15) can be defined such that ($\mathbb{P}\otimes \mathrm{d}t$-a.e.) $f(\cdot, z, u) = \operatorname{ess\,sup}_{(\gamma,\eta)\in\bar{C}} f^{(\gamma,\eta)}(\cdot, z, u)$ for all $(z, u) \in \mathbb{R}^d \times L^2(\zeta.\mathrm{d}\lambda)$ and $f$ is (a.e.) Lipschitz continuous in $(z, u) \in \mathbb{R}^d \times L^2(\lambda_t(\omega))$, with Lipschitz constant $K_f$. Indeed, such generator function $f$ can be defined at first (up to a $\mathbb{P}\otimes \mathrm{d}t$-nullset) for countably many $(z, u)$ with $z \in \mathbb{Q}^d$ and $u \in \{u^n,\ n \in \mathbb{N}\}$ dense subset of $L^2(\lambda)$ and, noting that $u\,\zeta_t(\omega)^{1/2}$ is in $L^2(\lambda)$ for $u$ in $L^2(\lambda_t(\omega))$, by Lipschitz-continuous extension for all $(z, u) \in \mathbb{R}^d \times L^2(\lambda_t(\omega))$. By setting $f(t, z, u) := 0$ elsewhere (for $u \in L^0(\mathcal{B}(E), \lambda) \setminus L^2(\lambda_t(\omega))$), one can define $f$ as Lipschitz continuous even for $(z, u) \in \mathbb{R}^d \times L^0(\mathcal{B}(E), \lambda)$.

By classical theory for Lipschitz-JBSDE, Eq. (5.15) thus has a unique solution $(Y, Z, U)$ in $\mathcal{S}^2 \times \mathcal{L}^2(B) \times \mathcal{L}^2(\widetilde{\mu})$ which by boundedness of X satisfies $Y \in \mathcal{S}^\infty$ (cf. e.g. [7, Proposition 3.2–3.3]). Note that for all $(\gamma, \eta) \in \bar{C}$, clearly $\beta := -\varphi + \eta$ is bounded and $\int_E |\gamma_t(e)|^2 \zeta_t(e)\lambda(\mathrm{d}e)$ is bounded uniformly in $t \le T$. Hence by Lemma 4.8, the BSDEs with generators $f^{\gamma,\eta}$ also have unique solutions $(Y^{\gamma,\eta}, Z^{\gamma,\eta}, U^{\gamma,\eta}) \in \mathcal{S}^\infty \times \mathcal{L}^2(B) \times \mathcal{L}^2(\widetilde{\mu})$, which satisfy $Y_t^{\gamma,\eta} = \mathbb{E}_{\mathbb{Q}^{\gamma,\eta}}(X|\mathcal{F}_t)$, $\mathbb{Q}^{\gamma,\eta}$-a.s., $t \le T$. Since $f = f^{\bar{\gamma},\bar{\eta}}$, we also have $Y_t = \mathbb{E}_{\bar{\mathbb{Q}}}(X|\mathcal{F}_t)$, $\bar{\mathbb{Q}}$-a.s. By Lemma 5.11 holds $\pi_t^u(X) \ge \mathbb{E}_{\bar{\mathbb{Q}}}(X|\mathcal{F}_t)$, $\bar{\mathbb{Q}}$-a.s., for all $t \le T$. To complete the proof, we show that $\pi_t^u(X) \le Y_t$. For all $(\gamma, \eta) \in C$ (defining $\mathbb{Q}^{\gamma,\eta} \in \mathcal{Q}^{\mathrm{ngd}}$) we have that $f_t(Z_t, U_t) = f_t^{\bar{\gamma},\bar{\eta}}(Z_t, U_t)$ dominates $f_t^{\gamma,\eta}(Z_t, U_t)$ for a.e. $t \le T$. Noting that $f^{\gamma,\eta}$ are Lipschitz in $(z, u)$ with (uniform) Lipschitz constant $K_f$ and

$$f_t^{\gamma,\eta}(Z_t^{\gamma,\eta}, U_t) - f_t^{\gamma,\eta}(Z_t^{\gamma,\eta}, U_t^{\gamma,\eta}) = \int_E \gamma_t(e)(U_t(e) - U_t^{\gamma,\eta}(e))\zeta_t(e)\lambda(\mathrm{d}e), \quad t \le T, \tag{5.17}$$

with $\mathcal{E}\left((-\varphi + \eta)\bullet B + \gamma * \widetilde{\mu}\right)$ being a martingale (see Example 3.3), one can apply comparison as in Proposition 3.1 to get $Y_t \ge Y_t^{\gamma,\eta}$, $\mathbb{P}$-a.s., $t \le T$, $(\gamma, \eta) \in C$. Hence $Y_t \ge \operatorname{ess\,sup}_{(\gamma,\eta)} Y_t^{\gamma,\eta} = \pi_t^u(X)$, $t \le T$, for $(\gamma, -\varphi + \eta)$ ranging over all Girsanov kernels of measures $\mathbb{Q} \in \mathcal{Q}^{\mathrm{ngd}}$. □

## References


1. Antonelli, F., Mancini, C.: Solutions of BSDEs with jumps and quadratic/locally lipschitz generator. Stochast. Process. Appl. **126**(10), 3124–3144 (2016)
2. Bandini, E.: Existence and uniqueness for backward stochastic differential equations driven by a random measure, possibly non quasi-left continuous. Electron. Commun. Probab. **20**, 13 pp. (2015)
3. Bandini, E., Confortola, F.: Optimal control of semi-markov processes with a backward stochastic differential equations approach. Math. Control Sign. Syst. **29**(1), 1 (2017)
4. Barles, G., Buckdahn, R., Pardoux, E.: BSDE's and integral-partial differential equations. Stochastics **60**, 57–83 (1997)
5. Barrieu, P., El Karoui, N.: Pricing, hedging and designing derivatives with risk measures. In: Carmona, R. (ed.) Indifference Pricing, Theory and Applications, pp. 77–146. Princeton Univ. Press (2009)
6. Barrieu, P., El Karoui, N.: Monotone stability of quadratic semimartingales with applications to unbounded general quadratic BSDEs. Ann. Probab. **41**(3B), 1831–1863 (2013)

7. Becherer, D.: Bounded solutions to backward SDEs with jumps for utility optimization and indifference hedging. Ann. Appl. Probab. **16**, 2027–2054 (2006)
8. Becherer, D.: Utility indifference valuation. In: Cont, R. (ed.) Encyclopedia of Quantitative Finance. Wiley, Chichester (2010)
9. Becherer, D., Kentia, K.: Good deal hedging and valuation under combined uncertainty about drift and volatility. Probab. Uncertain. Quant. Risk **2**(1), 13 (2017)
10. Becherer, D., Kentia, K.: Hedging under generalized good-deal bounds and model uncertainty. Math. Meth. Oper. Res. **86**(1), 171–214 (2017)
11. Björk, T., Slinko, I.: Towards a general theory of good-deal bounds. Rev. Financ. **10**, 221–260 (2006)
12. Carr, P., Geman, H., Madan, D., Yor, M.: The fine structure of asset returns: an empirical investigation. J. Bus. **75**(2), 305–332 (2002)
13. Cochrane, J., Saá Requejo, J.: Good deal asset price bounds in incomplete markets: beyond arbitrage. J. Polit. Econ. **108**, 79–119 (2000)
14. Cohen, S.N., Elliott, R.J.: Stochastic calculus and applications, 2nd edn. In: Probability and Its Applications. Springer, New York (2015)
15. Cohen, S.N., Elliott, R.J.: Comparisons for backward stochastic differential equations on Markov chains and related no-arbitrage conditions. Ann. Appl. Probab. **20**, 267–311 (2010)
16. Confortola, F., Fuhrman, M.: Backward stochastic differential equations associated to jump Markov processes and applications. Stochast. Process. Appl. **124**, 289–316 (2014)
17. Confortola, F., Fuhrman, M., Jacod, J.: Backward stochastic differential equations driven by a marked point process: an elementary approach with an application to optimal control. Ann. Appl. Probab. **26**(3), 1743–1773 (2016)
18. Dellacherie, C., Meyer, P.: Probabilities and potenial B: theory of martingales. In: Mathematics Studies. North Holland, Amsterdam, New York, Oxford (1982)
19. Delong, L., Pelsser, A.: Instantaneous mean-variance hedging and Sharpe ratio pricing in a regime-switching financial model. Stochast. Models **31**, 67–97 (2015)
20. Di Tella, P., Engelbert, H.-J.: The predictable representation property of compensated-covariation stable families of martingales. Theory Probab. Appl. **60**, 99–130 (2015)
21. Di Tella, P., Engelbert, H.-J.: On the predictable representation property of martingales associated with Lévy processes. Stochastics **87**(1), 1–15 (2015)
22. Eddahbi, M., Fakhouri, I., Ouknine, Y.: $\{L\}^p$ $(p \geq 2)$-solutions of generalized BSDEs with jumps and monotone generator in a general filtration. Modern Stochast.: Theor. Appl. **4**(1), 25–63 (2017)
23. El Karoui, N., Matoussi, A., Ngoupeyou, A.: Quadratic exponential semimartingales and application to BSDEs with jumps. arXiv preprint, arXiv:1603.06191 (2016)
24. Geiss, C., Steinicke, A.: Existence, uniqueness and comparison results for BSDEs with Lévy jumps in an extended monotonic generator setting. Probab. Uncertain. Quant. Risk. **3**(1), 9 (2018)
25. Geiss, C., Steinicke, A.: Malliavin derivative of random functions and applications to Lévy driven BSDEs. Electron. J. Probab. **21**, 28 pp. (2016)
26. Geiss, C., Labart, C.: Simulation of BSDEs with jumps by Wiener Chaos expansion. Stochast. Process. Appl. **126**(7), 2123–2162 (2016)
27. Geiss, C., Steinicke, A.: $L_2$-variation of Lévy driven BSDEs with non-smooth terminal conditions. Bernoulli **22**(2), 995–1025 (2016)
28. He, S., Wang, J., Yan, J.: Semimartingale Theory and Stochastic Calculus. Science Press, CRC Press, New York (1992)
29. Hu, Y., Liang, G., Tang, S.: Exponential utility maximization and indifference valuation with unbounded payoffs. arXiv preprint. arXiv:1707.00199 (2017)
30. Hu, Y., Imkeller, P., Müller, M.: Utility maximization in incomplete markets. Ann. Appl. Probab. **15**, 1691–1712 (2005)
31. Jacod, J., Shiryaev, A.: Limit Theorems for Stochastic Processes. Springer, Berlin (2003)
32. Kazamaki, N.: Continuous exponential martingales and BMO. In: Lecture Notes in Mathematics, vol. 1579. Springer, Berlin (1994)

33. Kazamaki, N.: A sufficient condition for the uniform integrability of exponential martingales. Math. Rep. Toyama University **2**, 1–11 (1979)
34. Kazi-Tani, N., Possamaï, D., Zhou, C.: Quadratic BSDEs with jumps: related nonlinear expectations. Stochast. Dyn. **16**(4), 1650012, 32 pp. (2016)
35. Kazi-Tani, N., Possamaï, D., Zhou, C.: Quadratic BSDEs with jumps: a fixed-point approach. Electron. J. Probab. **20**(66), 1–28 (2015)
36. Kentia, K.: Robust aspects of hedging and valuation in incomplete markets and related backward SDE theory. Ph.D. thesis, Humboldt-Universität zu Berlin. urn:nbn:de:kobv:11-100237580 (2015)
37. Kobylanski, M.: Backward stochastic differential equations and partial differential equations with quadratic growth. Ann. Appl. Probab. **28**, 558–602 (2000)
38. Kruse, T., Popier, A.: BSDEs with monotone generator driven by brownian and poisson noises in a general filtration. Stochastics **88**(4), 491–539 (2016)
39. Kruse, T., Popier, A.: Lp-solution for BSDEs with jumps in the case $p<2$. Stochastics **89**(8), 1201–1227 (2017)
40. Laeven, R.J.A., Stadje, M.A.: Robust portfolio choice and indifference valuation. Math. Oper. Res. **39**, 1109–1141 (2014)
41. Lepingle, D., Mémin, J.: Sur l'intégrabilité uniforme des martingales exponentielles. Z. Wahrscheinlichkeitstheor. verw. Geb. **42**, 175–203 (1978)
42. Mania, M., Chikvinidze, B.: New proofs of some results on BMO martingales using BSDEs. J. Theoret. Probab. **27**, 1213–1228 (2014)
43. Mania, M., Schweizer, M.: Dynamic exponential utility indifference valuation. Ann. Appl. Probab. **15**(3), 2113–2143 (2005)
44. Morlais, M.: Utility maximization in a jump market model. Stochastics **81**, 1–27 (2009)
45. Morlais, M.: A new existence result for quadratic BSDEs with jumps with application to the utility maximization problem. Stochast. Process. Appl. **120**, 1966–1995 (2010)
46. Nualart, D., Schoutens, W.: Chaotic and predictable representations for Lévy processes. Stochast. Process. Appl. **90**, 109–122 (2000)
47. Nualart, D., Schoutens, W.: Backward stochastic differential equations and Feynman-Kac formula for Lévy processes, with applications in finance. Bernoulli **7**(5), 761–776 (2001)
48. Nutz, M.: The opportunity process for optimal consumption and investment with power utility. Math. Financ. Econ. **3**(3–4), 139–159 (2010)
49. Papapantoleon, A., Possamaï, D., Saplaouras, A.: Existence and uniqueness results for BSDEs with jumps: the whole nine yards. Electron. J. Probab. 23, Paper no. 121, 68 pp (2018)
50. Pardoux, E., Peng, S.: Adapted solution of a backward stochastic differential equation. Syst. Control Lett. **14**, 55–61 (1990)
51. Protter, P., Shimbo, K.: No arbitrage and general semimartingales. Markov Processes and related Topics: A Festschrift for Thomas G. Kurtz **4**, 267–283 (2008)
52. Quenez, M.-C., Sulem, A.: BSDEs with jumps, optimization and applications to dynamic risk measures. Stochast. Process. Appl. **123**(8), 3328–3357 (2013)
53. Rockafellar, R.T.: Integral functionals, normal integrands and measurable selections. In: Waelbroeck, L. (ed.) Nonlinear Operators and Calculus of Variations. Lecture Notes in Mathematics 543, pp. 157–207. Springer, Berlin (1976)
54. Royer, M.: Backward stochastic differential equations with jumps and related non-linear expectations. Stochast. Process. Appl. **116**, 1358–1376 (2006)
55. Tang, S., Li, X.: Necessary conditions for optimal control for stochastic systems with random jumps. SIAM J. Control Optim. **32**, 1447–1475 (1994)
56. Tevzadze, R.: Solvability of backward stochastic differential equations with quadratic growth. Stochast. Process. Appl. **118**(3), 503–515 (2008)
57. Yao, S.: Lp solutions of backward stochastic differential equations with jumps. Stochast. Process. Appl. **127**(11), 3465–3511 (2017)

# On the Wellposedness of Some McKean Models with Moderated or Singular Diffusion Coefficient

**Mireille Bossy and Jean-François Jabir**

**Abstract** We investigate the well-posedness problem related to two models of nonlinear McKean Stochastic Differential Equations with some local interaction in the diffusion term. First, we revisit the case of the McKean-Vlasov dynamics with moderate interaction, previously studied by Méléard and Jourdain in [16], under slightly weaker assumptions, by showing the existence and uniqueness of a weak solution using a Sobolev regularity framework instead of a Hölder one. Second, we study the construction of a Lagrangian Stochastic model endowed with a conditional McKean diffusion term in the velocity dynamics and a nondegenerate diffusion term in the position dynamics.

**Keywords** Weak-strong wellposedness problems · McKean-Vlasov models · Singular McKean diffusions

## 1 Introduction

In this paper, we are interested in the wellposedness problem of some singular nonlinear McKean SDEs in the McKean-Vlasov sense in $\mathbb{R}^d$, in the particular situation where the diffusion term carries the singular McKean nonlinear dependency. General form of nonlinear McKean SDEs is given by

$$dX_t = b(X_t, Law(X_t))dt + \sigma(X_t, Law(X_t))dW_t, \quad \mu_0 = Law(X_0), \tag{1.1}$$

The second author has been supported by the Russian Academic Excellence Project "5–100".

M. Bossy (✉)
Université Côte d'Azur, Inria, France
e-mail: Mireille.Bossy@inria.fr

J.-F. Jabir
National Research University, Higher School of Economics, Moscow, Russia
e-mail: jjabir@hse.ru



S. N. Cohen et al. (eds.), *Frontiers in Stochastic Analysis - BSDEs, SPDEs and their Applications*, Springer Proceedings in Mathematics & Statistics 289,
https://doi.org/10.1007/978-3-030-22285-7_2

where $(W_t;\ t \geq 0)$ is a $\mathbb{R}^d$-valued standard Brownian motion, independent of $X_0$, and $\mu_0$ is a given probability measure on $\mathbb{R}^d$.

The class of singular McKean models we want to consider here are models for which the corresponding interacting particle system approximation (at least given formally, when the mean field limit is not yet established) gives rise to some singularity in the kernel function. In the context of the model (1.1), this means that for any $N \geq 2$, for any family of mollifiers $g_\varepsilon := \varepsilon^{-d} g(\frac{x}{\varepsilon})$ with parameter $\varepsilon > 0$, the mapping

$$x = (x^0, x^1, \ldots, x^N) \in \mathbb{R}^{(N+1)\times d} \mapsto b \text{ or } \sigma\left(x^0, \left(g_\varepsilon * \mu_x^N\right)\right) \in \mathbb{R}^d, \text{ with } \mu_x^N = \frac{1}{N}\sum_{i=1}^N \delta_{x^i}$$

attains its maximum norm on the subset $\bigcup_{i=1}^N \{x^0 = x^i\}$ which tends to $+\infty$ as $\varepsilon$ tends to zero. Well known examples of McKean SDEs with singular drift kernel $b$ are those of the probabilistic interpretation of Burgers' equation (Sznitman [32]), the stochastic vortex method model for fluid flow (see e.g. Chorin [9], Méléard [25] among others), or more recently the probabilistic interpretation of the Keller Segel equation for chemotaxis modeling (see e.g. Fournier and Jourdain [19]).

More precisely, we are interested in the wellposedness of the following coupled processes $(X_t, Y_t;\ t \geq 0)$ on a probability space $(\Omega, \mathcal{F}, \mathbb{P})$ satisfying the conditional McKean SDE

$$\begin{cases} X_t = X_0 + \int_0^t b(X_s, Y_s)ds + \int_0^t \sigma(X_s)dB_s, \\ Y_t = Y_0 + \int_0^t \mathbb{E}_{\mathbb{P}}[\ell(Y_s)|X_s]ds + \int_0^t \mathbb{E}_{\mathbb{P}}[\gamma(Y_s)|X_s]dW_s. \end{cases} \tag{1.2}$$

The initial condition $(X_0, Y_0)$ is distributed according to a given initial law $\mu_0$, $(W_t;\ t \geq 0)$ and $(B_t;\ t \geq 0)$ are two independent $\mathbb{R}^d$ standard Brownian motions, $x \mapsto \sigma(x)$ is a $\mathbb{R}^d \times \mathbb{R}^d$ valued function. Before briefly describing our hypotheses on the coefficients of (1.3), let us make some comments on such models.

Our particular interest for the study of singular dynamics as (1.2) is motivated by the wellposedness problem related to the class of Lagrangian stochastic models for turbulent flows. This class of models have been introduced in the general framework of the statistical description of fluid motions and aimed to describe the main characteristic properties (position, velocity, ...) of a generic particle of a particular fluid flow. From the turbulent modeling viewpoint, such SDEs are known as Lagrangian "fluid-particle" models and are translation in a Lagrangian point of view (SDE) of some Eulerian PDE turbulence models (see e.g. Pope [29, 30], Durbin and Speziale [10]). These models involve a particular family of nonlinear McKean-Vlasov SDEs where the McKean nonlinearities are of conditional form. Such particular form of nonlinearity models the influence of the macroscopic components of the flow on the particle motion. In some of our recent works, [4, 5], we have studied toy-version models of conditional McKean SDEs where the singularity is concentrated in the drift term. From a mathematical viewpoint, the wellposedness results obtained in [4, 5] are still far from covering the complexity of a meaningful 'fluid-particle' model, as such Lagrangian models contain conditional McKean nonlinearity in both drift and diffusion components. In this paper, we focus on singular McKean diffusive characteristic that motivates our interest in new wellposedness results in that direction.

In [4, 5] and for the construction of (numerical) approximation (we refer to Bossy et al. [6, 7] for some numerical description cases and experiments), we analyze the SDE (1.2) in the framework of an apriori existing density $\rho_t(x, y)dxdy = \mathbb{P}(X_t \in dx, Y_t \in dy)$. The model (1.2) thus becomes

$$\begin{cases} X_t = X_0 + \int_0^t b(X_s, Y_s)ds + \int_0^t \sigma(X_s)dB_s, \\ Y_t = Y_0 + \int_0^t \Lambda[X_s; \rho_s]ds + \int_0^t \Gamma[X_s; \rho_s]dW_s, \end{cases} \tag{1.3}$$

with $\Lambda$ and $\Gamma$ defined, for $(x, f)$ in $\mathbb{R}^d \times L^1(\mathbb{R}^d \times \mathbb{R}^d)$, as

$$\Lambda[x; f] = \frac{\int_{\mathbb{R}^d} \ell(y) f(x, y)dy}{\int_{\mathbb{R}^d} f(x, y)dy} \mathbb{1}_{\{\int_{\mathbb{R}^d} f(x,y)dy \neq 0\}} \quad \text{and} \quad \Gamma[x; f] = \frac{\int_{\mathbb{R}^d} \gamma(y) f(x, y)dy}{\int_{\mathbb{R}^d} f(x, y)dy} \mathbb{1}_{\{\int_{\mathbb{R}^d} f(x,y)dy \neq 0\}}.$$

In comparison, our wellposedness result for the solution of (1.2), presented in Sect. 3, uses a $L^2(\Omega)$-fixed point construction and a suitable Girsanov transformation that relies on the strong ellipticity assumption on $\sigma$. Essentially, our working hypotheses will be to assume boundedness and Lipschitz continuity of $b, \sigma, \ell$ and $\gamma$ for the wellposedness of a weak solution to (1.2), and some $L^p$ density condition on the initial distribution and a uniform elliptic property on $\gamma$ to handle pathwise uniqueness. At all time $t$, the time-marginal distributions $Law(X_t, Y_t)$ of this strong solution further admit a density function $\rho_t$, and so our constructed solution to (1.2) is also solution to (1.3).

In the context of complex flow modeling, we would like to emphasise that a targeted form of (1.2) is a coupled position-velocity $(X_t, U_t;\ 0 \leq t \leq T)$ kinetic process with degenerate diffusion in the $X$-component together with a linear drift $b(x, y) = y$:

$$\begin{cases} X_t = X_0 + \int_0^t U_s ds \\ U_t = Y_0 + \int_0^t \mathbb{E}_{\mathbb{P}}[\ell(U_s)|X_s]ds + \int_0^t \mathbb{E}_{\mathbb{P}}[\gamma(U_s)|X_s]dW_s. \end{cases}$$

But unbounded drift case, degenerate diffusion and singular McKean kernel are a mixture of difficulties that are quite hard to overcome jointly.

For future works, to overcome the strong ellipticity assumption on $\sigma$ in (1.2), we further investigate some weaker characterisation method based on mild-equation formulation as in [5]. In Sect. 2, we present a step further in that direction, applying such technique for our second study case on moderated McKean local diffusion equation:

$$\begin{cases} X_t = X_0 + \int_0^t \sigma(u(s, X_s))dW_s, \quad 0 \leq t \leq T, \\ d\,Law(X_t) = u(t, x)dx \text{ with } u \in L^\infty((0, T) \times \mathbb{R}^d) \cap L^2((0, T)\times\mathbb{R}^d), \\ u(0, x) = u_0(x),\ x \in \mathbb{R}^d, \text{ where } u_0 \text{ is a given probability density function on } \mathbb{R}^d, \end{cases} \tag{1.4}$$

for any arbitrary time horizon $0 < T < \infty$. Nevertheless, our existence proof based on approximation method needs some strict monotonicity assumption which still coincides with the strong ellipticity in the one dimensional framework. In [16], Jourdain and Méléard studied a moderately interacting model such as (1.4), extending a previous work from Oelschläger [28] on a moderately interacting model, where both the drift and diffusion coefficients depend locally on the time marginal densities of the law of the solution that are supposed to be smooth enough. Whenever the nonlinearity is reduced to the diffusion part, the model in [16] reduces to:

$$\begin{cases} X_t = X_0 + \int_0^t \sigma(p(s, X_s))dW_s, \quad 0 \leq t \leq T, \\ d\,Law(X_t) = p(t, x)\,dx \text{ with } p \in C_b^{1,2}([0, T] \times \mathbb{R}^d), \\ p(0, \cdot) \text{ is a given probability density that belongs in the Hölder space } H^{2+\alpha}(\mathbb{R}^d) \text{ with } 0 < \alpha < 1. \end{cases} \tag{1.5}$$

In Sect. 2, we prove the wellposdness of a strong solution of (1.4), mainly replacing the condition $p \in C_b^{1,2}([0, T] \times \mathbb{R}^d)$ by $p \in L^\infty((0, T) \times \mathbb{R}^d) \cap L^2((0, T) \times \mathbb{R}^d)$ (replacing the strong ellipticity condition on $\sigma$ needed for the Eq. (1.5) by a strict monotonicity condition). Our proof is focused on the simple case where the diffusion component is given by $\sigma(r)I_d$ for a scalar function $\sigma : [0, \infty) \to [0, \infty)$. Extensions to further multidimensional diffusion component are discussed at the end of Sect. 2.

Our main results are Theorem 2.3 in Sect. 2 which states the strong wellposedness for the moderated McKean local diffusion equation (1.4), and Theorem 3.3 in Sect. 3 for the strong wellposedness of conditional McKean SDEs (1.3). In the two cases, we obtained weak uniqueness of the solution with slighty weaker conditions.

We end this introductory section with a short review of results and approaches from the literature for SDEs with McKean diffusion term, in order give some insights to the reader with the two particular cases that we are addressing in this paper.

**Review of some wellposedness results for nonlinear SDEs with McKean diffusion term**

We consider McKean-Vlasov SDEs of the following specific form in $\mathbb{R}^d$,

$$X_t = X_0 + \int_0^t \sigma(X_s, Law(X_s))dW_s,\ 0 \leq t \leq T, \tag{1.6}$$

up to a (possibly infinite) horizon time $T$.

Under the assumptions that $\mathbb{E}[|X_0|^p] < \infty$, $1 \leq p < \infty$, and $\sigma$ is continuous on $\mathbb{R}^d \times \mathcal{P}_p(\mathbb{R}^d)$ for $\mathcal{P}_p(\mathbb{R}^d)$ being the space of probability measures with $p$-th finite moments, Funaki [13] showed the existence, on any arbitrary time interval, of a weak solution to (1.6) in terms of a martingale problem. Uniqueness of the solution to the martingale problem holds under the assumption that

$$\|\sigma(x,\mu) - \sigma(y,\nu)\| \leq C|x-y| + \kappa(\mathcal{W}_p(\mu,\nu)),$$

where $\mathcal{W}_p$ is the Wasserstein distance endowed with the cost function $|x-y|^p$, and $\kappa : [0,\infty) \to [0,\infty)$ is a strictly increasing function such that $\kappa(0) = 0$ and $\lim_{\epsilon\to 0^+} \int_\epsilon^\infty 1/\kappa^2(\sqrt{r})\,dr = \infty$. Oelschläger [27] considered the analog situation where $\sigma$ is bounded and Lipschitz for the metric

$$\|\mu - \nu\| = \sup\left\{\int f(x)\,(\mu(dx) - \nu(dx)) : \max_{x\in\mathbb{R}^d} |f(x)| \leq 1 \text{ and } \|f\|_{Lip} := \sup_{x\neq y} \frac{|f(x)-f(y)|}{|x-y|} \leq 1\right\},$$

and proved the existence of a solution in law, as well as a weak propagation of chaos result for the related stochastic particle system. Both cases include the particular situation when the interaction kernel has the form: $\sigma(x,\mu) = \int_{\mathbb{R}^d} \sigma(x,y)\mu(dy)$. Moreover, in this framework, Méléard [24] showed, through a fixed point argument in the space $(\mathcal{P}_2(\mathcal{C}([0,T];\mathbb{R}^d)), W_2)$, that whenever, $\sigma : \mathbb{R}^d \times \mathbb{R}^d \to \mathbb{R}^{d\times d}$ is Lipschitz continuous w.r.t. the two variables, the pathwise wellposedness and strong-pathwise propagation of chaos holds for the related stochastic particle system.

Jourdain and Méléard [16] extended the work of Oelschläger [28] on the moderately interacting drift term model and prove the wellposedness of (1.5) with the following assumptions:

- $p(0,x) = p_0(x)$ where $p_0$ belongs to the Hölder space $H^{2+\alpha}(\mathbb{R}^d)$ with $0 < \alpha < 1$;
- $\sigma : r \in \mathbb{R} \mapsto \sigma(r) \in \mathbb{R}^{d\times d}$ is a Lipschitz function, $\mathcal{C}^3$ on $\mathbb{R}$, with values in the space of symmetric non-negative matrices $d \times d$;
- Strong ellipticity holds for $\sigma$: there exists $m_\sigma > 0$ such that $\forall x \in \mathbb{R}^d, \forall r \in \mathbb{R},\ \ x^*\sigma(r)x \geq m_\sigma|x|^2$;
- Non negativity holds for the diffusion matrix leading to the Fokker-Planck equation written on divergence form:

$$\forall x \in \mathbb{R}^d, \forall r \in \mathbb{R},\quad x^*\left((\sigma\sigma^*)'(r)r + (\sigma\sigma^*)(r)\right)x \geq 0;$$

(This latest assumption is used to derive the uniqueness from the Fokker-Planck equation related to (1.5), written in divergence form:

$$\begin{cases} \dfrac{\partial p}{\partial t} = \displaystyle\sum_{i=1}^{d} \frac{\partial}{\partial x_i} \left( \frac{1}{2} \sum_{j=1}^{d} \Big( (\sigma\sigma^*)'_{ij}(p)p + (\sigma\sigma^*)_{ij}(p) \Big) \frac{\partial p}{\partial x_j} \right) \text{ on } (0,T) \times \mathbb{R}^d, \\ p(0,x) = p_0(x), \; x \in \mathbb{R}^d, \end{cases} \tag{1.7}$$

from maximum principle argument.)

- Strong ellipticity holds on the leading matrix: there exists $m_{\mathrm{div}} > 0$ such that

$$\forall x \in \mathbb{R}^d, \forall r \in \mathbb{R}, \quad x^* \left( (\sigma\sigma^*)'(r)r + (\sigma\sigma^*)(r) \right) x \geq m_{\mathrm{div}} |x|^2.$$

(With this additional assumption, the Cauchy problem (1.7) has a solution in $H^{1+\frac{\alpha}{2},2+\alpha}(\mathbb{R}^d)$, and the nonlinear SDE admits a unique strong solution.)

Kohatsu-Higa and Ogawa in [20] considered nonlinear McKean-Vlasov dynamic in convolution form

$$X_t = X_0 + \int_0^t A(X_s, \sigma * Law(X_s)) dW_s. \tag{1.8}$$

Assuming that $A$ and $\sigma$ are Lipschitz with at most linear growth, they prove the wellposedness of a strong solution and particle-time discrete approximation.

Jourdain and Reygner in [17] considered particular cases of scalar equation related to, and around, porus media equation which correspond to the case of Eq. (1.8) with $\sigma(x) = \mathbb{1}_{\{x \geq 0\}}$ and $A(x,u) = A(u) > 0$. The case $A(u) \geq 0$ is also studied using the limit of a reordered particle system.

Recently, Mishura and Veretennikov in [26] consider a model of the form

$$dX_t = \sigma[t, X_t; Law(X_t)] \, dW_t, \; X_0 \sim \mu_0,$$

where

$$\sigma[t, x; \mu] = \int \sigma(t, x, y) \mu(dy)$$

for $\sigma : (0, \infty) \times \mathbb{R}^d \times \mathbb{R}^d \to \mathbb{R}^{d \times d}$. Assuming that $X_0$ has finite fourth order moments, $(x, y) \mapsto \sigma(t, x, y)$ has linear growth, uniformly in $t$, and $(t, x, y) \mapsto \sigma(t, x, y)\sigma^*(t, x, y)$ is uniformly strongly elliptic, the SDE admits at least one weak solution.

We end this review by mentioning some recent works in the direction of the wellposedness of the following system of SDE

$$\begin{cases} \dfrac{dS_t}{S_t} = rdt + \dfrac{a(Y_t)}{\sqrt{\mathbb{E}[a^2(Y_t)|S_t]}}\sigma_{\text{Dup}}(t, S_t)S_t dW_t, \\ dY_t = \alpha(t, Y_t)dB_t + \xi(t)dt. \end{cases}$$

Such models arise in mathematical finance for the calibration of local and stochastic volatility models, and where $\sigma_{\text{Dup}}(t, y)$ is the Dupire's local volatility function (see Gyongy [15]). We emphasise that a major difference with the model (1.2) is in the conditioning variable which is the coupled variable $X_t$ in our case and the unknown $S_t$ in the volatility calibration model. This particular case generates different and yet hard difficulties compared to (1.2). The existence of a local-in-time solution to the Fokker-Planck equation associated to this model has been established by Abergel and Tachet in [1], while Jourdain and Zhou [18] recently obtained a first global-in-time wellposedness result in the case when $Y$ is a (constant in time) discrete valued random variable.

**Some Notations**

Hereafter, $\mathcal{C}$ denotes the space of continuous functions equipped with the uniform norm $\|f\|_\infty = \max|f|$ and $\mathcal{C}^k$ denotes the space of $k$ times continuously differentiable functions. $\mathcal{C}_c$ and $\mathcal{C}_c^k$ respectively refer to the corresponding compactly supported subsets. For $m \geq 1$, and $1 \leq p \leq \infty$, $L^p(\mathbb{R}^m)$ denotes the Lebesgue space of all Borel (measurable) functions $f : \mathbb{R}^m \to \mathbb{R}$ such that $\|f\|^p_{L^p(\mathbb{R}^m)} := \int |f(z)|^p \, dz < \infty$, and $W^{1,p}(\mathbb{R}^m)$ denotes the Sobolev space

$$W^{1,p}(\mathbb{R}^m) = \{f \in L^p(\mathbb{R}^m) \text{ s.t. } \|\nabla_z f\|_{L^p(\mathbb{R}^m)} < \infty\},$$

equipped with the norm $\|f\|_{W^{1,p}(\mathbb{R}^m)} := \|f\|_{L^p(\mathbb{R}^m)} + \|\nabla_z f\|_{L^p(\mathbb{R}^m)}$. As usual, $H^1(\mathbb{R}^m)$ stands for the particular space $W^{1,2}(\mathbb{R}^m)$. For $1 \leq q \leq \infty$, $L^q((0, T); W^{1,p}(\mathbb{R}^m))$ denotes the space of Borel functions $f$ defined on $(0, T) \times \mathbb{R}^m$ such that

$$\|f\|_{L^q((0,T);W^{1,p}(\mathbb{R}^m))} := \|f\|_{L^q((0,T);L^{1,p}(\mathbb{R}^m))} + \|\nabla f\|_{L^q((0,T);L^{1,p}(\mathbb{R}^m))}$$

is finite. Finally, the index *loc* will refer to local integrability property, namely $f$ belongs in $L^p_{loc}$ if for all $0 \leq R < \infty$, $f\mathbb{1}_{\{B(0,R)\}}$ is in $L^p(\mathbb{R}^m)$.

## 2 The Moderated McKean Local Diffusion Equation Revisited

In this section, we consider the wellposedness problem, up to an arbitrary finite horizon time $T > 0$, for the following SDE:

$$X_t = X_0 + \int_0^t \sigma(u(s, X_s))dW_s, \quad 0 \le t \le T, \tag{2.1a}$$

$$d\,Law(X_t) = u(t,x)dx \text{ where } u \text{ belongs in } L^\infty((0,T)\times\mathbb{R}^d)\cap L^2((0,T)\times\mathbb{R}^d), \tag{2.1b}$$

$$u(0,x) = u_0(x),\ x\in\mathbb{R}^d, \text{ where } u_0 \text{ is a given probability density function.} \tag{2.1c}$$

For the sake of simplicity, from now on, we restrict ourselves to the case of a diffusion matrix $\sigma$ mainly diagonal; that is $\sigma(r) = \sigma(r)I_d$ for $\sigma : \mathbb{R} \to \mathbb{R}$ and $I_d$ the identity $d \times d$ matrix. Extensions to more general diffusion matrices will be discussed at the end of this section (see Sect. 2.4).

Let us further point out that the notion of solution to (2.1) is intentionally restricted to the class of solutions satisfying (2.1b). We consider the class of solutions of continuous processes satisfying (2.1a) and whose time-marginal distributions admit a representant in $L^2(\mathbb{R}^d)$ for a.e. $0 \le t \le T$, and in $L^\infty((0,T)\times\mathbb{R}^d)$. The choice of working with this particular class is mainly motivated by the use of comparison principles and energy estimates techniques (see e.g. Evans [11] and Vasquez [34]) for the time-marginal distributions solution to the Fokker-Planck equation related to (2.1). Energy estimates will enable us to construct a suitable approximation to (2.1) and also to deduce the uniqueness of the marginal distributions $u(t)$, $0 \le t \le T$. We emphasise that the divergence form for the Fokker-Planck equation makes appear as a coefficient the map $r \mapsto \alpha(r)$ defined as

$$\alpha(r) := (\sigma^2(r)r)' = 2\sigma'(r)\sigma(r)r + \sigma^2(r), \text{ for } r\in\mathbb{R}^+, \tag{2.2}$$

which our main hypothesis is based on.
**Remark**: *In the case when $\sigma$ and $\alpha$ are bounded, our proof arguments and subsequent wellposedness results can be extended to the class of solutions to* (2.1a) *satisfying $u \in L^2((0,T)\times\mathbb{R}^d)$ in place of* (2.1b).

Throughout this section, Eq. (2.1) is considered under the following set of assumptions:

**Hypothesis 2.1**

($\boldsymbol{A}_0$) $u_0$ is a probability density function in $L^1(\mathbb{R}^d)\cap L^\infty(\mathbb{R}^d)$ such that $\int_{\mathbb{R}^d}|x|^2 u_0(x)dx < \infty$.

($\boldsymbol{A}_1$) The map $r \mapsto \sigma(r) \in \mathbb{R}$ is continuously differentiable on $\mathbb{R}^+$.

($\boldsymbol{A}_2$) The map $r \mapsto \alpha(r)$ is continuous on $\mathbb{R}^+$, and there exists some constant $\eta > 0$ such that

$$\alpha(r) \ge \eta > 0,\ \forall r \ge 0.$$

Assumption $(A_2)$ ensures that $\sigma$ is in $\mathcal{C}^1([0,\infty))$ and implies a classical assumption on the uniform positivity of $\sigma^2$:

$$\sigma^2(r) \ge \eta,\ \forall r \ge 0,$$

which further implies the uniform ellipticity of $\sigma^2 I_d$. Yet, most of the time, we will also make use of the following assumption which, together with a monotonic property of $\alpha$ (see Theorem 2.3 and Proposition 2.10 below), allows possible degeneracy of the diffusion $\sigma$ at point $r = 0$:

**Hypothesis 2.2**

**($A_2$-Weakend)** The map $r \mapsto \alpha(r)$ is continuous on $\mathbb{R}^+$, and

$$\alpha(r) \geq 0, \; \forall r \geq 0.$$

The main result of this section is the wellposedness for Eq. (2.1) given by the following theorem:

**Theorem 2.3** *Under Hypothesis 2.1, there exists a unique strong solution to (2.1). Uniqueness in law holds true under ($A_0$), ($A_1$), ($A_2$-weakened) with the additional hypothesis that $r \mapsto \alpha(r)$ is strictly increasing.*

**Main ingredients and steps of the proof.** The rest of this section is devoted to our proof of Theorem 2.3 that relies on the following three main ingredients:

1. An appropriate form of $\varepsilon$-nondegenerate approximation of the diffusion $\sigma$. In a first step, we show the wellposedness of a family of $\varepsilon$-approximation $\{(u^\varepsilon(t),\; X_t^\varepsilon;\; 0 \leq t \leq T), \varepsilon > 0\}$ to (2.1) where $\sigma$ is replaced by $\sigma_\varepsilon$ defined by

$$\sigma_\varepsilon^2(r) = \sigma^2(r) + \varepsilon, \; \forall r \geq 0. \tag{2.3}$$

   Notice that (2.3) produces a suitable approximation of the map $r \mapsto \alpha(r)$ by

$$\alpha_\varepsilon(r) = (\sigma_\varepsilon^2(r)r)' = (\sigma^2(r)r)' + \varepsilon = \alpha(r) + \varepsilon, \; \forall r \geq 0. \tag{2.4}$$

2. The construction of $(u^\varepsilon(t),\; X_t^\varepsilon;\; 0 \leq t \leq T)$ is then obtained from a preliminary existence result of a $L^\infty((0,T) \times \mathbb{R}^d) \cap L^2((0,T);\; H^1(\mathbb{R}^d))$-weak solution to the related Fokker-Planck equation, and some uniform energy estimates w.r.t $\varepsilon$ on this solution. Such estimates allow to deduce successively the relative compactness of $\{u^\varepsilon, \varepsilon > 0\}$ in $L^2((0,T);\; H^1(\mathbb{R}^d))$ and of $\{Law(X_t^\varepsilon;\; 0 \leq t \leq T), \varepsilon > 0\}$ in $\mathcal{P}(\mathcal{C}([0,T];\; \mathbb{R}^d))$. The corresponding limits of converging subsequences are then shown to jfbe a solution of the martingale problem related to (2.1). This main step is stated in Proposition 2.8 below.
3. Uniqueness in law is obtained from a mild-form equation in $L^2((0,T) \times \mathbb{R}^d)$ derived from the Fokker-Planck equation in Proposition 2.10. The mild approach used here allows us to get rid of the strong ellipticity hypothesis for $\sigma$, at least at point 0. The weak uniqueness result is then obtained under ($A_2$), but also under ($A_2$-weakened) with the adding of the strict monotonicity for $\alpha$. Uniqueness in the pathwise sense is stated in Proposition 2.12.

## 2.1 Nondegenerate Approximation of (2.1)

In this section, we construct a solution to the SDE

$$X_t^\varepsilon = X_0 + \int_0^t \sqrt{\sigma_\varepsilon^2(u^\varepsilon(s, X_s))} dW_s, \quad 0 \le t \le T, \tag{2.5a}$$

$$d\,Law(X_t^\varepsilon) = u^\varepsilon(t,x)dx, \text{ with } u^\varepsilon \in L^\infty((0,T)\times\mathbb{R}^d)\cap L^2((0,T)\times\mathbb{R}^d), \tag{2.5b}$$

$$u^\varepsilon(0,x) = u_0(x), x \in \mathbb{R}^d, \tag{2.5c}$$

where $\sigma_\varepsilon^2(r) = \sigma^2(r) + \varepsilon,\ \forall r \ge 0$, and we show some appropriate density estimates for the marginal densities $u^\varepsilon(t)$. Defined as such, the diffusion coefficient $\sigma_\varepsilon$ still satisfies $(A_1)$. Our existence proof is mainly deduced from a PDE analysis of the smoothed Fokker-Planck equation related to (2.5):

$$\begin{cases} \dfrac{\partial u^\varepsilon}{\partial t} - \dfrac{1}{2}\Delta_x(\sigma_\varepsilon^2(u^\varepsilon)u^\varepsilon) = 0, \text{ on } (0,T)\times\mathbb{R}^d, \\ u^\varepsilon(0,x) = u_0(x),\ x \in \mathbb{R}^d. \end{cases} \tag{2.6}$$

We proceed for the existence of a solution to (2.6), first by exhibiting the existence of a weak solution of a linearized version of (2.6), and next by a fixed point argument we deduce the existence. These results are given in the following two lemmas:

**Lemma 2.4** *Assume $(A_0)$, $(A_1)$ and ($A_2$-weakened). Let $v = v(t,x)$ be a non-negative given function belonging to $L^\infty((0,T)\times\mathbb{R}^d)\cap L^2((0,T)\times\mathbb{R}^d)$. Then there exists $\overline{u}^\varepsilon \in L^2((0,T); H^1(\mathbb{R}^d)) \cap C([0,T]; L^2(\mathbb{R}^d))$ such that, for all $f \in C_c^{1,2}([0,T]\times\mathbb{R}^d)$, for all $0 \le T_0 \le T$,*

$$\begin{aligned}&\int_{\mathbb{R}^d} \overline{u}^\varepsilon(T_0,x) f(T_0,x)\,dx - \int_{\mathbb{R}^d} u_0(x) f(0,x)\,dx \\ &= \int_{(0,T_0)\times\mathbb{R}^d} \overline{u}^\varepsilon(t,x)\partial_t f(t,x) - \frac{1}{2}\alpha_\varepsilon(v(t,x))\nabla_x\overline{u}^\varepsilon(t,x)\cdot\nabla_x f(t,x)\,dt\,dx,\end{aligned} \tag{2.7}$$

*for $\alpha_\varepsilon$ defined as in (2.4). In addition, $\overline{u}^\varepsilon$ has nonnegative values a.e. on $(0,T)\times\mathbb{R}^d$,*

$$\|\overline{u}^\varepsilon\|_{L^\infty((0,T)\times\mathbb{R}^d)} \le \|u_0\|_{L^\infty(\mathbb{R}^d)}, \tag{2.8}$$

*and*

$$\max_{t\in[0,T]} \|\overline{u}^\varepsilon(t)\|^2_{L^2(\mathbb{R}^d)} + \varepsilon\int_0^T \|\nabla_x\overline{u}^\varepsilon(t)\|^2_{L^2(\mathbb{R}^d)}dt \le \|u_0\|^2_{L^2(\mathbb{R}^d)}. \tag{2.9}$$

For the nonlinear PDE (2.6), we extend the notion of a $L^2((0,T); H^1(\mathbb{R}^d))$-weak solution, stated in this Lemma 2.4, as a function $u^\varepsilon \in L^2((0,T); H^1(\mathbb{R}^d))$ such that:

for all $0 < T_0 \leq T$, and $f \in C_c^\infty([0, T_0] \times \mathbb{R}^d)$,

$$\int_{\mathbb{R}^d} u^\varepsilon(T_0, x) f(T_0, x)\, dx - \int_{\mathbb{R}^d} u_0(x) f(0, x)\, dx$$
$$= \int\int_{(0,T_0)\times\mathbb{R}^d} u^\varepsilon(t, x)\partial_t f(t, x)\, dt\, dx$$
$$+ \int_{(0,T_0)\times\mathbb{R}^d} \frac{1}{2}\sigma_\varepsilon^2(u^\varepsilon(t, x))u^\varepsilon(t, x)\Delta_x f(t, x)\, dt\, dx.$$

**Lemma 2.5** *Assuming that* ($A_0$), ($A_1$) *and* ($A_2$-*weakened*) *hold, the nonlinear PDE (2.6) admits a unique nonnegative* $C([0, T]; L^2(\mathbb{R}^d))\cap L^2((0, T); H^1(\mathbb{R}^d))$-*weak solution* $u^\varepsilon$. *This solution is uniformly bounded with*

$$\|u^\varepsilon\|_{L^\infty((0,T)\times\mathbb{R}^d)} \leq \|u_0\|_{L^\infty(\mathbb{R}^d)}, \tag{2.10}$$

*and satisfies the energy inequality:*

$$\sup_{t\in[0,T]} \|u^\varepsilon(t)\|^2_{L^2(\mathbb{R}^d)} + \varepsilon \int_0^T \|\nabla_x u^\varepsilon(t)\|^2_{L^2(\mathbb{R}^d)} dt \leq \|u_0\|^2_{L^2(\mathbb{R}^d)}. \tag{2.11}$$

*In addition, we have, for all* $0 \leq T_0 \leq T$,

$$\int_{\mathbb{R}^d} \Psi_\varepsilon(u^\varepsilon(T_0, x))\, dx + \frac{1}{2}\int_{(0,T_0)\times\mathbb{R}^d} |\nabla_x \Phi_\varepsilon(u^\varepsilon(t, x))|^2\, dt\, dx = \int_{\mathbb{R}^d} \Psi_\varepsilon(u_0(x))\, dx, \tag{2.12}$$

*for* $\Phi_\varepsilon(r) = \sigma_\varepsilon^2(r)r$ *and* $\Psi_\varepsilon(r) = \int_0^r \Phi_\varepsilon(\theta)\, d\theta$.

Now the existence of a weak solution to (2.5) first could be classically reformulated into a martingale problem. Owing to the boundedness of $(t, x) \mapsto \sigma(u^\epsilon(t, x))$, this gives (see Theorem 2.6, in Figalli [12]) the following result.

**Proposition 2.6** *Under* ($A_0$), ($A_1$) *and* ($A_2$-*weakened*), *there exists a unique weak solution* $(X_t^\varepsilon;\ 0 \leq t \leq T)$ *to (2.5) such that the time marginal densities are given by* $(u^\varepsilon(t);\ 0 \leq t \leq T)$ *from Lemma 2.5.*

### 2.1.1 Proof of Lemma 2.4

For any $v \in L^2((0, T); H^1(\mathbb{R}^d))$, the identity (2.4) ensures that, for a.e. $t \in (0, T)$, the bilinear mapping

$$(u_1, u_2) \mapsto \mathcal{L}_t(u_1, u_2) = \int_{\mathbb{R}^d} \frac{1}{2}\alpha_\varepsilon(v(t, x))\, (\nabla_x u_1(x) \cdot \nabla_x u_2(x))\, dx$$

is continuous on $H^1(\mathbb{R}^d) \times H^1(\mathbb{R}^d)$, since $\mathcal{L}_t(u_1, u_2) \le \frac{1}{2}(\sup_{0 \le r \le \|v\|_{L^\infty}} \alpha_\varepsilon(r))\|\nabla_x u_1\|_{L^2(\mathbb{R}^d)}\|\nabla_x u_2\|_{L^2(\mathbb{R}^d)}$. Moreover, since $\mathcal{L}_t(u, u) \ge \frac{\epsilon}{2}\|\nabla_x u\|^2_{L^2(\mathbb{R}^d)}$, $\mathcal{L}_t$ satisfies the hypothesis of Theorem 1.1 in Lions [23, Chap. 4] : $\mathcal{L}_t(u, u) + \frac{\epsilon}{2}\|u^2\|_{L^2(\mathbb{R}^d)} \ge \frac{\epsilon}{2}\|u^2\|_{H^1(\mathbb{R}^d)}$ for all $u \in H^1(\mathbb{R}^d)$. Applying Theorem 1.1 and Lemma 1.1 in [23], we deduce the existence of a solution $\overline{u}^\varepsilon \in L^2((0, T); H^1(\mathbb{R}^d))$ to

$$\begin{aligned}
&\int_{\mathbb{R}^d} \overline{u}^\varepsilon(T_0, x) f(T_0, x)\, dx - \int_{\mathbb{R}^d} u_0(x) f(0, x)\, dx \\
&\quad = \int_{(0,T_0)\times\mathbb{R}^d} \overline{u}^\varepsilon(t, x)\frac{\partial f}{\partial t}(t, x) - \frac{1}{2}\alpha_\varepsilon(v(t, x))\,(\nabla_x \overline{u}^\varepsilon(t, x) \cdot \nabla_x f(t, x))\, dt\, dx, \\
&\qquad \forall\, 0 \le T_0 \le T, \forall\, f \in \mathcal{C}_c^\infty([0, T_0] \times \mathbb{R}^d).
\end{aligned}$$

The property of $u^\varepsilon \in \mathcal{C}([0, T]; L^2(\mathbb{R}^d))$ can be proved in the same way as in [23], Theorem 2.1, Chap. 4.

The energy estimate (2.9) is obtained by adapting some arguments of Ladyženskaja et al. [22], pp. 141–142. For $0 \le T_0 \le T$, for $h > 0$ such that $T_0 + h \le T$, and $\eta_h \in \mathcal{C}_c^\infty((-h, T_0) \times \mathbb{R}^d)$ such that $\eta_h(t, x) = 0$ whenever $t \le 0$ or $t \ge T_0 - h$, we define

$$\phi(t, x) = \frac{1}{h}\int_{t-h}^{t} \eta_h(s, x)\, ds = \int_{-1}^{0} \eta_h(t + sh, x)\, ds.$$

Plugging $\phi$ as a test function into the weak formulation (2.7) gives

$$0 = \int_{(0,T_0)\times\mathbb{R}^d} \overline{u}^\varepsilon(t, x)\partial_t\phi(t, x) - \frac{1}{2}\alpha_\varepsilon(v(t, x))(\nabla_x \overline{u}^\varepsilon(t, x) \cdot \nabla_x \phi(t, x))\, dt\, dx.$$

Since $\partial_t\phi(t, x) = \frac{1}{h}(\eta_h(t, x) - \eta_h(t - h, x))$,

$$\begin{aligned}
\int_{(0,T_0)\times\mathbb{R}^d} \overline{u}^\varepsilon(t, x)\partial_t\phi(t, x)\, dt\, dx &= \frac{1}{h}\left(\int_{(0,T_0)\times\mathbb{R}^d} \overline{u}^\varepsilon(t, x)\eta_h(t, x)\, dt\, dx \right.\\
&\qquad \left. - \int_{(0,T_0)\times\mathbb{R}^d} \overline{u}^\varepsilon(t, x)\eta_h(t - h, x)\, dt\, dx\right) \\
&= \frac{1}{h}\left(\int_{(0,T_0)\times\mathbb{R}^d} \overline{u}^\varepsilon(t, x)\eta_h(t, x)\, dt\, dx \right.\\
&\qquad \left. - \int_{(0,T_0)\times\mathbb{R}^d} \overline{u}^\varepsilon(t + h, x)\eta_h(t, x)\, dt\, dx\right) \\
&= \int_{(0,T_0)\times\mathbb{R}^d} \partial_t \overline{u}^\varepsilon_h(t, x)\eta_h(t, x)\, dt\, dx,
\end{aligned}$$

for $\overline{u}^\varepsilon_h(t, x) = \frac{1}{h}\int_t^{t+h} \overline{u}^\varepsilon(s, x)\, ds$. In the same manner, we have

$$\int_{(0,T_0)\times\mathbb{R}^d} \frac{1}{2}\alpha_\varepsilon(v(t,x))(\nabla_x \overline{u}^\varepsilon(t,x)\cdot\nabla_x\phi(t,x))\,dt\,dx = \int_{(0,T_0)\times\mathbb{R}^d} \frac{1}{2}\left(\alpha_\varepsilon(v)\nabla_x\overline{u}^\varepsilon\right)_h(t,x) \cdot \nabla_x\eta_h(t,x)\,dt\,dx$$

where

$$(\alpha_\varepsilon(v)\nabla_x\overline{u}^\varepsilon)_h(t,x) := \frac{1}{h}\int_t^{t+h} \alpha_\varepsilon(v(s,x))\nabla_x\overline{u}^\varepsilon(s,x)\,ds.$$

Therefore

$$0 = \int_{(0,T_0)\times\mathbb{R}^d} \partial_t\overline{u}^\varepsilon_h(t,x)\eta_h(t,x) + \frac{1}{2}(\alpha_\varepsilon(v)\nabla_x\overline{u}^\varepsilon)_h(t,x)\cdot\nabla_x\eta_h(t,x)\,dt\,dx.$$

Extending the previous equality from $\eta_h \in \mathcal{C}_c^\infty((0,T_0-h)\times\mathbb{R}^d)$ to $\eta \in L^2((0,T_0); H^1(\mathbb{R}^d))$ by density, it follows that

$$0 = \int_{(0,T_0)\times\mathbb{R}^d} \partial_t\overline{u}^\varepsilon_h(t,x)\eta(t,x) + \frac{1}{2}(\alpha_\varepsilon(v)\nabla_x\overline{u}^\varepsilon)_h(t,x)\cdot\nabla_x\eta(t,x)\,dt\,dx. \quad (2.13)$$

Next replacing $\eta = \overline{u}^\varepsilon_h$, we get

$$\begin{aligned}0 &= \int_{(0,T_0)\times\mathbb{R}^d} \partial_t\overline{u}^\varepsilon_h(t,x)\overline{u}^\varepsilon_h(t,x) + \frac{1}{2}(\alpha_\varepsilon(v)\nabla_x\overline{u}^\varepsilon)_h(t,x)\cdot\nabla_x\overline{u}^\varepsilon_h(t,x)\,dt\,dx \\ &= \int_{\mathbb{R}^d}\left(\overline{u}^\varepsilon_h(T_0,x)\right)^2dx - \int_{\mathbb{R}^d}\left(\overline{u}^\varepsilon_h(0,x)\right)^2dx \\ &\quad + \int_{(0,T_0)\times\mathbb{R}^d}\frac{1}{2}(\alpha_\varepsilon(v)\nabla_x\overline{u}^\varepsilon)_h(t,x)\cdot\nabla_x\overline{u}^\varepsilon_h(t,x)\,dt\,dx.\end{aligned}$$

Since

$$\begin{aligned}&\left|\int_{\mathbb{R}^d}\left(\overline{u}^\varepsilon_h(T_0,x)\right)^2dx - \int\left(\overline{u}^\varepsilon(T_0,x)\right)^2dx\right| \\ &= \left|\int_{\mathbb{R}^d}\left(\int_0^1\left(\overline{u}^\varepsilon(T_0+hs,x)\,ds - \overline{u}^\varepsilon(T_0,x)\right)\right)\left(\int_0^1\left(\overline{u}^\varepsilon(T_0+hs,x)\,ds + \overline{u}^\varepsilon(T_0,x)\right)\right)dx\right| \\ &\le \sqrt{\int_0^1 \|\overline{u}^\varepsilon(T_0+hs) - \overline{u}^\varepsilon(T_0)\|^2_{L^2(\mathbb{R}^d)}\,ds}\sqrt{\int_0^1 \|\overline{u}^\varepsilon(T_0+hs) + \overline{u}^\varepsilon(T_0)\|^2_{L^2(\mathbb{R}^d)}\,ds},\end{aligned}$$

we have

$$\lim_{h\to 0}\int_{\mathbb{R}^d}\left(\overline{u}^\varepsilon_h(T_0,x)\right)^2dx = \int_{\mathbb{R}^d}\left(\overline{u}^\varepsilon(T_0,x)\right)^2dx. \quad (2.14)$$

Similarly,

$$\lim_{h\to 0}\int_{\mathbb{R}^d}\left(\overline{u}^\varepsilon_h(0,x)\right)^2dx = \int_{\mathbb{R}^d}\left(\overline{u}^\varepsilon_0(x)\right)^2dx,$$

and

$$\lim_{h\to 0}\int_{(0,T_0)\times\mathbb{R}^d}\left(\frac{1}{2}\alpha_\varepsilon(v)\nabla_x\overline{u}^\varepsilon\right)_h(t,x)\cdot\nabla_x\overline{u}^\varepsilon_h(t,x)\,dt\,dx$$
$$=\int_{(0,T_0)\times\mathbb{R}^d}\frac{1}{2}\alpha_\varepsilon(v(t,x))\left|\nabla_x\overline{u}^\varepsilon_h(t,x)\right|^2\,dt\,dx,$$

From which we deduce (2.9).

The non-negativeness of $\overline{u}^\varepsilon$ and (2.8) follows from comparison principles: since $\overline{u}^\varepsilon$ is in $L^2((0,T);H^1(\mathbb{R}^d))$, its negative part $(\overline{u}^\varepsilon)^-=\max(0,-\overline{u}^\varepsilon))$ is also in $L^2((0,T);H^1(\mathbb{R}^d))$ and $\nabla_x(\overline{u}^\varepsilon)^-=-\nabla_x\overline{u}^\varepsilon\mathbb{1}_{\{\overline{u}^\varepsilon\leq 0\}}$. Taking $\eta=(\overline{u}^\varepsilon)^-$ in (2.13) yields

$$0=\int_{(0,T_0)\times\mathbb{R}^d}\partial_t\overline{u}^\varepsilon_h(t,x)(\overline{u}^\varepsilon)^-(t,x)+\frac{1}{2}\left(\alpha_\varepsilon(v)\nabla_x\overline{u}^\varepsilon\right)_h(t,x)\cdot\nabla_x(\overline{u}^\varepsilon)^-(t,x)\,dt\,dx$$
$$=-\int_{\mathbb{R}^d}\left(\left(\overline{u}^\varepsilon_h\right)^-(T_0,x)\right)^2dx+\int_{\mathbb{R}^d}\left(\left(\overline{u}^\varepsilon_h\right)^-(0,x)\right)^2dx$$
$$+\int_{(0,T_0)\times\mathbb{R}^d}\frac{1}{2}\left(\alpha_\varepsilon(v)\nabla_x\overline{u}^\varepsilon\right)_h(t,x)\cdot\nabla_x(\overline{u}^\varepsilon)^-(t,x)\,dt\,dx.$$

Replicating the same arguments as for (2.14),

$$\lim_{h\to 0}\int_{\mathbb{R}^d}\left(\left(\overline{u}^\varepsilon_h\right)^-(T_0,x)\right)^2dx=\int_{\mathbb{R}^d}\left((\overline{u}^\varepsilon)^-(T_0,x)\right)^2dx,$$
$$\lim_{h\to 0}\int_{\mathbb{R}^d}\left(\left(\overline{u}^\varepsilon_h\right)^-(0,x)\right)^2dx=\int_{\mathbb{R}^d}\left((u_0)^-(x)\right)^2dx,$$
$$\lim_{h\to 0}\int_{(0,T_0)\times\mathbb{R}^d}\frac{1}{2}\left(\alpha_\varepsilon(v)\nabla_x\overline{u}^\varepsilon\right)_h(t,x)\cdot\nabla_x\left(\overline{u}^\varepsilon_h\right)^-(t,x)\,dt\,dx$$
$$=\int_{(0,T_0)\times\mathbb{R}^d}\frac{1}{2}\alpha_\varepsilon(v(t,x))\nabla_x\overline{u}^\varepsilon(t,x)\cdot\nabla_x\left(\overline{u}^\varepsilon_h\right)^-(t,x)\,dt\,dx$$
$$=-\int_{(0,T_0)\times\mathbb{R}^d}\frac{1}{2}\alpha_\varepsilon(v(t,x))\left|\nabla_x\left(\overline{u}^\varepsilon_h\right)^-(t,x)\right|^2\,dt\,dx.$$

Consequently,

$$0=-\int_{\mathbb{R}^d}\left((\overline{u}^\varepsilon)^-(T_0,x)\right)^2dx+\int_{\mathbb{R}^d}\left((u_0)^-(x)\right)^2dx$$
$$-\int_{(0,T_0)\times\mathbb{R}^d}\frac{1}{2}\alpha_\varepsilon(v(t,x))\left|\nabla_x\left(\overline{u}^\varepsilon_h\right)^-(t,x)\right|^2\,dt\,dx.$$

Since $u_0\geq 0$ and $\alpha_\varepsilon$ is non-negative, we deduce immediately that $(\overline{u}^\varepsilon)^-=0$ on $(0,T)\times\mathbb{R}^d$.

For the proof of (2.8), we proceed similarly: we set $K = \|u_0\|_{L^\infty}$. Then the positive part of $\overline{u}^\varepsilon(t,x) - K$,

$$(\overline{u}^\varepsilon(t,x) - K)^+$$

is in $L^2((0,T); H^1(\mathbb{R}^d))$. This statement simply follows from the observation that

$$\left|(\overline{u}^\varepsilon(t,x) - K)^+\right|^2 \leq |\overline{u}^\varepsilon(t,x)|^2, \text{ a.e. on } (0,T) \times \mathbb{R}^d,$$

and that

$$\nabla_x (\overline{u}^\varepsilon - K)^+ = \nabla_x \overline{u}^\varepsilon \mathbb{1}_{\{\overline{u}^\varepsilon - K \geq 0\}}.$$

By plugging $\eta = (\overline{u}^\varepsilon - K)^+$ into (2.13), we get

$$\begin{aligned} 0 = \int_{(0,T_0)\times\mathbb{R}^d} & \partial_t \overline{u}^\varepsilon_h(t,x) (\overline{u}^\varepsilon(t,x) - K)^+ (t,x) \\ & + \frac{1}{2} (\alpha_\varepsilon(v) \nabla_x \overline{u}^\varepsilon)_h (t,x) \cdot \nabla_x (\overline{u}^\varepsilon(t,x) - K)^+ (t,x) \, dt \, dx. \end{aligned}$$

Taking the limit $h \to 0$ of the above expression yields

$$\begin{aligned} 0 = & \int_{\mathbb{R}^d} \left((\overline{u}^\varepsilon(t,x) - K)^+\right)^2 (t,x) \, dx - \int_{\mathbb{R}^d} \left((u_0(x) - K)^+\right)^2 (t,x) \, dx \\ & + \int_{(0,T_0)\times\mathbb{R}^d} \frac{1}{2} \alpha_\varepsilon(v(t,x)) \left|\nabla_x (\overline{u}^\varepsilon(t,x) - K)^+\right|^2 (t,x) \, dt \, dx. \end{aligned}$$

Since $(u_0(x) - K)^+ = 0$ for a.a. $x \in \mathbb{R}^d$, (2.8) follows.

### 2.1.2 Proof of Lemma 2.5

The existence of a weak solution to the non linear PDE (2.6) will be deduced from the following fixed point theorem that we apply to the mapping $A : v \in \mathcal{X} \mapsto A(v) \in \mathcal{X}$, for

$$\mathcal{X} = \left\{ v \in L^2((0,T); H^1(\mathbb{R}^d)) \; : \; v \geq 0 \text{ a.e. on } (0,T) \times \mathbb{R}^d \text{ and } \|v\|_{L^\infty((0,T)\times\mathbb{R}^d)} \leq \|u_0\|_{L^\infty(\mathbb{R}^d)} \right\}$$

equipped with the $\| \; \|_{L^2((0,T);H^1(\mathbb{R}^d))}$-norm, and where $A$ assigns to any nonnegative $v \in \mathcal{X}$, the weak solution $A(v)$ to the linear PDE (2.7) given by Lemma 2.4 with the estimates (2.8) and (2.9).

**Theorem 2.7** (Schaefer's fixed point Theorem, see [11], Theorem 4, Chap. 9, Sect. 2) *Let $\mathcal{X}$ be a Banach space and $A : \mathcal{X} \to \mathcal{X}$ be a continuous and compact mapping such that the set*

$$\mathcal{I} = \{u \in \mathcal{X} \text{ s.t. there exists } 0 \leq \lambda \leq 1 \text{ with } u = \lambda A(u)\}$$

*is bounded in $\mathcal{X}$. Then A has a fixed point in $\mathcal{X}$.*

**The continuity of $A$.** Let $\{v_n\}_n$ converge to $v_\infty$ in $L^2((0,T); H^1(\mathbb{R}^d))$. Then, since $A(v_n)$ and $A(v_\infty)$ are weak solution to (2.7) endowed with the diffusion coefficient $\sigma_\varepsilon(v_n)$ and $\sigma_\varepsilon(v_\infty)$ respectively, we have

$$0 = \int_{(0,T)\times\mathbb{R}^d} (A(v_n) - A(v_\infty))\, \partial_t f - \left(\frac{1}{2}\alpha_\varepsilon(v_n)\nabla_x A(v_n) - \frac{1}{2}\alpha_\varepsilon(v_\infty)\nabla_x A(v_\infty)\right) \nabla_x f,$$

for all $f \in C_c^\infty((0,T)\times\mathbb{R}^d)$. Replicating the arguments for the energy estimate (2.9) in Lemma 2.4, we deduce that

$$\begin{aligned}
&\int_{\mathbb{R}^d} |A(v_n)(T_0) - A(v_\infty)(T_0)|^2 + \int_{(0,T_0)\times\mathbb{R}^d} \frac{1}{2}\alpha_\varepsilon(v_n)\,|\nabla_x A(v_n) - \nabla_x A(v_\infty)|^2 \\
&= \int_{(0,T_0)\times\mathbb{R}^d} \frac{1}{2}\,(\alpha_\varepsilon(v_n) - \alpha_\varepsilon(v_\infty))\,(\nabla_x A(v_n) - \nabla_x A(v_\infty))\cdot\nabla_x A(v_n) \\
&\leq \frac{1}{\varepsilon}\int_{(0,T_0)\times\mathbb{R}^d} \frac{1}{2}\,|\alpha_\varepsilon(v_n) - \alpha_\varepsilon(v_\infty)|^2\,|\nabla_x A(v_n)|^2 + \frac{\varepsilon}{4}\int_{(0,T_0)\times\mathbb{R}^d} |\nabla_x A(v_n) - \nabla_x A(v_\infty)|^2.
\end{aligned}$$

Since $v_n \to v_\infty$ in $L^2((0,T)\times\mathbb{R}^d)$, there exists a subsequence $\{v_{n_k}\}_k$ such that $v_{n_k} \to v_\infty$ a.e. on $(0,T)\times\mathbb{R}^d$ and such that $\sup_k |v_{n_k}|$ is in $L^2((0,T)\times\mathbb{R}^d)$. Replacing $v_n$ by $v_{n_k}$ in the preceding inequality and applying the Lebesgue's dominated convergence theorem, we deduce that $\lim_k A(v_{n_k}) = A(v_\infty)$ in $L^2((0,T); H^1(\mathbb{R}^d))$. With the same reasoning, for any subsequence of $\{A(v_n)\}_n$, we can extract a subsequence which converges to $A(v_\infty)$. Since $\mathcal{X}$ is a closed subset of $L^2((0,T); H^1(\mathbb{R}^d))$, this implies the continuity of $A$.

**The compactness of $A$.** Owing to Lemma 2.4, for any converging sequence $\{v_n\}_n$ in $\mathcal{X}$, $\{A(v_n)\}_n$ is a bounded sequence in $L^2((0,T); H^1(\mathbb{R}^d))$. We can then extract a subsequence $\{A(v_{n_k})\}_k$ converging to some limit $A_\infty$ in the weak topology of $L^2((0,T); H^1(\mathbb{R}^d))$. In particular, since $\sup_k \|A(v_{n_k})\|_{L^2((0,T)\times\mathbb{R}^d)}$ is finite, $\|\nabla_x A(v_{n_k}) - A_\infty\|_{L^2((0,T)\times\mathbb{R}^d)}$ tends to 0 as $k$ tends to $\infty$. The uniform bound (2.8) naturally holds for $v_\infty$.

**The boundedness of $\mathcal{I}$.** Finally, let us consider the set

$$\mathcal{I} := \{u \in \mathcal{X} \text{ s.t. there exists } 0 \leq \lambda \leq 1 \text{ with } u = \lambda A(u)\}.$$

Excluding the trivial case $\lambda = 0$, one can check that for all $u_\lambda \in \mathcal{X}$ such that $u_\lambda = \lambda A(u_\lambda)$, $u_\lambda$ is the $L^2((0,T); H^1(\mathbb{R}^d))$-weak solution to

$$\begin{cases}
\dfrac{1}{\lambda}\dfrac{\partial u_\lambda}{\partial t} - \dfrac{1}{2}\Delta(\sigma_\varepsilon^2(u_\lambda)\dfrac{u_\lambda}{\lambda}) = 0, \text{ on } (0,T)\times\mathbb{R}^d, \\
\dfrac{1}{\lambda}u_\lambda(0,x) = u_0(x),\ x \in \mathbb{R}^d,
\end{cases}$$

given as in Lemma 2.4. Equivalently,

$$\int_{\mathbb{R}^d} u_\lambda(T_0,x)f(T_0,x)\,dx - \lambda\int_{\mathbb{R}^d} u_0(x)f(0,x)\,dx$$
$$= \int_{(0,T_0)\times\mathbb{R}^d} u_\lambda(t,x)\partial_t f(t,x) - \frac{1}{2}\alpha_\varepsilon(u_\lambda)\nabla_x u_\lambda(t,x)\cdot\nabla_x f(t,x)\,dt\,dx.$$

The energy estimate (2.9) from Lemma 2.4 then ensures that $\|u_\lambda\|^2_{L^2((0,T)\times\mathbb{R}^d)} \le \lambda^2 T\|u_0\|^2_{L^2(\mathbb{R}^d)}$ from which we conclude on the boundedness of $\mathcal{I}$.

The Schaefer Theorem ensures the existence of a $L^2((0,T);H^1(\mathbb{R}^d))$-weak solution to (2.6), for which the $L^2$ (and $L^\infty$) estimate established in Lemma 2.4 still hold true.

For the estimate (2.12), we replicate the proof arguments of Vasquez [34] (see Chap. 5). Owing to assumptions $(A_0)$ and $(A_1)$,

$$\int_{\mathbb{R}^d}\Psi_\varepsilon(u_0(x))\,dx = \int_{\mathbb{R}^d}\int_0^{u_0(x)}\Phi_\varepsilon(r)\,dr\,dx \le \frac{1}{2}\left(\sup_{0\le r\le\|u_0\|_{L^\infty(\mathbb{R}^d)}}(\sigma(r))^2+\varepsilon\right)\|u_0\|^2_{L^2(\mathbb{R}^d)}.$$

Replicating the proof arguments of (2.13), for all $\eta\in C_c^\infty((0,T)\times\mathbb{R}^d)$, for $h>0$, $T_0>0$ such that $T_0+h\le T$,

$$0 = \int_{(0,T_0)\times\mathbb{R}^d}\partial_t u_h^\varepsilon(t,x)\eta(t,x) + \frac{1}{2}\left(\alpha_\varepsilon(u^\varepsilon)\nabla_x u^\varepsilon\right)_h(t,x)\cdot\nabla_x\eta(t,x)\,dt\,dx, \tag{2.15}$$

with $u_h^\varepsilon(t,x) := \frac{1}{h}\int_t^{t+h} u^\varepsilon(s,x)\,ds$, and

$$\left(\alpha_\varepsilon(u^\varepsilon)\nabla_x u^\varepsilon\right)_h(t,x) := \frac{1}{h}\int_t^{t+h}\alpha_\varepsilon(u^\varepsilon(s,x))\nabla_x u^\varepsilon(s,x)\,ds.$$

Observing that

$$\|\Phi_\varepsilon(u_h^\varepsilon)\|_{L^2((0,T)\times\mathbb{R}^d)} = \|\left(\sigma^2(u^\varepsilon)+\varepsilon\right)u_h^\varepsilon\|_{L^2((0,T)\times\mathbb{R}^d)}$$
$$\le \frac{1}{2}\left(\sup_{0\le r\le\|u_0\|_{L^\infty(\mathbb{R}^d)}}(\sigma(r))^2+\varepsilon\right)\|u^\varepsilon\|_{L^2((0,T)\times\mathbb{R}^d)},$$

and that

$$\|\nabla_x\Phi_\varepsilon(u_h^\varepsilon)\|^2_{L^2((0,T)\times\mathbb{R}^d)} = \|\Phi'_\varepsilon(u_h^\varepsilon)\nabla_x u_h^\varepsilon\|^2_{L^2((0,T)\times\mathbb{R}^d)} = \|\alpha_\varepsilon(u^\varepsilon)\nabla_x u_h^\varepsilon\|^2_{L^2((0,T)\times\mathbb{R}^d)}$$
$$\le \left(\|\alpha(u^\epsilon)\|^2_{L^\infty(\mathbb{R}^d)}+\varepsilon^2\right)\|\nabla_x u^\varepsilon\|^2_{L^2((0,T)\times\mathbb{R}^d)},$$

$\Phi_\varepsilon(u_h^\varepsilon)$ is in $L^2((0,T);H^1(\mathbb{R}^d))$. Taking $\eta=\Phi_\varepsilon(u_h^\varepsilon)$ in (2.15)

$$
\begin{aligned}
0 =& \int_{(0,T_0)\times\mathbb{R}^d} \partial_t u_h^\varepsilon(t,x)\Phi_\varepsilon(u_h^\varepsilon)(t,x) + \frac{1}{2}\left(\alpha_\varepsilon(u^\varepsilon)\nabla_x u^\varepsilon\right)_h(t,x)\cdot\nabla_x\Phi_\varepsilon(u_h^\varepsilon)(t,x)\,dt\,dx\\
=& \int_{(0,T_0)\times\mathbb{R}^d} \partial_t \Psi_\varepsilon(u_h^\varepsilon)(t,x) + \frac{1}{2}\int_{(0,T_0)\times\mathbb{R}^d}\left(\nabla_x\Phi_\varepsilon(u_h^\varepsilon)\right)_h(t,x)\cdot\nabla_x\Phi_\varepsilon(u_h^\varepsilon)(t,x)\,dt\,dx\\
=& \|\Psi_\varepsilon(u_h^\varepsilon)(T_0)\|_{L^1(\mathbb{R}^d)} - \|\Psi_\varepsilon(u_h^\varepsilon)(0)\|_{L^1(\mathbb{R}^d)}\\
&+\frac{1}{2}\int_{(0,T_0)\times\mathbb{R}^d}\left(\nabla_x\Phi_\varepsilon(u_h^\varepsilon)\right)_h(t,x)\cdot\nabla_x\Phi_\varepsilon(u_h^\varepsilon)(t,x)\,dt\,dx.
\end{aligned}
$$

Taking the limit $h \to 0$ it follows that

$$
0 = \|\Psi_\varepsilon(u^\varepsilon)(T_0)\|_{L^1(\mathbb{R}^d)} - \|\Psi_\varepsilon(u_0)\|_{L^1(\mathbb{R}^d)} + \frac{1}{2}\int_{(0,T_0)\times\mathbb{R}^d}\left|\nabla_x\Phi_\varepsilon(u_h^\varepsilon)\right|^2(t,x)\,dt\,dx.
$$

Finally, since $u^\epsilon$ is a $L^2((0,T);H^1(\mathbb{R}^d))$-solution to (2.6), the property $u^\epsilon \in \mathcal{C}([0,T];L^2(\mathbb{R}^d))$ can be again obtained by following [23], Theorem 2.1, Chap. 4. This ends the proof.

## 2.2 *Existence Result for (2.1)*

The existence result in Theorem 2.3 will be deduced from the asymptotic behavior (up to a subsequence extraction) of the solution to (2.5) as $\varepsilon \to 0$. The precise result is the following:

**Proposition 2.8** *Assume $(A_0)$, $(A_1)$ and $(A_2)$. Consider $(X_t^\varepsilon;\ t \geq 0)$ solution to (2.5) given by Proposition 2.6. The sequence $\{(P^\varepsilon, u^\varepsilon)\}_{\varepsilon>0}$, defined by*

$$
P^\varepsilon = Law(X_t^\varepsilon;\ 0 \leq t \leq T),
$$

*and $u^\varepsilon$ given as in Lemma 2.5, admits a weakly converging subsequence $\{(P^{\varepsilon_k}, u^{\varepsilon_k})\}_k$ such that $u^0 = \lim_k u^{\varepsilon_k}$ in $L^2((0,T)\times\mathbb{R}^d)$, is a $L^2((0,T)\times\mathbb{R}^d)$-weak solution to*

$$
\begin{cases}
\dfrac{\partial u}{\partial t} - \dfrac{1}{2}\Delta_x(\sigma^2(u)u) = 0, \ \text{on } (0,T)\times\mathbb{R}^d,\\
u(0,x) = u_0(x),\ x \in \mathbb{R}^d,
\end{cases}
\tag{2.16}
$$

*and $P^0 = \lim_k P^{\varepsilon_k}$ is solution to the following martingale problem (MP): let $(x(t);\ 0 \leq t \leq T)$ denotes the canonical process on $\mathcal{C}([0,T];\mathbb{R}^d)$, we have*

***(MP)-(i)*** *For all $0 \leq t \leq T$, $P^0(x(t) \in dx) = u^0(t,x)\,dx$;*
***(MP)-(ii)*** *For all $f \in C_c^2(\mathbb{R}^d)$,*

$$
t \mapsto f(x(t)) - f(x(0)) - \frac{1}{2}\int_0^t \sigma^2(u^0(s,x))\Delta_x f(x(s))\,ds
$$

*is a continuous martingale under* $P^0$.

From Lemma 2.5, the $L^2((0,T)\times\mathbb{R}^d)$-convergence of $u^\varepsilon$ to $u^0$ ensures that

$$\|u^0\|_{L^\infty((0,T)\times\mathbb{R}^d)} < \infty, \text{ and } \int_{\mathbb{R}^d} \left|\nabla\Phi_\varepsilon(u^0(t,x))\right|^2 dt\,dx < \infty.$$

In particular under $(A_2)$, the control of $\|\nabla\Phi_\varepsilon(u^0(t,x))\|_{L^2((0,T)\times\mathbb{R}^d)}$ yields to the estimate:

$$\eta\|\nabla_x u^0\|_{L^2((0,T)\times\mathbb{R}^d)} \le \|\alpha_\varepsilon(u^0)\nabla_x u^0\|_{L^2((0,T)\times\mathbb{R}^d)} \le \|\nabla_x\Phi_\varepsilon(u^0)\|_{L^2((0,T)\times\mathbb{R}^d)} < \infty.$$

Therefore

**Corollary 2.9** *The time marginal densities* $u^0(t)$ *of* $(X^0_t;\, 0\le t\le T)$ *given in Proposition 2.8 are in* $L^\infty((0,T)\times\mathbb{R}^d)\cap L^2((0,T);H^1(\mathbb{R}^d))$.

*Proof of Proposition* 2.8. Owing to (2.11), we deduce that $\{u^\varepsilon\}_\varepsilon$ is relatively compact for the weak topology in $L^2((0,T)\times\mathbb{R}^d)$. Denote by $\{u^{\varepsilon_k}\}_k$ a (weakly) converging subsequence and $u^0$ its limit. Under the assumption $(A_2)$, the estimate (2.12) ensures that

$$\sup_\varepsilon \|\nabla_x u^\varepsilon\|_{L^2((0,T)\times\mathbb{R}^d)} \le \frac{1}{\eta}\sup_\varepsilon \|\alpha(u^\varepsilon)\nabla_x u^\varepsilon\|_{L^2((0,T)\times\mathbb{R}^d)} = \frac{1}{\eta}\sup_\varepsilon \|\nabla_x\Phi(u^\varepsilon)\|_{L^2((0,T)\times\mathbb{R}^d)} < \infty,$$

so that $\lim_k \|u^{\varepsilon_k} - u^0\|_{L^2((0,T)\times\mathbb{R}^d)} = 0$. We can further extract a converging subsequence such that the convergence holds a.e. on $(0,T)\times\mathbb{R}^d$ and $\sup_k |u^{\varepsilon_k}| \in L^2((0,T)\times\mathbb{R}^d)$. Since $\sigma_\varepsilon(u^\varepsilon)$ is bounded and by $(A_0)$, according to the Kolmogorov-Centov criterion, the sequence $\{P^\varepsilon\}_\varepsilon$ is tight on $(\mathcal{C}([0,T];\mathbb{R}^d), \mathcal{B}(\mathcal{C}([0,T];\mathbb{R}^d)))$. Denote for simplicity by $\{(P^{\varepsilon_k}, u^{\varepsilon_k})\}_k$ a converging pair of $\{(P^\varepsilon, u^\varepsilon)\}_\varepsilon$ and by $(P^0, u^0)$ its limit. Since $P^{\varepsilon_k}(x(t)\in dx) = u^{\varepsilon_k}(t,x)\,dx$ the $L^2$-convergence of $u^{\varepsilon_k}$ to $u^0$ and the convergence of the time marginal distributions of $P^{\varepsilon_k}$ to $P^0$ ensure that $P^0(x(t)\in dx) = u^0(t,x)\,dx$ for a.e. $0\le t\le T$.

Coming back to (2.6) and taking the limit $k\to 0$ in the expression,

$$0 = \int_{(0,T)\times\mathbb{R}^d} u^\varepsilon(t,x)\partial_t f(t,x) - \frac{1}{2}\alpha_\varepsilon(u^\varepsilon(t,x))\nabla_x u^\varepsilon(t,x)\cdot\nabla_x f(t,x)\,dt\,dx,$$

for $f\in\mathcal{C}_c^\infty((0,T)\times\mathbb{R}^d)$, and owing to the continuity of $\sigma$, we deduce that $u^0$ is a weak $L^2((0,T)\times\mathbb{R}^d)$-solution to (2.16).

In order to identify $P^0$ as the solution of the martingale problem *(MP)*, it is sufficient to show that, for all $0\le s<t\le T$, $\psi:\mathcal{C}([0,s];\mathbb{R}^d)\to\mathbb{R}$ bounded and continuous, $f\in\mathcal{C}_c^2(\mathbb{R}^d)$,

$$\begin{aligned}
&\lim_k \mathbb{E}_{P^{\varepsilon_k}}\left[\psi(x(r); 0 \le r \le s)\int_s^t \sigma_\varepsilon^2(u^\varepsilon(\theta, x(\theta)))\Delta_x f(x(\theta))\, d\theta\right] \\
&= \mathbb{E}_{P^0}\left[\psi(x(r); 0 \le r \le s)\int_s^t \sigma^2(u^0(\theta, x(\theta)))\Delta_x f(x(\theta))\, d\theta\right].
\end{aligned} \tag{2.17}$$

To this end, let us introduce a smooth approximation of $\sigma_\varepsilon^2(u^{\varepsilon_k})$ and $\sigma^2(u^0)$ with

$$\left(\sigma_\varepsilon^2(u^{\varepsilon_k}(t))\right)_\beta(x) := \left(\phi_\beta * \sigma_\varepsilon^2(u^{\varepsilon_k}(t))\right)(x), \text{ and } \left(\sigma^2(u^0(t))\right)_\beta(x) := \left(\phi_\beta * \sigma^2(u^0(t))\right)(x),$$

for $*$ denoting the convolution product on the variable $x \in \mathbb{R}^d$ and $\{\phi_\beta\}_{\beta>0}$ a sequence of mollifiers on $\mathbb{R}^d$ given by $\phi_\beta(y) = \frac{1}{\beta^d}\phi(\frac{y}{\beta})$ with $\phi \ge 0$, $\phi \in C_c^\infty(\mathbb{R}^d)$ and $\int \phi(y)\, dy = 1$.

Then, we can consider

$$\begin{aligned}
&\Bigg|\mathbb{E}_{P^{\varepsilon_k}}\left[\psi(x(r); 0 \le r \le s)\int_s^t \sigma_\varepsilon^2(u^\varepsilon(\theta, x(\theta)))\Delta_x f(x(\theta))\, d\theta\right] \\
&\qquad -\mathbb{E}_{P^0}\left[\psi(x(r); 0 \le r \le s)\int_s^t \sigma^2(u^0(\theta, x(\theta)))\Delta_x f(x(\theta))\, d\theta\right]\Bigg| \\
&\le \left|\mathbb{E}_{P^{\varepsilon_k}}\left[\psi(x(r); 0 \le r \le s)\int_s^t \left(\sigma_\varepsilon^2(u^\varepsilon(\theta, x(\theta))) - \left(\sigma_\varepsilon^2(u^{\varepsilon_k}(\theta))\right)_\beta(x(\theta))\right)\Delta_x f(x(\theta))\, d\theta\right]\right| \\
&\quad + \Bigg|\mathbb{E}_{P^{\varepsilon_k}}\left[\psi(x(r); 0 \le r \le s)\int_s^t \left(\sigma_\varepsilon^2(u^{\varepsilon_k}(\theta))\right)_\beta(x(\theta))\Delta_x f(x(\theta))\, d\theta\right] \\
&\qquad -\mathbb{E}_{P^0}\left[\psi(x(r); 0 \le r \le s)\int_s^t \left(\sigma^2(u^0(\theta))\right)_\beta(x(\theta))\Delta_x f(x(\theta))\, d\theta\right]\Bigg| \\
&\quad + \left|\mathbb{E}_{P^0}\left[\psi(x(r); 0 \le r \le s)\int_s^t \left(\left(\sigma^2(u^0(\theta))\right)_\beta(x(\theta)) - \sigma^2(u^0(\theta, x(\theta)))\Delta_x f(x(\theta))\right) d\theta\right]\right| \\
&=: I_1^{\varepsilon,\beta} + I_2^{\varepsilon,\beta} + I_3^\beta.
\end{aligned}$$

By the weak convergence of $P^{\varepsilon_k}$ and since $\left(\sigma_\varepsilon^2(u^{\varepsilon_k}(t))\right)_\beta(x)$ converges locally to $\left(\sigma^2(u^0(t))\right)_\beta(x)$, it follows that $\lim_{\beta\to 0}\lim_{\varepsilon\to 0} I_2^{\varepsilon,\beta} = 0$.

For $I_1^{\varepsilon,\beta}$, observe that

$$\begin{aligned}
I_1^{\varepsilon,\beta} &\le \|\psi\|_\infty \mathbb{E}_{P^{\varepsilon_k}}\left[\int_0^T \left|\sigma_\varepsilon^2(u^\varepsilon(\theta, x(\theta))) - \left(\sigma_\varepsilon^2(u^{\varepsilon_k}(\theta))\right)_\beta(x(\theta))\right| |\Delta_x f(x(\theta))|\, d\theta\right] \\
&\le \|\psi\|_\infty \sup_k \|u^{\varepsilon_k}\|_{L^2((0,T)\times\mathbb{R}^d)}\sqrt{\int_{(0,T)\times\mathbb{R}^d}\left|\sigma_\varepsilon^2(u^\varepsilon(\theta, x)) - \left(\sigma_\varepsilon^2(u^{\varepsilon_k}(\theta))\right)_\beta(x)\right|^2 |\Delta_x f(x)|^2\, dx\, d\theta}.
\end{aligned}$$

Since

$$\left|\sigma_{\varepsilon_k}^2(u^{\varepsilon_k}(\theta, x)) - \left(\sigma_{\varepsilon_k}^2(u^{\varepsilon_k}(\theta))\right)_\beta(x)\right| \le \int_{\mathbb{R}^d}\phi(y)\left|\sigma_{\varepsilon_k}^2(u^{\varepsilon_k}(\theta, x)) - \sigma_{\varepsilon_k}^2(u^{\varepsilon_k}(\theta, x - \beta y))\right| dy,$$

we have

$$\int_{(0,T)\times\mathbb{R}^d} \left|\sigma_\varepsilon^2(u^\varepsilon(\theta,x)) - \left(\sigma_\varepsilon^2(u^{\varepsilon_k}(\theta))\right)_\beta(x)\right|^2 |\Delta_x f(x)|^2 \, dx \, d\theta$$
$$\leq \int_{\mathbb{R}^d} \phi(y) \left(\int_{(0,T)\times\mathbb{R}^d} \left|\sigma_{\varepsilon_k}^2(u^{\varepsilon_k}(\theta,x)) - \sigma_{\varepsilon_k}^2(u^{\varepsilon_k}(\theta,x-\beta y))\right|^2 |\Delta_x f(x)|^2 \, d\theta \, dx\right) dy.$$

Then we observe that, for all $y \in \mathbb{R}^d$,

$$\begin{aligned}
&\int_{(0,T)\times\mathbb{R}^d} \left|\sigma_{\varepsilon_k}^2(u^{\varepsilon_k}(\theta,x)) - \sigma_{\varepsilon_k}^2(u^{\varepsilon_k}(\theta,x-\beta y))\right|^2 |\Delta_x f(x)|^2 \, d\theta \, dx \\
&\leq \int_{(0,T)\times\mathbb{R}^d} \left|\sigma_{\varepsilon_k}^2(u^{\varepsilon_k}(\theta,x)) - \sigma_{\varepsilon_k}^2(u^0(\theta,x))\right|^2 |\Delta_x f(x)|^2 \, d\theta \, dx \\
&\quad + \int_{(0,T)\times\mathbb{R}^d} \left|\sigma_{\varepsilon_k}^2(u^0(\theta,x)) - \sigma_{\varepsilon_k}^2(u^0(\theta,x-\beta y))\right|^2 |\Delta_x f(x)|^2 \, d\theta \, dx \\
&\quad + \int_{(0,T)\times\mathbb{R}^d} \left|\sigma_{\varepsilon_k}^2(u^{\varepsilon_k}(\theta,x-\beta y)) - \sigma_{\varepsilon_k}^2(u^{\varepsilon_k}(\theta,x-\beta y))\right|^2 |\Delta_x f(x)|^2 \, d\theta \, dx.
\end{aligned} \tag{2.18}$$

By continuity of $\sigma$, as $k$ tends to $\infty$, $\sigma(u^{\varepsilon_k})$ tends to $\sigma(u^0)$ a.e. on $(0,T)\times\mathbb{R}^d$. Therefore, by Lebesgue's dominated convergence theorem, the first expression in the right hand side of (2.18) tends to 0. In the same way

$$\begin{aligned}
&\lim_k \int_{(0,T)\times\mathbb{R}^d} \left|\sigma_{\varepsilon_k}^2(u^{\varepsilon_k}(\theta,x-\beta y)) - \sigma_{\varepsilon_k}^2(u^{\varepsilon_k}(\theta,x-\beta y))\right|^2 |\Delta_x f(x)|^2 \, d\theta \, dx \\
&= \lim_k \int_{(0,T)\times\mathbb{R}^d} \left|\sigma_{\varepsilon_k}^2(u^{\varepsilon_k}(\theta,x)) - \sigma_{\varepsilon_k}^2(u^{\varepsilon_k}(\theta,x))\right|^2 |\Delta_x f(x+\beta y)|^2 \, d\theta \, dx = 0.
\end{aligned}$$

For the remaining component in (2.18), assuming that the support of $f$ is included in the open ball $B(0,R)$ for some radius $R<\infty$,

$$\begin{aligned}
&\int_{(0,T)\times\mathbb{R}^d} \left|\sigma_{\varepsilon_k}^2(u^0(\theta,x)) - \sigma_{\varepsilon_k}^2(u^0(\theta,x-\beta y))\right|^2 |\Delta_x f(x)|^2 \, d\theta \, dx \\
&\leq \|\Delta_x f\|_\infty \int_{(0,T)\times B(0,R)} \left|\sigma_{\varepsilon_k}^2(u^0(\theta,x)) - \sigma_{\varepsilon_k}^2(u^0(\theta,x-\beta y))\right|^2 \, d\theta \, dx
\end{aligned}$$

Since $\sigma_{\varepsilon_k}^2(u^0)$ is bounded, the continuity of

$$z \in \mathbb{R}^d \mapsto \int_{(0,T)\times B(0,R)} \left|\sigma_{\varepsilon_k}^2(u^0(\theta,x)) - \sigma_{\varepsilon_k}^2(u^0(\theta,x-z))\right|^2 \, d\theta \, dx$$

ensures that

$$\lim_{\beta\to 0}\int_{(0,T)\times\mathbb{R}^d}\left|\sigma^2_{\varepsilon_k}(u^0(\theta,x))-\sigma^2_{\varepsilon_k}(u^0(\theta,x-\beta y))\right|^2|\Delta_x f(x)|^2\,d\theta\,dx=0.$$

Coming back to (2.18), we deduce that

$$\lim_{\beta\to 0}\lim_{\varepsilon\to 0}\int_{(0,T)\times\mathbb{R}^d}\left|\sigma^2_{\varepsilon_k}(u^{\varepsilon_k}(\theta,x))-\sigma^2_{\varepsilon_k}(u^{\varepsilon_k}(\theta,x-\beta y))\right|^2|\Delta_x f(x)|^2\,d\theta\,dx=0,$$

and by extension that $\lim_{\beta\to 0}\lim_{\varepsilon\to 0} I_1^{\varepsilon,\beta}=0$.

Finally, for $I_3^{\beta}$, replicating the arguments for $I_1^{\varepsilon_k,\beta}$, we have

$$\begin{aligned}
&\left|\mathbb{E}_{P^0}\left[\psi(x(r);0\le r\le s)\int_s^t\left(\left(\sigma^2(u^0(\theta))\right)_\beta(x(\theta))-\sigma^2(u^0(\theta,x(\theta)))\Delta_x f(x(\theta))\right)d\theta\right]\right|\\
&\le\|\psi\|_\infty\sup_k\|u^{\varepsilon_k}\|_{L^2((0,T)\times\mathbb{R}^d)}\sqrt{\int_{(0,T)\times\mathbb{R}^d}\left|\sigma^2(u^0(\theta,x))-\left(\sigma^2(u^0(\theta))\right)_\beta(x)\right|^2|\Delta_x f(x)|^2\,dx\,d\theta}\\
&\le\|\psi\|_\infty\sup_k\|u^{\varepsilon_k}\|_{L^2((0,T)\times\mathbb{R}^d)}\sqrt{\int_{\mathbb{R}^d}\phi(y)\int_{(0,T)\times\mathbb{R}^d}\left|\sigma^2(u^0(\theta,x))-\sigma^2(u^0(\theta))(x-\beta y)\right|^2|\Delta_x f(x)|^2\,dx\,d\theta\,dy}
\end{aligned}$$

where the last upper bound tends to 0 as $k\to\infty$. We then conclude on (2.17).

## 2.3 Uniqueness Result for (2.1)

Let us first start by showing that the time marginal distribution of (2.1) are unique.

**Proposition 2.10** *Assume that* $(A_0)$, $(A_1)$ *and* $(A_2)$ *hold true. Let* $(X^1_t, u^1_t;\ t\in[0,T])$ *and* $(X^2_t, u^2_t;\ t\in[0,T])$ *be two weak solutions to (2.1). Then, for all* $0\le t\le T$, $u^1_t=u^2_t$ *a.e. on* $\mathbb{R}^d$. *This conclusion holds true also under* $(A_0)$, $(A_1)$ *and* ($A_2$*-weakened) plus the assumption that* $\alpha$ *is strictly increasing.*

*Proof* We give the proof assuming $(A_0)$, $(A_1)$ and ($A_2$-weakened) plus the assumption that $\alpha$ is strictly increasing, the other case can be easily to deduce from the following arguments. Consider two weak solutions $(X^1_t, u^1_t;\ t\in[0,T])$ and $(X^2_t, u^2_t;\ t\in[0,T])$ to (2.1). Given some $\gamma>0$, that will be chosen later, define

$$G^\gamma_{s,t}(x,y)=G^\gamma_{t-s}(x-y)=\left(2\pi\gamma^2(t-s)\right)^{-\frac{d}{2}}\exp\left(-\frac{|x-y|^2}{2\gamma^2(t-s)}\right).\qquad(2.19)$$

Since the kernel $G^\gamma_t$ gives the fundamental solution related to the parabolic operator $\partial_s+\frac{\gamma^2}{2}\Delta_x$; that is

$$\partial_s G^\gamma_{s,t}+\frac{\gamma^2}{2}\Delta_x G^\gamma_{s,t}=0,\ \lim_{s\to t^-}G^\gamma_{s,t}=\delta_{\{y\}},\ \delta_{\{y\}}\text{ the Dirac measure in }y,$$

the function $v(s,x) = G^{\gamma}_{s,t}(f)(x) = G^{\gamma}_{t-s} * f(x)$, $s \leq t, x \in \mathbb{R}^d$, $f \in \mathcal{C}^{\infty}_c(\mathbb{R}^d)$ ($*$ denoting the convolution product on $\mathbb{R}^d$) is a $\mathcal{C}^{\infty}([0,t] \times \mathbb{R}^d)$ function satisfying:

$$\begin{cases} \partial_s v(s,x) + \frac{\gamma^2}{2}\Delta_x v(s,x) = 0, \ 0 \leq s < t, x \in \mathbb{R}^d, \\ v(t,x) = f(x), \ x \in \mathbb{R}^d. \end{cases}$$

Applying Itô formula to $v(s, X^i_s)$ for $i = 1, 2$,

$$\begin{aligned} \int f u^i_t \, dx = \mathbb{E}\left[v(t, X^i_t)\right] &= \mathbb{E}\left[v(0, X_0)\right] + \mathbb{E}\left[\int_0^t \frac{\partial v}{\partial s}(s, X^i_s) ds\right] + \frac{1}{2}\mathbb{E}\left[\int_0^t \sigma^2(u^i_s(X^i_s))\Delta_x v(s, X^i_s) ds\right] \\ &= \mathbb{E}\left[v(0, X_0)\right] + \frac{1}{2}\mathbb{E}\left[\int_0^t \left(\sigma^2(u^i_s(X^i_s)) - \gamma^2\right)\Delta_x v(s, X^i_s) ds\right] \end{aligned}$$

so that for $f \in \mathcal{C}^{\infty}_c(\mathbb{R}^d)$, $G^{\gamma}_{0,t} = G^{\gamma}_t$,

$$\int f u^i_t \, dx = \int f \left( G^{\gamma}_t(\mu_0) + \frac{1}{2}\sum_{k=1}^d \int_0^t \frac{\partial^2}{\partial x_k^2} G^{\gamma}_{t-s}\left(\left(\sigma^2(u^i_s) - \gamma^2\right) u^i_s\right) ds \right) dx.$$

We then obtain that

$$\begin{aligned} &\int f \left(u^1_t - u^2_t\right) dx \\ &= \int f \left( \frac{1}{2}\sum_{k=1}^d \int_0^t \frac{\partial^2}{\partial x_k^2} G^{\gamma}_{t-s}\left(\left(\sigma^2(u^1_s) - \gamma^2\right) u^1_s - \left(\sigma^2(u^2_s) - \gamma^2\right) u^2_s\right) ds \right) dx. \end{aligned}$$

Next, we take the supremum over all $f \in \mathcal{C}^{\infty}_c(\mathbb{R}^d)$ such that $\|f\|_{L^2(\mathbb{R}^d)} = 1$ and we integrate the resulting expression over $(0, T)$. It follows that

$$\begin{aligned} &\|u^1 - u^2\|_{L^2((0,T)\times\mathbb{R}^d)} \\ &= \|\frac{1}{2}\sum_{k=1}^d \int_0^{\cdot} \frac{\partial^2}{\partial x_k^2} G^{\gamma}_{\cdot-s}\left(\left(\sigma^2(u^1_s) - \gamma^2\right) u^1_s - \left(\sigma^2(u^2_s) - \gamma^2\right) u^2_s\right) ds\|_{L^2((0,T)\times\mathbb{R}^d)}. \end{aligned} \tag{2.20}$$

Now, let us recall that for all $f \in L^2((0,T) \times \mathbb{R}^d)$, (see e.g. Stroock and Varadhan [31], Appendix A.2, Lemmas A.2.1 and A.2.2)

$$\|\int_0^{\cdot} \partial^2_{x_k x_l} G^{\gamma}_{\cdot-s}(f(s)) \, ds\|_{L^2((0,T)\times\mathbb{R}^d)} \leq \frac{2}{\gamma^2}\|f\|_{L^2((0,T)\times\mathbb{R}^d)}, \ 1 \leq k, l \leq d.$$

Thank to the boundedness of $\sigma^2(u^i),\ i = 1, 2$, the preceding estimate ensures that the r.h.s. of (2.20) is well defined. A closer investigation of the proof arguments in [31] enables to slightly improve the preceding estimate with

$$\begin{aligned}
&\| \int_0^{\cdot} \triangle_x G^{\gamma}_{\cdot-s}(f(s))\, ds \|_{L^2((0,T)\times\mathbb{R}^d)} \\
&= \| \sum_{k=1}^{d} \int_0^{\cdot} \frac{\partial^2}{\partial x_k^2} G^{\gamma}_{\cdot-s}(f(s))\, ds \|_{L^2((0,T)\times\mathbb{R}^d)} \leq \frac{2}{\gamma^2} \| f \|_{L^2((0,T)\times\mathbb{R}^d)}
\end{aligned} \tag{2.21}$$

(see the appendix section, for a short proof). Applying this estimate to (2.20), we obtain

$$\begin{aligned}
\|u^1 - u^2\|_{L^2((0,T)\times\mathbb{R}^d)} &\leq \frac{1}{\gamma^2} \| \left(\sigma^2(u^1) - \gamma^2\right) u^1 - \left(\sigma^2(u^2) - \gamma^2\right) u^2 \|_{L^2((0,T)\times\mathbb{R}^d)} \\
&= \frac{1}{\gamma^2} \| \left(\sigma^2(u^1)u^1 - \sigma^2(u^2)u^2 - \gamma^2(u^1 - u^2)\right) \|_{L^2((0,T)\times\mathbb{R}^d)}.
\end{aligned}$$

For $\alpha(r)$ as in (2.2), by observing that the first order Taylor expansion writes

$$\sigma^2(r_2)r_2 - \sigma^2(r_1)r_1 = \int_0^1 (r_1 - r_2)\alpha(r_1 + \theta(r_2 - r_1))\, d\theta,$$

and choosing $\gamma > 0$ so that

$$\gamma^2 > \max_{i=1,2} | \int_0^1 \alpha(u^i + \theta(u^i - u^j))\, d\theta |,$$

we get

$$| \left(\sigma^2(u^1)u^1 - \sigma^2(u^2)u^2 - \gamma^2(u^1 - u^2)\right) | = \left(\gamma^2 - \int_0^1 \alpha(u_1 + \theta(u_2 - u_1))\, d\theta\right) |u^1 - u^2|,$$

and deduce that

$$\|u^1 - u^2\|^2_{L^2((0,T)\times\mathbb{R}^d)} \leq \frac{1}{\gamma^4} \| \left(\gamma^2 - \int_0^1 \alpha(u_1 + \theta(u_2 - u_1))\, d\theta\right) |u^1 - u^2| \|^2_{L^2((0,T)\times\mathbb{R}^d)}. \tag{2.22}$$

Splitting $\|u^1 - u^2\|^2_{L^2((0,T)\times\mathbb{R}^d)}$ into the sum

$$\|(u^1 - u^2)\mathbb{1}_{\{u^2 - u^1 \geq \kappa\}}\|^2_{L^2((0,T)\times\mathbb{R}^d)} + \|(u^1 - u^2)\mathbb{1}_{\{u^2 - u^1 < \kappa\}}\|^2_{L^2((0,T)\times\mathbb{R}^d)}$$

for some arbitrary $\kappa > 0$, (2.22) reduces to

$$
\begin{aligned}
&\|(u^1-u^2)\mathbb{1}_{\{u^2-u^1\geq\kappa\}}\|^2_{L^2((0,T)\times\mathbb{R}^d)}\\
&\leq \frac{1}{\gamma^4}\|\left(\gamma^2-\int_0^1\alpha(u_1+\theta(u_2-u_1))\,d\theta\right)|u^1-u^2|\,\|^2_{L^2((0,T)\times\mathbb{R}^d)}-\|(u^1-u^2)\mathbb{1}_{\{u^2-u^1<\kappa\}}\|^2_{L^2((0,T)\times\mathbb{R}^d)}\\
&\leq \frac{1}{\gamma^4}\|\left(\gamma^2-\int_0^1\alpha(u_1+\theta(u_2-u_1))\,d\theta\right)|u^1-u^2|\,\mathbb{1}_{\{u^2-u^1\geq\kappa\}}\|^2_{L^2((0,T)\times\mathbb{R}^d)}
\end{aligned}
$$

Fixing $\kappa>0$, and setting

$$\zeta^2(\kappa):=\sup\left\{\beta>0\,;\int_0^1\alpha(r\kappa)\,dr>\beta\right\}$$

which is (strictly) positive by ($A_2$-weakened) and the monotone assumption of $\alpha$,

$$\gamma^2-\int_0^1\alpha(u_1+\theta(u_2-u_1))\,d\theta\leq\gamma^2-\int_0^1\alpha(\theta\kappa)\,d\theta\leq\gamma^2-\zeta^2(\kappa),$$

which implies that $\|\,|u^1-u^2|\mathbb{1}_{\{u^2-u^1\geq\kappa\}}\|_{L^2((0,T)\times\mathbb{R}^d)}<(\gamma^2-\zeta^2)\|\,|u^1-u^2|\,\mathbb{1}_{\{u^2-u^1\geq\kappa\}}\|_{L^2((0,T)\times\mathbb{R}^d)}/\gamma^2<\|\,|u^1-u^2|\mathbb{1}_{\{u^2-u^1\geq\kappa\}}\|_{L^2((0,T)\times\mathbb{R}^d)}$. Since $\kappa$ is arbitrary, $u_2\leq u_1$ for a.e. on $(0,T)\times\mathbb{R}^d$. By symmetry, we can also exchange $u^1$ and $u^2$, and deduce that $u_1\leq u_2$ for a.e. on $(0,T)\times\mathbb{R}^d$, from which we conclude that $u_1=u_2$ for a.a. $(t,x)\in(0,T)\times\mathbb{R}^d$. This conclude the claim.

To conclude on the strong uniqueness of the solution to (2.8), let us recall the following result due to Champagnat and Jabin [8].

**Theorem 2.11** (Theorems 1.1 and 1.2, [8]) *Let $(Z^1_t;\,t\geq0)$ and $(Z^2_t;\,t\geq0)$ be two solutions to the SDE*

$$dZ_t=\Sigma(t,Z_t)\,dW_t,\ Z_0=\xi,$$

*with one-dimensional time marginal $u_{Z_1}(t,z)\,dz$ and $u_{Z_2}(t,z)\,dz$ in $L^{2q}_{loc}((0,\infty);W^{1,2p}(\mathbb{R}^d))$. If $\Sigma:(0,\infty)\times\mathbb{R}^d\to\mathbb{R}^{d\times d}$ is in $L^\infty((0,\infty)\times\mathbb{R}^d)\cap L^{q'}_{loc}((0,\infty);W^{1,p'}(\mathbb{R}^d))$, for $1/p+1/p'=1$, $1/q+1/q'=1$, then one has pathwise uniqueness: a.s. $\sup_{t\geq0}|Z^1_t-Z^2_t|=0$.*

According to Proposition 2.10 and Corollary 2.9, for any solution to (2.1), $u$ is in $L^\infty((0,T)\times\mathbb{R}^d)$. Hence, a direct application of Theorem 2.11 (with $p=q=1$, $p'=q'=\infty$) gives

**Proposition 2.12** *Under ($A_0$), ($A_1$) and ($A_2$), (2.1) admits at most one strong solution.*

## 2.4 Generalization to Matrix Valued Diffusion

We may remark that the main ideas for the proof of Theorem 2.3 can be extended to obtain the existence and uniqueness of a weak solution to (2.1) in the situation where the diffusion component is a $d \times d$-matrix valued function; namely $\sigma : [0, \infty) \to \mathbb{R}^{d\times d}$. Such extension holds provided that the assumption $(A_0)$ remains unchanged meanwhile $(A_1)$, $(A_2)$ and $(A_2$-weakened) are respectively replaced by the following:

**Hypothesis 2.13**

**$(A_1')$** The map $r \mapsto \sigma(r) \in \mathbb{R}^{d\times d}$ is continuously differentiable on $\mathbb{R}^+$.

**$(A_2')$** For $a(r) = \sigma\sigma^*(r)$, the map $r \in [0, \infty) \mapsto \alpha(r) \in \mathbb{R}^{d\times d}$ given by

$$\alpha(r) = \{\alpha^{i,j}(r) := (a^{i,j}(r)r)' = (a^{i,j})'(r)r + a^{i,j}(r),\ 1 \le i, j \le d\},$$

is continuous and strongly elliptic in the sense that, for some $\eta_a > 0$,

$$\xi \cdot \alpha(r)\xi \ge \eta_a |\xi|^2,\ \forall r \ge 0,\ \forall \xi \in \mathbb{R}^d.$$

**$(A_2'$-weakend)** For $a(r) = \sigma\sigma^*(r)$, the map $r \in [0, \infty) \mapsto \alpha(r) \in \mathbb{R}^d \times \mathbb{R}^d$ given by

$$\alpha(r) = \{\alpha^{i,j}(r) := (a^{i,j}(r)r)' = (a^{i,j})'(r)r + a^{i,j}(r),\ 1 \le i, j \le d\}$$

is continuous and positive semi-definite:

$$\xi \cdot \alpha(r)\xi \ge 0,\ \forall \xi \in \mathbb{R}^d.$$

The particular strict monotone assumption in Proposition 2.10 can be replaced by the assumption that

$$\xi \cdot \alpha(r)\xi < \xi \cdot \alpha(r')\xi,\ \forall \xi \in \mathbb{R}^d,\ \forall r, r' \ge 0 \text{ such that } r < r'. \tag{2.23}$$

Under $(A_0)$, $(A_1')$ and $(A_2')$, Theorem 2.3 can be extended to the existence and uniqueness of a strong solution to

$$\begin{cases} X_t = X_0 + \int_0^t \sigma(u(s, X_s))dW_s, \quad 0 \le t \le T, \\ dLaw(X_t) = u(t, x)dx \text{ with } u \in L^\infty((0, T) \times \mathbb{R}^d) \cap L^2((0, T) \times \mathbb{R}^d), \\ u(0, x) = u_0(x),\ x \in \mathbb{R}^d,\ \sigma : \mathbb{R}^+ \to \mathbb{R}^{d\times d}. \end{cases} \tag{2.24}$$

**For the Existence of a Weak Solution to** (2.24)
Assumptions $(A_0)$, $(A_1')$ and $(A_2')$ are enough to replicate the proof arguments of the estimates 2.10 and 2.11 in Lemma 2.5 and enables to construct, as in Proposition 2.6, a weak solution to

$$\begin{cases} X_t^\varepsilon = X_0 + \int_0^t \sqrt{a_\varepsilon(u^\varepsilon(s, X_s))} dW_s, \quad 0 \le t \le T, \\ d\,Law(X_t^\varepsilon) = u^\varepsilon(t, x)dx, \text{ with } u^\varepsilon \in L^\infty((0, T) \times \mathbb{R}^d) \cap L^2((0, T) \times \mathbb{R}^d), \\ u^\varepsilon(0, x) = u_0(x), \ \sigma : \mathbb{R}^+ \to \mathbb{R}^d \times \mathbb{R}^d, \end{cases}$$

where $\sqrt{a_\varepsilon(r)}$ is the square root matrix of $a(r) + \varepsilon I_d$, $a'(r) = \{(a^{i,j})'(r),\ 1 \le i, j \le d\}$. Although the identity (2.12) doesn't have any trivial multidimensional extension, and since $u^\varepsilon$ satisfies the analogous of (2.7): for all for all $0 \le T_0 \le T$, $f \in C_c^{1,2}([0, T_0] \times \mathbb{R}^d)$,

$$\begin{aligned} &\int_{\mathbb{R}^d} u^\varepsilon(T_0, x) f(T_0, x)\, dx - \int_{\mathbb{R}^d} u_0(x) f(0, x)\, dx \\ &= \int_{(0,T_0)\times\mathbb{R}^d} u^\varepsilon(t, x)\partial_t f(t, x) - \nabla_x u^\varepsilon(t, x) \cdot a_\varepsilon'(u^\varepsilon(t, x))\nabla_x f(t, x)\, dt\, dx, \end{aligned} \tag{2.25}$$

the convergence (up to a subsequence) of $\{(X_t^\varepsilon, u^\varepsilon(t);\ 0 \le t \le T)\}_{\varepsilon > 0}$ to a weak solution to (2.24) can still be derived from the energy estimate:

$$\max_{0 \le t \le T} \|u^\varepsilon(t)\|^2_{L^2(\mathbb{R}^d)} + \eta_a \int_0^T \|\nabla_x u^\varepsilon(t)\|^2_{L^2(\mathbb{R}^d)}\, dt \le \|u_0\|^2_{L^2(\mathbb{R}^d)},$$

which follows from (2.25).

**For the Uniqueness of a Strong Solution to** (2.24)

Under ($A_1'$) and ($A_2'$-weakened), the main arguments of Proposition 2.10 can be extended, replacing $G_{s,t}^\gamma$ by the fundamental solution to $G_{s,t}^\Gamma$ related to the parabolic operator $L(f) = \partial_s f + \frac{1}{2}\text{Trace}\left(\Gamma \nabla_x^2 f\right)$ where $\Gamma$ is a (constant) positive definite matrix such that $\xi \cdot \Gamma\xi \ge \gamma^2|\xi|^2$, $\gamma \neq 0$. Taking two weak solutions $(X_t^1, u_t^1;\ t \in [0, T])$ and $(X_t^2, u_t^2;\ t \in [0, T])$ to (2.25), and by replicating the first proof steps of Proposition 2.10, we get the analog of (2.20):

$$\begin{aligned} &\|u^1 - u^2\|_{L^2((0,T)\times\mathbb{R}^d)} \\ &= \|\frac{1}{2}\sum_{k,l=1}^d \int_0^\cdot \partial^2_{x_k x_l} G^\gamma_{\cdot - s}\left(\left(a(u_s^1) - \Gamma\right)^{k,l} u_s^1 - \left(a(u_s^2) - \Gamma\right)^{k,l} u_s^2\right) ds\|_{L^2((0,T)\times\mathbb{R}^d)}. \end{aligned}$$

Using the following generalization of (2.21) (see Appendix section):

$$\begin{aligned} &\|\sum_{i,j=1}^d \int_0^\cdot \partial^2_{x_i x_j} G^\Gamma_{\cdot - s}(F^{i,j}(s))\, ds\|_{L^2((0,T)\times\mathbb{R}^d)} \\ &\le \frac{2}{\gamma^2}\sum_{i,j=1}^d \|F^{i,j}\|_{L^2((0,T)\times\mathbb{R}^d)}, \quad F^{i,j} \in L^2((0, T) \times \mathbb{R}^d), \end{aligned}$$

we deduce the analog of (2.22)

$$\|u^1 - u^2\|_{L^2((0,T)\times\mathbb{R}^d)} \leq \frac{1}{\gamma^2}\sum_{k,l=1}^{d} \| \left(\Gamma - \int_0^1 \alpha(u_1+\theta(u_2-u_1))\,d\theta\right)^{k,l} |u^1-u^2| \,\|_{L^2((0,T)\times\mathbb{R}^d)}.$$

Taking $\Gamma$ large enough so that $\xi \cdot (\Gamma - \alpha(r))\xi < 0$, for all $\xi \in \mathbb{R}^d$, $r \geq 0$, the strict elliptic assumption in $(A_2')$ or (2.23) are sufficient to ensure that $u^1 = u^2$. The uniqueness of a strong solution to (2.24) still follows from Champagnat and Jabin [8].

## 3 Conditional Nonlinear Diffusion Case

**Hypothesis 3.1**

**(H0)** The initial law $\mu_0$ admits a density $\rho_0$ such that $\int_{\mathbb{R}^d\times\mathbb{R}^d} (|x|^2+|y|^2)\rho_0(x,y)\,dx\,dy < \infty$.

**(H1)** The coefficients $b$ and $\sigma$ are bounded Lipschitz continuous functions.

**(H2)** The kernels $\ell$ and $\gamma$ are bounded and Lipschitz continuous on $\mathbb{R}^d$.

**(H3)** Strong ellipticity is assumed for $\sigma$: there exists $a_* > 0$ such that, for all $x \in \mathbb{R}^d$,

$$a_*|\xi|^2 \leq \xi\sigma(x)\xi, \quad \forall \xi \in \mathbb{R}^d.$$

**Hypothesis 3.2**

**(H4)** The initial marginal density $\rho_X(0,x) = \int_{\mathbb{R}^d} \rho_0(x,y)dy$ is in $L^1(\mathbb{R}^d) \cap L^p(\mathbb{R}^d)$ for some $p \geq 2d+2$.
Moreover, for all $R > 0$, for all $x \in B(0,R)$, there exists a constant $\mu_R > 0$ such that $\rho_X(0,x) \geq \mu_R$.

**(H5)** $\sigma$ and $\gamma$ are in $C^2(\mathbb{R}^d)$ with bounded derivatives up to second order.

**(H6)** Strong ellipticity is assumed for $\gamma\gamma^*$: there exists $\alpha_* > 0$ such that, for all $(x,y) \in \mathbb{R}^d \times \mathbb{R}^d$,

$$\alpha_*|\xi|^2 \leq \xi\gamma(y)\gamma(y)^*\xi, \quad \forall \xi \in \mathbb{R}^d.$$

Our main result concerns the wellposedness (in the weak and strong sense) of a solution to (1.2). More precisely, we have

**Theorem 3.3** *Under Hypothesis 3.1, there exists a unique weak solution to (1.2). With the addition of Hypothesis 3.2, pathwise uniqueness holds for the solution of (1.2) and* $Law(X_t, Y_t)$ *admits a density function at all time* $0 \leq t \leq T$.

Before entering in the details of the proof, let us point out an important remark for the construction of the solution of (1.2). Consider for a while the case when $(X_t;\, t \geq 0)$ doesn't depend on $(Y_t;\, t \geq 0)$, namely when $b$ does not depend on $Y$, or the simpler situation when $b = 0$. Hypothesis 3.1 ensure the existence of a unique strong solution for

$$X_t = X_0 + \int_0^t \sigma(X_s)\,dB_s. \tag{3.1}$$

Then, based on the fact that $(X_t;\ t \geq 0)$ is now an exogenous process, we can consider the following fixed point construction, similar to those in Sznitman [33] and Méléard [24]. From now on, we fix an arbitrary time horizon $0 \leq T < +\infty$ and we consider the classical Hilbert space $M^2(0, T)$ of real adapted continuous processes $\zeta = (\zeta_t; 0 \leq t \leq T)$ such that $\mathbb{E}_{\mathbb{P}}[\int_0^T \zeta_s^2 ds] < +\infty$ ($\mathbb{E}_{\mathbb{P}}$ denoting the expectation under $\mathbb{P}$), and endowed with the following scalar product and norm

$$(\zeta, \xi)_c = \mathbb{E}_{\mathbb{P}}\left[\int_0^T \exp(-cs)\zeta_s\xi_s ds\right], \quad \|\zeta\|_c^2 = \mathbb{E}_{\mathbb{P}}\left[\int_0^T \exp(-cs)|\zeta_s|^2 ds\right],$$

where $c$ is a positive constant that will be chosen later. Given $(\Omega, \mathcal{F}, (\mathcal{F}_t;\ 0 \leq t \leq T), \mathbb{P})$ a filtered probability space under which are defined $(W_t;\ 0 \leq t \leq T)$ and $(B_t;\ 0 \leq t \leq T)$, two independent $\mathbb{R}^d$-Brownian motions, and $(X_0, Y_0) \sim \mu_0$ (independent of $(W_t;\ 0 \leq t \leq T)$ and $(B_t;\ 0 \leq t \leq T)$).

From any element $\zeta$ in $M^2(0, T)$, we construct the application $\zeta \mapsto Y(\zeta)$ taking values in $M^2(0, T)$ and defined as

$$Y(\zeta)_t = Y_0 + \int_0^t \mathbb{E}_{\mathbb{P}}\left[\ell(\zeta_s) \mid X_s\right] ds + \int_0^t \mathbb{E}_{\mathbb{P}}\left[\gamma(\zeta_s) \mid X_s\right] dW_s,\ 0 \leq t \leq T. \tag{3.2}$$

Owing to **(H0)** and **(H2)**, it is clear that $\|Y(\zeta)\|_c < +\infty$. Now, for $\zeta$ and $\xi$ in $M^2(0, T)$, $0 \leq t \leq T$, we have

$$\begin{aligned}
\mathbb{E}_{\mathbb{P}}\left[|Y(\zeta)_t - Y(\xi)_t|^2\right] &\leq 2\int_0^t \mathbb{E}_{\mathbb{P}}\left[\left|\mathbb{E}_{\mathbb{P}}\left[\ell(\zeta_s) \mid X_s\right] - \mathbb{E}_{\mathbb{P}}\left[\ell(\xi_s) \mid X_s\right]\right|^2\right] ds \\
&\quad + 2\int_0^t \mathbb{E}_{\mathbb{P}}\left[\left\|\mathbb{E}_{\mathbb{P}}\left[\gamma(\zeta_s) \mid X_s\right] - \mathbb{E}_{\mathbb{P}}\left[\gamma(\xi_s) \mid X_s\right]\right\|^2\right] ds \\
&\leq 2\left(\|\ell\|_{Lip}^2 + \|\gamma\|_{Lip}^2\right)\int_0^t \mathbb{E}_{\mathbb{P}}\left[\mathbb{E}_{\mathbb{P}}\left[|\zeta_s - \xi_s|^2 \mid X_s\right]\right] ds \\
&\leq 2\left(\|\ell\|_{Lip}^2 + \|\gamma\|_{Lip}^2\right)\int_0^t \mathbb{E}_{\mathbb{P}}\left[|\zeta_s - \xi_s|^2\right] ds.
\end{aligned}$$

Multiplying both sides by $\exp(-ct)$ and integrating (in time) the resulting expression over the interval $(0, T)$ gives

$$\begin{aligned}
&\int_0^T \exp(-ct)\mathbb{E}_{\mathbb{P}}\left[|Y(\zeta)_t - Y(\xi)_t|^2\right] dt \\
&\leq 2\left(\|\ell\|_{Lip}^2 + \|\gamma\|_{Lip}^2\right)\int_0^T \exp(-ct)\int_0^t \mathbb{E}_{\mathbb{P}}\left[|\zeta_s - \xi_s|^2\right] ds\, dt.
\end{aligned}$$

An integration by part in time then yields

$$\begin{aligned}\|Y(\zeta)-Y(\xi)\|_c^2 &\leq -\frac{\exp(-cT)}{c}\int_0^T \mathbb{E}_{\mathbb{P}}\left[|\zeta_s-\xi_s|^2\right]ds\\ &\quad+\frac{2}{c}\left(\|\ell\|_{Lip}^2+\|\gamma\|_{Lip}^2\right)\int_0^T \exp(-ct)\mathbb{E}_{\mathbb{P}}\left[|\zeta_t-\xi_t|^2\right]dt\\ &\leq \frac{2}{c}\left(\|\ell\|_{Lip}^2+\|\gamma\|_{Lip}^2\right)\|\zeta-\xi\|_c^2.\end{aligned}$$

Choosing $c>2(\|\ell\|_{Lip}^2+\|\gamma\|_{Lip}^2)$, we get the existence of a unique fixed point solution of Eq. (1.2), when $b=0$. In the case $b(x,y)=b(x)$, the same arguments lead to the same result.

The rest of this section is dedicated to the proof of Theorem 3.3, which essentially relies on a Girsanov transform to go back to a situation similar to the previous case. First, in Sect. 3.1, we collect some preliminary remarks on the Girsanov transform that we use to remove the drift in (1.2), and deduce some apriori controls on the associated change of probability measure.

Second, in Sect. 3.2, we use the fixed point technique for the $L^2$-existence and weak uniqueness for solution to (1.2).

Finally in Sect. 3.3, assuming some stronger regularity on the kernels $\ell$ and $\gamma$ and strong ellipticity on $\gamma$ (see Hypothesis 3.2, we obtain some apriori regularity on the nonlinear coefficients using averaging lemma technique, and then deduce the strong uniqueness property.

## *3.1 Preliminary Remarks on (1.2)*

Fix an arbitrary $0\leq T<+\infty$ and let $(X_t,Y_t;\,0\leq t\leq T)$ be a solution to (1.2) up to $T$, defined on $(\Omega,\mathcal{F},(\mathcal{F}_t;\,0\leq t\leq T),\mathbb{P})$. Then, define $(Z_t;\,0\leq t\leq T)$ as

$$Z_t=\exp\left\{-\int_0^t\left(\sigma^{-1}b\right)(X_s,Y_s)\,dB_s-\frac{1}{2}\int_0^t\left|\sigma^{-1}b\right|^2(X_s,Y_s)\,ds\right\},$$

for $\sigma^{-1}(x)$ the inverse matrix of $\sigma(x)$, $(\sigma^{-1}b)(x,y)=\sigma^{-1}(x)b(x,y)$ and $\left|\sigma^{-1}b\right|^2(x,y)=\left(\sigma^{-1}b\right)(x,y)\cdot\left(\sigma^{-1}b\right)(x,y)$. Then, under the probability measure $\mathbb{Q}$ defined on $(\Omega,\mathcal{F}_T,(\mathcal{F}_t;\,0\leq t\leq T))$ by

$$\left.\frac{d\mathbb{Q}}{d\mathbb{P}}\right|_{\mathcal{F}_T}=Z_T,$$

the process

$$\widehat{B}_t=\int_0^t(\sigma^{-1}b)(X_s,Y_s)\,ds+B_t,\;0\leq t\leq T,$$

is a $\mathbb{R}^d$-Brownian motion (by means of Girsanov transformation). Observing that the covariation between $(W_t;\ 0 \leq t \leq T)$ and $(\widehat{B}_t;\ 0 \leq t \leq T)$ is zero, $(\widehat{B}_t;\ 0 \leq t \leq T)$ is independent of $(W_t;\ 0 \leq t \leq T)$.

In addition, for all $\mathcal{F}_t$-adapted process $(\theta_t;\ 0 \leq t \leq T)$ such that $\mathbb{E}_{\mathbb{P}}[|\theta_t|] < +\infty$ for all $t$, the characterization of the conditional expectation ensures that, $\mathbb{P}$-a.s. (or equivalently $\mathbb{Q}$-a.s.),

$$\mathbb{E}_{\mathbb{P}}[\theta_t \mid X_t] = Z_t \mathbb{E}_{\mathbb{Q}}[(Z_t)^{-1}\theta_t \mid X_t],\ 0 \leq t \leq T. \tag{3.3}$$

Following this change of probability measure, under $\mathbb{Q}$, Eq. (1.2) formulate as the following self-contained SDE:

$$\begin{cases} X_t = X_0 + \int_0^t \sigma(X_s)\, d\widehat{B}_s,\ 0 \leq t \leq T, \\ Y_t = Y_0 + \int_0^t Z_s \mathbb{E}_{\mathbb{Q}}\left[Z_s^{-1}\ell(Y_s) \mid X_s\right] ds + \int_0^t Z_s \mathbb{E}_{\mathbb{Q}}\left[Z_s^{-1}\gamma(Y_s) \mid X_s\right] dW_s, \\ Z_t = \exp\left\{-\int_0^t \left(\sigma^{-1}b\right)(X_s, Y_s)\, d\widehat{B}_s + \frac{1}{2}\int_0^t \left|\sigma^{-1}b\right|^2 (X_s, Y_s)\, ds\right\}, \\ (X_0, Y_0) \sim \mu_0. \end{cases} \tag{3.4}$$

Conversely, starting from (3.4), defined on $(\Omega, \mathcal{F}_T, (\mathcal{F}_t,\ 0 \leq t \leq T), \mathbb{Q})$ endowed with two independent Brownian motions $(\widehat{B}_t;\ 0 \leq t \leq T)$ and $(W_t;\ 0 \leq t \leq T)$, independent to $(X_0, Y_0)$, one can easily check that $(X_t, Y_t;\ 0 \leq t \leq T)$ is a solution to (1.2) on $(\Omega, \mathcal{F}_T, (\mathcal{F}_t,\ 0 \leq t \leq T), \widehat{\mathbb{P}})$ where $\widehat{\mathbb{P}}$ is given by

$$\left.\frac{d\widehat{\mathbb{P}}}{d\mathbb{Q}}\right|_{\mathcal{F}_T} = Z_T^{-1} = \exp\left\{\int_0^t \left(\sigma^{-1}b\right)(X_s, Y_s)\, d\widehat{B}_s - \frac{1}{2}\int_0^t \left|\sigma^{-1}b\right|^2 (X_s, Y_s)\, ds\right\},\ 0 \leq t \leq T.$$

The existence and uniqueness of a weak solution to (1.2) is then an immediate consequence of the existence and uniqueness of a weak solution to (3.4).

Let $\zeta$ in $M^2(0, T)$. We consider the linearized system

$$\begin{cases} X_t = X_0 + \int_0^t \sigma(X_s)\, d\widehat{B}_s,\ 0 \leq t \leq T, \\ Z(\zeta)_t = \exp\left\{-\int_0^t \left(\sigma^{-1}b\right)(X_s, \zeta_s)\, d\widehat{B}_s + \frac{1}{2}\int_0^t \left|\sigma^{-1}b\right|^2 (X_s, \zeta_s) ds\right\}, \\ (X_0, Y_0) \sim \mu_0. \end{cases} \tag{3.5}$$

**Lemma 3.4** *Assume Hypothesis 3.1. Let $\zeta$ and $\xi$ in $M^2(0, T)$.*

*(i) There exits a positive constant $C$ depending only on $T$ and on $\|\sigma^{-1}b\|_\infty$ such that*

$$\mathbb{E}_{\mathbb{Q}}\big[Z(\zeta)_t^2 | X_t\big](Z(\zeta)_t)^{-2} \leq C.$$

*(ii) There exits a positive constant $C'$ depending only on $T$, $\|\sigma^{-1}b\|_\infty$ and $\|\ell\|_{Lip} + \|\gamma\|_{Lip}$ such that*

$$\mathbb{E}_{\mathbb{Q}}\left[\max_{0\le s\le t}\frac{|Z(\zeta)_s^2 - Z(\xi)_s|^2}{Z(\zeta)_s^2}\right] \le C'\int_0^t \mathbb{E}_{\mathbb{Q}}\left[\frac{|Z(\zeta)_s - Z(\xi)_s|^2}{Z(\zeta)_s^2} + |\zeta_s - \xi_s|^2\right]ds. \tag{3.6}$$

*Proof* For *(i)*, we fix $\zeta \in M^2(0,T)$. We denote by $\mathcal{E}$ the exponential martingale (under $\mathbb{Q}$) defined as

$$\mathcal{E}_t = \exp\left\{-\int_0^t 2\left(\sigma^{-1}b\right)(X_s,\zeta_s)\,d\widehat{B}_s - \int_0^t 2\left|\sigma^{-1}b\right|^2(X_s,\zeta_s)ds\right\}.$$

From (3.5), we have

$$Z(\zeta)_t^2 = \mathcal{E}_t \exp\left\{+3\int_0^t \left|\sigma^{-1}b\right|^2(X_s,\zeta_s)ds\right\}$$

and then

$$\begin{aligned}\mathbb{E}_{\mathbb{Q}}[(Z(\zeta))_t^2|X_t](Z(\zeta)_t)^{-2} &= \mathcal{E}_t^{-1}\mathbb{E}_{\mathbb{Q}}\left[\mathcal{E}_t \exp\left\{+3\int_0^t \left|\sigma^{-1}b\right|^2(X_s,\zeta_s)ds\right\}|X_t\right]\exp\left\{-3\int_0^t\left|\sigma^{-1}b\right|^2(X_s,\zeta_s)ds\right\}\\ &\le \mathcal{E}_t^{-1}\mathbb{E}_{\mathbb{Q}}\left[\mathcal{E}_t\exp\left\{+3\int_0^t\left|\sigma^{-1}b\right|^2(X_s,\zeta_s)ds\right\}|X_t\right].\end{aligned}$$

Let us define the probability measure $\widetilde{\mathbb{Q}}$ on $(\Omega, \mathcal{F}_T, (\mathcal{F}_t;\ 0 \le t \le T))$ by

$$\left.\frac{d\widetilde{\mathbb{Q}}}{d\mathbb{Q}}\right|_{\mathcal{F}_T} = \mathcal{E}_T.$$

Then, as in (3.3), we obtain that for all $\mathcal{F}_t$-adapted process $(\theta_t;\ 0 \le t \le T)$ such that $\mathbb{E}_{\widetilde{\mathbb{Q}}}[|\theta_t|] < +\infty$ for all $t$ in $[0,T]$, the characterization of the conditional expectation ensures that, $\mathbb{P}$-a.s. (or equivalently $\mathbb{Q}$-a.s.),

$$\mathbb{E}_{\widetilde{\mathbb{Q}}}[\theta_t \mid X_t] = \mathcal{E}_t^{-1}\mathbb{E}_{\mathbb{Q}}[\mathcal{E}_t\theta_t \mid X_t],\ 0 \le t \le T,$$

from which we immediately deduce that

$$\mathcal{E}_t^{-1}\mathbb{E}_{\mathbb{Q}}\left[\mathcal{E}_t\exp\left\{+3\int_0^t\left|\sigma^{-1}b\right|^2(X_s,\zeta_s)ds\right\}|X_t\right] \le \exp(3T\|\sigma^{-1}b\|_\infty).$$

For *(ii)*, we fix again a $\zeta \in M^2(0,T)$. From (3.5), we have

$$dZ(\zeta)_t = Z(\zeta)_t\left(-(\sigma^{-1}b)(X_t,\zeta_t)\,d\widehat{B}_t + |\sigma^{-1}b|^2(X_t,\zeta_t)\,dt\right),\ Z(\zeta)_0 = 1,$$

and

$$dZ(\zeta)_t^{-1} = Z(\zeta)_t^{-1}(\sigma^{-1}b)(X_t, \zeta_t)\, d\widehat{B}_t,\ Z(\zeta)_0 = 1,$$

from which we compute, using the Itô formula, for a $\xi \in M^2(0, T)$

$$\begin{aligned}
&|Z(\zeta)_t - Z(\xi)_t|^2 \\
&= 2\int_0^t (Z(\zeta)_s - Z(\xi)_s)\left\{Z(\zeta)_s(\sigma^{-1}b)(X_s, \zeta_s) - Z(\xi)_s(\sigma^{-1}b)(X_s, \xi_s)\right\} d\widehat{B}_s \\
&\quad + 2\int_0^t (Z(\zeta)_s - Z(\xi)_s)\left\{Z(\zeta)_s|\sigma^{-1}b|^2(X_s, \zeta_s) - Z(\xi)_s|\sigma^{-1}b|^2(X_s, \xi_s)\right\} ds \\
&\quad + \int_0^t \left(Z(\zeta)_s(\sigma^{-1}b)(X_s, \zeta_s) - (Z(\xi)_s\sigma^{-1}b)(X_s, \xi_s)\right)^2 ds
\end{aligned}$$

and

$$d(Z(\zeta)_t^{-1})^2 = (Z(\zeta)_t^{-1})^2\left(2(\sigma^{-1}b)(X_t, \zeta_t)\, d\widehat{B}_t + |\sigma^{-1}b|^2(X_t, \zeta_t)dt\right).$$

Applying again the Itô formula,

$$\begin{aligned}
&\frac{|Z(\zeta)_t - Z(\xi)_t|^2}{Z(\zeta)_t^2} \\
&= \int_0^t \frac{|Z(\zeta)_s - Z(\xi)_s|^2}{Z(\zeta)_s^2}\left(2(\sigma^{-1}b)(X_s, \zeta_s)\, d\widehat{B}_s + |\sigma^{-1}b|^2(X_s, \zeta_s)ds\right) \\
&\quad + 2\int_0^t \frac{(Z(\zeta)_s - Z(\xi)_s)}{Z(\zeta)_s^2}\left\{Z(\zeta)_s(\sigma^{-1}b)(X_s, \zeta_s) - Z(\xi)_s(\sigma^{-1}b)(X_s, \xi_s)\right\} d\widehat{B}_s \\
&\quad + 2\int_0^t \frac{(Z(\zeta)_s - Z(\xi)_s)}{Z(\zeta)_s^2}\left\{Z(\zeta)_s|\sigma^{-1}b|^2(X_s, \zeta_s) - Z(\xi)_s|\sigma^{-1}b|^2(X_s, \xi_s)\right\} ds \\
&\quad + \int_0^t \frac{1}{Z(\zeta)_s^2}\left|Z(\zeta)_s(\sigma^{-1}b)(X_s, \zeta_s) - (Z(\xi)_s\sigma^{-1}b)(X_s, \xi_s)\right|^2 ds \\
&\quad + 4\int_0^t \frac{(Z(\zeta)_s - Z(\xi)_s)}{Z(\zeta)_s^2}\Big(Z(\zeta)_s|\sigma^{-1}b|^2(X_t, \zeta_s) - Z(\xi)_s(\sigma^{-1}b)(X_s, \xi_s)(\sigma^{-1}b)(X_s, \zeta_s)\Big)ds.
\end{aligned}$$

$Z(\cdot)$ being an exponential martingale, the $L^2$-integrability of each integrands in the right-hand side of the preceding expression derive from the boundedness of $\sigma^{-1}b$. For any terms of the form

$$\frac{1}{Z(\zeta)_s^2}\left\{Z(\zeta)_s g(X_s, \zeta_s) - Z(\xi)_s g(X_s, \xi_s)\right\},$$

for $g$ equal to $\sigma^{-1}b$ or $|\sigma^{-1}b|^2$, we add and subtract the same element $Z(\zeta)_s g(X_s, \zeta_s)$ to get

$$\frac{1}{Z(\zeta)_s}\Big(g(X_s,\zeta_s)-g(X_s,\xi_s)\Big)+\frac{1}{Z(\zeta)_s^2}(Z(\zeta)_s-Z(\zeta)_s)g(X_s,\xi_s).$$

Noticing that $\sigma^{-1}b$ and $|\sigma^{-1}b|^2$ are bounded Lipschitz, by taking the expectation, and by introducing the appropriate pivots in the three last integrals, we get

$$\begin{aligned}\mathbb{E}_{\mathbb{Q}}\left[\frac{|Z(\zeta)_t-Z(\xi)_t|^2}{Z(\zeta)_t^2}\right]&\leq C\|\sigma^{-1}b\|_\infty^2\int_0^t\mathbb{E}_{\mathbb{Q}}\left[\frac{|Z(\zeta)_s-Z(\xi)_s|^2}{Z(\zeta)_s^2}\right]ds\\&+C\min\left(\|\sigma^{-1}b\|_{Lip},2\|\sigma^{-1}b\|_\infty\right)\int_0^t\mathbb{E}_{\mathbb{Q}}\left[\frac{|Z(\zeta)_s-Z(\xi)_s||\zeta_s-\xi_s|}{Z(\zeta)_s}\right]ds.\end{aligned}$$

We end the proof of *(ii)* by applying Young's inequality.

### 3.2 $L^2$-Existence and Weak Uniqueness

$$\begin{cases}X_t=X_0+\displaystyle\int_0^t\sigma(X_s)\,d\widehat{B}_s,\ 0\leq t\leq T,\\ Y(\zeta)_t=Y_0+\displaystyle\int_0^t Z(\zeta)_s\mathbb{E}_{\mathbb{Q}}\left[Z(\zeta)_s^{-1}\ell(\zeta_s)\,|\,X_s\right]ds+\int_0^t Z(\zeta)_s\mathbb{E}_{\mathbb{Q}}\left[Z(\zeta)_s^{-1}\gamma(\zeta_s)\,|\,X_s\right]dW_s,\\ Z(\zeta)_t=\exp\left\{-\displaystyle\int_0^t\left(\sigma^{-1}b\right)(X_s,\zeta_s)\,d\widehat{B}_s+\frac{1}{2}\int_0^t\left|\sigma^{-1}b\right|^2(X_s,\zeta_s)\,ds\right\},\\ (X_0,Y_0)\sim\mu_0.\end{cases}\tag{3.7}$$

Remark that, from (3.3), for any bounded Borel function $g:\mathbb{R}^d\to\mathbb{R}$,

$$Z(\zeta)_t\mathbb{E}_{\mathbb{Q}}[(Z(\zeta)_t)^{-1}|g(\zeta_t)|\,|\,X_t]=\mathbb{E}_{\mathbb{P}}[|g(\zeta_t)|\,|\,X_t]\leq\|g\|_{L^\infty},\ 0\leq t\leq T,\tag{3.8}$$

so that, $(Y(\zeta)_t;\ 0\leq t\leq T)$ is in $M^2(0,T)$. In addition, we have

**Proposition 3.5** *There exists $0<C<\infty$ depending only on $T$, $\|(\sigma^{-1}b)(\sigma^{-1}b)^*\|_{L^\infty}$, $\|\ell\|_{L^\infty}$ such that, for all $\zeta,\xi\in M^2(0,T)$, for all $0\leq t\leq T$,*

$$\mathbb{E}_{\mathbb{Q}}\left[|Y(\zeta)_t-Y(\xi)_t|^2\right]\leq C\int_0^t\mathbb{E}_{\mathbb{Q}}\left[\frac{|Z(\zeta)_s-Z(\xi)_s|^2}{|Z(\xi)_s|^2}+\frac{|Z(\zeta)_s-Z(\xi)_s|^2}{|Z(\zeta)_s|^2}+|\zeta_s-\xi_s|^2\right]ds.$$

*Proof* Applying Itô's formula, we get that

$$\begin{aligned}&|Y(\zeta)_t-Y(\xi)_t|^2\\&=2\int_0^t(Y(\zeta)_s-Y(\xi)_s)\left(Z(\zeta)_s\mathbb{E}_{\mathbb{Q}}\left[Z(\zeta)_s^{-1}\ell(\zeta_s)\,|\,X_s\right]-Z(\xi)_s\mathbb{E}_{\mathbb{Q}}\left[Z(\xi)_s^{-1}\ell(\xi_s)\,|\,X_s\right]\right)ds\\&\quad+\int_0^t\left(Z(\zeta)_s\mathbb{E}_{\mathbb{Q}}\left[Z(\zeta)_s^{-1}\gamma(\zeta_s)\,|\,X_s\right]-Z(\xi)_s\mathbb{E}_{\mathbb{Q}}\left[Z(\xi)_s^{-1}\gamma(\xi_s)\,|\,X_s\right]\right)dW_s\end{aligned}$$

$$+\int_0^t \operatorname{Trace}\left(\left(Z(\zeta)_s\mathbb{E}_{\mathbb{Q}}\left[Z(\zeta)_s^{-1}\gamma(\zeta_s)\,|\,X_s\right]-Z(\xi)_s\mathbb{E}_{\mathbb{Q}}\left[Z(\xi)_s^{-1}\gamma(\xi_s)\,|\,X_s\right]\right)\right.$$
$$\left.\times\left(Z(\zeta)_s\mathbb{E}_{\mathbb{Q}}\left[Z(\zeta)_s^{-1}\gamma(\zeta_s)\,|\,X_s\right]-Z(\xi)_s\mathbb{E}_{\mathbb{Q}}\left[Z(\xi)_s^{-1}\gamma(\xi_s)\,|\,X_s\right]\right)^*\right)ds.$$

Taking the expectation on both sides of the preceding equality ($L^2$ integrability is again ensured from the boundedness of the coefficients combined with Identity (3.3)), it follows

$$\begin{aligned}
&\mathbb{E}_{\mathbb{Q}}\left[|Y(\zeta)_t-Y(\xi)_t|^2\right]\\
&=2\int_0^t \mathbb{E}_{\mathbb{Q}}\left[(Y(\zeta)_s-Y(\xi)_s)\left(Z(\zeta)_s\mathbb{E}_{\mathbb{Q}}\left[Z(\zeta)_s^{-1}\ell(\zeta_s)\,|\,X_s\right]-Z(\xi)_s\mathbb{E}_{\mathbb{Q}}\left[Z(\xi)_s^{-1}\ell(\xi_s)\,|\,X_s\right]\right)\right]ds\\
&\quad+\int_0^t \mathbb{E}_{\mathbb{Q}}\left[\operatorname{Trace}\left(\left(Z(\zeta)_s\mathbb{E}_{\mathbb{Q}}\left[Z(\zeta)_s^{-1}\gamma(\zeta_s)\,|\,X_s\right]-Z(\xi)_s\mathbb{E}_{\mathbb{Q}}\left[Z(\xi)_s^{-1}\gamma(\xi_s)\,|\,X_s\right]\right)\right.\right.\\
&\qquad\left.\left.\times\left(Z(\zeta)_s\mathbb{E}_{\mathbb{Q}}\left[Z(\zeta)_s^{-1}\gamma(\zeta_s)\,|\,X_s\right]-Z(\xi)_s\mathbb{E}_{\mathbb{Q}}\left[Z(\xi)_s^{-1}\gamma(\xi_s)\,|\,X_s\right]\right)^t\right)\right]ds.
\end{aligned}\tag{3.9}$$

By Young's inequality, for the first integral on the r.h.s., we have

$$\begin{aligned}
&\int_0^t \mathbb{E}_{\mathbb{Q}}\left[(Y(\zeta)_s-Y(\xi)_s)\left(Z(\zeta)_s\mathbb{E}_{\mathbb{Q}}\left[Z(\zeta)_s^{-1}\ell(\zeta_s)\,|\,X_s\right]-Z(\xi)_s\mathbb{E}_{\mathbb{Q}}\left[Z(\xi)_s^{-1}\ell(\xi_s)\,|\,X_s\right]\right)\right]ds\\
&\le\frac{1}{2}\int_0^t \mathbb{E}_{\mathbb{Q}}\left[|Y(\zeta)_s-Y(\xi)_s|^2\right]ds\\
&\quad+\frac{1}{2}\int_0^t \mathbb{E}_{\mathbb{Q}}\left[\left|Z(\zeta)_s\mathbb{E}_{\mathbb{Q}}\left[Z(\zeta)_s^{-1}\ell(\zeta_s)\,|\,X_s\right]-Z(\xi)_s\mathbb{E}_{\mathbb{Q}}\left[Z(\xi)_s^{-1}\ell(\xi_s)\,|\,X_s\right]\right|^2\right]ds.
\end{aligned}$$

In the last integral, adding and subtracting $Z(\xi)_s\mathbb{E}_{\mathbb{Q}}\left[Z(\zeta)_s^{-1}\ell(\zeta_s)\,|\,X_s\right]$ yields

$$\begin{aligned}
&\mathbb{E}_{\mathbb{Q}}\left[\left|Z(\zeta)_s\mathbb{E}_{\mathbb{Q}}\left[Z(\zeta)_s^{-1}\ell(\zeta_s)\,|\,X_s\right]-Z(\xi)_s\mathbb{E}_{\mathbb{Q}}\left[Z(\xi)_s^{-1}\ell(\xi_s)\,|\,X_s\right]\right|^2\right]\\
&\le 2\mathbb{E}_{\mathbb{Q}}\left[(Z(\zeta)_s-Z(\xi)_s)^2\left(\mathbb{E}_{\mathbb{Q}}\left[Z(\zeta)_s^{-1}\ell(\zeta_s)\,|\,X_s\right]\right)^2\right]\\
&\quad+2\mathbb{E}_{\mathbb{Q}}\left[Z(\zeta)_s^2\left|\mathbb{E}_{\mathbb{Q}}\left[Z(\xi)_s^{-1}\ell(\xi_s)-Z(\zeta)_s^{-1}\ell(\zeta_s)\,|\,X_s\right]\right|^2\right].
\end{aligned}$$

The identity (3.3) then ensures that

$$\begin{aligned}
&\mathbb{E}_{\mathbb{Q}}\left[(Z(\zeta)_s-Z(\xi)_s)^2\left(\mathbb{E}_{\mathbb{Q}}\left[Z(\zeta)_s^{-1}\ell(\zeta_s)\,|\,X_s\right]\right)^2\right]\\
&=\mathbb{E}_{\mathbb{Q}}\left[\frac{(Z(\zeta)_s-Z(\xi)_s)^2}{Z(\zeta)_s^2}\left(Z(\zeta)_s\mathbb{E}_{\mathbb{Q}}\left[Z(\zeta)_s^{-1}\ell(\zeta_s)\,|\,X_s\right]\right)^2\right]\\
&\le\|\ell\|_{L^\infty}\mathbb{E}_{\mathbb{Q}}\left[\frac{(Z(\zeta)_s-Z(\xi)_s)^2}{Z(\zeta)_s^2}\right].
\end{aligned}$$

Using the properties of the conditional expectation, we also have

$$
\begin{aligned}
&\mathbb{E}_{\mathbb{Q}}\left[Z(\zeta)_s^2\left|\mathbb{E}_{\mathbb{Q}}\left[Z(\xi)_s^{-1}\ell(\xi_s)-Z(\zeta)_s^{-1}\ell(\zeta_s)\,|\,X_s\right]\right|^2\right]\\
&\leq \mathbb{E}_{\mathbb{Q}}\left[Z(\zeta)_s^2\mathbb{E}_{\mathbb{Q}}\left[\left|Z(\xi)_s^{-1}\ell(\xi_s)-Z(\zeta)_s^{-1}\ell(\zeta_s)\right|^2\,|\,X_s\right]\right]\\
&\leq 2\|\ell\|_{L^\infty}^2\mathbb{E}_{\mathbb{Q}}\left[Z(\zeta)_s^2\mathbb{E}_{\mathbb{Q}}\left[\left(Z(\xi)_s^{-1}-Z(\zeta)_s^{-1}\right)^2\,\middle|\,X_s\right]\right]\\
&\quad+2\mathbb{E}_{\mathbb{Q}}\left[Z(\zeta)_s^2\mathbb{E}_{\mathbb{Q}}\left[Z(\zeta)_s^{-2}\left|\ell(\xi_s)-\ell(\zeta_s)\right|^2\,\middle|\,X_s\right]\right]
\end{aligned}
\tag{3.10}
$$

where, by Lemma 3.4-*(i)*,

$$
\begin{aligned}
\mathbb{E}_{\mathbb{Q}}\left[Z(\zeta)_s^2\mathbb{E}_{\mathbb{Q}}\left[\left(Z(\xi)_s^{-1}-Z(\zeta)_s^{-1}\right)^2\,\middle|\,X_s\right]\right]&=\mathbb{E}_{\mathbb{Q}}\left[\mathbb{E}_{\mathbb{Q}}\left[Z(\zeta)_s^2\,|\,X_s\right]\left(Z(\xi)_s^{-1}-Z(\zeta)_s^{-1}\right)^2|\ell(\xi_s)|^2\right]\\
&=\mathbb{E}_{\mathbb{Q}}\left[\mathbb{E}_{\mathbb{Q}}\left[Z(\zeta)_s^2\,|\,X_s\right]\frac{(Z(\xi)_s-Z(\zeta)_s)^2}{Z(\xi)_s^2Z(\zeta)_s^2}\right]\\
&\leq C\mathbb{E}_{\mathbb{Q}}\left[\frac{|Z(\xi)_s-Z(\zeta)_s|^2}{Z(\xi)_s^2}\right]
\end{aligned}
$$

and

$$
\begin{aligned}
\mathbb{E}_{\mathbb{Q}}\left[Z(\xi)_s^2\mathbb{E}_{\mathbb{Q}}\left[Z(\xi)_s^{-2}|\ell(\xi_s)-\ell(\zeta_s)|^2\,|\,X_s\right]\right]&=\mathbb{E}_{\mathbb{Q}}\left[\mathbb{E}_{\mathbb{Q}}\left[Z(\xi)_s^2\,|\,X_s\right]Z(\xi)_s^{-2}|\ell(\xi_s)-\ell(\zeta_s)|^2\right]\\
&\leq C\|\ell\|_{Lip}^2\mathbb{E}_{\mathbb{Q}}\left[|\xi_s-\zeta_s|^2\right].
\end{aligned}
$$

Putting the two last upper bounds together, we obtain the following bound for the l.h.s of (3.10),

$$
\begin{aligned}
&\mathbb{E}_{\mathbb{Q}}\left[Z(\zeta)_s^2\left|\mathbb{E}_{\mathbb{Q}}\left[Z(\xi)_s^{-1}\ell(\xi_s)-Z(\zeta)_s^{-1}\ell(\zeta_s)\,|\,X_s\right]\right|^2\right]\\
&\leq C\|\ell\|_{L^\infty}^2\mathbb{E}_{\mathbb{Q}}\left[\frac{|Z(\xi)_s-Z(\zeta)_s|^2}{Z(\xi)_s^2}\right]+C\|\ell\|_{Lip}^2\mathbb{E}_{\mathbb{Q}}\left[|\xi_s-\zeta_s|^2\right].
\end{aligned}
\tag{3.11}
$$

For the second integral in (3.9), again by Young's inequality, we have

$$
\begin{aligned}
&\mathbb{E}_{\mathbb{Q}}\Big[\mathrm{Trace}\Big(\big(Z(\zeta)_s\mathbb{E}_{\mathbb{Q}}\left[Z(\zeta)_s^{-1}\gamma(\zeta_s)\,|\,X_s\right]-Z(\xi)_s\mathbb{E}_{\mathbb{Q}}\left[Z(\xi)_s^{-1}\gamma(\xi_s)\,|\,X_s\right]\big)\\
&\qquad\times\big(Z(\zeta)_s\mathbb{E}_{\mathbb{Q}}\left[Z(\zeta)_s^{-1}\gamma(\zeta_s)\,|\,X_s\right]-Z(\xi)_s\mathbb{E}_{\mathbb{Q}}\left[Z(\xi)_s^{-1}\gamma(\xi_s)\,|\,X_s\right]\big)^t\Big)\Big]\\
&=\sum_{i,j,k=1}^{d}\mathbb{E}_{\mathbb{Q}}\Big[\Big(\Big(Z(\zeta)_s\mathbb{E}_{\mathbb{Q}}\left[Z(\zeta)_s^{-1}\gamma^{i,j}(\zeta_s)\,|\,X_s\right]-Z(\xi)_s\mathbb{E}_{\mathbb{Q}}\left[Z(\xi)_s^{-1}\gamma^{i,j}(\xi_s)\,|\,X_s\right]\Big)\\
&\qquad\times\Big(Z(\zeta)_s\mathbb{E}_{\mathbb{Q}}\left[Z(\zeta)_s^{-1}\gamma^{i,k}(\zeta_s)\,|\,X_s\right]-Z(\xi)_s\mathbb{E}_{\mathbb{Q}}\left[Z(\xi)_s^{-1}\gamma^{i,k}(\xi_s)\,|\,X_s\right]\Big)\Big)\Big]\\
&\leq d\sum_{i,j=1}^{d}\mathbb{E}_{\mathbb{Q}}\left[\left|Z(\zeta)_s\mathbb{E}_{\mathbb{Q}}\left[Z(\zeta)_s^{-1}\gamma^{i,j}(\zeta_s)\,|\,X_s\right]-Z(\xi)_s\mathbb{E}_{\mathbb{Q}}\left[Z(\xi)_s^{-1}\gamma^{i,j}(\xi_s)\,|\,X_s\right]\right|^2\right].
\end{aligned}
$$

Each component of the above sum can be bounded in the same manner than (3.11), replacing $\|\ell\|$ by some $\|\gamma^{i,j}\|$. Putting all together, we get, for some positive constant $C$,

$$\mathbb{E}_{\mathbb{Q}}\left[|Y(\zeta)_t - Y(\xi)_t|^2\right] \leq C\int_0^t \mathbb{E}_{\mathbb{Q}}\left[\frac{|Z(\xi)_s - Z(\zeta)_s|^2}{Z(\xi)_s^2} + \frac{|Z(\xi)_s - Z(\zeta)_s|^2}{Z(\zeta)_s^2}\right] ds$$
$$+ C\int_0^t \mathbb{E}_{\mathbb{Q}}\left[|\xi_s - \zeta_s|^2\right] ds.$$

Combining the result of Proposition 3.5 with Lemma 3.4-$(ii)$, and following the same procedure as for (3.1)–(3.2), we deduce with an appropriate choice of the constant $c$ that

$$\|Y(\zeta) - Y(\xi)\|_c < \|\zeta - \xi\|_c, \ \forall \zeta, \xi \in M^2(0, T).$$

This ensures that the mapping $\zeta \in M^2(0, T) \mapsto Y(\zeta) \in M^2(0, T)$ which assigns to each element $\xi \in M^2(0, T)$, the solution $(Y(\zeta)_t;\ 0 \leq t \leq T)$ given by (3.7) is contracting in $(M^2(0, T), \|\ \|_c)$. This enable us to conclude on the existence and uniqueness of a strong solution to (3.4).

By Girsanov transformation, this also enable us to conclude on the wellposedness of a weak solution to (1.2).

## 3.3 Strong Uniqueness

The strong wellposedness of (1.2) will be given by a direct application of the following theorem due to Veretennikov [35]:

**Theorem 3.6** (Theorem 1, [35]) *Let $b : [0, \infty) \times \mathbb{R}^d \to \mathbb{R}^d$ be a bounded measurable function. Let $\sigma : [0, \infty) \times \mathbb{R}^d \to \mathbb{R}^d \times \mathbb{R}^d$ be such that $a : (t, x) \mapsto a(t, x) = \sigma\sigma^*(t, x)$ is continuous, $x \mapsto a(t, x)$ is uniformly continuous in each compact $K \subset \mathbb{R}^d$, for any $t \in (0, T]$, $0 < T < \infty$, and for some positive $\lambda$*

$$(\xi \cdot a(t, x)\xi) \geq \lambda|\xi|^2, \quad \textit{for all } \xi \in \mathbb{R}^d,\ (t, x) \in [0, \infty) \times \mathbb{R}^d.$$

*Moreover, assume that there exist some Borel functions $\sigma_d \in W^{1,2d+2}_{loc}(\mathbb{R}^n)$, $\sigma_{d+1} \in L^{2d+2}_{loc}((0, \infty); W^{1,2d+2}_{loc}(\mathbb{R}^n))$ and $\sigma_L : (0, t) \times \mathbb{R}^n \times \mathbb{R}^n \to \mathbb{R}^d \times \mathbb{R}^d$, $\mathbb{R}^n \times \mathbb{R}^n$-Lipschitz continuous uniformly for $t \in (0, T]$, $0 < T < \infty$, such that*

$$\sigma(t, x) = \sigma_L(t, \sigma_d(x), \sigma_{d+1}(t, x)).$$

*Then, given a $\mathbb{R}^d$-valued standard Brownian motion $(w_t;\ t \geq 0)$, the stochastic differential equation:*

$$x_t = x + \int_0^t b(s, x_s)\, ds + \int_0^t \sigma(s, x_s) dw_s, \quad x \in \mathbb{R}^d,\ t \geq 0,$$

*has a unique strong solution.*

We are going to show that the nonlinear diffusion coefficient

$$(t, x) \mapsto \mathbb{E}_{\mathbb{P}}[\gamma(Y_t) \mid X_t = x]$$

is continuous and admits a derivative (in the Sobolev sense) w.r.t. $x$ such that $\nabla_x \mathbb{E}_{\mathbb{P}}[\gamma(Y_t) \mid X_t = x]$ is locally in $L^{2d+2}$-integrable on $(0, T) \times \mathbb{R}^d$. Before that, as a preliminary remark, let us point out that owing to **(H0)** and **(H1)**, for all $t \geq 0$, the law of $X_t$ admits a density function $\rho_X(t, x)$. In addition, since $\rho_X(t, x)$ is a weak solution to

$$\partial_t \rho_X + \nabla_x \left(\rho_X B_{\rho_X}\right) - \frac{1}{2} \text{Trace}\left(\nabla_x^2 \times \left(\rho_X \sigma \sigma^*\right)\right) = 0,$$

where $B_{\rho_X} = B_{\rho_X}(t, x)$ is the bounded Borel measurable $\mathbb{R}^d$-vector field given by

$$B_{\rho_X}(t, x) = \mathbb{E}_{\mathbb{P}}\left[b(X_t, Y_t) \mid X_t = x\right],$$

we have the following bounds (see e.g. Aronson [2])

$$\begin{aligned} c \int G_t^{1/\kappa}(x - x_0) \rho_X(0, x_0)\, dx_0 &\leq \rho_X(t, x) \\ \leq C \int G_t^{\kappa}(x - x_0) \rho_X(0, x_0)\, dx_0, &\quad x \in \mathbb{R}^d,\ 0 \leq t \leq T, \end{aligned} \tag{3.12}$$

where $G_t^\kappa$ is the centered Gaussian kernel with variance $\kappa^2 t$ and $\kappa$, $c$, $C$ are some finite positive constants depending only on $T$, $d$, $a_*$ and $a*$. Then, under the assumption **(H4)**, for all $0 < R < \infty$, we have

$$\begin{aligned} \inf_{x \in B(0,R)} \rho_X(t, x) &\geq c \inf_{x \in B(0,R)} \int G_t^{1/\kappa}(x - x_0) \rho_X(0, x_0)\, dx_0 \\ &\geq c \inf_{x \in B(0,R)} \int G_1^{1/\kappa}(x_0) \rho_X(0, x + t x_0)\, dx_0 \\ &\geq c \inf_{x \in B(0,R)} \int_{B(0,R)} G_1^{1/\kappa}(x_0) \rho_X(0, x + t x_0)\, dx_0 \\ &\geq c \inf_{x \in B(0,R), x_0 \in B(0,R), 0 \leq t \leq T} \rho_X(0, x + t x_0) \left( \int_{B(0,R)} G_1^{1/\kappa}(x_0)\, dx_0 \right), \end{aligned}$$

that leads to the following lower bound for $\rho_X$:

$$\inf_{x\in B(0,R)} \rho_X(t,x) \geq m_R > 0, \quad \text{with} \quad m_R := c \inf_{z\in B(0,R+TR)} \rho_X(0,z) \left( \int_{B(0,R)} G_1^{1/\kappa}(x_0)\,dx_0 \right). \tag{3.13}$$

The positiveness of $\rho_X$ ensures that the component $\mathbb{E}_{\mathbb{P}}\left[\gamma(Y_t) \mid X_t = x\right]$ is defined a.e. on $(0,T) \times \mathbb{R}^d$ and writes as a Borel measurable function:

$$(t,x) \mapsto \mathbb{E}_{\mathbb{P}}\left[\gamma(Y_t) \mid X_t = x\right] = \int \gamma(y)\mu^{Y \mid X=x}(t,dy,x), \text{ for a.a. } (t,x) \in (0,T) \times \mathbb{R}^d,$$

where $\int \gamma(y)\mu^{Y \mid X=x}(t,dy,x)$ is the disintegration of $\int \gamma(y)\mu(t,dx,dy)$ for $\mu(t) = Law(X_t,Y_t)$ w.r.t. $\rho_X(t,x)\,dx$. To exhibit the smoothness of $(t,x) \mapsto \mathbb{E}_{\mathbb{P}}\left[\gamma(Y_t) \mid X_t = x\right]$, we prove below a general result showing that any distribution of the form

$$(t,x) \mapsto \int m(y)\mu(t,dx,dy), \quad \text{for any bounded Borel function } m : \mathbb{R}^d \to \mathbb{R},$$

is absolutely continuous w.r.t. the Lebesgue measure on $\mathbb{R}^d$ and its related density is smooth in a suitable Sobolev sense. Such property is precisely given by the following lemma:

**Lemma 3.7** *Assume that Assumptions 3.1 and 3.2 hold. For $(X_t,Y_t;\ 0 \leq t \leq T)$ the weak solution to (1.2), let $\mu(t)$ denote the joint law of $(X_t,Y_t)$ and let $\rho_X(t)$ be the density function of $Law(X_t)$ for $0 \leq t \leq T$. Then, for any Borel measurable function $m : \mathbb{R}^d \to [0,\infty)$ not-identically equal to 0, bounded, of class $C^2(\mathbb{R}^d)$ on $\mathbb{R}^d$ such that its derivatives up to second order are bounded, the family of distributions*

$$\int m(y)\mu(t,dx,dy)$$

*admits a representant denoted $\overline{m\mu}$ in $L^p((0,T);\ W^{2,p}_{loc}(\mathbb{R}^d))$ for any $p \geq d+2$.*

Splitting $y \mapsto \gamma(y)$ into its positive part $(\gamma)^+$ and negative part $(\gamma)^-$, Lemma 3.7 ensures that, $\mathbb{E}_{\mathbb{P}}\left[\gamma(Y_t) \mid X_t = x\right]$ rewrites according to

$$\mathbb{E}_{\mathbb{P}}\left[\gamma(Y_t) \mid X_t = x\right] = \frac{\overline{\gamma\mu}(t,x)}{\rho_X(t,x)} \left( = \frac{\overline{(\gamma)^+\mu}(t,x)}{\rho_X(t,x)} + \frac{\overline{(\gamma)^-\mu}(t,x)}{\rho_X(t,x)} \right).$$

Since Lemma 3.7 also guarantees that $\rho_X$ and $\overline{\gamma\mu}$ are both in $L^p((0,T);\ W^{2,p}_{loc}(\mathbb{R}^d))$ for any $p$ large enough, the lower bound (3.13) further ensures that $(t,x) \mapsto \frac{\overline{\gamma\mu}(t,x)}{\rho_X(t,x)}$ is in $L^p((0,T);\ W^{2,p}_{loc}(\mathbb{R}^d))$. We then conclude on the strong uniqueness of $(X_t,Y_t;\ 0 \leq t \leq T)$ solution to (1.2).

*Proof of Lemma* 3.7. As a preliminary step, let us point out that, since $\rho_X(0)$ is in $L^1(\mathbb{R}^d)\cap L^p(\mathbb{R}^d)$ for $p\geq 2d+2$, then $\rho_X(0)\in L^r(\mathbb{R}^d)$ for all $1\leq r\leq p$. Combined with the Gaussian upper-bound in (3.12), this estimate ensures that, whenever $g:\mathbb{R}^d\mapsto\mathbb{R}$ is bounded, for all $0\leq t\leq T$,

$$\left|\int_{\mathbb{R}^d}\psi(x)\int_{\mathbb{R}^d}g(y)\mu(t,dx,dy)\right|\leq C\|g\|_{L^\infty(\mathbb{R}^d)}\|\psi\|_{L^q(\mathbb{R}^d)},\ 1\leq\frac{q}{q-1}\leq 2d+2,\ \psi\in\mathcal{C}_c^\infty(\mathbb{R}^d).$$

Riesz's representation theorem then implies that

$$\int_{\mathbb{R}^d}g(y)\mu(t,dx,dy)=\overline{g\mu}(t,x)\,dx, \tag{3.14}$$

for some $\overline{g\mu}$ in $L^\infty((0,T);L^r(\mathbb{R}^d))$.

By Itô formula, for all $\phi\in\mathcal{C}_c^\infty((0,T)\times\mathbb{R}^d)$, we have

$$\begin{aligned}&\mathbb{E}_\mathbb{P}\left[\int_0^T m(Y_t)\left(\partial_t\phi(t,X_t)+b(X_t,Y_t)\cdot\nabla_x\phi(t,X_t)+\frac{1}{2}\text{Trace}\left(\sigma\sigma^*(X_t)\nabla_x^2\phi(t,X_t)\right)\right)dt\right]\\&+\mathbb{E}_\mathbb{P}\left[\int_0^T\phi(t,X_t)\left(\mathbb{E}_\mathbb{P}\left[\ell(Y_t)\,|\,X_t\right]\cdot\nabla_y m(Y_t)+\frac{1}{2}\text{Trace}\left(\nabla_y^2 m(Y_t)\left(\mathbb{E}_\mathbb{P}\left[\gamma(Y_t)\,|\,X_t\right]\mathbb{E}_\mathbb{P}\left[\gamma^*(Y_t)\,|\,X_t\right]\right)\right)\right)dt\right]=0.\end{aligned}$$

Rewriting the preceding expression into

$$\begin{aligned}&\int_{(0,T)\times\mathbb{R}^{2d}}m(y)\left(\partial_t\phi(t,x)+b(x,y)\cdot\nabla_x\phi(t,x)+\frac{1}{2}\text{Trace}\left(\sigma\sigma^*(x)\nabla_x^2\phi(t,x)\right)\right)\mu(t,dx,dy)\,dt\\&+\int_{(0,T)\times\mathbb{R}^{2d}}\phi(t,x)\left(\mathbb{E}_\mathbb{P}\left[\ell(Y_t)\,|\,X_t=x\right]\cdot\nabla_y m(y)\right)\mu(t,dx,dy)\,dt\\&+\frac{1}{2}\int_{(0,T)\times\mathbb{R}^{2d}}\phi(t,x)\text{Trace}\left(\nabla_y^2 m(y)\mathbb{E}_\mathbb{P}\left[\gamma(Y_t)\,|\,X_t=x\right]\mathbb{E}_\mathbb{P}\left[\gamma^*(Y_t)\,|\,X_t=x\right]\right)\mu(t,dx,dy)\,dt=0,\end{aligned}$$

we deduce that

$$\overline{m\mu}(t,dx):=\int m(y)\mu(t,dx,dy),$$

satisfies

$$\begin{aligned}&\int_{(0,T)\times\mathbb{R}^d}\left(\partial_t\phi(t,x)+\frac{1}{2}\text{Trace}\left(\sigma\sigma^*(x)\nabla_x^2\phi(t,x)\right)\right)\overline{m\mu}(t,dx)\,dt\\&+\int_{(0,T)\times\mathbb{R}^d}\left(\nabla_x\phi(t,x)\cdot\int m(y)b(x,y)\mu(t,dx,dy)\right)dt\\&+\int_{(0,T)\times\mathbb{R}^d}\left(\phi(t,x)\mathbb{E}_\mathbb{P}\left[\ell(Y_t)\,|\,X_t=x\right]\cdot\int\nabla_y m(y)\mu(t,dx,dy)\right)dt\\&+\frac{1}{2}\int_{(0,T)\times\mathbb{R}^{2d}}\phi(t,x)\text{Trace}\left(\mathbb{E}_\mathbb{P}\left[\gamma(Y_t)\,|\,X_t=x\right]\mathbb{E}_\mathbb{P}\left[\gamma^*(Y_t)\,|\,X_t=x\right]\nabla_y^2 m(y)\right)\mu(t,dx,dy)\,dt=0.\end{aligned}$$

Since $b$, $\ell$, $\gamma$, $m$, $\nabla_y m$ and $\nabla_y^2 m$ are all bounded, we have

$$\int_{(0,T)\times\mathbb{R}^d}\left(\partial_t\phi(t,x)+\frac{1}{2}\text{Trace}\left(\sigma\sigma^*(x)\nabla_x^2\phi(t,x)\right)\right)\overline{m\mu}(t,dx)\,dt$$
$$+\int_{(0,T)\times\mathbb{R}^d}(E_1[t,x]+E_2[t,x])\,\phi(t,x)\overline{m\mu}(t,dx)\,dt \tag{3.15}$$
$$=-\int_{(0,T)\times\mathbb{R}^d}(B[t,x]\cdot\nabla_x\phi(t,x))\,\overline{m\mu}(t,dx)\,dt,$$

where, following the convention in (3.14), $B[t,x]=(B^{(i)}[t,x];\,1\le i\le d)$, $E_1[t,x]$ and $E_1[t,x]$ are given by

$$B^{(i)}[t,x]=\frac{\overline{mb^{(i)}\mu}}{\overline{m\mu}}(t,x)\mathbb{1}_{\{\overline{m\mu}(t,x)\neq 0\}},$$
$$E_1[t,x,\rho(t)]=\sum_{i=1}^{d}\mathbb{E}_{\mathbb{P}}\left[\ell^i(Y_t)\,|\,X_t=x\right]\frac{\overline{(\partial_{y_i}m)\mu}}{\overline{m\mu}}(t,x)\mathbb{1}_{\{\overline{m\mu}(t,x)\neq 0\}},$$
$$E_2[t,x]=\sum_{i,j=1}^{d}\left(\mathbb{E}_{\mathbb{P}}\left[\gamma(Y_t)\,|\,X_t=x\right]\mathbb{E}_{\mathbb{P}}\left[\gamma^*(Y_t)\,|\,X_t=x\right]\right)^{i,j}\frac{\overline{(\partial^2_{y_iy_j}m)\mu}}{\overline{m\mu}}(t,x)\mathbb{1}_{\{\overline{m\mu}(t,x)\neq 0\}}.$$

Since $m$ is not identically equal to 0 on $\mathbb{R}^d$, (3.12) implies that $\overline{m\mu}(t,x)\neq 0$ a.e. on $(0,T)\times\mathbb{R}^d$ and that the fraction in each of the above functions are defined a.e. on $(0,T)\times\mathbb{R}^d$.) Noticing that $B$, $E_1$, $E_2$ are all locally bounded, the continuity of $\overline{m\mu}$ on $(0,T)\times\mathbb{R}^d$ and its local Sobolev regularity then follow from the application of the following results from Bogachev et al. [3].

**Proposition 3.8** [Corollaries 6.4.3 and 6.4.4 in [3]] *Let $\mathcal{D}\subset\mathbb{R}^d$ be an open set. Assume that $a:(0,T)\times\mathbb{R}^d\to\mathbb{R}^{d\times d}$ is uniformly elliptic such that $a$ and $a^{-1}$ are locally bounded on $(0,T)\times\mathbb{R}^d$ and that, for $p>d+2$,*

$$\sup_{1\le i,j\le d}\sup_t\|a^{i,j}\|_{W^{1,p}(B(x_0,R))}<\infty\ \ \textit{for all}\ x_0\in\mathcal{D},\ 0<R<\infty.$$

*Assume that $b^1,b^2,\cdots,b^d$, $f^1,f^2,\cdots,f^d$ are in $L^p_{loc}((0,T)\times\mathcal{D})$, $c$ is in $L^{p/2}_{loc}((0,T)\times\mathcal{D})$, and assume that $\mu$ is a locally finite Borel measure on $(0,T)\times\mathbb{R}^d$ such that*

$$\int_{(0,T)\times\mathcal{D}}\left(\partial_t\phi+Trace(a\nabla^2\phi)+b\cdot\nabla\phi+c\phi\right)\mu(dt,dx)=\int_{(0,T)\times\mathcal{D}}f\cdot\nabla\phi,\ \forall\psi\in C_c^\infty((0,T)\times\mathcal{D}).$$

*Then $\mu$ has a locally Hölder continuous density that belongs to the space $L^p(J;W^{1,p}$
$(V))$ for all $J\subset(0,T)$, $V\subset\mathcal{D}$ such that $J\times V$ has compact closure in $(0,T)\times\mathcal{D}$.*

**Acknowledgements** The second author has been supported by the Russian Academic Excellence Project "5-100".

## 4 Appendix

The proof of the estimate (2.21) relies on the original arguments exhibited in [31] (pp. 304–306) in a one-dimensional setting. We simply extend the result to a multi-dimensional case: For $\delta > 0$, define

$$\beta_\delta(t) = \mathbb{1}_{\{\delta \le t \le 1/\delta\}}, \; G_t^{\gamma,\delta}(x, y) = \beta_\delta(t) G_t^\gamma(x, y).$$

Let $f$ be a $\mathcal{C}_c^\infty((0, T) \times \mathbb{R}^d)$-function so that $(t, x) \mapsto \int_0^t \Delta_x G_{t-s}^\gamma(f)(s, x)\, ds = \int_0^t \Delta_x G_{t-s}^\gamma * f(s, x)\, ds$ (for simplicity $G_t^\gamma$ denotes the $\mathcal{N}(0, \gamma^2 t)$-Gaussian density function/kernel) is $\mathcal{C}^\infty((0, T) \times \mathbb{R}^d) \cap L^2((0, T) \times \mathbb{R}^d)$. By Parseval's equality: $\|h\|_{L^2(\mathbb{R}^m)} = \frac{1}{(2\pi)^m} \|\mathcal{F}(h)\|_{L^2(\mathbb{R}^m)}$, $h \in L^2(\mathbb{R}^m)$, $m \ge 1$, we get

$$\Big\| \int_0^\cdot \Delta_x G_{\cdot-s}^{\gamma,\delta} * f(s, x)\, ds \Big\|_{L^2((0,T)\times\mathbb{R}^d)} = \frac{1}{(2\pi)^{d+1}} \big\| \mathcal{F}(\Delta_x G^{\gamma,\delta}) \mathcal{F}(f \mathbb{1}_{\{[0,T]\}}) \big\|_{L^2(\mathbb{R}^{d+1})}$$

where $\mathcal{F}$ denote the Fourier transformation along the variables $t$ and $x$:

$$\mathcal{F}(f \mathbb{1}_{\{[0,T]\}})(\tau, \xi) = \int_{\mathbb{R}} \int_{\mathbb{R}^d} \exp\{-\mathbf{i} t\tau - \mathbf{i} x \cdot \xi\} f(t, x) \mathbb{1}_{[0,T]}\, dt\, dx.$$

Since

$$\mathcal{F}(\Delta_x G^{\gamma,\delta})(\tau, \xi) = -|\xi|^2 \mathcal{F}(G^{\gamma,\delta})(\tau, \xi)$$

with

$$\begin{aligned} \mathcal{F}(G^{\gamma,\delta})(\tau, \xi) &= \int_0^T \beta_\delta(t) \exp\{-\mathbf{i} t\tau\} \left( \int_{\mathbb{R}^d} \exp\{-\mathbf{i}\xi \cdot x\} G^\gamma(t, x)\, dx \right) dt \\ &= \int_0^T \beta_\delta(t) \exp\{-\mathbf{i} t\tau\} \exp\{-t\gamma^2 |\xi|^2/2\}\, dt \\ &= \frac{2}{2\mathbf{i}\tau - \gamma^2 |\xi|^2} \left( \exp\{-\delta(\mathbf{i}\tau - \gamma^2|\xi|^2/2)\} - \exp\{-\frac{(\mathbf{i}\tau - \gamma^2|\xi|^2/2)}{\delta}\} \right), \end{aligned}$$

we get

$$\begin{aligned} &|\xi|^2 \left| \mathcal{F}(\Delta_x G^{\gamma,\delta})(\tau, \xi) \mathcal{F}(f \mathbb{1}_{[0,T]})(\tau, \xi) \right| \\ &\le \left| \frac{-2|\xi|^2}{2\mathbf{i}\tau - \gamma^2|\xi|^2} \right| |\mathcal{F}(f \mathbb{1}_{[0,T]})|(\tau, \xi) \le \frac{2}{\gamma^2} |\mathcal{F}(f \mathbb{1}_{[0,T]})|(\tau, \xi). \end{aligned}$$

Integrating both sides of the preceding inequality over $\mathbb{R}^{d+1}$, it follows that

$$\frac{1}{(2\pi)^{d+1}}\Big\|\int_0^{\cdot}\mathcal{F}(\Delta_x G^{\gamma,\delta})\mathcal{F}(f\mathbb{1}_{[0,T]})\Big\|_{L^2((0,T)\times\mathbb{R}^d)} \leq \frac{2}{\gamma^2(2\pi)^{d+1}}\big\|\mathcal{F}(f\mathbb{1}_{[0,T]})\big\|_{L^2((0,T)\times\mathbb{R}^d)}$$
$$\leq \frac{2}{\gamma^2}\|f\|_{L^2((0,T)\times\mathbb{R}^d)}.$$

Since $\int_0^{\cdot}\Delta_x G^{\gamma}_{\cdot-s}(f(s))\,ds = \lim_{\delta\to 0^+}\int_0^{\cdot}\Delta_x G^{\gamma,\delta}_{\cdot-s}(f(s))\,ds$, we deduce, by extension, that

$$\Big\|\int_0^{\cdot}\Delta_x G^{\gamma}_{\cdot-s}(f(s))\,ds\Big\|_{L^2((0,T)\times\mathbb{R}^d)} \leq \frac{2}{\gamma^2}\|f\|_{L^2((0,T)\times\mathbb{R}^d)},\ \forall f\in\mathcal{C}^\infty_c((0,T)\times\mathbb{R}^d).$$

Since $\mathcal{C}^\infty_c((0,T)\times\mathbb{R}^d)$ is dense in $L^2((0,T)\times\mathbb{R}^d)$, we conclude (2.21).

In the same way, for any given positive definite $\mathbb{R}^{d\times d}$-matrix $\Gamma$ satisfying $\xi\cdot\Gamma\xi\geq\gamma^2|\xi|^2$ for some $\gamma\neq 0$, let $G^\Gamma_t$ be the $\mathcal{N}(0,\Gamma t)$-Gaussian density function and define, for $\beta_\delta$ as above,

$$G^{\Gamma,\delta}_t = \beta_\delta(t)G^\Gamma_t.$$

Observing that

$$\Big|\mathcal{F}(\partial^2_{x_i x_j}G^{\Gamma,\delta})(\tau,\xi)\Big| = |\xi_i\xi_j|\,|\mathcal{F}(G^{\Gamma,\delta})(\tau,\xi)|$$
$$= |\xi_i\xi_j|\Big|\int_0^T\beta_\delta(t)\exp\{-\mathbf{i}t\tau\}\exp\{-t(\xi\cdot\Gamma\xi)/2\}\,dt\Big| \leq \frac{2\,|\xi_i\xi_j|}{|2\mathbf{i}\tau-(\xi\cdot\Gamma\xi)|}\leq\frac{2}{\gamma^2},$$

then, for any family $F^{i,j}\in\mathcal{C}^\infty_c((0,T)\times\mathbb{R}^d)$, $1\leq i,j\leq d$, we have

$$|\xi|^2\Big|\mathcal{F}(\partial^2_{x_i x_j}G^{\Gamma,\delta})(\tau,\xi)\mathcal{F}(F^{i,j}\mathbb{1}_{[0,T]})(\tau,\xi)\Big| \leq \frac{2}{\gamma^2}|\mathcal{F}(F^{i,j}\mathbb{1}_{[0,T]})|(\tau,\xi)$$

from which we deduce, as previously, that

$$\Big\|\sum_{i,j=1}^d\int_0^{\cdot}\partial^2_{x_i x_j}G^\Gamma_{\cdot-s}(F^{i,j}(s))\,ds\Big\|_{L^2((0,T)\times\mathbb{R}^d)} = \Big\|\lim_{\delta\to 0^+}\sum_{i,j=1}^d\int_0^{\cdot}\partial^2_{x_i x_j}G^{\Gamma,\delta}_{\cdot-s}(F^{i,j}(s))\,ds\Big\|_{L^2((0,T)\times\mathbb{R}^d)}$$
$$\leq\frac{2}{\gamma^2}\sum_{i,j=1}^d\|F^{i,j}\|_{L^2((0,T)\times\mathbb{R}^d)}.$$

## References


1. Abergel, F., Tachet, R.: A nonlinear partial integro-differential equation from mathematical Finance. Discrete Continuous Dyn. Syst.-Series A (DCDS-A) **27**(3), 907–917 (2010)
2. Aronson, D.G.: Bounds for the fundamental solution of a parabolic equation. Bull. Amer. Math. Soc. **73**, 890–896 (1967)
3. Bogachev, V.I., Krylov, N.V., Röckner, M., Shaposhnikov, S.: Fokker-Planck-Kolmogorov Equations. American Mathematical Society (2015)

4. Bossy, M., Jabir, J.-F., Talay, D.: On conditional McKean Lagrangian stochastic models. Probab. Theor. Relat. Fields **151**(1–2), 319–351 (2011)
5. Bossy, M., Jabir, J.-F.: Lagrangian stochastic models with specular boundary condition. J. Funct. Anal. **268**(6), 1309–1381 (2015)
6. Bossy, M., Espina, J., Morice, J., Paris, C., Rousseau, A.: Modeling the wind circulation around mills with a Lagrangian stochastic approach. SMAI-J. Comput. Math. **2**, 177–214 (2016)
7. Bossy, M., Dupré, A., Drobinski, P., Violeau, L., Briard, C.: Stochastic Lagrangian approach for wind farm simulation (2018). hal.inria.fr
8. Champagnat, N., Jabin, P.-E.: Strong solutions to stochastic differential equations with rough coefficients. To appear in Annals of Probability (2018)
9. Chorin, A.J.: Numerical study of slightly viscous flows. J. Fluid Mech. **57**, 785–796 (1973)
10. Durbin, P.A., Speziale, C.G.: Realizability of second moment closure via stochastic analysis. J. Fluid Mech. **280**, 395–407 (1994)
11. Evans, L.C.: Partial Differential Equations. American Mathematical Institute (1997)
12. Figalli, A.: Existence and uniqueness of martingale solutions for SDEs with rough or degenerate coefficients. J. Funct. Anal. **254**(1), 109–153 (2008)
13. Funaki, T.: A certain class of diffusion processes associated with nonlinear parabolic equations. Z. Wahrsch. Verw. Gebiete **67**(3), 331–348 (1984)
14. Guyon, J., Henry-Labordère, P.: The smile calibration problem solved. SSRN Electron. J. (2011)
15. Gyöngy, I.: Mimicking the one-dimensional marginal distributions of processes having an Itô Differential. Probab. Theor. Relat. Fields **71**, 501–516 (1986)
16. Jourdain, B., Méléard, S.: Propagation of chaos and fluctuations for a moderate model with smooth initial data. Ann. Inst. H. Poincaré Probab. Statist. **34**(6), 726–766 (1998)
17. Jourdain, B., Reygner, J.: Propagation of chaos for rank-based interacting diffusions and long time behaviour of a scalar quasilinear parabolic equation. Stochast. Partial Differ. Eq. Anal. Comput. **1**(3), 455–506 (2013)
18. Jourdain, B., Zhou, A.: Existence of a calibrated regime switching local volatility model and new fake Brownian motions. Preprint (2016)
19. Fournier, N., Jourdain, B.: Stochastic particle approximation of the Keller-Segel equation and two-dimensional generalization of Bessel processes. Ann. Appl. Probab. **27**(5), 2807–2861 (2017)
20. Kohatsu-Higa, A., Ogawa, S.: Weak rate of convergence for an Euler scheme of nonlinear SDE's. Monte Carlo Meth. Appl. **3**(4), 327–345 (1997)
21. Krylov, N.V.: Lecture on Elliptic and Parabolic Equations in Sobolev Spaces. American Mathematical Society (2008)
22. Ladyženskaja, O.A., Solonnikov, V.A., Uralćeva, N.: Linear and Quasi-linear Equations of Parabolic Type. American Mathematical Society, coll. Translations of Mathematical Monographs (1968)
23. Lions, J.-L.: Equations Différentielles Opérationnelles et Problèmes aux Limites. Grundlehren der mathematischen Wissenschaften (1961)
24. Méléard, S.: Asymptotic behaviour of some interacting particle systems; McKean-Vlasov and Boltzmann models. In: Probabilistic Models for Nonlinear Partial Differential Equations (Montecatini Terme, 1995). Lecture Notes in Mathematics, vol. 1627, pp. 42–95 (1996)
25. Méléard, S.: Monte-Carlo approximation for 2d Navier-Stokes equations with measure initial data. Probab. Theory Relat. Fields **121**, 367–388 (2001)
26. Mishura, Y.S., Veretennikov, A.Y.: Existence and uniqueness theorems for solutions of McKean-Vlasov stochastic equations. Preprint (2017)
27. Oelschläger, K.: A martingale approach to the law of large numbers for weakly interacting stochastic processes. Ann. Probab. **12**(2), 458–497 (1984)
28. Oelschläger, K.: A law of large numbers for moderately interacting diffusion processes. Zeitschrift für Wahrscheinlichkeitstheorie und verwandte Gebiete **69**(2), 279–322 (1985)
29. Pope, S.B.: Lagrangian pdf methods for turbulent flows. Annu. Rev. Fluid Mech. **26**, 23–63 (1994)

30. Pope, S.B.: Turbulent Flows, 11th edn. Cambridge University Press (2011)
31. Stroock, D., Varadhan, S.R.: Multidimensional Diffusion Processes. Springer-Verlag (1979)
32. Sznitman, A.S.: A propagation of chaos result for Burgers' equation. Probab. Theor. Relat. Fields **71**(4), 581–613 (1986)
33. Sznitman, A.S.: Topics in Propagation of Chaos. In: École d'Été de Probabilités de Saint-Flour XIX-1989, pp. 165–251. In: Lecture Notes in Mathematics, 1464. Springer (1989)
34. Vasquez, J.L.: The Porous Medium Equation. Oxford University Publications (2006)
35. Veretennikov, A., Yu: On strong solutions and explicit formulas for solutions of stochastic integral equations. Mat. Sb. (N.S.) **111**(3), 434–452 (1980)

# On the Uniqueness of Solutions to Quadratic BSDEs with Non-convex Generators

**Philippe Briand and Adrien Richou**

**Abstract** In this paper we prove some uniqueness results for quadratic backward stochastic differential equations without any convexity assumptions on the generator. The bounded case is revisited while some new results are obtained in the unbounded case when the terminal condition and the generator depend on the path of a forward stochastic differential equation. Some of these results are based on strong estimates on $Z$ that are interesting on their own and could be applied in other situations.

**Keywords** Backward stochastic differential equations · Generator of quadratic growth · Unbounded terminal condition · Uniqueness result

**AMS Subject Classifications** 60H10

## 1 Introduction

In this paper, we consider the following quadratic backward stochastic differential equation (BSDE in short for the remaining of the paper)

$$Y_t = \xi + \int_t^T f(s, Y_s, Z_s)ds - \int_t^T Z_s dW_s, \quad 0 \leqslant t \leqslant T, \tag{1.1}$$

where the generator $f$ has a quadratic growth with respect to $z$. In [11] Kobylanski studied the case where $\xi$ and the random part of $f$ are bounded. She proved the existence of a solution $(Y, Z)$ such that $Y$ is bounded and she get that this solution is unique amongst solutions $(\tilde{Y}, \tilde{Z})$ such that $\tilde{Y}$ is bounded. The unbounded case

P. Briand
Université Grenoble Alpes, Université Savoie Mont Blanc, CNRS, LAMA, 73000 Chambéry, France
e-mail: philippe.briand@univ-smb.fr

A. Richou (✉)
Université Bordeaux, IMB,UMR 5251, 33400 Talence, France
e-mail: adrien.richou@math.univ-bordeaux.fr





S. N. Cohen et al. (eds.), *Frontiers in Stochastic Analysis - BSDEs, SPDEs and their Applications*, Springer Proceedings in Mathematics & Statistics 289,
https://doi.org/10.1007/978-3-030-22285-7_3

was investigated in [3] where authors obtained an existence result. The problem of uniqueness in the unbounded framework was tackled in [4–6] by assuming that $f$ is a convex function with respect to $z$. The case of a non-convex generator $f$ was treated in [12] but uniqueness results where obtained in some classes involving bounds on $Z$.

The main contribution of this paper is to strengthen these uniqueness results. Concerning the bounded case, we are able to expand the class of uniqueness: the bounded solution obtained by Kobylanski is unique amongst solutions $(\tilde{Y}, \tilde{Z})$ such that $\tilde{Y}$ has a specific exponential moment. In the unbounded framework, we are able to relax the convexity assumption on the generator by assuming that the terminal condition and the random part of the generator depend on the path of a forward stochastic differential equation

$$X_t = x + \int_0^t b(X_s)ds + \int_0^t \sigma(s, X_s)dW_s.$$

Moreover, the class of uniqueness only involves the process $Y$. To get into the details, two different situations are investigated.

- When $\sigma$ only depends on $s$, we can deal with a terminal condition and a generator that are locally Lipschitz functions of the path of $X$. This uniqueness result relies on a strong estimate on $Z$ given by

$$|Z_t| \leqslant C(1 + \sup_{s\in[0,t]} |X_s|^r), \quad d\mathbb{P} \otimes dt \text{ a.e.}$$

  This estimate is a generalization of an estimate obtained in [12] in the Markovian framework and is interesting on its own.
- When $\sigma$ depends on $X$, we start by the case of a terminal condition and a generator that are Lipschitz functions of the path of $X$. In this case, we are able to show that $Z$ is bounded $d\mathbb{P} \otimes dt$ a.e. which is also a new estimate interesting on its own.

Let us emphasize that, in these two situations, we are able to get a uniqueness result, even if we add a bounded random variable to the terminal condition and a bounded process to the generator.

The paper is organized as follows. In Sect. 2, we prove some elementary theoretical uniqueness results that will be usefull in the following of the article. Finally, Sect. 3 is devoted to the different frameworks detailed previously: the bounded case and the two different unbounded cases.

Let us close this introduction by giving the notations that we will use in all the paper. For the remaining of the paper, let us fix a nonnegative real number $T > 0$. First of all, $(W_t)_{t\in[0,T]}$ is a standard Brownian motion with values in $\mathbb{R}^d$ defined on some complete probability space $(\Omega, \mathcal{F}, \mathbb{P})$. $(\mathcal{F}_t)_{t\geqslant 0}$ is the natural filtration of the Brownian motion $W$ augmented by the $\mathbb{P}$-null sets of $\mathcal{F}$. The sigma-field of predictable subsets of $[0, T] \times \Omega$ is denoted by $\mathcal{P}$.

By a solution to the BSDE (1.1) we mean a pair $(Y_t, Z_t)_{t\in[0,T]}$ of predictable processes with values in $\mathbb{R}\times\mathbb{R}^{1\times d}$ such that $\mathbb{P}$-a.s., $t\mapsto Y_t$ is continuous, $t\mapsto Z_t$ belongs to $L^2(0,T)$, $t\mapsto f(t,Y_t,Z_t)$ belongs to $L^1(0,T)$ and $\mathbb{P}$-a.s. $(Y,Z)$ verifies (1.1). The terminal condition $\xi$ is $\mathcal{F}_T$-measurable.

For any real $p\geqslant 1$, $\mathcal{S}^p$ denotes the set of real-valued, adapted and càdlàg processes $(Y_t)_{t\in[0,T]}$ such that

$$\|Y\|_{\mathcal{S}^p} := \mathbb{E}\left[\sup_{0\leqslant t\leqslant T}|Y_t|^p\right]^{1/p} < +\infty.$$

$\mathcal{M}^p$ denotes the set of (equivalent classes of) predictable processes $(Z_t)_{t\in[0,T]}$ with values in $\mathbb{R}^{1\times d}$ such that

$$\|Z\|_{\mathcal{M}^p} := \mathbb{E}\left[\left(\int_0^T |Z_s|^2\,ds\right)^{p/2}\right]^{1/p} < +\infty.$$

We will use the notation $Y^* := \sup_{0\leqslant t\leqslant T}|Y_t|$ and by $\mathcal{S}^\infty$ we denote the set of adapted càdlàg processes such that $Y^*$ belongs to $L^\infty$.

Let us recall that a continuous local martingale is bounded in mean oscillations if

$$||M||_{\mathrm{BMO}_2} = \sup_\tau \left|\left|\mathbb{E}[\langle M\rangle_T - \langle M\rangle_\tau|\mathcal{F}_\tau]^{1/2}\right|\right|_\infty < \infty$$

where the supremum is taken over all stopping time $\tau\leq T$. We refer to [10] for further details on BMO-martingales.

Finally, $\mathbb{D}^{1,2}$ stands for the set of random variables $X$ which are differentiable in the Malliavin sense and such that

$$\mathbb{E}\left[|X|^2 + \int_0^T |D_sX|^2\,ds\right] < \infty.$$

Moreover, $\mathbb{L}_{1,2}$ denote the set of real-valued progressively measurable processes $(u_t)_{t\in[0,T]}$ such that

- for a.e. $t\in[0,T]$, $u_t\in\mathbb{D}^{1,2}$,
- $(t,\omega)\mapsto Du_t(\omega)\in L^2([0,T])$ admits a progressively measurable version,
- $\mathbb{E}\left[\left(\int_0^T |u_t|^2dt\right)^{1/2} + \left(\int_0^T\int_0^T |D_\theta u_t|^2 d\theta dt\right)^{1/2}\right] < +\infty.$

## 2 Some Elementary Uniqueness Results

We are looking for a uniqueness result for the BSDE

$$Y_t = \xi + \int_t^T f(s, Y_s, Z_s)ds - \int_t^T Z_s dW_s, \quad 0 \le t \le T, \tag{2.1}$$

where we assume the following assumptions: **(B1)** $f : [0, T] \times \Omega \times \mathbb{R} \times \mathbb{R}^{1\times d} \to \mathbb{R}$ is a measurable function with respect to $\mathcal{P} \otimes \mathcal{B}(\mathbb{R}) \otimes \mathcal{B}(\mathbb{R}^{1\times d})$. There exist two constants $K_y > 0$ and $K_z > 0$ such that, for all $t \in [0, T]$, $y, y' \in \mathbb{R}$, $z, z' \in \mathbb{R}^{1\times d}$

1. $|f(t, y, z) - f(t, y', z)| \leqslant K_y|y - y'|$ a.s.,
2. $\tilde{z} \mapsto f(s, y, \tilde{z})$ is $C^1$ and

$$|\nabla_z f(s, y, z) - \nabla_z f(s, y, z')| \leqslant K_z|z - z'| \quad \text{a.s.}$$

*Remark 2.1* Since we have

$$f(s, 0, z) - f(s, 0, 0) = z \cdot \nabla_z f(s, 0, 0) + |z|^2 \int_0^1 \frac{z\,(\nabla_z f(s, 0, uz) - \nabla_z f(s, 0, 0))}{|z|^2} \mathbb{1}_{z\neq 0} du,$$

we can remark that assumption (B1) implies the following upper bound: for all $\eta > 0$, for all $s \in [0, T]$, $y \in \mathbb{R}$, $z \in \mathbb{R}^{1\times d}$, we have

$$|f(s, y, z)| \leqslant |f(s, 0, 0)| + \frac{|\nabla_z f(s, 0, 0)|^2}{4\eta} + K_y|y| + \left(\frac{K_z}{2} + \eta\right)|z|^2 \quad \text{a.s.}$$

**Theorem 2.2** *Let $p > 1$ and $\varepsilon > 0$ and let us assume the existence of a solution $(Y, Z)$ to* (2.1) *such that*

$$\mathcal{E}_T := e^{\int_0^T \nabla_z f(s,Y_s,Z_s)dW_s - \frac{1}{2}\int_0^T |\nabla_z f(s,Y_s,Z_s)|^2 ds} \in L^p \tag{2.2}$$

*and*

$$\mathbb{E}\left[e^{\frac{2p}{p-1}K_z(1+\varepsilon)|Y^*|}\right] < +\infty. \tag{2.3}$$

*Then, this solution is unique amongst solutions to* (2.1) *such that the exponential integrability* (2.3) *holds true.*

*Proof* Let us consider $(\tilde{Y}, \tilde{Z})$ a solution of (2.1) such that

$$\mathbb{E}\left[e^{\frac{2p}{1-p}K_z(1+\varepsilon)|\tilde{Y}^*|}\right] < +\infty$$

and let us denote $\delta Y := \tilde{Y} - Y$ and $\delta Z := \tilde{Z} - Z$. We get

$$\delta Y_t = 0 + \int_t^T \left[f(s, \tilde{Y}_s, \tilde{Z}_s) - f(s, Y_s, Z_s)\right] ds - \int_t^T \delta Z_s dW_s$$

and we can write

$$f(s, \tilde{Y}_s, \tilde{Z}_s) - f(s, Y_s, Z_s) = b_s \delta Y_s + \delta Z_s \nabla_z f(s, Y_s, Z_s) + a_s |\delta Z_s|^2$$

with

$$b_s := \frac{f(s, \tilde{Y}_s, \tilde{Z}_s) - f(s, Y_s, \tilde{Z}_s)}{\delta Y_s} \mathbb{1}_{|\delta Y_s|>0}$$

and

$$a_s := \int_0^1 \frac{\delta Z_s \left(\nabla_z f(s, Y_s, Z_s + u\delta Z_s) - \nabla_z f(s, Y_s, Z_s)\right)}{|\delta Z_s|^2} \mathbb{1}_{|\delta Z_s|>0} du.$$

Thanks to assumptions (B1) we know that $|b_s| \leqslant K_y$ and $|a_s| \leqslant \frac{K_z}{2}$ for all $s \in [0, T]$. Moreover, since (2.2) is fulfilled, we are allowed to apply Girsanov's theorem: There exists a new probability $\mathbb{Q}$ under which $W^{\mathbb{Q}} := (W_t - \int_0^t \nabla_z f(s, Y_s, Z_s) ds)_{t\in[0,T]}$ is a Brownian motion. Thus, we get

$$\delta Y_t = 0 + \int_t^T \left(b_s \delta Y_s + a_s |\delta Z_s|^2\right) ds - \int_t^T \delta Z_s dW_s^{\mathbb{Q}}, \quad 0 \le t \le T.$$

For any stopping time $\sigma \le T$, setting

$$B_s = e^{\int_0^s b_u \mathbb{1}_{u\ge\sigma} du},$$

we have, from Itô's formula, for any real number $r$,

$$de^{rB_s\delta Y_s} = re^{rB_s\delta Y_s} B_s \delta Z_s \, dW_s^{\mathbb{Q}} + e^{rB_s\delta Y_s} B_s |\delta Z_s|^2 \left(\frac{r^2}{2} B_s - ra_s\right) ds.$$

In particular, if $\tau \ge \sigma$, since $ra_s \le |r| K_z/2$,

$$\begin{aligned} e^{r\delta Y_\sigma} &= e^{rB_\tau \delta Y_\tau} + \int_\sigma^\tau e^{rB_s\delta Y_s} B_s |\delta Z_s|^2 \left(ra_s - \frac{r^2}{2} B_s\right) ds - \int_\sigma^\tau re^{rB_s\delta Y_s} B_s \delta Z_s \, dW_s^{\mathbb{Q}}, \\ &\le e^{rB_\tau \delta Y_\tau} + \frac{|r|}{2} \int_\sigma^\tau e^{rB_s\delta Y_s} B_s |\delta Z_s|^2 \left(K_z - |r| B_s\right) ds - r \int_\sigma^\tau e^{rB_s\delta Y_s} B_s \delta Z_s \, dW_s^{\mathbb{Q}}. \end{aligned} \tag{2.4}$$

For the remaining of the proof we set $\eta = \left((4K_y)^{-1} \log(1+\varepsilon)\right) \wedge T$ which implies in particular that $e^{-K_y\eta} \geqslant (1+\varepsilon)^{-1/4}$. For any $n \in \mathbb{N}^*$ we define the stopping time

$$\tau_n := \inf\left\{t \in [T-\eta, T] \;\middle|\; \int_{T-\eta}^t |\delta Z_s|^2 ds > n\right\}.$$

Let $t \in [T-\eta, T]$ and let us use the inequality (2.4) with $\sigma = t \wedge \tau_n$, $\tau = \tau_n$ and $r = K_z(1+\varepsilon)^{1/2}$. For $\sigma \leq s \leq \tau$,

$$(1+\varepsilon)^{1/4} \geqslant e^{K_y \eta} \geqslant B_s \geqslant e^{-K_y \eta} \geqslant (1+\varepsilon)^{-1/4}.$$

Thus $|r| B_s \geqslant K_z(1+\varepsilon)^{1/4} \geqslant K_z$ and (2.4) gives

$$e^{K_z \sqrt{1+\varepsilon}\, \delta Y_{t\wedge\tau_n}} \leqslant \mathbb{E}_t^{\mathbb{Q}}[e^{K_z \sqrt{1+\varepsilon} B_{\tau_n} \delta Y_{\tau_n}}] \leqslant \mathbb{E}_t^{\mathbb{Q}}[e^{K_z (1+\varepsilon)^{3/4} |\delta Y_{\tau_n}|}]. \tag{2.5}$$

By applying Hölder inequality and by using (2.3) for $Y$ and $\tilde{Y}$, we can remark that

$$\begin{aligned} \mathbb{E}^{\mathbb{Q}}\left[e^{K_z(1+\varepsilon)|\delta Y_{\tau_n}|}\right] &= \mathbb{E}\left[\mathcal{E}_T e^{K_z(1+\varepsilon)|\delta Y_{\tau_n}|}\right] \\ &\leqslant \mathbb{E}\left[\mathcal{E}_T^p\right]^{1/p} \mathbb{E}\left[e^{\frac{p}{p-1} K_z(1+\varepsilon)|\delta Y_{\tau_n}|}\right]^{\frac{p-1}{p}} \\ &\leqslant \mathbb{E}\left[\mathcal{E}_T^p\right]^{1/p} \mathbb{E}\left[e^{\frac{2p}{p-1} K_z(1+\varepsilon)|Y^*|}\right]^{\frac{p-1}{2p}} \mathbb{E}\left[e^{\frac{2p}{p-1} K_z(1+\varepsilon)|\tilde{Y}^*|}\right]^{\frac{p-1}{2p}} < +\infty. \end{aligned} \tag{2.6}$$

Thus, $(e^{K_z(1+\varepsilon)^{3/4}|\delta Y_{\tau_n}|})_{n\in\mathbb{N}}$ is uniformly integrable under $\mathbb{Q}$. Since we clearly have that $\tau_n \to T$ a.s. and $\delta Y_{\tau_n} \to 0$ a.s. when $n \to +\infty$, we get

$$\mathbb{E}_t^{\mathbb{Q}}[e^{K_z(1+\varepsilon)^{3/4}|\delta Y_{\tau_n}|}] \to 1 \quad \text{a.s.}$$

By taking $n \to +\infty$ in (2.5) we finally obtain that $\tilde{Y}_t \leqslant Y_t$ a.s. for all $t \in [T-\eta, T]$. By the same argument (the quadratic term in (2.4) depends on $|r|$), we can also derive the inequality

$$e^{-K_z\sqrt{1+\varepsilon}\delta Y_{t\wedge\tau_n}} \leqslant \mathbb{E}_t^{\mathbb{Q}}[e^{K_z(1+\varepsilon)^{3/4}|\delta Y_{\tau_n}|}], \quad \forall t \in [T-\eta, T],$$

which gives us that $\tilde{Y}_t \geqslant Y_t$ a.s. for all $t \in [T-\eta, T]$. Finally, $\mathbb{E}[\sup_{s\in[T-\eta,T]} |\delta Y_s|^2] = 0$ since $Y$ and $\tilde{Y}$ are continuous a.s. It is clear that we can iterate the proof on intervals $[T-(k+1)\eta, T-k\eta] \cap [0, T]$ for $k \in \mathbb{N}^*$ to get that $\mathbb{E}[\sup_{s\in[0,T]} |\delta Y_s|^2] = 0$, As usual it is sufficient to apply Itô formula to $\delta Y$ to obtain that $\mathbb{E}\left[\int_0^T |\delta Z_s|^2 ds\right] = 0$ which concludes the proof. □

By using same arguments we can also obtain two other versions of this result.

**Theorem 2.3** *We assume the existence of a solution* $(Y, Z)$ *to* (2.1) *such that*

$$\mathcal{E}_T := e^{\int_0^T \nabla_z f(s,Y_s,Z_s) dW_s - \frac{1}{2}\int_0^T |\nabla_z f(Z_s)|^2 ds} \in \bigcap_{p>1} L^p \tag{2.7}$$

*and*

$$e^{K_z|Y^*|} \in \bigcap_{p>1} L^p. \tag{2.8}$$

*Then, this solution is unique amongst solutions* $(Y, Z)$ *to* (2.1) *such that*

$$e^{K_z|Y^*|} \in \bigcup_{p>1} L^p. \tag{2.9}$$

**Theorem 2.4** *We assume the existence of a solution* $(Y, Z)$ *to* (2.1) *such that*

$$\mathcal{E}_T = e^{\int_0^T \nabla_z f(s,Y_s,Z_s)dW_s - \frac{1}{2}\int_0^T |\nabla_z f(Z_s)|^2 ds} \in \bigcup_{p>1} L^p \tag{2.10}$$

*and*

$$e^{K_z|Y^*|} \in \bigcap_{p>1} L^p. \tag{2.11}$$

*Then, this solution is unique amongst solutions* $(Y, Z)$ *to* (2.1) *for which*

$$e^{K_z|Y^*|} \in \bigcap_{p>1} L^p. \tag{2.12}$$

*Proof* The proof of Theorems 2.3 and 2.4 are overall similar to the previous one. We only sketch the proof of Theorem 2.3, the proof of Theorem 2.4 following same lines: we consider $(\tilde{Y}, \tilde{Z})$ a solution of (2.1) such that

$$e^{K_z|\tilde{Y}^*|} \in \bigcup_{p>1} L^p, \tag{2.13}$$

and we show that $Y = \tilde{Y}$ a.s. The only difference is in the inequality (2.6): instead of applying Cauchy-Schwarz inequality, we use Hölder inequality to get, for any $r > 1$ and $p > 1$,

$$\mathbb{E}^{\mathbb{Q}}\left[e^{K_z(1+\varepsilon)\delta Y_{\tau_n}}\right] \leqslant \mathbb{E}\left[\mathcal{E}_T^p\right]^{1/p} \mathbb{E}\left[e^{\frac{rp}{(r-1)(p-1)} K_z(1+\varepsilon)|Y^*|}\right]^{\frac{(r-1)(p-1)}{rp}} \mathbb{E}\left[e^{\frac{rp}{p-1} K_z(1+\varepsilon)|\tilde{Y}^*|}\right]^{\frac{p-1}{rp}}.$$

Then, by taking $p > 1$ large enough, $r > 1$ small enough and $\varepsilon > 0$ small enough we obtain that

$$\mathbb{E}\left[\mathcal{E}_T^p\right]^{1/p} \mathbb{E}\left[e^{\frac{rp}{(r-1)(p-1)} K_z e^{K_y T}(1+\varepsilon)|Y^*|}\right]^{\frac{(r-1)(p-1)}{rp}} \mathbb{E}\left[e^{\frac{rp}{p-1} K_z e^{K_y T}(1+\varepsilon)|\tilde{Y}^*|}\right]^{\frac{p-1}{rp}} < +\infty$$

thanks to (2.7), (2.8) and (2.13). The remaining of the proof stays the same. □

# 3 Applications to Particular Frameworks

## 3.1 The Bounded Case

Since the seminal paper of Kobylanski [11] it is now well known that we have existence and uniqueness of a solution $(Y, Z) \in \mathcal{S}^\infty \times \mathcal{M}^2$ to (2.1) when $\xi$ and $(f(s, 0, 0))_{s\in[0,T]}$ are bounded. We are now able to extend the uniqueness to a larger class of solution.

**Proposition 3.1** *We assume that*

$$M := |\xi|_{L^\infty} + \left| \int_0^T |f(s, 0, 0)| + \sup_{y\in\mathbb{R}} |\nabla_z f(s, y, 0)| ds \right|_{L^\infty} < +\infty.$$

*Then there exists $q > 1$ that depends only on $M$, $K_y$ and $K_z$ such that the BSDE* (2.1) *admits a unique solution $(Y, Z)$ satisfying*

$$\mathbb{E}\left[e^{2K_z q|Y^*|}\right] < +\infty.$$

*In particular, the BSDE* (2.1) *admits a unique solution $(Y, Z)$ satisfying*

$$e^{K_z|Y^*|} \in \bigcap_{p>1} L^p.$$

*Proof* Thanks to Kobylanski [11] we know that the BSDE (2.1) admits a unique solution $(Y, Z) \in \mathcal{S}^\infty \times \mathcal{M}^2$ and this solution satisfies

$$|Y^*|_{L^\infty} + \left| \int_0^\cdot Z_s dW_s \right|_{BMO} < +\infty.$$

It implies that

$$\left| \int_0^\cdot \nabla_z f(s, Y_s, Z_s) dW_s \right|_{BMO} = \left| \sup_{\tau\in[0,T]} \mathbb{E}_\tau \left[ \int_\tau^T |\nabla_z f(s, Y_s, Z_s)|^2 ds \right]^{1/2} \right|_{L^\infty}$$
$$\leqslant \left| \int_0^T \sup_{y\in\mathbb{R}} |\nabla_z f(s, y, 0)| ds \right|_{L^\infty} + K_z \left| \int_0^\cdot Z_s dW_s \right|_{BMO} < +\infty.$$

Then, the reverse Hölder inequality (see e.g. [10]) implies that there exists $p^* > 1$ such that

$$\mathcal{E}_T \in \bigcap_{1\leqslant p<p^*} L^p.$$

Finally we just have to apply Theorem 2.2: for any $\varepsilon > 0$ we have the uniqueness of the solution amongst solutions $(Y, Z)$ that satisfy

$$\mathbb{E}\left[e^{\frac{2p^*}{p^*-1}K_z(1+\varepsilon)|Y^*|}\right] < +\infty. \tag{3.1}$$

*Remark 3.2* It is possible to have an estimate of the exponent $q$ appearing in Proposition 3.1. Indeed, following the proof, the exponent $q$ is a function of $p^*$ and, using the proof of Theorem 3.1 in [10], this exponent $p^*$ is given by

$$p^* := \phi^{-1}\left(\left|\int_0^\cdot \nabla_z f(s, Y_s, Z_s) dW_s\right|_{BMO}\right), \quad \text{with } \phi : p \mapsto \left(1 + \frac{1}{q^2}\log\frac{2q-1}{2q-2}\right)^{1/2} - 1.$$

Moreover, $\left|\int_0^\cdot \nabla_z f(s, Y_s, Z_s) dW_s\right|_{BMO}$ is bounded by an explicit function of $\left|\int_0^\cdot Z_s dW_s\right|_{BMO}$ and we have some estimates of this last quantity, see for example [2].

## 3.2 A First Unbounded Case

In this subsection we consider an SDE with an additive noise

$$X_t = x + \int_0^t b(X_s)ds + \int_0^t \sigma(s)dW_s, \quad 0 \leqslant t \leqslant T, \tag{3.2}$$

where $b$ and $\sigma$ satisfy classical assumptions: **(F1)**

1. $b : \mathbb{R}^d \to \mathbb{R}^d$ is a Lipschitz function: for all $(x, x') \in \mathbb{R}^d \times \mathbb{R}^d$ we have $|b(x) - b(x')| \leqslant K_b|x - x'|$.
2. $\sigma : [0, T] \to \mathbb{R}^{d\times d}$ is a bounded measurable function.

We want to study the following BSDE

$$Y_t = \xi + h((X_s)_{s\in[0,T]}) + \int_t^T f(s, Y_s, Z_s) + g((X_{u\wedge s})_{u\in[0,T]}, Y_s, Z_s)ds - \int_t^T Z_s dW_s, \tag{3.3}$$

with $h : \mathcal{C}([0, T], \mathbb{R}^d) \to \mathbb{R}$, $f : [0, T] \times \Omega \times \mathbb{R} \times \mathbb{R}^{1\times d} \to \mathbb{R}$ and $g : \mathcal{C}([0, T], \mathbb{R}^d) \times \mathbb{R} \times \mathbb{R}^{1\times d} \to \mathbb{R}$ some measurable functions with respect to $\mathcal{B}(\mathcal{C}([0, T], \mathbb{R}^d))$, $\mathcal{P} \otimes \mathcal{B}(\mathbb{R}^d) \otimes \mathcal{B}(\mathbb{R}^{1\times d})$ and $\mathcal{B}(\mathcal{C}([0, T], \mathbb{R}^d)) \otimes \mathcal{B}(\mathbb{R}^d) \otimes \mathcal{B}(\mathbb{R}^{1\times d})$ . We will assume following assumptions: **(B2)**

1.

$$|\xi|_{L^\infty} + \left|\int_0^T |f(s, 0, 0)|ds\right|_{L^\infty} + \left|\sup_{s\in[0,T], y\in\mathbb{R}, \mathbf{x}\in\mathcal{C}([0,T],\mathbb{R}^d)} |\nabla_z f(s, y, 0)| + |\nabla_z g(\mathbf{x}, y, 0)|\right|_{L^\infty} < +\infty$$

2. there exists $C > 0$ such that, for all $t \in [0, T]$, $y \in \mathbb{R}$ and $z \in \mathbb{R}^{1\times d}$,

$$|f(t, y, z)| \leqslant C.$$

3. There exist $K_h > 0$, $K_g > 0$ and $r \in [0,1)$ such that, for all $\mathbf{x}, \tilde{\mathbf{x}} \in C([0,T], \mathbb{R}^d)$, $y \in \mathbb{R}$, $z \in \mathbb{R}^{1\times d}$,

$$|h(\mathbf{x}) - h(\tilde{\mathbf{x}})| \leqslant K_h(1 + |\mathbf{x}|_\infty^r + |\tilde{\mathbf{x}}|_\infty^r)|\mathbf{x} - \tilde{\mathbf{x}}|_\infty,$$

$$|g(\mathbf{x}, y, z) - g(\tilde{\mathbf{x}}, y, z)| \leqslant K_g(1 + |\mathbf{x}|_\infty^r + |\tilde{\mathbf{x}}|_\infty^r)|\mathbf{x} - \tilde{\mathbf{x}}|_\infty,$$

4. (B1) holds true for $(s, y, z) \mapsto f(s, y, z)$ and $(s, y, z) \mapsto g((X_{u\wedge s})_{u\in[0,T]}, y, z)$.

**Proposition 3.3** *We consider the path-dependent framework and so we assume that* $\xi = 0$ *and* $f = 0$. *We also assume that Assumptions (F1)-(B2) hold true. Then there exists a solution* $(Y, Z)$ *of the path-dependent BSDE* (3.3) *in* $\mathcal{S}^2 \times \mathcal{M}^2$ *such that,*

$$|Z_t| \leqslant C(1 + \sup_{s\in[0,t]} |X_s|^r) \qquad d\mathbb{P} \otimes dt \ a.e. \tag{3.4}$$

*Proof* The Markovian case was already treated in [12]. The idea is to generalize this result to the discrete path dependent case, as in [9], and then pass to the limit to obtain the general path dependent case. Since the only novelty is the gathering of known methods and results, we will only sketch the proof.

1. First of all, we start by localizing the generator $g$ to obtain a Lipschitz continuous generator. Let us consider $\rho_N$ a regularized version of the projection on the centered Euclidean ball of radius $N$ in $\mathbb{R}^{1\times d}$ such that $|\rho_N| \leqslant N$, $|\nabla\rho_N| \leqslant 1$ and $\rho_N(x) = x$ when $|x| \leqslant N-1$. We denote $(Y^N, Z^N) \in \mathcal{S}^2 \times \mathcal{M}^2$ the unique solution of the BSDE

$$Y_t^N = h((X_s)_{s\in[0,T]}) + \int_t^T g^N((X_{u\wedge s})_{u\in[0,T]}, Y_s^N, Z_s^N)ds - \int_t^T Z_s^N dW_s$$

with $g^N = g(.,., \rho_N(.))$. In the remaining of the proof we will see how to prove that (3.4) is satisfied by $Z^N$ with a constant $C$ that does not depend on $N$. Let us remark that this is sufficient to conclude since it is quite standard to show that $(Y^N, Z^N)$ is a Cauchy sequence in $\mathcal{S}^2 \times \mathcal{M}^2$ and that the limit is solution of (3.3), by using for example a linearization argument and the uniform estimate on $Z^N$. For the reading convenience we will skip the superscript $^N$ in the following.

2. We approximate $h$ and the random part of $g$ by some discrete functions: by a mere generalization of [13] there exists a family $\Pi = \{\pi\}$ of partitions of $[0,T]$ and some families of discrete functionals $\{h^\pi\}$, $\{g^\pi\}$ such that, for any $\pi \in \Pi$, assuming $\pi : 0 = t_0 < ... < t_n = T$, we have

- $h^\pi \in \mathcal{C}_b^\infty(\mathbb{R}^{d(n+1)})$ and $g^\pi(.,y,z) \in \mathcal{C}_b^\infty(\mathbb{R}^{d(n+1)})$ for all $(y,z) \in \mathbb{R} \times \mathbb{R}^{1\times d}$,
- $\sum_{i=0}^n |\partial_{x_i} h^\pi(x)| \leqslant K_h(1 + 2\sup_{0\leqslant i\leqslant n} |x_i|^r)$ for all $x = (x_0, ..., x_n) \in \mathbb{R}^{d(n+1)}$,
- $\sum_{i=0}^n |\partial_{x_i} g^\pi(x,y,z)| \leqslant K_g(1 + 2\sup_{0\leqslant i\leqslant n} |x_i|^r)$ for all $x = (x_0, ..., x_n) \in \mathbb{R}^{d(n+1)}$ and $(y,z) \in \mathbb{R} \times \mathbb{R}^{1\times d}$,
- $\lim_{|\pi|\to 0} |h^\pi(x_{t_0}, ..., x_{t_n}) - h(x)| = 0$, for all $x \in \mathcal{C}([0,T], \mathbb{R}^d)$,

- $\lim_{|\pi|\to 0} |g^\pi(x_{t_0}, ..., x_{t_n}, y, z) - g(x, y, z)| = 0$, for all $x \in \mathcal{C}([0,T], \mathbb{R}^d)$ and $(y, z) \in \mathbb{R} \times \mathbb{R}^{1\times d}$.

Let us emphasize that $K_h$ and $K_g$ do not depend on $N$ and $\pi$. We firstly assume that $g$ is smooth enough with respect to $y, z$ and $b$ is smooth enough with respect to $x$, then we have the representation

$$Z_t^\pi = \nabla Y_t^\pi \nabla X_t^{-1} \sigma(t) \qquad \forall t \in [0, T],$$

where

$$\nabla Y_t^\pi = \sum_{i=1}^n \nabla^i Y_t^\pi \mathbb{1}_{[t_{i-1}, t_i)}(t) + \nabla^n Y_{T^-}^\pi \mathbb{1}_{\{T\}}(t)$$

$$\nabla^i Y_t^\pi = \sum_{j\geqslant i} \partial_{x_j} h^\pi \nabla X_{t_j} + \int_t^T \sum_{j\geqslant i} \partial_{x_j} g^\pi \nabla X_s + \partial_y g^\pi \nabla^i Y_s^\pi + \partial_z g^\pi \nabla^i Z_s^\pi ds - \int_t^T \nabla^i Z_s^\pi dW_s$$

$$\nabla X_t = I_d + \int_0^t \partial_x b \nabla X_s ds.$$

Thanks to this representation of the process $Z$, we can now apply the same strategy than in [12] to show that

$$|Z_t^\pi| \leqslant C(1 + \sup_{s\in[0,t]} |X_s|^r) \qquad d\mathbb{P} \otimes dt \text{ a.e.} \tag{3.5}$$

where $C$ only depends on constants appearing in (F1)-(B2) and does not depend on $\pi$ nor on $N$. We emphasize the fact that this is possible due to the uniform (in $N$ and $\pi$) bound on $\sum_{i=0}^n |\partial_{x_i} h^\pi(x)|$ and $\sum_{i=0}^n |\partial_{x_i} g^\pi(x, y, z)|$. When $g$ and $b$ are not smooth we can obtain the same result by a standard smooth approximation.

3. Since $h^\pi$ tends to $h$ and $g^\pi$ tends to $g$, recalling we have a Lipschitz generator, we can use a standard stability result to get that $(Y^\pi, Z^\pi) \to (Y, Z)$ in $\mathcal{S}^2 \times \mathcal{M}^2$ and so

$$|Z_t| \leqslant C(1 + \sup_{s\in[0,t]} |X_s|^r) \qquad d\mathbb{P} \otimes dt \text{ a.e.} \tag{3.6}$$

*Remark 3.4*
- The case $r = 1$ can be also tackled with extra assumptions as in [12]. More precisely, we have to assume that $K_h$, $K_g$ and $T$ are small enough to ensure exponential integrability of the terminal condition and the random part of the generator. These extra assumptions are natural when we are looking for the existence of a solution, see e.g. [3].
- The estimate (3.4) is interesting in itself and can be useful in many situations. For example, we can adapt the proof to obtain the same kind of estimate in a super-quadratic setting, as in [12], and then obtain an existence and uniqueness result for path-dependent super-quadratic BSDEs. We can also use this estimate to get an explicit error bound when we consider a truncated (in $z$) approximation of the BSDE in order to deal with BSDE numerical approximation schemes (see Sect. 5

in [12]). See also [1] for a possible application of this kind of estimate to BSDEs driven by Gaussian Processes.

**Proposition 3.5** *We assume that Assumptions (F1) and (B2) hold. Then the BSDE* (3.3) *admits a unique solution* $(Y, Z)$ *satisfying*

$$e^{K_z|Y^*|} \in \bigcap_{p>1} L^p.$$

*Proof* We start by considering the BSDE

$$Y_t^1 = h((X_s)_{s\in[0,T]}) + \int_t^T g((X_{u\wedge s})_{u\in[0,T]}, Y_s^1, Z_s^1)ds - \int_t^T Z_s^1 dW_s. \tag{3.7}$$

Using Proposition 3.3, we have the existence of a solution $(Y^1, Z^1) \in \mathcal{S}^2 \times \mathcal{M}^2$ to Eq. (3.7) such that

$$|Z_t^1| \leqslant C(1 + \sup_{s\in[0,t]} |X_s|^r), \quad d\mathbb{P} \otimes dt \text{ a.e.} \tag{3.8}$$

Moreover, by using this estimate, assumption (B2) and Remark 2.1 we can remark that

$$|g((X_{u\wedge s})_{u\in[0,T]}, Y_s^1, Z_s^1)| \leqslant C(1 + |Y_s^1| + \sup_{u\in[0,s]} |X_u|^{1+r}), \quad d\mathbb{P} \otimes dt \text{ a.e.}$$

and

$$|h((X_s)_{s\in[0,T]})| \leqslant C(1 + \sup_{s\in[0,T]} |X_s|^{1+r}).$$

Then, a classical estimate on solutions of BSDE (3.7) gives us that

$$|(Y^1)^*| \leqslant C(1 + \sup_{s\in[0,T]} |X_s|^{1+r}). \tag{3.9}$$

Now we introduce a new BSDE

$$\begin{aligned} Y_t^2 &= \xi + \int_t^T f(s, Y_s^1 + Y_s^2, Z_s^1 + Z_s^2)ds - \int_t^T Z_s^2 dW_s^{\mathbb{Q}} \\ &+ \int_t^T \left\{ g((X_{u\wedge s})_{u\in[0,T]}, Y_s^1 + Y_s^2, Z_s^1 + Z_s^2) - g((X_{u\wedge s})_{u\in[0,T]}, Y_s^1, Z_s^1) \right\} ds \\ &- \int_t^T Z_s^2 \nabla_z g((X_{u\wedge s})_{u\in[0,T]}, Y_s^1, Z_s^1)ds, \end{aligned} \tag{3.10}$$

where $dW_s^{\mathbb{Q}} = dW_s - \nabla_z g((X_{u\wedge s})_{u\in[0,T]}, Y_s^1, Z_s^1)ds$. By using Novikov's condition, there exists a probability $\mathbb{Q}$ under which $W^{\mathbb{Q}}$ is a Brownian motion. Then, [11] gives us the existence of a solution $(Y^2, Z^2) \in \mathcal{S}^\infty(\mathbb{Q}) \times \mathcal{M}^2(\mathbb{Q})$ to the previous BSDE such that $Z^2 \in BMO(\mathbb{Q})$. Now we can remark that $(Y, Z) := (Y^1 + Y^2, Z^1 + Z^2)$ is a solution of (3.3). We denote

$$F(t, y, z) := f(t, y, z) + g((X_{u\wedge t})_{u\in[0,T]}, y, z)$$

and we get

$$\mathcal{E}_T := e^{\int_0^T \nabla_z F(s,Y_s,Z_s)dW_s - \frac{1}{2}\int_0^T |\nabla_z F(s,Y_s,Z_s)|^2 ds} = e_1 e_2 e_3,$$

with

$$\begin{aligned}
e_1 &= e^{\int_0^T (\nabla_z F(s,Y_s,Z_s)-\nabla_z F(s,Y_s,Z_s^1))dW_s - \frac{1}{2}\int_0^T |\nabla_z F(s,Y_s,Z_s)-\nabla_z F(s,Y_s,Z_s^1)|^2 ds},\\
e_2 &= e^{\int_0^T \nabla_z F(s,Y_s,Z_s^1)dW_s - \frac{1}{2}\int_0^T |\nabla_z F(s,Y_s,Z_s^1)|^2 ds},\\
e_3 &= e^{-\int_0^T \langle \nabla_z F(s,Y_s,Z_s), \nabla_z F(s,Y_s,Z_s)-\nabla_z F(s,Y_s,Z_s^1)\rangle ds}.
\end{aligned}$$

We will study the integrability of these terms. First of all, we can remark that

$$\int_0^T |\nabla_z F(s, Y_s, Z_s^1)|^2 ds \leqslant C(1 + \sup_{s\in[0,T]} |X_s|^{2r}) \tag{3.11}$$

due to (B2)-1, (B2)-4 and (3.8). By using Novikov's condition and classical estimates on exponential moments of SDEs, it implies that

$$e_2 \in \bigcap_{p\geqslant 1} L^p. \tag{3.12}$$

For same reasons we have

$$e^{\int_0^T \nabla_z g((X_{u\wedge s})_{u\in[0,T]}, Y_s^1, Z_s^1)dW_s - \frac{1}{2}\int_0^T |\nabla_z g((X_{u\wedge s})_{u\in[0,T]}, Y_s^1, Z_s^1)|^2 ds} \in \bigcap_{p\geqslant 1} L^p. \tag{3.13}$$

Since $|\nabla_z F(s, Y_s, Z_s) - \nabla_z F(s, Y_s, Z_s^1)| \leqslant 2K_z|Z_s^2|$, then we obtain

$$\int_0^\cdot \left(\nabla_z F(s, Y_s, Z_s) - \nabla_z F(s, Y_s, Z_s^1)\right) dW_s \in BMO(\mathbb{Q})$$

and so there exists $\ell > 1$ such that $e_1 \in L^\ell(\mathbb{Q})$. By using (3.13) and Hölder inequality we get that $e_1 \in L^{1+\frac{\ell-1}{2}}$. We can also observe that,

$$e_3 \leqslant e^{-\int_0^T \langle \nabla_z F(s,Y_s,Z_s^1), \nabla_z F(s,Y_s,Z_s)-\nabla_z F(s,Y_s,Z_s^1)\rangle ds}$$

and, by using Young inequality,

$$|\langle \nabla_z F(s, Y_s, Z_s^1), \nabla_z F(s, Y_s, Z_s) - \nabla_z F(s, Y_s, Z_s^1)\rangle| \leqslant \frac{1}{4\varepsilon}|\nabla_z F(s, Y_s, Z_s^1)|^2 + \varepsilon K_z^2 |Z_s^2|^2,$$

for all $\varepsilon > 0$. It implies that

$$e_3 \leqslant e^{\frac{C}{\varepsilon}(1+\sup_{s\in[0,T]}|X_s|^{2r})+\varepsilon K_z^2 \int_0^T |Z_s^2|^2 ds}.$$

Since $\int_0^\cdot Z_s^2 dW_s$ is BMO we can apply the John-Nirenberg inequality (see [10]) and by using Cauchy-Schwarz inequality and classical estimates on exponential moments of SDEs we get

$$e_3 \in \bigcap_{p>1} L^p. \tag{3.14}$$

Finally, by using (3.12), (3.14) and the estimate $e_1 \in \cup_{p>1} L^p$, we get that

$$\mathcal{E}_T \in \bigcup_{p>1} L^p.$$

Recalling (3.9) and the fact that $Y^2$ is bounded, we just have to apply Theorem 2.4 to conclude.

### 3.3 A Second Unbounded Case

In this subsection we consider a more general SDE

$$X_t = x + \int_0^t b(X_s)ds + \int_0^t \sigma(X_s)dW_s, \quad 0 \leqslant t \leqslant T, \tag{3.15}$$

where $b$ and $\sigma$ satisfies classical assumptions: **(F2)**

1. $b : \mathbb{R}^d \to \mathbb{R}^d$ and $\sigma : \mathbb{R}^d \to \mathbb{R}^{d\times d}$ are Lipschitz functions: for all $(x, x') \in \mathbb{R}^d \times \mathbb{R}^d$ we have $|b(x) - b(x')| \leqslant K_b|x - x'|$ and $|\sigma(x) - \sigma(x')| \leqslant K_\sigma |x - x'|$.
2. $\sigma$ is bounded by $|\sigma|_\infty$.

Now we want to study the same BSDE (3.3) under following assumptions: **(B3)**

1.

$$|\xi|_{L^\infty} + \left|\int_0^T |f(s, 0, 0)|ds\right|_{L^\infty} + \left|\sup_{s\in[0,T], y\in\mathbb{R}, \mathbf{x}\in C([0,T],\mathbb{R}^d)} |\nabla_z f(s, y, 0)| + |\nabla_z g(\mathbf{x}, y, 0)|\right|_{L^\infty} < +\infty$$

2. there exists $C > 0$ such that, for all $s \in [0, T]$, $y \in \mathbb{R}$ and $z \in \mathbb{R}^{1\times d}$,

$$|f(t, y, z)| \leqslant C.$$

3. There exist $K_h > 0$ and $K_g > 0$ such that, for all $\mathbf{x}, \tilde{\mathbf{x}} \in C([0,T], \mathbb{R}^d)$, $y \in \mathbb{R}$, $z \in \mathbb{R}^{1\times d}$,
$$|h(\mathbf{x}) - h(\tilde{\mathbf{x}})| \leqslant K_h |\mathbf{x} - \tilde{\mathbf{x}}|_\infty,$$
$$|g(\mathbf{x}, y, z) - g(\tilde{\mathbf{x}}, y, z)| \leqslant K_g |\mathbf{x} - \tilde{\mathbf{x}}|_\infty,$$
4. (B1) holds true for $(s, y, z) \mapsto f(s, y, z)$ and $(s, y, z) \mapsto g((X_{u\wedge s})_{u\in[0,T]}, y, z)$.

Firstly we give a general lemma.

**Lemma 3.6** *We assume that (B1) is in force and*

- $\mathbb{E}\left[|\xi|^2 + \int_0^T |f(t,0,0)|^2 dt\right] < +\infty$,
- $\xi \in \mathbb{D}^{1,2}$ *and for all* $(y,z) \in \mathbb{R} \times \mathbb{R}^{1\times d}$, $f(.,y,z) \in \mathbb{L}_{1,2}$,
- $\left|\sup_{y\in\mathbb{R}, s\in[0,T]} |\nabla_z f(s,y,0)|\right|_{L^\infty} \leqslant M_z$.

*Let us consider an auxiliary BSDE*

$$R_t = \xi + \int_t^T f(s, R_s, 0)ds - \int_t^T S_s dW_s, \tag{3.16}$$

*with a unique solution* $(R,S) \in \mathcal{S}^2 \times \mathcal{M}^2$. *If*

$$\left| \sup_{t\in[0,T]} \mathbb{E}_t \left[ |D_t\xi| + \int_t^T |D_t f(s,y,0)|_{y=R_s} ds \right] \right|_{L^\infty} < +\infty, \tag{3.17}$$

*then $S$ is $d\mathbb{P}\otimes dt$ a.e. bounded and there exists a solution $(Y,Z) \in \mathcal{S}^2 \times \mathcal{M}^2$ of (2.1) such that $\int_0^\cdot Z_s dW_s$ is BMO.*

*If moreover we have, for all $p > 1$,*

$$\left| \sup_{t\in[0,T]} \mathbb{E}_t \left[ |D_t\xi|^p + \left( \int_t^T |D_t f(s,y,z)|_{y=Y_s, z=Z_s} ds \right)^p \right] \right| < +\infty,$$

*then there exists a solution $(Y,Z) \in \mathcal{S}^2 \times \mathcal{M}^2$ of (2.1) such that $Z$ is $d\mathbb{P}\otimes dt$ a.e. bounded.*

*Proof* Let us assume that $f$ and $\xi$ satisfy assumptions of Proposition 5.3 in [7] (smoothness and integrability assumptions). Then we can differentiate (in the Malliavin sense) BSDE (3.16): We obtain, for all $t \in [0,T]$,

$$D_t R_t = \mathbb{E}_t \left[ e^{\int_t^T \nabla_y f(s,R_s,0)ds} D_t\xi + \int_t^T e^{\int_t^s \nabla_y f(r,R_r,0)dr} (D_t f(s,y,0))_{y=R_s} ds \right]$$

and a version of $S$ is given by $(D_t R_t)_{t\in[0,T]}$. Thus we get that there exists $C > 0$ such that, for all $t \in [0,T]$,

$$|S_t| = |D_t R_t| \leqslant e^{K_y T}\mathbb{E}_t\left[|D_t\xi| + \int_t^T |D_t f(s,y,0)|_{y=R_s} ds\right] \leqslant C. \tag{3.18}$$

When $f$ is not smooth enough and $f, \xi$ are not enough integrable, we can show by a standard approximation procedure that inequality (3.18) stays true $d\mathbb{P}\otimes dt$ a.e. Now we consider the following BSDE:

$$U_t = \int_t^T f(s, U_s + R_s, V_s + S_s) - f(s, R_s, 0)ds - \int_t^T V_s dW_s. \tag{3.19}$$

If we set $\Psi(s,u,v) := f(s,u+R_s, v+S_s) - f(s,R_s,0)$, then, by using (3.18) and assumptions of the Lemma on $f$, we have, for all $s \in [0,T]$, $u, u' \in \mathbb{R}$, $v, v' \in \mathbb{R}^{1\times d}$,

- $|\Psi(s,u,v) - \Psi(s,u',v)| \leqslant |v - v'|$,
- $|\Psi(s,u,v) - \Psi(s,u,v')| \leqslant \big(M_z + K_z(|z| + |z'| + 2C)\big)\,|z - z'|$,
- $|\Psi(s,u,v)| \leqslant K_y|u| + (M_z + K_z|v+S_s|)|v+S_s| \leqslant C(1 + |u| + |v|^2)$.

By applying results of [11] we obtain a unique solution $(U,V) \in \mathcal{S}^\infty \times \mathcal{M}^2$ and moreover $\int_0^\cdot V_s dW_s$ is BMO. Finally, we can remark that $(Y,Z) := (U+R, V+S)$ is a solution of BSDE (2.1). So, since $S$ is bounded, $\int_0^\cdot Z_s dW_s$ is BMO.

Concerning the boundedness of $Z$, we just have to adapt the proof of Theorem 3.6 in [12] in a non Markovian framework which does not create any difficulty. For the reader convenience, we only sketch the proof and we refer to [12] for further details. We start by assuming that $f$ and $\xi$ satisfy assumptions of Proposition 5.3 in [7] (smoothness and integrability assumptions). Then we can differentiate (in the Malliavin sense) BSDE (2.1): We obtain, for all $t \in [0,T]$, $u \in [0,T]$,

$$\begin{aligned} D_u Y_t &= D_u\xi + \int_t^T (D_u f(s,y,z))_{y=Y_s, z=Z_s} + D_u Y_s \nabla_y f(s,Y_s,Z_s) + D_u Z_s \nabla_z f(s,Y_s,Z_s) ds \\ &\quad - \int_t^T D_u Z_s dW_s \\ &= D_u\xi e^{\int_t^T \nabla_y f(s,Y_s,Z_s)ds} + \int_t^T e^{\int_s^T \nabla_y f(r,Y_r,Z_r)dr}(D_u f(s,y,z))_{y=Y_s,z=Z_s} ds - \int_t^T D_u Z_s dW_s^{\mathbb{Q}}, \end{aligned}$$

where $dW_s^{\mathbb{Q}} := dW_s - \nabla_z f(s,Y_s,Z_s)ds$ and a version of $Z$ is given by $(D_t Y_t)_{t\in[0,T]}$. Thanks to assumptions on the growth of $\nabla_z f$ and the fact that $\int_0^\cdot Z_s dW_s$ is BMO, we know that there exists a probability $\mathbb{Q}$ under which $W^{\mathbb{Q}}$ is a Brownian motion. It implies the following estimate

$$|Z_t| = |D_t Y_t| \leqslant e^{K_y T}\mathbb{E}_t^{\mathbb{Q}}\left[|D_t\xi| + \int_t^T |D_t f(s,y,z)|_{y=Y_s,z=Z_s} ds\right].$$

Then, we use once again BMO properties of $\int_0^\cdot Z_s dW_s$: thanks to the reverse Hölder inequality (see Kazamaki [10]), we can apply Hölder inequality to the previous estimate to obtain the existence of $C > 0$ and $p > 1$ (that depend only on constants appearing in assumptions of the Lemma) such that

$$|Z_t| = |D_t Y_t| \leqslant Ce^{K_y T}\mathbb{E}_t\left[|D_t\xi|^p + \left(\int_t^T |D_t f(s,y,z)|_{y=Y_s,z=Z_s} ds\right)^p\right]^{1/p}. \tag{3.20}$$

We use (3.17) to conclude. When $f$ is not smooth enough and $f$, $\xi$ are not enough integrable, we can show by a standard approximation procedure that inequality (3.20) stays true $d\mathbb{P}\otimes dt$ a.e.

**Corollary 3.7** *We consider the path-dependent framework and so we assume that $\xi = 0$ and $f = 0$. We also assume that Assumptions (F2)-(B3) hold true. Then there exists a solution $(Y, Z)$ of the path-dependent BSDE* (3.3) *in $\mathcal{S}^2 \times \mathcal{M}^2$ such that,*

$$|Z_t| \leqslant C, \quad d\mathbb{P}\otimes dt \text{ a.e.}$$

*Proof* 1. Let us start by the Markovian framework. Without lost of generality we can assume that $X_t^1 = t$ for all $t \in [0, T]$. We assume that, for all $\mathbf{x} \in C([0,T], \mathbb{R}^d)$, $y \in \mathbb{R}$, $z \in \mathbb{R}^{1\times d}$, we have $h(\mathbf{x}) = \tilde{h}(\mathbf{x}_T)$ and $g(\mathbf{x}, y, z) = \tilde{g}(\mathbf{x}_{\sup_{t\in[0,T]}\mathbf{x}_t^1}, y, z)$ with $\tilde{h} : \mathbb{R} \to \mathbb{R}$ a $K_h$-Lipschitz function and $\tilde{g} : \mathbb{R}^d \times \mathbb{R} \times \mathbb{R}^{1\times d} \to \mathbb{R}$ a $K_g$-Lipschitz function with respect to the first variable (uniformly in $y$ and $z$). If $\tilde{h}$, $\tilde{g}$, $b$ and $\sigma$ are smooth enough then $\tilde{h}(X_T)$ and $\tilde{g}(X_s, y, z)$ are Malliavin differentiable and the chain rule gives us

$$D_t\tilde{h}(X_T) = \nabla_x\tilde{h}(X_T)\nabla X_T(\nabla X_t)^{-1}\sigma(X_t), \quad D_t\tilde{g}(X_s,y,z) = \nabla_x\tilde{g}(X_s,y,z)\nabla X_s(\nabla X_t)^{-1}\sigma(X_t)\mathbb{1}_{t\leqslant s}.$$

So we get, for all $p \geqslant 1$,

$$\begin{aligned}\mathbb{E}_t\left[|D_t\tilde{h}(X_T)|^p + \left(\int_t^T |D_t\tilde{g}(X_s,y,z)|ds\right)^p\right] &\leqslant (K_h^p + K_g^p T^p)|\sigma|_\infty^p \mathbb{E}_t\left[\sup_{s\in[t,T]} |\nabla X_s(\nabla X_t)^{-1}|^p\right]\\ &\leqslant (K_h^p + K_g^p T^p)|\sigma|_\infty^p C_p\end{aligned}$$

with $C_p$ that only depends on $p$, $T$, $K_b$ and $K_\sigma$, thanks to classical estimates on SDEs. Then we just have to apply the Lemma 3.6 to obtain that $Z$ is bounded with a bound that only depends on constants appearing in assumptions. When $\tilde{h}$, $\tilde{g}$, $b$ and $\sigma$ are not smooth enough we can show that this result stays true by a standard approximation procedure.

2. To deal with the general path-dependent framework we just have to apply the same strategy than in Proposition 3.3, we firstly consider the discrete path-dependent case and then we pass to the limit. We refer to this proof for further details.

*Remark 3.8* Corollary 3.7 answers an open question in the Sect. 3 of [12]. In light to this result a new question arise: what happens when $g$ and $h$ are only locally Lipschitz? More precisely, does Proposition 3.3 stays true when we replace assumption (F1) by assumption (F2)? Let us remark that the answer is not clear even when $f(z) = \frac{|z|^2}{2}$, see [12].

**Proposition 3.9** *We assume that Assumptions (F2)-(B3) hold true. Then the BSDE* (3.3) *admits a unique solution* $(Y, Z)$ *satisfying*

$$e^{K_z|Y^*|} \in \bigcap_{p>1} L^p.$$

*Proof* The proof follows the same lines than the proof of Proposition 3.5.

*Remark 3.10* Let us remark that we assume some boundedness assumptions on $\xi$ and $f$ to be able to use BMO martingales tools. In particular, these tools are essential to get (2.2). As pointed out by a referee, it is known, see e.g. [8], that we can sometimes keep the BMO property on $\int_0^\cdot Z_s dW_s$ by assuming only that

$$\left(\mathbb{E}\left[\xi + \int_0^T f(s,0,0)ds \mid \mathcal{F}_t\right]\right)_{t\in[0,T]} \in BMO. \tag{3.21}$$

Nevertheless, contrary to [8], our generator depends on $Y$ which implies some extra difficulties. More precisely, if we consider BSDE (3.16) under assumptions (B1) and (3.21) then $\int_0^\cdot S_s dW_s$ is not necessarily a BMO martingale except if we add a boundedness assumption on $f$:

$$|f(t,y,z)| \leqslant \alpha_t + C|z|^2, \quad \forall (t,y,z) \in [0,T] \times \mathbb{R} \times \mathbb{R}^{1\times d}$$

where $\alpha : [0,T] \to \mathbb{R}^+$ satisfies $\int_0^T |\alpha_t|^2 dt < +\infty$. In this case we can construct a solution $(Y, Z)$ in the space $\mathcal{S}^2 \times \mathcal{M}^2$ of BSDE (2.1) such that $\int_0^\cdot Z_s dW_s$ is a BMO martingale by using the solution $(R, S) \in \mathcal{S}^2 \times \mathcal{M}^2$ of BSDE (3.16) up to some extra assumption on the smallness of the BMO martingale (3.21). This assumption is necessary to get some exponential integrability for $Y$. But to avoid making this article more cumbersome we do not enter more into details.

Finally, let us point out that we also run into extra difficulties if we want to relax boundedness assumptions on $\xi$ and $f$ in the framework of Sect. 3.2. Indeed, we have to study the BSDE (3.10) under a new probability $\mathbb{Q}$ and so we should replace boundedness assumptions on $\xi$ and $f$ by

$$\left(\mathbb{E}\left[\xi + \int_0^T f(s,0,0)ds \mid \mathcal{F}_t\right]\right)_{t\in[0,T]} \in BMO(\mathbb{Q}) \tag{3.22}$$

which is not necessarily equivalent to (3.21) since the change of probability does not come from the Doleans-Dade exponential of a BMO martingale.

## References

1. Bender, C.: Backward SDEs driven by Gaussian processes. Stochast. Process. Appl. **124**(9), 2892–2916 (2014)
2. Briand, P., Elie, R.: A simple constructive approach to quadratic BSDEs with or without delay. Stochast. Process. Appl. **123**(8), 2921–2939 (2013)
3. Briand, P., Hu, Y.: BSDE with quadratic growth and unbounded terminal value. Probab. Theory Related Fields **136**(4), 604–618 (2006)
4. Briand, P., Hu, Y.: Quadratic BSDEs with convex generators and unbounded terminal conditions. Probab. Theory Related Fields **141**(3–4), 543–567 (2008)
5. Delbaen, F., Hu, Y., Richou, A.: On the uniqueness of solutions to quadratic BSDEs with convex generators and unbounded terminal conditions: the critical case. arXiv:1303.4859v1, to appear in Discrete and Continuous Dynamical Systems. Series A
6. Delbaen, F., Hu, Y., Richou, A.: On the uniqueness of solutions to quadratic BSDEs with convex generators and unbounded terminal conditions. Ann. Inst. Henri Poincaré Probab. Stat. **47**(2), 559–574 (2011)
7. El Karoui, N., Peng, S., Quenez, M.C.: Backward stochastic differential equations in finance. Math. Financ. **7**(1), 1–71 (1997)
8. Frei, P.: Splitting multidimensional BSDEs and finding local equilibria. Stochast. Process. Appl. **124**(8), 2654–2671 (2014)
9. Hu, Y., Ma, J.: Nonlinear Feynman-Kac formula and discrete-functional-type BSDEs with continuous coefficients. Stochast. Process. Appl. **112**(1), 23–51 (2004)
10. Kazamaki, N.: Continuous exponential martingales and BMO. In: Lecture Notes in Mathematics, vol. 1579. Springer-Verlag, Berlin (1994)
11. Kobylanski, M.: Backward stochastic differential equations and partial differential equations with quadratic growth. Ann. Probab. **28**(2), 558–602 (2000)
12. Richou, A.: Markovian quadratic and superquadratic BSDEs with an unbounded terminal condition. Stochast. Process. Appl. **122**(9), 3173–3208 (2012)
13. Zhang, J.: A numerical scheme for BSDEs. Ann. Appl. Probab. **14**(1), 459–488 (2004)

# An Example of Martingale Representation in Progressive Enlargement by an Accessible Random Time

**Antonella Calzolari and Barbara Torti**

**Abstract** Given two martingales on the same probability space, both enjoying the predictable representation property with respect to their own filtrations, it can happens that their quadratic covariation process enters in the martingale representation of the filtration obtained as the union of the original ones. This fact on one hand influences the multiplicity of the enlarged filtration and on the other hand it is linked to the behavior of the sharp brackets of the martingales. Here we illustrate these arguments presenting an elementary example of martingale representation in the context of progressive enlargement by an accessible random time.

**Keywords** Semi-martingales · Predictable representations property · Enlargement of filtration · Completeness of a financial market

**AMS 2010** 60G48 · 60G44 · 60H05 · 60H30 · 91G99

## 1 Introduction

In stochastic analysis, given a filtration $\mathbb{G} = (\mathcal{G}_t)_{t\geq 0}$ on a probability space $(\Omega, \mathcal{F}, P)$, the problem arises to represent every random variable in $L^2(\Omega, \mathcal{G}_\infty, P)$ as the final value of an element of the stable subspace generated by a given number of square integrable $(P, \mathbb{G})$-martingales. Many papers up to the seventies deal with the *martingale representation problem* and among the others we recall the fundamental contribution of Davis and Varaiya in [6] and the recent contribution of Song in [19]. In [6] the authors suggest how to derive the *multiplicity* of $\mathbb{G}$, that is the minimal number of pairwise orthogonal martingales needed to generate via stochastic integration $L^2(\Omega, \mathcal{G}_\infty, P)$. In [19] the focus is on some important properties common to any set of representing martingales for a given filtration $\mathbb{G}$. Inspired by practical situa-

A. Calzolari (✉) · B. Torti
Dipartimento di Matematica, Università di Roma "Tor Vergata",
via della Ricerca Scientifica 1, I 00133 Roma, Italy
e-mail: calzolar@axp.mat.uniroma2.it



S. N. Cohen et al. (eds.), *Frontiers in Stochastic Analysis - BSDEs, SPDEs and their Applications*, Springer Proceedings in Mathematics & Statistics 289,
https://doi.org/10.1007/978-3-030-22285-7_4

tions, many other papers consider the opposite point of view. More precisely, given a fixed number of $\mathbb{G}$-adapted semi-martingales, the question is if they are enough to represent $L^2(\Omega, \mathcal{G}_\infty, Q)$, where $Q$ is a suitable probability measure equivalent to $P$.

A well-known result is due to Kusuoka. Let $B$ be a Brownian motion and let $\mathbb{I}_{\tau\leq\cdot}$ be the *occurrence process* of a random time $\tau$. Let $\mathbb{G}$ be the *progressive enlargement* of the Brownian filtration $\mathbb{F}^B$ by the occurrence of $\tau$, that is $\mathcal{G}_t := \cap_{s>t}\mathcal{F}_s^B \vee \sigma(\tau \wedge s)$. As well known, $B$ is not necessarily a $\mathbb{G}$-Brownian motion. In [18] Kusuoka gives conditions under which $B$ is still a $\mathbb{G}$-Brownian motion, and, together with the Doob-Meyer martingale part of $\mathbb{I}_{\tau\leq\cdot}$, represents $L^2(\Omega, \mathcal{G}_\infty, P)$.

A key hypothesis in [18] is the *density hypothesis* on $\tau$, namely the absolute continuity for all $t$ of the $\mathcal{F}_t^B$-conditional law of $\tau$ with respect to a deterministic measure without atoms. Under the last condition $\tau$ has no atoms and therefore it is *totally inaccessible* with respect to the natural filtration $\mathbb{H}$ of $\mathbb{I}_{\tau\leq\cdot}$. Moreover, under the density hypothesis $\tau$ avoids $\mathbb{F}^B$-stopping times, that is $P(\tau = \sigma) = 0$ for any finite $\mathbb{F}^B$-stopping time $\sigma$ (see IV-107 in [7] and Proposition 1 in [13]). It is natural to ask how to represent $L^2(\Omega, \mathcal{G}_\infty, P)$ when no density hypothesis on $\tau$ is assumed and in particular when $\tau$ is allowed to have a law with atoms or even a discrete law. In the last case $\tau$ turns out to be *accessible* with respect to $\mathbb{H}$ and obviously also with respect to $\mathbb{G}$.

Many papers are devoted to possible extensions of Kusuoka's result, either by considering more general local martingales than $B$ or by changing hypotheses on $\tau$ (see e.g. [1, 2, 13]). In particular we point out the main contribution given by Jeanblanc and Song in [14], where the authors propose a general methodology to find solutions in great generality to enlargement problems, namely, to establish whether a given $\mathbb{F}$-local martingale is a $\mathbb{G}$-semimartingale and, if it is the case, what is the canonical decomposition of that $\mathbb{G}$-semimartingale. Indeed they recover all the results obtained in the past literature by using different techniques according to the kind of enlargement to be considered. In particular they work out the martingale representation problem for a Brownian filtration enlarged by the occurrence of an honest default time (see Definition 5.9.4.9 in [15]).

In [4], given a probability space $(\Omega, \mathcal{F}, P)$ and fixed a finite time horizon $T$, we deal with the progressive enlargement on $[0, T]$ of a filtration $\mathbb{F}$, such that $\mathcal{F}_T \subset \mathcal{F}$ and $L^2(\Omega, \mathcal{F}_T, P)$ is represented by a local martingale $M$, by the occurrence of a general random time $\tau$. A fundamental hypothesis in our setting is the existence of an equivalent probability measure decoupling $\mathbb{F}$ and $\mathbb{H}$. This hypothesis is compensated by the fact that no conditions on M and $\tau$ are required. We prove that the multiplicity of $\mathbb{G}$ in the sense of Davis and Varaiya is at most three and we propose a martingale representation formula on the interval $[0, T]$.

In the present paper we face the martingale representation problem for an elementary model which, in spite of its simplicity, allows to fully illustrate the results of [4]. More precisely we consider a standard Brownian motion $B$ independent of a random time $\eta$ with values in the set $\{1, 2, 3\}$ and we call $M$ the martingale part of the Doob-Meyer decomposition of $B + \mathbb{I}_{\eta\leq\cdot}$ and $\mathbb{F}$ its natural filtration. We assume $\tau$ to be a binary random variable in the set $\{2, 4\}$ and as usual we put $\mathcal{G}_t := \cap_{s>t}\mathcal{F}_s \vee \sigma(\tau \wedge s)$. In Sect. 2 we discuss precisely the model and we give conditions under which the multiplicity of $\mathbb{G}$ is exactly three and we compute explicitly a $\mathbb{G}$-*basis*, that is three orthogonal martingales giving rise to the representation of $L^2(\Omega, \mathcal{G}_T, P)$. In Sect. 3 we state the necessary theoretical results.

The interest of our example is linked to the application of martingale representation results to reduced form approach of credit risk theory, in view of the fact that recent financial literature stress the role of predictable random times in modeling markets with critical announced random dates (see e.g. [9, 16, 17]). In fact the process $B + \mathbb{I}_{\eta\leq\cdot}$ can be interpreted as the risky asset price of a complete market and the random time $\tau$ can be taught as a default time, actually the simplest example of a non trivial accessible random time.

## 2 The Toy Example

Let $T$ be a finite time horizon. On a probability space $(\Omega, \mathcal{F}, P)$ we consider the process $M = (M_t)_{t\in[0,T]}$ defined as follows

$$M := B + H^\eta,$$

where $B = (B_t)_{t\in[0,T]}$ is a standard Brownian motion with natural filtration $\mathbb{F}^B = (\mathcal{F}_t^B)_{t\in[0,T]}$, independent of a random time $\eta$ in the set $\{1, 2, 3\}$. Let $\mathbb{F}^\eta = (\mathcal{F}_t^\eta)_{t\in[0,T]}$ be the natural filtration of the occurrence process $\mathbb{I}_{\eta\leq\cdot} = (\mathbb{I}_{\eta\leq t})_{t\in[0,T]}$ and $H^\eta = (H_t^\eta)$ the $(P, \mathbb{F}^\eta)$-*compensated occurrence process of* $\eta$, that is

$$H_t^\eta := \mathbb{I}_{\eta\leq t} - A_t^{\eta,P,\mathbb{F}^\eta}, \tag{1}$$

where $A^{\eta,P,\mathbb{F}^\eta}$ is the $(P, \mathbb{F}^\eta)$-*compensator* of $\mathbb{I}_{\eta\leq\cdot}$ (see Definition VI-77 in [8]). Then, by the independence hypothesis, the filtration

$$\mathbb{F} := \mathbb{F}^B \vee \mathbb{F}^\eta$$

satisfies usual conditions (see Theorem 1 in [10]). Trivially $M$ turns out to be a $(P, \mathbb{F})$-martingale.

Moreover we assume $\tau$ to be a random time with values in the set $\{2, 4\}$. In the next we will call $\mathbb{H} = (\mathcal{H}_t)_{t\in[0,T]}$ the natural filtration of the occurrence process $\mathbb{I}_{\tau\leq\cdot} = (\mathbb{I}_{\tau\leq t})_{t\in[0,T]}$ and, similarly to (1), we will denote by $H = (H_t)_{t\in[0,T]}$ the $(P, \mathbb{H})$-*compensated occurrence process of* $\tau$ defined by

$$H_t := \mathbb{I}_{\tau\leq t} - A_t^{\tau,P,\mathbb{H}}. \tag{2}$$

As already observed in the introduction, as all the discrete non negative random variables, $\tau$ turns out to be an accessible random time with respect to $\mathbb{H}$ and then with respect to any bigger filtration (see Definition 3.34 in [11] and Theorem IV-105 in [7]) .

Finally we set

$$\mathbb{G} := \mathbb{F} \vee \mathbb{H}$$

and we denote by $H' = (H'_t)_{t\in[0,T]}$ the $(P, \mathbb{G})$-compensated occurrence process of $\tau$, that is

$$H'_t := \mathbb{I}_{\tau\le t} - A_t^{\tau,P,\mathbb{G}}. \tag{3}$$

We recall that a local martingale (possibly multidimensional) enjoys the *predictable representation property (p.r.p.)* if any one-dimensional local martingale can be represented as (vector) stochastic integral with respect to it ( see e.g. Definition 13.1 in [11]).

It is well-known that $B$ enjoys the $(P, \mathbb{F}^B)$-p.r.p. Moreover the $(P, \mathbb{H})$-martingale $H$ enjoys the $(P, \mathbb{H})$-p.r.p. and the $(P, \mathbb{F}^\eta)$-martingale $H^\eta$ enjoys the $(P, \mathbb{F}^\eta)$-p.r.p. (see Proposition 2 in [5]).

**Lemma 2.1** *$M$ enjoys the $(P, \mathbb{F})$-p.r.p.*

*Proof* The result follows by using Remark 3.2. In fact $B$ and $\eta$ are $P$-independent and the random measures induced by the sharp brackets $\langle B\rangle^{P,\mathbb{F}^B}$, $\langle H^\eta\rangle^{P,\mathbb{F}^\eta}$ are $P$-a.s. mutually singular (see e.g. [4] for the computation of $\langle H^\eta\rangle^{P,\mathbb{F}^\eta}$).

Before stating the main result of this note we recall the following definition.

**Definition 2.2** [6] The multiplicity of $\mathbb{G}$ is the minimal number of orthogonal real martingales needed to represent by stochastic integration all $\mathbb{G}$-local martingales, i.e. the minimal number of elements of a basis for $\mathbb{G}$.

It is to note that the multiplicity of a filtration is invariant under equivalent (regular enough) changes of measure (see Remark 3.12 in [4]).

Let us announce the main theorem of our paper.

**Theorem 2.3** *In the above model assume that $B$ is $P$-independent of $(\eta, \tau)$ and the joint law $p_{\eta,\tau}$ of $(\eta, \tau)$ is strictly positive on the set $\{1, 2, 3\} \times \{2, 4\}$. Then*

*(i) the multiplicity of $\mathbb{G}$ is equal to three;*

*(ii) $M$ and $H'$ can be explicitly written as*

$$M_t = B_t + \mathbb{I}_{\eta\le t} - P(\eta = 1)\mathbb{I}_{1\le t} - P(\eta = 2 \mid \sigma\{\eta = 1\})\mathbb{I}_{2\le t} - \mathbb{I}_{\eta=3}\mathbb{I}_{3\le t}, \tag{4}$$

$$H'_t = \mathbb{I}_{\tau\le t} - P(\tau = 2 \mid \sigma\{\eta = 1\})\mathbb{I}_{2\le t} - \mathbb{I}_{\tau=4}\mathbb{I}_{4\le t}. \tag{5}$$

*(iii) If moreover*

$$P(\tau = 2 \mid \eta = 2) = P(\tau = 2 \mid \eta \neq 1) = P(\tau = 2 \mid \eta = 3), \tag{6}$$

*the triplet* $(M, H', K)$, *where the process* $K = (K_t)_{t\in[0,T]}$ *is defined by*

$$K_t := \left(\mathbb{I}_{\eta=2,\tau=2} - \gamma\mathbb{I}_{\eta=2} - P(\eta = 2 \mid \eta \neq 1)\mathbb{I}_{\eta\neq 1,\tau=2} + \gamma P(\eta = 2 \mid \eta \neq 1)\mathbb{I}_{\eta\neq 1}\right)\mathbb{I}_{2\leq t} \tag{7}$$

*with*

$$\gamma := P(\tau = 2 \mid \eta = 2), \tag{8}$$

*is a* $(P, \mathbb{G})$*-basis.*

The proof of this theorem requires the following lemma.

**Lemma 2.4** *Under the hypotheses of Theorem 2.3 the triplet* $\big(M, H', [M, H]\big)$ *enjoys the* $(P, \mathbb{G})$*-p.r.p.*

*Proof* We apply Theorem 3.3 with $P^*$ defined on $(\Omega, \mathcal{G}_T)$, up to a standard extension procedure, by the rule

$$P^*(A \cap C \cap D) := P(A)P(C)P(D), \quad \text{for any } A \in \mathcal{F}^B_T, C \in \mathcal{F}^\eta_T, D \in \mathcal{H}_T. \tag{9}$$

$P^*$ is the unique *equivalent decoupling martingale preserving probability measure* and in particular $M$ and $H$ are independent $(P^*, \mathbb{G})$-martingales. In fact: $\mathbb{F}$ and $\mathbb{H}$ turn out to be $P^*$-independent by construction; $M$ is a $(P^*, \mathbb{F})$-martingale since $P^*|_{\mathcal{F}_T} = P|_{\mathcal{F}_T}$ and, by the independence of $\mathbb{F}$ and $\mathbb{H}$, $M$ is also a $(P^*, \mathbb{G})$-martingale; in the same way we can argue that $H$ is a $(P^*, \mathbb{G})$-martingale; the equivalence between $P$ and $P^*$ follows by the fact that the joint law $p_{\eta,\tau}$ of $(\eta, \tau)$ is strictly positive on the set $\{1, 2, 3\} \times \{2, 4\}$.

Let us show that condition (6) implies that $P$ is the *minimal martingale measure* for $H'$ on $(\Omega, \mathcal{F}, \mathbb{G}, P^*)$, that is any $(P^*, \mathbb{G})$-local martingale orthogonal to $\mathbb{H}$ is a $(P, \mathbb{G})$-local martingale.

By Theorem 4.5 in [3] the subset of $\mathcal{M}^2_0(P^*, \mathbb{G})$ of the martingales orthogonal to $H$ coincides with $\mathcal{Z}^2(M) \oplus \mathcal{Z}^2([M, H])$ (here as usual, we denote by $\mathcal{M}^2_0(P^*, \mathbb{G})$ the set of centred square integrable real valued $(P^*, \mathbb{G})$-martingales, and by $\mathcal{Z}^2(M)$ and $\mathcal{Z}^2([M, H])$ the *stable subspaces* generated by $M$ and $[M, H]$ respectively). This reduces our problem to prove that under condition (6) $M$ and $[M, H]$ are $(P, \mathbb{G})$-martingales.

Let's start by observing that

$$\mathcal{F}^\eta_t = \begin{cases} \{\phi, \Omega\} \text{ if } t < 1 \\ \sigma\{\eta = 1\} \text{ if } t \in [1, 2) \\ \sigma(\eta) \text{ if } t \geq 2 \end{cases} \qquad \mathcal{H}_t = \begin{cases} \{\phi, \Omega\} \text{ if } t < 2 \\ \sigma(\tau) \text{ if } t \geq 2 \end{cases}$$

so that

$$\mathcal{F}_t^\eta \vee \mathcal{H}_t = \begin{cases} \{\phi, \Omega\} \text{ if } t < 1 \\ \sigma\{\eta = 1\} \text{ if } t \in [1, 2) \\ \sigma(\eta) \vee \sigma(\tau) \text{ if } t \geq 2. \end{cases}$$

Note that $M$ is a $(P, \mathbb{G})$-martingale without any further condition, since $M$ is a sum of two $(P, \mathbb{G})$-martingales, $B$ and $H^\eta$. The fact that $H^\eta$ is a $(P, \mathbb{G})$-martingale follows by observing that

$$A^{\eta, P, \mathbb{G}} = A^{\eta, P^*, \mathbb{G}}.$$

In fact by the $P$-independence of $B$ of $(\eta, \tau)$ we get

$$A^{\eta, P, \mathbb{G}} = A^{\eta, P, \mathbb{F}^\eta \vee \mathbb{H}},$$

and by the expressions of $\mathcal{F}_t^\eta \vee \mathcal{H}_t$ and of $\mathcal{F}_t^\eta$

$$A^{\eta, P, \mathbb{F}^\eta \vee \mathbb{H}} = A^{\eta, P, \mathbb{F}^\eta}$$

(see Proposition 3.4). Moreover since $B$ and $\eta$ are $P$-independent

$$A^{\eta, P, \mathbb{F}^\eta} = A^{\eta, P, \mathbb{F}},$$

and finally since by construction $P|_{\mathcal{F}_T} = P^*|_{\mathcal{F}_T}$ and $\mathbb{F}$ and $\tau$ are $P^*$-independent

$$A^{\eta, P, \mathbb{F}} = A^{\eta, P^*, \mathbb{G}}.$$

Let us now prove that under condition (6) the $(P^*, \mathbb{G})$-martingale $[M, H]$ is also a $(P, \mathbb{G})$-martingale. From the general formula

$$[M, H]_t = \langle M^c, H^c \rangle_t^{P^*, \mathbb{G}} + \sum_{s \leq t} \Delta H_s \Delta H_s,$$

recalling that $H^c \equiv 0$, we derive that $P^*$-a.s. and therefore $P$-a.s.

$$[M, H]_t = \sum_{s \leq t} \Delta M_s \Delta H_s.$$

Observe that the only common jump between $M$ and $H$ is $t = 2$ and therefore

$$[M, H]_t = \Delta M_2 \Delta H_2 \mathbb{I}_{2 \leq t}. \tag{10}$$

Then we need to explicitly compute the $(P^*, \mathbb{G})$-martingales $M$ and $H$, that is to compute the explicit expression of the involved compensators. Applying Proposition 3.4 we easily get

$$A_t^{\eta,P,\mathbb{F}} = A_t^{\eta,P^*,\mathbb{G}} = \begin{cases} 0 & \text{if } t < 1 \\ P(\eta = 1) & \text{if } t \in [1, 2) \\ P(\eta = 1) + P\big(\eta = 2 \mid \sigma\{\eta = 1\}\big) & \text{if } t \in [2, 3) \\ P(\eta = 1) + P\big(\eta = 2 \mid \sigma\{\eta = 1\}\big) + \mathbb{I}_{\eta=3} & \text{if } t \geq 3 \end{cases}$$

and

$$A_t^{\eta,P,\mathbb{H}} = A_t^{\tau,P^*,\mathbb{G}} = \begin{cases} 0 & \text{if } t < 2 \\ P(\tau = 2) & \text{if } t \in [2, 4) \\ P(\tau = 2) + \mathbb{I}_{\tau=4} & \text{if } t \geq 4 \end{cases}$$

so that

$$H_t^\eta = \mathbb{I}_{\eta \leq t} - P(\eta = 1)\mathbb{I}_{1\leq t} - P\big(\eta = 2 \mid \sigma\{\eta = 1\}\big)\mathbb{I}_{2\leq t} - \mathbb{I}_{\eta=3}\mathbb{I}_{3\leq t} \tag{11}$$

$$H_t = \mathbb{I}_{\tau\leq t} - P(\tau = 2)\mathbb{I}_{2\leq t} - \mathbb{I}_{\tau=4}\mathbb{I}_{4\leq t}.$$

In particular

$$\Delta M_2 = \Delta H_2^\eta = \mathbb{I}_{\eta=2} - P\big(\eta = 2 \mid \sigma\{\eta = 1\}\big), \tag{12}$$

$$\Delta H_2 = \mathbb{I}_{\tau=2} - P(\tau = 2). \tag{13}$$

Therefore $\Delta M_2 \Delta H_2$ coincides with

$$\mathbb{I}_{\tau=2,\eta=2} - P(\tau = 2)\mathbb{I}_{\eta=2} - P\big(\eta = 2 \mid \sigma\{\eta = 1\}\big)\mathbb{I}_{\tau=2} + P(\tau = 2)P\big(\eta = 2 \mid \sigma\{\eta = 1\}\big).$$

Observe that, for $s \leq t$, $E\big[[M, H]_t \mid \mathcal{G}_s\big] = E\big[[M, H]_t \mid \mathcal{F}_s^\eta \vee \mathcal{H}_s\big]$ and then

$$E\big[[M, H]_t \mid \mathcal{G}_s\big] - [M, H]_s = \begin{cases} 0 & \text{if } 0 \leq s < t < 2 \\ E\big[[M, H]_t\big] & \text{if } 0 < s < 1 < 2 \leq t \\ E\big[[M, H]_t \mid \sigma\{\eta = 1\}\big] & \text{if } 1 \leq s < 2 \leq t \\ 0 & \text{if } 2 \leq s < t. \end{cases}$$

It follows that the process $[M, H]$ is a $(P, \mathbb{G})$-martingale if and only if

$$E\big[[M, H]_t \mid \sigma\{\eta = 1\}\big] = 0.$$

Since the left hand side in the above equality for $t \geq 2$ is equal to

$$P\big(\tau = 2, \eta = 2 \mid \sigma\{\eta = 1\}\big) - P\big(\tau = 2|\sigma\{\eta = 1\}\big)P\big(\eta = 2 \mid \sigma\{\eta = 1\}\big),$$

previous equality holds if and only if

$$P\big(\tau = 2, \eta = 2|\sigma\{\eta = 1\}\big) = P\big(\tau = 2 \mid \sigma\{\eta = 1\}\big)P\big(\eta = 2 \mid \sigma\{\eta = 1\}\big),$$

or equivalently, if and only if condition (6) holds.

Before proving Theorem 2.3 let us observe that the triplet $(M, H', [M, H])$ is not a $(P, \mathbb{G})$-basis of martingales, since $[M, H]$ is not null and then it cannot be orthogonal to $M$ and $H'$ (see Remark 4.8 in [4]).

*Proof of Theorem* 2.3 (i) This point follows by statement (iii) of Theorem 3.1, since formula (10) says that the quadratic covariation process $[M, H]$ is not identically zero and therefore by Proposition 3.6 in [4], the random measures induced by $\langle M^{dp}\rangle^{P,\mathbb{F}}$ and $\langle H^{dp}\rangle^{P,\mathbb{H}}$ are not $P$-a.s. singular.

(ii) The explicit expression (4) of $M$ follows immediately by (11). As far as $H'$ is concerned, by using Proposition 3.4 we compute

$$A_t^{\tau,P,\mathbb{G}} = \begin{cases} 0 & \text{if } t < 2 \\ P\big(\tau = 2 \mid \sigma\{\eta = 1\}\big) & \text{if } t \in [2, 4) \\ P\big(\tau = 2 \mid \sigma\{\eta = 1\}\big) + \mathbb{I}_{\tau=4} & \text{if } t \geq 4 \end{cases} \tag{14}$$

and (5) follows.

(iii) By Theorem 3.3, we already know that $M$ and $H'$ are orthogonal $(P, \mathbb{G})$-martingales. Then, following Proposition 4.41 in [12] with $k = 3$, $(M_1, M_2, M_3) = (M, H', [M, H])$, we get as a $(P, \mathbb{G})$-basis the triplet $(M, H', K)$, where $K$ is the projection of the $(P, \mathbb{G})$-martingale $[M, H]$ on the subspace orthogonal to $\mathcal{Z}^2(M, H') = \mathcal{Z}^2(M) \oplus \mathcal{Z}^2(H')$.

By the expression of $[M, H]$ (see (10)), in order to get $K$ it is enough to project the random variable $\Delta M_2 \Delta H_2$ into $L^2(\Omega, \sigma\big(\Delta M_2), P\big) \oplus L^2\big(\Omega, \sigma(\Delta H_2'), P\big)$. Actually

$$\sigma(\Delta M_2) = \sigma(\eta), \quad \sigma(\Delta H_2') = \sigma(\tau) \vee \sigma\{\eta = 1\},$$

since

$$\Delta M_2 = \begin{cases} 1 - \frac{P(\eta=2)}{P(\eta=2)+P(\eta=3)} & \text{if } \eta = 2, \eta \neq 1 \Leftrightarrow \eta = 2 \\ -\frac{P(\eta=2)}{P(\eta=2)+P(\eta=3)} & \text{if } \eta \neq 2, \eta \neq 1 \Leftrightarrow \eta = 3 \\ 0 & \text{if } \eta = 1, \eta \neq 2 \Leftrightarrow \eta = 1 \end{cases}$$

and

$$\Delta H_2' = \begin{cases} 1 - \frac{P(\tau=2,\eta=1)}{P(\eta=1)} & \text{if } \eta = 1, \tau = 2 \\ 1 - \frac{P(\tau=2,\eta\neq1)}{P(\eta=2)+P(\eta=3)} & \text{if } \eta \neq 1, \tau = 2 \\ -\frac{P(\tau=4,\eta=1)}{P(\eta=1)} & \text{if } \eta = 1, \tau = 4 \\ -\frac{P(\tau=4,\eta\neq1)}{P(\eta=2)+P(\eta=3)} & \text{if } \eta \neq 1, \tau = 4 \end{cases}$$

(see (12) and (5)).

We start by projecting $\Delta M_2 \Delta H_2$ into $L^2\big(\Omega, \sigma(\eta), P\big)$. By (13) we derive

$$E\big[\Delta M_2 \Delta H_2 \mid \sigma(\eta)\big] = \Delta M_2\Big(P\big(\tau = 2 \mid \sigma(\eta)\big) - P(\tau = 2)\Big). \qquad (15)$$

By using condition (6) we get

$$\begin{aligned} P\big(\tau = 2 \mid \sigma(\eta)\big) =& P(\tau = 2 \mid \eta = 1)\mathbb{I}_{\eta=1} + P(\tau = 2 \mid \eta = 2)\mathbb{I}_{\eta=2} + P(\tau = 2 \mid \eta = 3)\mathbb{I}_{\eta=3} \\ =& P(\tau = 2 \mid \eta = 1)\mathbb{I}_{\eta=1} + \gamma \mathbb{I}_{\eta\neq 1}, \end{aligned}$$

with $\gamma$ as in (8). It follows that $E\big[\Delta M_2 \Delta H_2 \mid \sigma(\eta)\big]$ is equal to

$$\begin{aligned} &\gamma \mathbb{I}_{\eta=2} - P(\tau = 2)\mathbb{I}_{\eta=2} - \gamma P(\eta = 2 \mid \eta \neq 1)\mathbb{I}_{\eta\neq 1} + P(\tau = 2)P(\eta = 2 \mid \eta \neq 1)\mathbb{I}_{\eta\neq 1} \\ &= \big(\gamma - P(\tau = 2)\big)\mathbb{I}_{\eta=2} - \big(\gamma - P(\tau = 2)\big)P(\eta = 2 \mid \eta \neq 1)\mathbb{I}_{\eta\neq 1} \ \text{ if } t \geq 2. \end{aligned}$$

Consequently $\Delta M_2 \Delta H_2 - E\big[\Delta M_2 \Delta H_2 | \sigma(\eta)\big]$ coincides with

$$\mathbb{I}_{\eta=2,\tau=2} - \gamma \mathbb{I}_{\eta=2} - P(\eta = 2 \mid \eta \neq 1)\mathbb{I}_{\eta\neq 1,\tau=2} + \gamma P(\eta = 2 \mid \eta \neq 1)\mathbb{I}_{\eta\neq 1},$$

and, since by easy computations

$$E\Big[\Delta M_2 \Delta H_2 - E\big[\Delta M_2 \Delta H_2 | \sigma(\eta)\big] \mid \sigma(\tau) \vee \sigma\{\eta = 1\}\Big] = 0,$$

then

$$\Delta M_2 \Delta H_2 - E\big[\Delta M_2 \Delta H_2 \mid \sigma(\eta)\big] \in \Big(L^2(\Omega, \sigma(\Delta M_2), P) \oplus L^2(\Omega, \sigma(\Delta H_2'), P)\Big)^{\perp}.$$

Therefore

$$K_t = \Big(\Delta M_2 \Delta H_2 - E\big[\Delta M_2 \Delta H_2 \mid \sigma(\eta)\big]\Big)\mathbb{I}_{2\leq t}, \qquad (16)$$

that is (7) holds. □

*Remark 2.5* Observe that $P\big(\tau = 2 \mid \sigma(\eta)\big)$ is measurable with respect to $\sigma\{\eta = 1\} = \mathcal{G}_{2-}$, so that the process

$$\Big(P\big(\tau = 2 \mid \sigma(\eta)\big) - P(\tau = 2)\Big)\mathbb{I}_{2\leq \cdot}$$

is $\mathbb{G}$-predictable and then by (15)

$$E\big[\Delta M_2 \Delta H_2 \mid \sigma(\eta)\big]\mathbb{I}_{2\leq t} = \int_0^t \big(P(\tau = 2 \mid \sigma(\eta)) - P(\tau = 2)\big)\mathbb{I}_{2\leq s} dM_s. \quad (17)$$

(10), (16) and (17) give

$$[M, H]_t = \int_0^t \Big(P\big(\tau = 2 \mid \sigma(\eta)\big) - P(\tau = 2)\Big)\mathbb{I}_{2\leq s} dM_s + K_t, \quad (18)$$

that is the Kunita-Watanabe representation of $[M, H]$ with respect to the $(P, \mathbb{G})$-martingale $M$. Actually (18) coincides also with the Kunita-Watanabe representation of $[M, H]$ with respect to the $(P, \mathbb{G})$-martingale $(M, H')$.

*Remark 2.6* Trivially the value $\gamma$ (see (8)) and the marginal laws of $\tau$ and $\eta$, let us say $(p, 1-p)$ and $(q_1, q_2, q_3)$, determine $p_{\eta,\tau}$. In fact as soon as

$$p - \gamma\,(q_2 + q_3) \in (0, 1), \quad q_1 - p + \gamma\,(q_2 + q_3) \in (0, 1)$$

that is

$$\frac{p - q_1}{q_2 + q_3} \vee 0 < \gamma < \frac{p}{q_2 + q_3} \wedge 1,$$

one gets

$$p_{\eta,\tau}(2, 2) = \gamma\, q_2, \quad p_{\eta,\tau}(3, 2) = \gamma\, q_3, \quad p_{\eta,\tau}(1, 2) = p - \gamma\,(q_2 + q_3)$$

$$p_{\eta,\tau}(2, 4) = (1-\gamma)\, q_2, \quad p_{\eta,\tau}(3, 4) = (1-\gamma)\, q_3, \quad p_{\eta,\tau}(1, 4) = q_1 - p + \gamma\,(q_2 + q_3).$$

Then, fixed the marginal laws of $\tau$ and $\eta$, depending on the choice of $\gamma$ there is an infinite number of probability measures $P$ associated to the same probability decoupling measure $P^*$. This do not contradict the uniqueness of the minimal martingale measure for $H'$, as different choices of $\gamma$ lead to different expressions of $A^{\tau,P,\mathbb{G}}$. More precisely, recalling (14), with the notation just introduced

$$A_t^{\tau,P,\mathbb{G}} = \begin{cases} 0 & \text{if } t < 2 \\ \frac{p-\gamma\,(q_2+q_3)}{q_1}\mathbb{I}_{\eta=1} + \gamma\mathbb{I}_{\eta\neq 1} & \text{if } t \in [2, 4) \\ \frac{p-\gamma\,(q_2+q_3)}{q_1}\mathbb{I}_{\eta=1} + \gamma\mathbb{I}_{\eta\neq 1} + \mathbb{I}_{\tau=4} & \text{if } t \geq 4. \end{cases}$$

## 3 Some Theoretical Results

Let $T$ be a finite time horizon and let $(\Omega, \mathcal{F}, P)$ be a given probability space. Let $\mathbb{F} = (\mathcal{F}_t)_{t\in[0,T]}$ a filtration satisfying usual conditions and let $M = (M_t)_{t\in[0,T]}$ be a square integrable $(P, \mathbb{F})$-martingale enjoying the $(P, \mathbb{F})$-p.r.p. Let also $\mathbb{H} = (\mathcal{H}_t)_{t\in[0,T]}$ be the natural filtration of $(\mathbb{I}_{\{\tau\leq t\}})_{t\in[0,T]}$ where $\tau$ is a random time.

As in the previous section let us denote by $\mathbb{G}$ the union filtration $\mathbb{F} \vee \mathbb{H}$. Here we assume the following hypothesis (**D**).

(**D**) there exists $Q$ on $(\Omega, \mathcal{G}_T)$ such that $Q$ is equivalent to $P|_{\mathcal{G}_T}$ and $\mathbb{F}$ and $\mathbb{H}$ are $Q$-independent (*decoupling assumption*).

Finally we denote by $H = (H_t)_{t\in[0,T]}$ and $H' = (H'_t)_{t\in[0,T]}$ the $(P, \mathbb{H})$-martingale and the $(P, \mathbb{G})$-martingale respectively, defined like in (2) and (3).

Let $P^*$ be defined on $(\Omega, \mathcal{G}_T, P)$ by

$$dP^* := \frac{dP}{dQ}\bigg|_{\mathcal{F}_T} \frac{dP}{dQ}\bigg|_{\mathcal{H}_T} dQ.$$

Then $M$ and $H$ are independent $(P^*, \mathbb{G})$-martingales.

Let $M^{dp}$ and $H^{dp}$ be the $\mathbb{G}$-accessible martingale part of $M$ and $H$ respectively (see Theorem 1.4 in [20]). Then Theorem 3.11 in [4] implies next result.

**Theorem 3.1** *The multiplicity of* $\mathbb{G}$ *is*

*(i) 1, if and only if the random measures induced by* $\langle M\rangle^{P,\mathbb{F}}$ *and* $\langle H\rangle^{P,\mathbb{H}}$ *are mutually singular. In this case* $M + H$ *is a* $(P^*, \mathbb{G})$*-basis;*
*(ii) 2, if and only if the random measures induced by* $\langle M^{dp}\rangle^{P,\mathbb{F}}$ *and* $\langle H^{dp}\rangle^{P,\mathbb{H}}$ *are mutually singular and those induced by* $\langle M\rangle^{P,\mathbb{F}}$ *and* $\langle H\rangle^{P,\mathbb{H}}$ *are not mutually singular. In this case* $(M, H)$ *is a* $(P^*, \mathbb{G})$*-basis;*
*(iii) 3, if and only if the random measures induced by* $\langle M^{dp}\rangle^{P,\mathbb{F}}$ *and* $\langle H^{dp}\rangle^{P,\mathbb{H}}$ *are not mutually singular. In this case* $(M, H, [M, H])$ *is a* $(P^*, \mathbb{G})$*-basis.*

*Remark 3.2* We stress that the statement of Theorem 3.11 in [4] is more general than the previous one. In fact when $M$ stays for any martingale enjoying the $(P, F^M)$-p.r.p. and $H$ is replaced by any martingale $N$ enjoying the $(P, F^N)$-p.r.p. the above result still holds changing $\mathbb{F}$ in $F^M$ and $\mathbb{H}$ in $F^N$.

Note that, since $H' = H + A^{\tau,P,\mathbb{H}} - A^{\tau,P,\mathbb{G}}$, then $H$ is the $(P^*, \mathbb{G})$-martingale part of the $(P^*, \mathbb{G})$-semi-martingale $H'$. Theorem 4.5 in [4] can be resumed as follows.

**Theorem 3.3** *Assume that any* $(P^*, \mathbb{G})$*-local martingale orthogonal to* $H$ *is a* $(P, \mathbb{G})$*-local martingale, that is that* $P$ *is the minimal martingale measure for* $H'$ *on* $(\Omega, \mathcal{F}, \mathbb{G}, P^*)$.

*Then* $M$ *and* $H'$ *are strongly orthogonal* $(P, \mathbb{G})$*-martingales and the* $\mathbb{R}^3$*-valued* $(P, \mathbb{G})$*-martingale* $\big(M, H', [M, H]\big)$ *enjoys the* $(P, \mathbb{G})$*-p.r.p.*

In order to explicitly compute the elements of the triplet $\big(M, H', [M, H]\big)$ it can be useful next result (see Proposition 3.3 in [4]).

**Proposition 3.4** *Let* $\mathbb{A}$ *be a filtration on* $(\Omega, \mathcal{F}, P)$*. For any* $\mathbb{A}$*-accessible random time* $\theta$ *the* $(P, \mathbb{A})$*-compensator of* $\mathbb{I}_{\theta\leq\cdot}$ *is*

$$A_t^{\theta,P,\mathbb{A}} = \sum_{n\in\mathbb{N}} E\left[\mathbb{I}_{\{\theta=\theta_n\}} \mid \mathcal{A}_{\theta_n-}\right] \mathbb{I}_{[0,t]}(\theta_n),$$

*where* $(\theta_n)_{n\in\mathbb{N}}$ *is the sequence of* $\mathbb{A}$*-predictable stopping times with disjoint graphs enveloping* $\theta$.

**Acknowledgements** The authors acknowledge the MIUR Excellence Department Project awarded to the Department of Mathematics, University of Rome Tor Vergata, CUP E83C18000100006.

## References


1. Aksamit, A., Jeanblanc, M., Rutkowski, M.: Predictable representation property for progressive enlargements of a Poisson filtration. Tech. rep., arXiv:1512.03992v1 (2015)
2. Callegaro, G., Jeanblanc, M., Zargari, B.: Carthaginian enlargement of filtrations. ESAIM Probab. Stat. **17**, 550–566 (2013)
3. Calzolari, A., Torti, B.: Enlargement of filtration and predictable representation property for semi-martingales. Stochastics **88**(5), 680–698 (2016)
4. Calzolari, A., Torti, B.: Martingale representation in dynamic enlargement setting: the role of accessible random times. Tech. rep., arXiv:1708.05858v2 (2019)
5. Chou, C.S., Meyer, P.A.: Sur la représentation des martingales comme intégrales stochastiques dans les processus ponctuels. 226–236. In: Lecture Notes in Mathematics, vol. 465
6. Davis, M.H.A., Varaiya, P.: The multiplicity of an increasing family of $\sigma$-fields. Ann. Probab. **2**, 958–963 (1974)
7. Dellacherie, C., Meyer, P.-A.: Probabilities and potential. A. In: North-Holland Mathematics Studies, pp. xvii+463, vol. 29 North-Holland Publishing Co., Amsterdam 91978)
8. Dellacherie, C., Meyer, P.-A.: Probabilities and potential. B. In: North-Holland Mathematics Studies, pp. xvii+463, vol. 72. North-Holland Publishing Co., Amsterdam (1982). Theory of martingales, Translated from the French by J. P. Wilson
9. Fontana, C., Schmidt, T.: General dynamics term structures under default risk. Tech. rep., arXiv: 1603.03198v2 (2017)
10. He, S.W., Wang, J.G.: The property of predictable representation of the sum of independent semimartingales. Z. Wahrsch. Verw. Gebiete **61**(1), 141–152 (1982)
11. He, S.W., Wang, J.G., Yan, J.A.: Semimartingale theory and stochastic calculus, pp. xiv+546. Kexue Chubanshe (Science Press), Beijing; CRC Press, Boca Raton, FL (1992)
12. Jacod, J.: Calcul stochastique et problèmes de martingales. In: Lecture Notes in Mathematics, pp. x+539, vol. 714. Springer, Berlin (1979)
13. Jeanblanc, M., Le Cam, Y.: Immersion property and credit risk modelling. In: Optimality and Risk—Modern Trends in Mathematical Finance, pp. 99–131. Springer, Berlin (2009)
14. Jeanblanc, M., Song, S.: Martingale representation property in progressively enlarged filtrations. Stochast. Process. Appl. **125**(11), 4242–4271 (2015)
15. Jeanblanc, M., Yor, M., Chesney, M.: Mathematical methods for financial markets, pp. xxvi+732. Springer Finance. Springer-Verlag London Ltd., London (2009)
16. Jiao, Y., Li, S.: The generalized density approach in progressive enlargement of filtrations. Electron. J. Probab. **20**(85), 21 (2015)
17. Jiao, Y., Li, S.: Modeling sovereign risks: from a hybrid model to the generalized density approach. Math. Financ. **28**(1), 240–267 (2018)
18. Kusuoka, S.A.: Remark on default risk models. In: Advances in Mathematical Economics. Tokyo,: vol. 1 of Adv. Math. Econ. Springer, Tokyo **1999**, 69–82 (1997)
19. Song, S.: Drift operator in a viable expansion of information flow. Stochast. Process. Appl. **126**(8), 2297–2322 (2016)

20. Yoeurp, C.: Décompositions des martingales locales et formules exponentielles. In: Séminaire de Probabilités, X (Seconde partie: Théorie des intégrales stochastiques, Univ. Strasbourg, Strasbourg, année universitaire 1974/1975). Lecture Notes in Mathematics, vol. 511, pp. 432–480. Springer, Berlin (1976)

# European Option Pricing with Stochastic Volatility Models Under Parameter Uncertainty

**Samuel N. Cohen and Martin Tegnér**

**Abstract** We consider stochastic volatility models under parameter uncertainty and investigate how model derived prices of European options are affected. We let the pricing parameters evolve dynamically in time within a specified region, and formalise the problem as a control problem where the control acts on the parameters to maximise/minimise the option value. Through a dual representation with backward stochastic differential equations, we obtain explicit equations for Heston's model and investigate several numerical solutions thereof. In an empirical study, we apply our results to market data from the S&P 500 index where the model is estimated to historical asset prices. We find that the conservative model-prices cover 98% of the considered market-prices for a set of European call options.

**Keywords** Option pricing · Stochastic volatility · Model uncertainty

## 1 Introduction

In this paper, we consider the problem of European-option pricing when the underlying assets follow a stochastic volatility model in a setting that accommodates for *parameter uncertainty*, and in particular, how this transfers to *conservative bounds* for derived option prices.

Stochastic volatility models feature an instantaneous variance of the asset price, the *volatility*, that evolves stochastically in time. It is a natural generalisation of the seminal constant-volatility model of [15], and examples include the models introduced by [10, 38–40, 54] to mention a few. Evidence supporting this generalisation

S. N. Cohen · M. Tegnér
Mathematical Institute, University of Oxford, Oxford, UK
e-mail: samuel.cohen@maths.ox.ac.uk

M. Tegnér (✉)
Department of Engineering Science, University of Oxford, Oxford, UK
e-mail: martin.tegner@eng.ox.ac.uk

Department of Mathematical Sciences, University of Copenhagen, Copenhagen, Denmark



S. N. Cohen et al. (eds.), *Frontiers in Stochastic Analysis - BSDEs, SPDEs and their Applications*, Springer Proceedings in Mathematics & Statistics 289,
https://doi.org/10.1007/978-3-030-22285-7_5

in terms of empirical asset-return behaviour goes back to [14], while for instance [55] highlights the prediction mismatch of a constant volatility and option prices observed from the market. Stochastic volatility serves as an attractive alternative and numerous studies are available from the literature in their favour.

Being a parametric model immediately implies that the stochastic volatility model has to be fitted with data before it is employed for pricing or hedging market instruments. At least two approaches are conventional for this purpose: either estimation from historical asset-prices, or calibration from market option-prices by matching the model derived price (or a combination of the two, see for example [1]). Regardless of which approach is used, one is exposed to *parameter uncertainty* since point-estimates from either are subject to errors. An estimator based on observed time-series of asset prices has an inherent variance (and could potentially be biased), while the calibration problem might be ill-posed—the minimum obtained from numerical optimisation can be local, and several parameter settings might give the same model-to-market matching.

A notion of parameter uncertainty naturally arises with statistical estimation from asset-prices since the error of the point-estimate can be quantified by an inferred confidence interval. The confidence interval thus defines an uncertainty set which contains the true value of the model parameters, at a given confidence level. In this case, inferred uncertainty and estimated parameters will be associated with the *real-world* probability measure, as opposed to the *risk-neutral* measure(s) used for no-arbitrage pricing. On the other hand, calibration from option-prices will give a set of parameters associated with the risk-neutral measure. In this case, however, there is no obvious way of how to deduce an uncertainty set for the parameters which quantifies the errors that stem from the calibration.

The question remains how the parameter uncertainty affects option prices outputted by the stochastic volatility model. In the case of statistical estimation, one needs to establish the relation between the parameters under the statistical measure and under a risk-neutral pricing measure. In financial terms, this is accommodated by the market price of risk, and typically in such a way that the model remains form-invariant. The uncertainty may then be propagated to the risk-neutral parameters which are used for option pricing. We consider uncertainty in drift- and jump parameters[1] to offer an interpretation of the parameter uncertainty as representative for the *incompleteness* of the stochastic volatility model: there exists a space of equivalent pricing measures as given by the span of risk-neutral parameter values in the uncertainty set (we elaborate on this in the introducing discussion of Sect. 3).

We immediately look at model pricing from a best/worst case point of view, and aim to obtain conservative *pricing bounds* inferred from the parameter uncertainty. Two approaches are fair: either optimising the pricing function over the parameters

---

[1] A supporting case for this assumption is the fact pointed out for instance by [53]: while volatilities may be estimated within reasonable confidence with a few years of data, drift estimation requires data from much longer time periods.

constrained by the uncertainty set, or treating the parameters as dynamical components of a *control process* which acts to optimise the option value. The former is thus a special case of the latter where the control process is restricted to take constant values only. We formalise the problem as a control problem and since all pricing measures are equivalent, this can be seen as change of measure problem. Following the results due to [52], the optimal value function of the option price may then by expressed as a backward stochastic differential equation.

The postulation of parameter uncertainty, or more generally *model uncertainty*, as an inherent model feature is certainly not novel and its importance in finance was early acknowledged by [26]. Conceptually, model uncertainty draws on the principles due to [43, 45] of the *unknown unknown* as distinguished from the *known unknown*. Following the overview by [8], model uncertainty—the unknown unknown—refers to the situation when a whole family of models is available for the financial market, but the *likelihood* of each individual model is unknown. Parameter uncertainty is thus the special case where the family of models may be parametrized. Further, if a probability measure is attributed to the model (parameter) family—the known unknown—one is in the situation of *model (parameter) risk*.

When it comes to option pricing, Bayesian methods offer a fruitful way of inferring parameter and model risk, and take it into account by model averaging, see for example [19, 36, 41] and the non-parametric approach to local volatility by [56]. Considering the situation of model uncertainty, the worst-case approach taken here was pioneered in the works of [6, 7, 28, 47]. Our control-theoretic approach is similar to that of [6], but in contrast to their unspecified volatility, we place ourself in a "within-model" setting of parametrised volatility models where the parameters are controlled instead of the volatility itself. We thus account for a case where the uncertain family of volatility models gives a more detailed description of the financial market. Arguably, this implies conservative prices which are more realistic. Since we also suggest how to infer the uncertainty set for the parameters, our approach should be particularly appealing for stochastic-volatility inclined practitioners.

**Overview** The model proposed by [38] will be the working model of our study, and we present the risk-neutral pricing of European options in Sect. 2 along with the BSDE representation of the controlled value process. We show how to derive the optimal driver that generates the BSDE of the optimally controlled value processes, which gives pricing bounds for options under parameter uncertainty. To obtain actual values for the pricing bounds, we resort to numerical solutions for the BSDE that governs the optimal value. In Appendix A, we detail some simulation schemes for this purpose, and demonstrate the methods in a controlled setting to be able to compare and evaluate their performance. With a suggested numerical scheme in hand, we proceed in Sect. 3 to illustrate our method empirically on real-world market data. For a set of market quotes of European call options on the S&P 500 index, we investigate how well the (numerically calculated) model bounds actually cover observed market prices. We also compare the results with the corresponding constant-parameter optimal price.

For completion, we finally treat the general multi-asset case of a Markovian stochastic volatility model with jumps in Sect. 4. Section 5 concludes.

## 2 The Heston Stochastic Volatility Model

To set the scene, we consider a financial market consisting of a risk-free money account and a risky asset over a fixed time period $[0, T]$. We assume the standard assumptions of a frictionless market: short selling is permitted and assets may be held in arbitrary amounts, there are no transaction costs and borrowing and lending are made at the same interest rate. The prices of the assets will be modelled as adapted stochastic processes on a filtered probability space, the notion of which will be formalised in the following section.

### *2.1 European Option Pricing*

Let $(\Omega, \mathcal{F}, \{\mathcal{F}_t\}_{t\geq 0}, P)$ be a filtered probability space where $\{\mathcal{F}_t\}_{t\geq 0}$ is the natural filtration generated by two independent Wiener processes $W^1$ and $W^2$, augmented to satisfy the usual conditions of $P$-completeness and right continuity. We assume that the asset price $S$ and variance $V$ follow the model by [38], with real-world dynamics (under the objective probability measure $P$) given by

$$
\begin{aligned}
dS_t &= \mu(V_t)S_t dt + \sqrt{V_t}S_t(\rho dW_t^1 + \sqrt{1-\rho^2}dW_t^2),\\
dV_t &= \kappa(\theta - V_t)dt + \sigma\sqrt{V_t}dW_t^1,
\end{aligned}
$$

for nonnegative constants $\kappa$, $\theta$, $\sigma$ and instantaneous correlation $\rho \in (-1, 1)$. The variance follows a square root process[2] and it is bounded below by zero. If Feller's condition is satisfied, $2\kappa\theta \geq \sigma^2$, the boundary cannot be achieved. Furthermore, the relative rate of return $\mu$ is taken to be a deterministic function of the variance. In addition to the risky asset, the market contains a risk-free money account which value processes is denoted $B$. The money account pays a constant rate of return $r$, which means that $B$ obeys the deterministic dynamics $dB_t = rB_t dt$.

The market price of risk processes $(\gamma^1, \gamma^2)$ associated with $W^1$ and $W^2$ are assumed to be specified such that

$$
\frac{\mu(V) - r}{\sqrt{V}} = \left(\rho\gamma^1 + \sqrt{1-\rho^2}\gamma^2\right) \tag{1}
$$

[2]Also know as a CIR process from its use as a model for short-term interest rates by [24]. The square root process goes back to [29].

and as suggested by [38], we let $\gamma^1 \equiv \lambda\sqrt{V}$ for some constant $\lambda$.[3] We then have that the stochastic exponential of $-(\gamma^1, \gamma^2) \bullet (W^1, W^2)$ is given by[4]

$$\mathcal{E}(-\gamma \bullet W) = \exp\left(-\int_0^{\cdot} \lambda\sqrt{V_s} dW_s^1 - \int_0^{\cdot} \gamma_s^2 dW_s^2 - \frac{1}{2}\int_0^{\cdot} (\lambda^2 V_s + (\gamma_s^2)^2) ds\right)$$

and if we define the measure $Q$ on $\mathcal{F}_T$ for a fixed deterministic time $T$ by

$$\frac{dQ}{dP} = \mathcal{E}(-\gamma \bullet W)_T$$

we have that $Q$ is equivalent to $P$ (provided the stochastic exponential is a martingale, i.e. $\mathbb{E}[\mathcal{E}(-\gamma \bullet W)_t] = 1$ for all $t \in [0, T]$, for which Novikov's and Kazamaki's conditions are sufficient. [57] express this explicitly in terms of the parameters). Further, by the Girsanov theorem, $\{\tilde{W}_t^1\}_{t\in[0,T]}$ and $\{\tilde{W}_t^2\}_{t\in[0,T]}$ defined by

$$d\tilde{W}_t^1 = \lambda\sqrt{V_t}dt + dW_t^1,$$
$$d\tilde{W}_t^2 = \gamma_t^2 dt + dW_t^2,$$

are independent Wiener processes under $Q$. By virtue of Eq. (1), this gives the $Q$-dynamics of the model

$$\begin{aligned} dS_t &= rS_t dt + \sqrt{V_t} S_t(\rho d\tilde{W}_t^1 + \sqrt{1-\rho^2} d\tilde{W}_t^2), \\ dV_t &= (\kappa\theta - [\kappa + \sigma\lambda] V_t) dt + \sigma\sqrt{V_t} d\tilde{W}_t^1, \end{aligned} \tag{2}$$

for $t \in [0, T]$ and we note that the variance dynamics are form invariant under the measure change: $V$ also follows a square root process under $Q$ with "risk-neutral" parameters $\tilde{\kappa}, \tilde{\theta}, \sigma$ where

$$\tilde{\kappa} = \kappa + \sigma\lambda \quad \text{and} \quad \tilde{\theta} = \frac{\kappa\theta}{\kappa + \sigma\lambda}.$$

We also see that the discounted asset price $B^{-1}S$ will be a $Q$-martingale (i.e. $Q$ is an equivalent martingale measure) such that the financial market model $(B, S)$ is arbitrage-free. However, as $(\gamma^1, \gamma^2)$ may be arbitrarily chosen as long as (1) is satisfied, the model is incomplete. This means that $\lambda$ could be determined by a single exogenously given asset (with a volatility dependent price) to complete the market,

---

[3] Heston motivates this choice from the model of [18] under the assumption that the equilibrium consumption process also follows a square-root process; the risk premium is then proportional to variance. Aggregate risk preferences aside, a consequence is that the pricing equation (3) conveniently allows for Heston's pricing formula.

[4] We use $\bullet$ to denote the stochastic integral of $d$-dimensional processes: $H \bullet M = \sum_{i=1}^d \int_0^{\cdot} H_t^i dM_t^i$ for $H, M$ taking values in $\mathbb{R}^d$.

and $\gamma^2$ is uniquely determined by Eq. (1). Any other contingent claim will then be uniquely priced.

For a European option with payoff $g(S_T)$ at maturity $T$, we have that the $\mathcal{C}^{1,2}$ function $D(t, s, v)$ of the pricing rule $D_t = D(t, S_t, V_t)$, $t \in [0, T]$, for the option satisfies the following partial differential equation

$$\frac{\partial D}{\partial t} + rs\frac{\partial D}{\partial s} + \{\kappa\theta - v(\kappa + \sigma\lambda)\}\frac{\partial D}{\partial v} + \frac{1}{2}s^2 v\frac{\partial^2 D}{\partial s^2} + \rho\sigma vs\frac{\partial^2 D}{\partial v \partial s} + \frac{1}{2}\sigma^2 v\frac{\partial^2 D}{\partial v^2} = rD, \quad (3)$$

with terminal condition $D(T, s, v) = g(s)$. Notice that the expression in curly brackets can be equivalently written $\tilde{\kappa}(\tilde{\theta} - v)$ with the risk-neutral specification of the parameters. Equivalently, by Feynman–Kac, this is the usual risk-neutral pricing formula

$$D(t, s, v) = \mathbb{E}^Q\left[e^{-r(T-t)}g(S_T)\middle| (S_t, V_t) = (s, v)\right]$$

where $(S, V)$ follows the $Q$-dynamics with initial value $(S_t, V_t) = (s, v)$ at the initial time $t \in [0, T]$.

The pricing equation (3) is the same as in Heston's original paper if we let $\lambda_v = \sigma\lambda$ and $\lambda(t, S_t, V_t) = \lambda_v V_t$ the price of volatility risk used in his exposition. The equation is solved by considering the Fourier transform of the price and the resulting "semi-closed" pricing formula is obtained by the inverse transform. In practice, however, the inverse transform has to be calculated by numerical integration methods.

## 2.2 *Conservative Pricing Under Parameter Uncertainty*

Heston's model (and any other stochastic volatility model) is fundamentally a model for the underlying financial market even if it is predominantly used for option pricing. The pricing measure is often taken as being fixed for convenience, for instance through model-to-market calibration of option prices, and the connection to the objective measure is not important for the analysis; and hence not necessarily made explicit.

Although we are dealing with option pricing as well, we take a slightly converse approach in the case when the pricing measure inherits its uncertainty from the objective measure. Here we infer uncertainty of pricing parameters from statistical estimation of objective parameters, and the relation between the measures will thus play an integral role. On the other hand when uncertainty is deduced from a calibration method of pricing parameters directly, there is no need to make an explicit connection between the measures. We will handle both cases simultaneously and for this purpose, we assume a pricing measure $Q$ to be given momentarily just to be able to replace it with another pricing measure that is subject to uncertainty.

To this end, we introduce parameter uncertainty in our model by modifying our reference measure with the effect of a control that governs the parameter processes. Namely, we replace the risk-neutral measure $Q$ with an equivalent measure $Q^u$ under which we have the controlled dynamics

$$\begin{aligned} dS_t &= r^u(u_t)S_t dt + \sqrt{V_t}S_t(\rho dW_t^{u1} + \sqrt{1-\rho^2}dW_t^{u2}), \\ dV_t &= \kappa^u(u_t)\left(\theta^u(u_t) - V_t\right)dt + \sigma\sqrt{V_t}dW_t^{u1}, \end{aligned} \tag{4}$$

for $t \in [0, T]$. The control process $\{u_t\}_{t\geq 0}$ is an $\mathcal{F}_t$-predictable process that takes values in a compact set $U \subset \mathbb{R}^3$, which we call the parameter uncertainty set. We write $\mathcal{U}$ for the space of admissible control processes (that is, predictable processes taking values in $U$ with sufficient integrability) and under $Q^u$, we have that $W^{u1}$ and $W^{u2}$ are independent Wiener processes, as will be explained in a moment. The control process realises its paths stochastically, and we simply do not know beforehand which $\{u_t\}_{t\geq 0} \in \mathcal{U}$ will be governing (4): the uncertainty is tantamount to this choice.

Furthermore, we denote the components of the control $\{u_t\}_{t\geq 0} = \{r_t, \kappa_t, \theta_t\}_{t\geq 0}$ and let the controlled drift-functions of (4), all $f : U \to \mathbb{R}^+$, be defined as

$$r^u(u_t) = r_t, \qquad \kappa^u(u_t) = \kappa_t + \sigma\lambda \quad \text{and} \quad \theta^u(u_t) = \frac{\kappa_t\theta_t}{\kappa_t + \sigma\lambda}. \tag{5}$$

Notice that this specification of the controlled drift relies on the premise that the $Q$-parameters $r, \tilde{\kappa}, \tilde{\theta}$ are subject to parameter uncertainty by their replacement with $r^u, \kappa^u, \theta^u$. The uncertainty is in turn taken to be inferred from statistical estimation of the objective $P$-parameters, represented by $(r_t, \kappa_t, \theta_r) \in U$ where $U$ is the statistical uncertainty set, and transferred to the pricing parameters by the map

$$U \ni u_t \mapsto \left(r^u(u_t), \kappa^u(u_t), \theta^u(u_t)\right) \in U^\lambda$$

as given by (5). Here $U^\lambda$ is the uncertainty set for the controlled parameters, induced by the range of the mapping. The parameter $\lambda$ associated with $Q$ thus plays an instrumental role in facilitating the uncertainty transfer and it determines the set $U^\lambda$ where uncertain price-parameters live. In practice, we forcefully set $\lambda = 0$ to obtain that the uncertainty in price-parameters is exactly that of the uncertainty in estimated real-world parameters, i.e. $U^\lambda \equiv U$. However, note that this does not imply $P \equiv Q$ nor $\mu = r$, cf. Eq. (1).

Similarly, when a calibration approach is employed to give pricing parameters and an associated uncertainty set directly, the identities

$$r^u(u_t) = r_t, \qquad \kappa^u(u_t) = \kappa_t, \qquad \theta^u(u_t) = \theta_t \tag{6}$$

will facilitate the replacement of $r, \tilde{\kappa}, \tilde{\theta}$ with the control $(r_t, \kappa_t, \theta_t) \in U$, now representing uncertain pricing parameters restricted to lie within a risk-neutral uncertainty set $U$.[5]

With the controlled dynamics of $(S, V)$ representing the model under parameter uncertainty, we proceed to define the upper and lower boundary for the price of a

---

[5]Here we could be a bit more finical on notation, for instance with $(r_t, \tilde{\kappa}_t, \tilde{\theta}_t) \in \tilde{U}$ representing the pricing uncertainty deduced from calibration. For brevity, we refrain from such a notional distinction.

European option. For an option written on $S$ with terminal payoff at time $T$ given by a square-integrable $\mathcal{F}_T$-measurable random variable $G$, we take the most conservative prices from valuation under the controlled pricing measures (i.e. under parameter uncertainty) as given by the control problems

$$D_t^- = \operatorname*{ess\,inf}_{\{u_t\}\in\mathcal{U}} \mathbb{E}_u\left[e^{-\int_t^T r_s ds} G \middle| \mathcal{F}_t\right] \quad \text{and} \quad D_t^+ = \operatorname*{ess\,sup}_{\{u_t\}\in\mathcal{U}} \mathbb{E}_u\left[e^{-\int_t^T r_s ds} G \middle| \mathcal{F}_t\right] \tag{7}$$

for $t \in [0, T]$, where $\mathbb{E}_u(\cdot|\mathcal{F})$ denotes the conditional expectation under $Q^u$. In a sense, we thus consider the super-replication costs of selling a long/short position in the option when the uncertain parameters evolve stochastically in the uncertainty set, in an optimal way.[6]

In order to find a pricing PDE that corresponds to Eq. (3) of the previous section, we henceforth consider payoffs given by $G = g(S_T)$ for some non-negative function $g$. Due to the Markovian structure of the problem, we then have that the optimal value processes will be functions of the current asset price and variance state

$$D_t^- = D^-(t, S_t, V_t),$$
$$D_t^+ = D^+(t, S_t, V_t),$$

for some continuous functions $D^\pm : [0, T] \times \mathbb{R}^+ \times \mathbb{R}^+ \to \mathbb{R}$. As we will see later, these functions will satisfy a semilinear version of the standard pricing equation for European options. However, before we arrive at more precise expressions for the optimally controlled value processes (and their generating functions) we take one step backwards: we will first consider the value process for a fixed control and its link to a backward stochastic differential equation. Following the approach due to [52] as outlined in [22], we then consider the optimally controlled value process as the solution to a closely related BSDE.

To find pricing-boundaries through a dual formulation with BSDEs, consider the following representation from Girsanov's theorem

$$\begin{aligned} dW_t^{1u} &= d\tilde{W}_t^1 - \alpha_1(t, S_t, V_t)dt, \\ dW_t^{2u} &= d\tilde{W}_t^2 - \alpha_2(t, S_t, V_t)dt. \end{aligned} \tag{8}$$

For its kernel $\alpha = (\alpha_1, \alpha_1)^\top$, straightforward algebra then shows that

$$\alpha(S_t, V_t, u_t) = \frac{1}{\sigma\sqrt{V_t}} \begin{bmatrix} \kappa^u(u_t)\theta^u(u_t) - \tilde{\kappa}\tilde{\theta} - (\kappa^u(u_t) - \tilde{\kappa})V_t \\ \frac{-\rho\left(\kappa^u(u_t)\theta^u(u_t) - \tilde{\kappa}\tilde{\theta} - (\kappa^u(u_t) - \tilde{\kappa})V_t\right) + \sigma(r_t - r)}{\sqrt{1-\rho^2}} \end{bmatrix} \tag{9}$$

[6]This draws on the interpretation that $\{Q^u : u \in \mathcal{U}\}$ is the set of equivalent martingale measures of an incomplete market model, such that the most conservative risk-neutral price of an option equals the super-replication cost of a short position in the same: with $\Pi_t(G) = \inf_\phi\{\tilde{V}_t(\phi) : V_T(\phi) \geq G, \text{ a.s.}\}$ being the discounted portfolio value of the (cheapest) admissible strategy $\phi$ that super-replicates $G$, then $\Pi_t(G) = \operatorname{ess\,sup}_{u\in\mathcal{U}} \mathbb{E}_u[\tilde{G}|\mathcal{F}_t]$ and the supremum is attained. See for instance [23], Sect. 10.2.

ties together the dynamics (4) and (2) associated with $W^u$ and $\tilde{W}$, respectively. Formally, if the stochastic exponential of the process $\alpha(S, V, u)^\top \bullet (\tilde{W}^1, \tilde{W}^2)$ defines the measure change $Q \to Q^u$ on $\mathcal{F}_T$,

$$\frac{dQ^u}{dQ} = \mathcal{E}\left( \int_0^\cdot \alpha_1(S_t, V_t, u_t) d\tilde{W}_t^1 + \int_0^\cdot \alpha_2(S_t, V_t, u_t) d\tilde{W}_t^2 \right)_T,$$

(provided $\mathcal{E}(\alpha^\top \bullet \tilde{W})$ is a martingale), then $W^{1u}$ and $W^{2u}$ defined in (8) are two independent Wiener processes under $Q^u$ by Girsanov's theorem.

Next, define the linear driver function $f : (0, \infty) \times (0, \infty) \times \mathbb{R} \times \mathbb{R}^{1\times 2} \times U \to \mathbb{R}$ as

$$f(s, v, y, z, u) = \varrho(s, v, u)y + z\alpha(s, v, u) \tag{10}$$

where $\varrho(s, v, u) \equiv -r$, that is, the (negative) first component of the control which represents the risk-free interest rate. With the driver defined by (9)–(10), we obtain the following representation of the expected value under $Q^u$: for a given, fixed control process $u = \{u_t\}_{t\geq 0} \in \mathcal{U}$, the controlled value process given by

$$J_t(u) = \mathbb{E}_u\left[ e^{-\int_t^T r_s ds} g(S_T) \middle| \mathcal{F}_t \right], \quad t \in [0, T],$$

is the unique solution to the linear Markovian backward stochastic differential equation

$$\begin{aligned} dJ_t(u) &= -f(S_t, V_t, J_t(u), Z_t, u_t)dt + Z_t d\tilde{W}_t, \\ J_T(u) &= g(S_T), \end{aligned} \tag{11}$$

where $Z = (Z^1, Z^2)^\top$—the martingale representation part of $J(u)$—is a process taking values in $\mathbb{R}^{1\times 2}$ (being a part of the solution to the BSDE). To see this, consider the process $J(u)$ that solves (11) and let $\mathcal{E}(\Gamma)$ be the stochastic exponential of

$$\Gamma = \int_0^\cdot -r_t dt + \int_0^\cdot \alpha(S_t, V_t, u_t)^\top d\tilde{W}_t.$$

Apply Itô's product rule to $\mathcal{E}(\Gamma)J(u)$ to obtain

$$d\left(\mathcal{E}(\Gamma)_t J_t(u)\right) = \mathcal{E}(\Gamma)_t \left( Z_t + J_t(u)\alpha(S_t, V_t, u_t)^\top \right) d\tilde{W}_t$$

and thus, since $\mathcal{E}(\Gamma)J(u)$ is a martingale under $Q$,

$$\begin{aligned} J_t(u) &= \frac{1}{\mathcal{E}(\Gamma)_t} \mathbb{E}^Q \left[ \mathcal{E}(\Gamma)_T\, g(S_T) \middle| \mathcal{F}_t \right] \\ &= e^{\int_0^t r_s ds} \frac{1}{\mathcal{E}(\alpha^\top \bullet \tilde{W})_t} \mathbb{E}^Q \left[ e^{-\int_0^T r_s ds} \mathcal{E}(\alpha^\top \bullet \tilde{W})_T\, g(S_T) \middle| \mathcal{F}_t \right] \\ &= \mathbb{E}_u \left[ e^{-\int_t^T r_s ds} g(S_T) \middle| \mathcal{F}_t \right] \end{aligned}$$

as $\mathcal{E}(\alpha^\top \bullet \tilde{W})$ is the density for the measure change $Q \to Q^u$.

The BSDE (11) governs the value process under the impact of a fixed control process that evolves in the uncertainty set $U$. To obtain the lowest (highest) value scenario, the value process is minimised (maximised) over all admissible controls in $\mathcal{U}$, and as we detail in the following, this is done through pointwise optimisation with respect to $u \in U$ of the driver function for the value process. Hence, define the following drivers optimised over the parameter uncertainty set

$$H^-(s, v, y, z) = \operatorname*{ess\,inf}_{u \in U} f(s, v, y, z, u) \quad \text{and} \quad H^+(s, v, y, z) = \operatorname*{ess\,sup}_{u \in U} f(s, v, y, z, u),$$

and note that as $U$ is compact, the infimum and supremum are both attained. We then have the main result due to the comparison principle for BSDEs: the lower/upper optimally controlled value processes

$$D_t^- = \operatorname*{ess\,inf}_{u \in \mathcal{U}} J_t(u) \quad \text{and} \quad D_t^+ = \operatorname*{ess\,sup}_{u \in \mathcal{U}} J_t(u)$$

for $t \in [0, T]$, have cadlag modifications that are the unique solutions of the BSDEs

$$\begin{aligned} dD_t^\pm &= -H^\pm(S_t, V_t, D_t^\pm, Z_t)dt + Z_t d\tilde{W}_t, \\ D_T^\pm &= g(S_T). \end{aligned} \tag{12}$$

In particular, the processes are deterministic functions of $(t, S_t, V_t)$, that is, $D_t^\pm = D^\pm(t, S_t, V_t)$ for some continuous functions $D^\pm : [0, T] \times \mathbb{R}^+ \times \mathbb{R}^+ \to \mathbb{R}$. As the infimum (supremum) of $H$ is attained, we further have that there exists optimal controls $\{u_t^{\pm *}\}_{t \in [0,T]} \in \mathcal{U}$ which are *feedback* controls. This means that the processes

$$u_t^{\pm *} = u^{\pm *}(t, S_t, V_t), \quad t \in [0, T],$$

are the optimal controls among all predictable controls for some deterministic functions $u^{\pm *} : [0, T] \times \mathbb{R}^+ \times \mathbb{R}^+ \to U$. Finally, by the semilinear Feynman-Kac formula (provided a solution exists), we have that $D^-(t, s, v)$ satisfies the following semilinear parabolic PDE

$$\frac{\partial D}{\partial t} + \frac{1}{2}s^2 v \frac{\partial^2 D}{\partial s^2} + \rho\sigma v s \frac{\partial^2 D}{\partial v \partial s} + \frac{1}{2}\sigma^2 v \frac{\partial^2 D}{\partial v^2} \tag{13}$$

$$+ \operatorname*{ess\,inf}_{(r,\kappa,\theta) \in U} \left\{ -rD + rs\frac{\partial D}{\partial s} + \kappa^u(\kappa)\left(\theta^u(\theta) - v\right)\frac{\partial D}{\partial v} \right\} = 0 \tag{14}$$

with terminal value $D^-(T, s, v) = g(s)$. In the corresponding equation for $D^+(t, s, v)$ we have a supremum substituted for the infimum.

*Proof* For the first part of the result, since $H^-(s, v, y, z) \leq f(s, v, y, z, u)$ by definition, we have that the (unique) solution[7] $Y$ to the BSDE with data $(g(S_T), H^-)$ satisfies $Y_t \leq J_t(u)$ for all controls $u \in \mathcal{U}$ (up to indistinguishability). This is a consequence of the comparison theorem for BSDEs ([50], see also [27]). Further, by Filippov's implicit function theorem ([12], see also [48] and [30]), for each $\epsilon > 0$ there exists a predictable control $u^\epsilon \in \mathcal{U}$ such that $f(s, v, y, z, u^\epsilon) \leq H^-(s, v, y, z) + \epsilon$. Since $Y_t + \epsilon(T - t)$ solves the BSDE with driver $H^-(s, v, y, z) + \epsilon$, the comparison theorem yields $J_t(u^\epsilon) \leq Y_t + \epsilon(T - t)$ (up to indistinguishability) and we have the inclusion

$$Y_t \leq J_t(u^\epsilon) \leq Y_t + \epsilon(T - t).$$

Letting $\epsilon \to 0$ we have that $Y_t = \operatorname{ess\,inf}_{u \in \mathcal{U}} J_t(u) = D_t^-$ for every $t$ which is to say that $Y$ is a version of the optimal value process. That $D_t^-$ can be written as a continuous function of $(t, S_t, V_t)$ is due to the fact that (12) is a Markovian BSDE. Further, as the optimal control is attainable, Filippov's theorem gives that it is a function of $(t, S_t, V_t, D_t^-, Z_t)$ where $Z_t = z(t, S_t, V_t)$—due to the Markovian BSDE[8]—and we have the result that $u_t^{-*} = u^{-*}(t, S_t, V_t)$ for a deterministic function. □

To obtain an expression for the optimised driver $H^\pm$, we note that the driver of the value function is conveniently expressed in terms of divergence of the controlled drift from the original parameters; by rearrangement of (10)

$$f(S_t, V_t, Y_t, Z_t, u_t) = (r_t - r)\left(\frac{Z_t^2}{\sqrt{1-\rho^2}\sqrt{V_t}} - Y_t\right) + (\kappa^u(u_t) - \tilde{\kappa})\left(\frac{-Z_t^1\sqrt{V_t}}{\sigma} + \frac{\rho Z_t^2\sqrt{V_t}}{\sigma\sqrt{1-\rho^2}}\right) \tag{15}$$

$$+ (\kappa^u(u_t)\theta^u(u_t) - \tilde{\kappa}\tilde{\theta})\left(\frac{Z_t^1}{\sigma\sqrt{V_t}} - \frac{\rho Z_t^2}{\sigma\sqrt{1-\rho^2}\sqrt{V_t}}\right) - rY_t. \tag{16}$$

[7] As we assume the driver $f$ to be sufficiently integrable for the $J(u)$-BSDE to admit a unique solution (i.e. it is a stochastic Lipschitz driver) the integrability carries over to $H$ such that the $Y$-BSDE admits a unique solution as well.

[8] The function for the martingale representation $Z$ is obtained explicitly by applying Itô's lemma to $D_t = D(t, S_t, V_t)$ and using the semilinear pricing PDE (14), which gives

$$dD(t, S_t, V_t) = -H(S_t, V_t, D_t, Z_t)dt + \partial_x D(t, S_t, V_t)\sigma(S_t, V_t)d\tilde{W}_t$$

where $\partial_x f \equiv (\partial_s f, \partial_v f)$ and $\sigma(s, v)$ should be understood as the diffusion matrix of (2). Hence, by uniqueness of the BSDE solution, $z(t, s, v) \equiv \partial_x D(t, s, v)\sigma(s, v)$ is the deterministic generating function for $Z$.

Alternatively, this can be expressed as

$$f(S_t, V_t, Y_t, Z_t, u_t) = (r_t - r)\left(\frac{Z_t^2}{\sqrt{1-\rho^2}\sqrt{V_t}} - Y_t\right) + (\kappa_t - \kappa)\left(\frac{-Z_t^1\sqrt{V_t}}{\sigma} + \frac{\rho Z_t^2\sqrt{V_t}}{\sigma\sqrt{1-\rho^2}}\right) \tag{17}$$

$$+ (\kappa_t\theta_t - \kappa\theta)\left(\frac{Z_t^1}{\sigma\sqrt{V_t}} - \frac{\rho Z_t^2}{\sigma\sqrt{1-\rho^2}\sqrt{V_t}}\right) - rY_t \tag{18}$$

since $\kappa^u(u_t) - \tilde{\kappa} = \kappa_t - \kappa$ and $\kappa^u(u_t)\theta^u(u_t) - \tilde{\kappa}\tilde{\theta} = \kappa_t\theta_t - \kappa\theta$, which is due to the linear form of the drift (and the simple form of the parameter change under $P \to Q$, regardless of the value of $\lambda$). If we let $\beta_t \equiv \kappa_t\theta_t$ and use the parametrisation $(r_t, \kappa_t, \beta_t) \mapsto (r_t, \kappa_t, \theta_t)$, we thus have that the driver is a linear function of the divergence

$$\tilde{u}_t = (r_t - r, \kappa_t - \kappa, \beta_t - \beta).$$

Hence, the optimal drivers $H^\pm$ are obtained by minimising/maximising a linear objective subject to the constraint given by the compact uncertainty set $U$—Eq. (17) in case of statistical uncertainty for $P$-parameters, or similarly, (15) with the identities (6) when dealing with $Q$-parameters under uncertainty deduced from calibration.

To this end, we consider elliptical uncertainty sets as given by the quadratic form

$$U = \left\{u : \tilde{u}^\top \Sigma^{-1}\tilde{u} \le \chi\right\}$$

for some positive semi-definite matrix $\Sigma$ and positive constant $\chi$. In particular, from statistical inference, we have that the $1-\alpha$ confidence ellipse

$$\tilde{u}^\top \Sigma_{r,\kappa,\beta}^{-1}\tilde{u} \le \chi_3^2(1-\alpha) \tag{19}$$

represents $u \in U$ (for a significance level $\alpha$) where $\Sigma_{r,\kappa,\beta}$ is the covariance matrix of the parameters and $\chi_3^2(1-\alpha)$ is the quantile of the chi-squared distribution with three degrees of freedom (see further Sect. 3.1). The formal justification for elliptical uncertainty sets is that (19) is a level set of the asymptotic likelihood, due to large-sample normality of the maximum likelihood estimator of the parameters. This is the same form of uncertainty as given by the confidence region under Wald's hypothesis test, and approximates the optimal confidence set, by Neyman–Pearson lemma (see [46]).

As $\tilde{u} \mapsto f(\tilde{u})$ is linear, it has no internal stationary points and the quadratic problems

$$H^- = \inf f(\tilde{u}) \quad \text{and} \quad H^+ = \sup f(\tilde{u})$$
$$\text{subject to } \tilde{u}^\top \Sigma^{-1}\tilde{u} = \chi$$

give the optimised drivers. The solutions are (obtained e.g. by a Lagrange multiplier)

$$\begin{aligned} H^\pm(S_t, V_t, Z_t, Y_t) &= \pm\sqrt{\chi n_t^\top \Sigma^\top n_t} - rY_t \\ \tilde{u}^\pm(S_t, V_t, Z_t, Y_t) &= \pm\sqrt{\frac{\chi}{n_t^\top \Sigma^\top n_t}}\,\Sigma\, n_t \end{aligned} \tag{20}$$

where $n_t$ is the $3 \times 1$ vector of coefficients to the parameter deviances of Eq. (17) given by

$$n_t = \left[ \left( \frac{Z_t^2}{\sqrt{1-\rho^2}\sqrt{V_t}} - Y_t \right), \left( \frac{-Z_t^1\sqrt{V_t}}{\sigma} + \frac{\rho Z_t^2\sqrt{V_t}}{\sigma\sqrt{1-\rho^2}} \right), \left( \frac{Z_t^1}{\sigma\sqrt{V_t}} - \frac{\rho Z_t^2}{\sigma\sqrt{1-\rho^2}\sqrt{V_t}} \right) \right]^\top .$$

The optimal drivers in (20) conclude our analysis since we now have an explicit form for the stochastic differential equation (12) that describes the evolution of the pricing boundaries. Before we proceed to approximative solutions of these equation by numerical methods, a few remarks are in order. Firstly, the approach applies unchanged to a portfolio of options with time-$T$ terminal payoff $\sum_i g_i(S_T)$. Due to the non-linearity of the pricing boundaries (7), we further have $D^+(\sum w_i g_i(S_T)) \leq \sum w_i D^+(g_i(S_T))$ for weights $\sum w_i = 1$, such that the super-replication cost for individual hedging might be lowered by hedging the portfolio as a whole. Secondly, for a general payoff represented by $G \in \mathcal{F}_T$, for instance a path-dependent European options on $S$, we have a value process equation corresponding to (12) with terminal condition $D_T^{\pm} = G$. However, this problem do no longer yield a Markovian structure, and we do not have $D^{\pm}$ (nor $Z$) being generated by deterministic functions, neither do the numerical methods of Appendix A apply. Thirdly, we deliberately impose parameter uncertainty by replacing $Q \to Q^u$ in contrast to replacing $P \to Q^u$ directly (which would yield the same form of the effect $\alpha(t, S_t, V_t)$ that governs the measure change, but with $r$ replaced by $\mu$ in (9)). The reason is that the governing BSDEs (12) will have a terminal condition $g(S_T)$ where $S_T \sim Q$, for which we have accessible parameters (in particular, we may directly observe the $Q$-drift $r$ instead of estimating the $P$-drift $\mu$).

We conclude this section with a technical remark.

*Remark 1* So far, we have not expressed any integrability conditions on the process $\alpha(S_t, V_t, u_t)$ in order to guarantee that (i) the density $dQ^u/dQ$ and (ii) the driver $f(S_t, V_t, Y_t, Z_t, u_t)$ are well defined, i.e. for the measure change $Q \to Q^u$ to be eligible and to certify that $f$ (and hence $H^{\pm}$) yields a BSDE which admits a unique solution. For this purpose, Novikov's condition

$$\mathbb{E}^Q \left[ e^{\frac{1}{2}\int_0^T ||\alpha(S_t, V_t, u_t)||^2 dt} \right] < \infty \tag{21}$$

is sufficient for both (i) and (ii) since then we have that the driver is stochastic Lipschitz in $y$ and $z$ (note that $r_t$ is bounded in $U$), i.e.

$$|f(S_t, V_t, y, z, u_t) - f(S_t, V_t, y', z', u_t)| \leq ||\alpha(S_t, V_t, u_t)|| \left( |y - y'| + ||z - z'|| \right)$$

where $||\alpha(S_t, V_t, u_t)||$ is predictable and such that (21) holds. With a stochastic Lipschitz driver, the concerned BSDE admits a unique solution which is bounded if the terminal condition $g(S_T)$ is bounded (see e.g. [22], Appendix A.9.2).

For Novikov's condition in (21) note that the integrand of the exponent can be written

$$||\alpha(S_t, V_t, u_t)||^2 = a(u_t)V_t + b(u_t)\frac{1}{V_t} + c(u_t)$$

with

$$\begin{aligned} a(u_t) &= \frac{(\kappa_t - \kappa)^2}{\sigma^2(1-\rho^2)} \\ b(u_t) &= \frac{\sigma^2(r_t - r)^2 + (\kappa_t\theta_t - \kappa\theta)^2 + 2\rho\sigma(r - r_t)(\kappa_t\theta_t - \kappa\theta)}{\sigma^2(1-\rho^2)} \\ c(u_t) &= -2\frac{(\sigma\rho(r - r_t) + \kappa_t\theta_t - \kappa\theta)(\kappa_t - \kappa)}{\sigma^2(1-\rho^2)} \end{aligned}$$

such that for the expectation in (21) we have

$$\mathbb{E}^Q\left[e^{\frac{1}{2}\int_0^T a(u_t)V_t dt} e^{\frac{1}{2}\int_0^T b(u_t)\frac{1}{V_t}dt} e^{\frac{1}{2}\int_0^T c(u_t)dt}\right] \le k\sqrt{\mathbb{E}^Q\left[e^{\int_0^T a(u_t)V_t dt}\right]\mathbb{E}^Q\left[e^{\int_0^T b(u_t)\frac{1}{V_t}dt}\right]} \tag{22}$$

since $e^{\frac{1}{2}\int_0^T c(u_t)dt}$ is bounded by a constant $k$ (for $u_t \in U$) and where we have used the Cauchy-Schwarz inequality. If we begin with the first expectation on the right hand side of (22) we have $a(u_t) \le \bar{a}$ for a constant $\bar{a}$. As the Laplace transform of the integrated CIR process[9] is finite for $\bar{a} \le \tilde{\kappa}^2/(2\sigma^2)$, we end up with the condition

$$|\kappa_t - \kappa| \le \tilde{\kappa}\frac{\sqrt{1-\rho^2}}{\sqrt{2}}. \tag{23}$$

For the second expectation of (22), we use that $b(u_t) \le \bar{b}$ for a constant $\bar{b}$ and that the Laplace transform of the integrated inverse-CIR process[10] is finite for

---

[9]The Laplace transform of the integrated variance $\mathbb{E}[\exp(-\beta\int_0^T V_t dt)]$ goes back to [24] and is well defined for $-\beta \le \kappa^2/(2\sigma^2)$, see also [20].

[10][21] gives an expression for the joint transform of the log-price and integrated variance of a 3-over-2 process. Applying Itô's formula to $1/V_t$ we find that the inverse-CIR $(\kappa, \theta, \sigma)$ process is a 3-over-2 process with parameters $(\hat{\kappa} \equiv \kappa\theta - \sigma^2, \hat{\theta} \equiv \kappa/(\kappa\theta - \sigma^2), \hat{\sigma} \equiv -\sigma)$. Using their transform, provided $\hat{\kappa} > -\hat{\sigma}^2/2$,

$$\mathbb{E}\left[e^{-\lambda\int_0^T \frac{1}{V_t}dt}\right] = \frac{\Gamma(\gamma - \alpha)}{\Gamma(\gamma)}\left(\frac{2}{\hat{\sigma}^2 y(0, 1/V_0)}\right)^{\alpha} M\left(\alpha, \gamma, -\frac{2}{\hat{\sigma}^2 y(0, 1/V_0)}\right)$$

where

$$\begin{aligned} y(t, x) &\equiv x(e^{\hat{\kappa}\hat{\theta}(T-t)} - 1)/(\hat{\kappa}\hat{\theta}) = x(e^{\kappa(T-t)} - 1)/\kappa \\ \alpha &\equiv -(1/2 + \hat{\kappa}/\sigma^2) + \sqrt{(1/2 + \hat{\kappa}/\sigma^2)^2 + 2\lambda/\sigma^2} \\ \gamma &\equiv 2(\alpha + 1 + \hat{\kappa}/\sigma^2) = 1 + 2\sqrt{(1/2 + \hat{\kappa}/\sigma^2)^2 + 2\lambda/\sigma^2} \end{aligned}$$

and $M$ is the confluent hypergeometric function. From this, we see that

$$\bar{b} \leq \left( \frac{2\tilde{\kappa}\tilde{\theta} - \sigma^2}{2\sqrt{2}\sigma} \right)^2. \tag{24}$$

Rearranging this condition, we have that

$$\sigma^2 (r_t - r)^2 + (\kappa_t \theta_t - \kappa\theta)^2 + 2\rho\sigma(r - r_t)(\kappa_t\theta_t - \kappa\theta) \leq \frac{1-\rho^2}{2} \left( \kappa\theta - \sigma^2/2 \right)^2$$

together with (23) are sufficient conditions for (21) to hold.

## 3 The Empirical Perspective

In this section we investigate how the conservative pricing method carries over to market data from the S&P 500 index. We perform an empirical experiment with the following rationale.

We let statistical parameters estimated from historical observations of the index price and variance represent the financial market model under the objective measure $P$. Hence, $(S, V)$ is assumed to evolve with $P$-dynamics according to Heston's model specified with the estimated parameters. We consider $(B, S)$ exclusively as the traded assets in the financial market model driven by two random sources, $W^1$ and $W^2$, and refrain from the assumption that there exists an additional, exogenously given, (volatility dependent) asset which would complete the model. On the other hand, we exclude arbitrage opportunities and affirm the existence of a space of risk-neutral measures: $Q$ exists (not necessarily unique) in a set $\mathcal{Q}$ of probability measures equivalent to $P$, such that the discounted asset price is a martingale under any measure in $\mathcal{Q}$. This implies that $(S, V)$ will have the same diffusion matrix under every $Q \in \mathcal{Q}$ as given by the diffusion matrix of the $P$-dynamics. This follows from the notion that the quadratic variation (continuous part) of a semimartingale is invariant under equivalent probability measures on a complete filtered space. Further, by Girsanov's theorem, we have that the law of the driving random sources is invariant: they will be (independent) Wiener process under all equivalent measures in $\mathcal{Q}$.

With this in mind, fix the diffusion matrix of $(S, V)$ to be as given by the estimated diffusion parameters from historical data. We then take the space of equivalent risk-neutral measures to be equal the space spanned by the controlled measure $Q^u$ over all admissible controls: $\mathcal{Q} \equiv \{Q^u : u \in \mathcal{U}\}$ where $\mathcal{U}$ represents the space of predictable control processes $u = \{u_t\}_{t\geq 0}$ that lives in the compact uncertainty set $U \subset \mathbb{R}^d$. In particular, we deduce the uncertainty set from statistical estimation and define $U$ to be represented by the elliptical confidence region derived from the observed Fisher

---

$$\lambda \geq -\left( \frac{2\hat{\kappa} + \sigma^2}{2\sqrt{2}\sigma} \right)^2 = -\left( \frac{2\kappa\theta - \sigma^2}{2\sqrt{2}\sigma} \right)^2$$

is a sufficient condition for the transform to being well defined.

information (which asymptotically approximates the sampling covariance of our estimates). The question we ask is then if market option prices are covered by the pricing rules in $\mathcal{Q}$ as given by the corresponding model pricing-boundaries.

Since the volatility process of an asset is latent by nature it has to be measured with some method. In the following section we briefly present the realized volatility measure which gives a commonly used nonparametric estimator of the variance process. We then detail some estimation methods that we employ for point estimation of the model parameters and for drawing inference thereof. The empirical study based on S&P 500 data finally follows.

## 3.1 Measured Variance and Statistical Inference

We use variance of the S&P 500 price as estimated with the *realised volatility measure* from high-frequency observations ($\sim$5 min) of the index returns (see for instance [5]). If $s = (s_{t_1}, s_{t_2}, \dots)$ is a time-series with daily prices, the realized variance measure is

$$RV([t_{i-1}, t_i]) = \sum_{k: s_k, s_{k+1} \in [t_{i-1}, t_i]} (y_{s_{k+1}} - y_{s_k})^2,$$

where $[t_{i-1}, t_i]$ is the duration of the day and $s_k \in [t_{i-1}, t_i]$ are intra-day time points over which $y_{s_k} = \log(s_{s_k}) - \log(s_{s_{k-1}})$ are log-returns. The realized variance approximates the integrated variance: $RV([t_{i-1}, t_i]) \xrightarrow{P} \int_{t_{i-1}}^{t_i} V_s ds$, (for a continuous return process, the quadratic variation for a general semimartingale), and the measured variance at time $t_i$ is taken to be

$$v_{t_i} = \frac{1}{t_i - t_{i-1}} RV([t_{i-1}, t_i])$$

such that a time-series of daily variances $v = (v_{t_1}, v_{t_2}, \dots)$ is obtained. For convenience, we use precomputed variance estimates from the Oxford-Man Institute's realised library[11] shown in Fig. 1 together with the closing price of the S&P 500 index from the period January 3rd, 2000 to February 29th, 2016.

Next, consider the estimation of the parameters $\Theta = (\kappa, \theta, \sigma)$ from $n + 1$ observations $v = (v_0, \dots, v_n)$ of the variance. Here, $v_i$ is treated as the observed value of $V_{t_i}$ for a set of discrete time-points $(t_0, \dots, t_n)$ at which the observations are made, and we denote with $\Delta_i = t_{i+1} - t_i$ the length of the $i$th time-interval between two consecutive observations. In general, inference on the parameters $\Theta$ of a Markov process with transition density $f(y; x, \delta, \Theta)$ for $V_{t+\delta} | V_t = x$ can be made with the likelihood function

[11]The Realised Library version 0.2 by Heber, Gerd, Lunde, Shephard and Sheppard (2009), http://realized.oxford-man.ox.ac.uk.

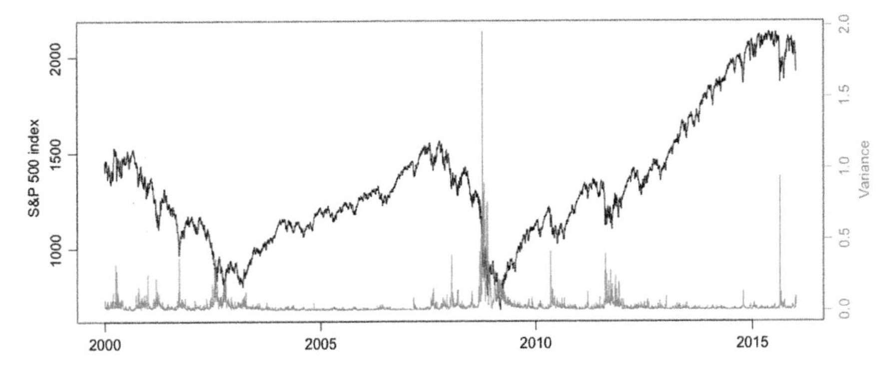


**Fig. 1** Historical closing prices and realized variances of the S&P 500 index, 4,035 daily observations from January 3rd, 2000 to February 29th, 2016

$$L_n(\Theta) = \prod_{i=0}^{n-1} f(v_{i+1}; v_i, \Delta_i, \Theta)$$

and the maximising argument of the (log-) likelihood is the maximum likelihood estimator $\hat{\Theta}$ of $\Theta$. If the transition density is unknown in closed form, or, as in the case for the square root process, of a kind that is impenetrable for optimisation (both analytically and by numerical schemes), one might consider alternatives based on approximations of the likelihood. A direct way is to consider the time-discrete approximation $V^\pi$ of the process $V$ given by a Euler-Maruyama scheme. For the square root process[12]

$$V^\pi_{t+\delta} = V^\pi_t + \kappa(\theta - V^\pi_t)\delta + \sigma\sqrt{V^\pi_t}\,(W_{t+\delta} - W_t) \tag{25}$$

which give an approximative Gaussian log-likelihood

$$l_n(\Theta) \equiv \log L_n(\Theta) = -\frac{1}{2}\sum_{i=0}^{n-1} \frac{(v_{i+1} - v_i - \kappa(\theta - v_i)\Delta_i)^2}{\sigma^2 v_i \Delta_i} + \log(2\pi\sigma^2 v_i \Delta_i). \tag{26}$$

A function on the same form as above was considered for least-squares estimation of drift parameters in [51]. For processes with ergodic property, [42] considered the joint estimation of drift and diffusion parameters with a Gaussian approximation to the transition density (26) and showed that under general conditions, the estimator is asymptotically normal and efficient. Their approach addresses the case when the mean and variance of the transition density are unknown and uses approximations in their place. For the square root process, the explicit expressions

[12]Note that (25) may generate negative outcomes of $V^\pi_{t+\delta}$ and is thus not suitable for simulation in its standard form. Alternative schemes are discussed in Appendix A.3. Here we use (25) for an approximative Gaussian likelihood—the Euler contrast (26)—which is well defined.

$$\mathbb{E}[V_{t+\delta}|V_t = v] = \theta + (v - \theta)e^{-\kappa\delta} \equiv \mu(v, \delta),$$
$$\mathrm{Var}(V_{t+\delta}|V_t = v) = v\frac{\sigma^2}{\kappa}(e^{-\kappa\delta} - e^{-2\kappa\delta}) + \theta\frac{\sigma^2}{2\kappa}(1 - e^{-\kappa\delta})^2 \equiv s^2(v, \delta),$$

in place of the approximations $\mu(v, \delta) \approx v + \kappa(\theta - v)\delta$ and $s^2(v, \delta) \approx \sigma^2 v\delta$ in (26) give that

$$l_n(\Theta) = -\frac{1}{2}\sum_{i=0}^{n-1} \frac{(v_{i+1} - \mu(v_i, \Delta_i))^2}{s^2(v_i, \Delta_i)} + \log(2\pi s^2(v_i, \Delta_i)) \tag{27}$$

forms an approximative Gaussian likelihood with exact expressions for the conditional mean and variance. While the Euler contrast (26) is known to have substantial bias when the time between observations is large, estimators such as (27) are consistent for any value of $\delta$. The reason is that they form quadratic martingale estimating functions, see [13] and [35].

An approximation for the variance of the maximum likelihood estimator $\hat{\Theta} = \arg\max l_n(\Theta)$ is given by the observed information matrix

$$I_o = -\left.\frac{\partial^2 l_n(\Theta)}{\partial\Theta^\top \partial\Theta}\right|_{\Theta=\hat{\Theta}}$$

which may be calculated by numerical differentiation of the log-likelihood function at estimated values. The approximative covariance matrix of $\hat{\Theta}$ is given by the inverse $\Sigma_{\hat{\Theta}} = I_o^{-1}$ and an estimated standard error of the $j$th parameter by $\sqrt{(I_o^{-1})_{jj}}$. Hence, an approximate $1 - \alpha$ confidence region for $\Theta$ is given by the ellipse

$$\left\{\Theta : (\Theta - \hat{\Theta})\Sigma_{\hat{\Theta}}^{-1}(\Theta - \hat{\Theta})^\top \leq \chi_d^2(1 - \alpha)\right\}$$

where $d$ is the dimension of the row vector $\Theta$ and $\chi_d^2(1 - \alpha)$ the $1 - \alpha$ quantile of the chi-square distribution with $d$ degrees of freedom.

Finally we note that several additional approaches for statistical estimation of Heston's model are known in the literature. For example, with access to high-frequency returns data, one may consult [9] or [25].

## 3.2 Empirical Study

We base the empirical study on market data from the Oxford-Man Institute and Wharton Research Data Services.[13] Data from Oxford-Man is used for the price of the S&P 500 index, both historical closing prices with a daily frequency and high-frequency (~5 min) returns in the (pre-calculated) estimation of historical volatility.

[13] http://realized.oxford-man.ox.ac.uk and https://wrds-web.wharton.upenn.edu/wrds/.

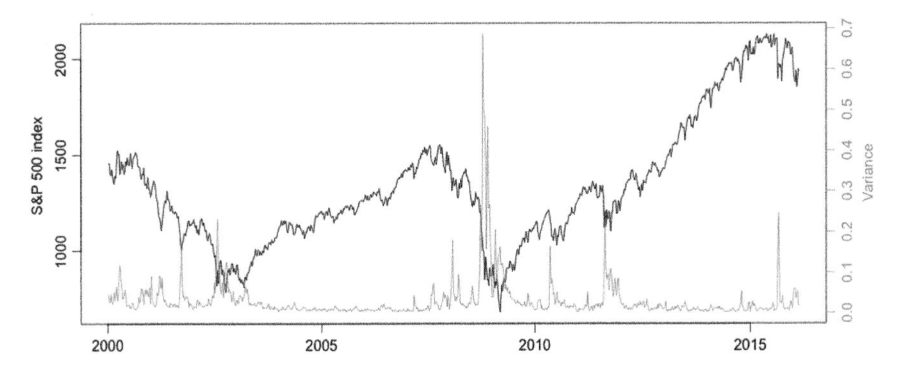


**Fig. 2** Historical closing prices and realized variances of the S&P 500 index, 843 weekly observations from January 3rd, 2000 to February 29th, 2016

Data from Wharton Research, is used for the price of European call options written on the S&P 500 index. Here we use historical quotes of bid and offer prices from options with different strike-prices and maturities along with the relevant dividend yield paid by the underlying index and the risk-free interest rate that corresponds to the maturity of each option.

Prior to the numerical calculation of pricing bounds for the call options, we estimate the parameters of Heston's model from the S&P 500 price and variance according to the following steps:

1. First, we decimate the observation frequency of the variance to weekly observations by calculating the realized variance measure over week-long intervals, see Fig. 2. This operation smooths the measured variance process and, in particular, it removes the extreme variance spikes (cf. Fig. 1) which might cause non-robust parameter estimates.
2. We estimate $(\kappa, \beta, \sigma)$ from the weekly variance with the parametrisation $(\kappa, \beta, \sigma) \mapsto (\kappa, \theta, \sigma)$ of the model. We employ the approximative likelihood based on Euler conditional moments[14] and calculate the approximative covariance matrix accordingly by numerical differentiation. Results are given in Table 1 (with squared elements of the covariance matrix for a notion of standard errors) together with estimation results from the daily variance. From the results in Table 1 note that the daily variance yields relatively high estimates of the mean-reversion speed to accommodate extreme observations and also large standard errors of both drift parameters, which indicate that the square root process is a poorly fitting model for the daily variance data. With the weekly variance we obtain more sensible results and lower standard errors.

[14]Alternatively, we may employ the (approximative) likelihood with exact conditional moments. For daily observations, the numerical optimisation does not converge while for weekly data, this yields very similar parameter estimates and standard errors as with approximative moments.

**Table 1** Parameters and standard errors estimated from historical data of the S&P 500 index. **Left table**: results based on 4,035 daily observations. **Right table**: results based on 843 weekly observations decimated from the original data. All estimates from numerical optimisation and differentiation of the approximative likelihood function based on Euler moments

Estimates, daily data

| $\kappa$ | $\theta$ | $\sigma$ |
|---|---|---|
| 29.6 | 0.0315 | 2.58 |

Estimates, weekly data

| $\kappa$ | $\theta$ | $\sigma$ |
|---|---|---|
| 4.59 | 0.0307 | 0.775 |

Standard errors (daily data)

| | $\kappa$ | $\beta$ | $\sigma$ |
|---|---|---|---|
| $\kappa$ | 4.314 | 0.544 | 2.37e-06 |
| $\beta$ | 0.544 | 0.074 | 1.79e-06 |
| $\sigma$ | 2.37e-06 | 1.79e-06 | 0.0288 |

Standard errors (weekly data)

| | $\kappa$ | $\beta$ | $\sigma$ |
|---|---|---|---|
| $\kappa$ | 1.395 | 0.621 | 1.91e-06 |
| $\beta$ | 0.621 | 0.0269 | 4.84e-06 |
| $\sigma$ | 1.91e-06 | 4.84e-06 | 0.0189 |

3. In addition to estimated parameters of the variance process, we estimate the correlation coefficient of the model with a realised covariation measure.[15] This gives an estimate $\rho = -0.274$ from the weekly variance and closing price of the S&P 500 index.

With estimated model parameters and standard errors in hand, we continue to the calculation of upper and lower pricing bounds for European options. We proceed according to the following:

1. We consider call options on the S&P 500 index with historical market prices from the 3-year period of August 31st, 2012 to August 31st, 2015. We select the dates from this period that coincides with the weekly index data (i.e. dates for which both the option and the S&P 500 closing price are quoted). This results in 157 dates and a total of 244,239 option quotes for different strikes and maturities.
2. For each of the 157 dates during the time period, we chose a single option with strike price and time to maturity as shown in the right pane of Fig. 3. Here, the "initial" option of each period is selected with a medium-sized maturity and a strike-price as close as possible to being at-the-money. We then retain the same maturity and strike price (the same option) as far as there is market quotes available. This gives us four options in total. We further record the relevant risk-free rate as given by the zero-coupon interest rate with corresponding maturity, and the current dividend yield. The left pane of Fig. 3 shows the resulting bid/offer quotes after they have been converted to zero-dividend prices. This is done for each quote by first calculating the Black-Scholes implied volatility (with the effective dividend yield) and then recalculating the Black-Scholes price with zero dividend.

[15]The quadratic covariation of logarithmic data gives $\frac{1}{t}[\log S, \frac{1}{\sigma}\log V]_t = \frac{1}{t}\int_0^t \sqrt{V_s}\frac{1}{\sqrt{V_s}}d[\rho W^1 + \sqrt{1-\rho^2}W^2, W^1]_s = \rho$ and we use a realized covariation estimate thereof.

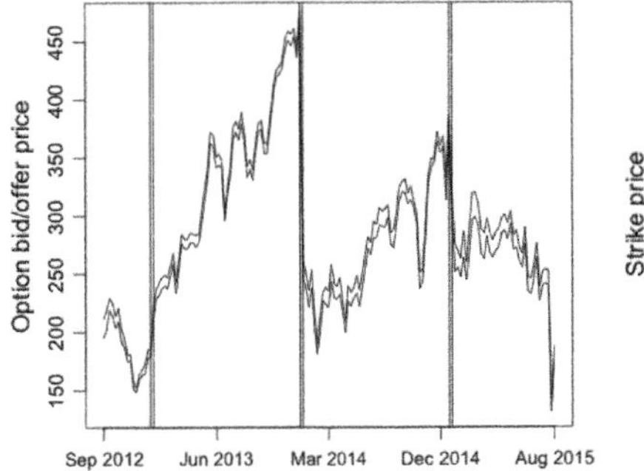


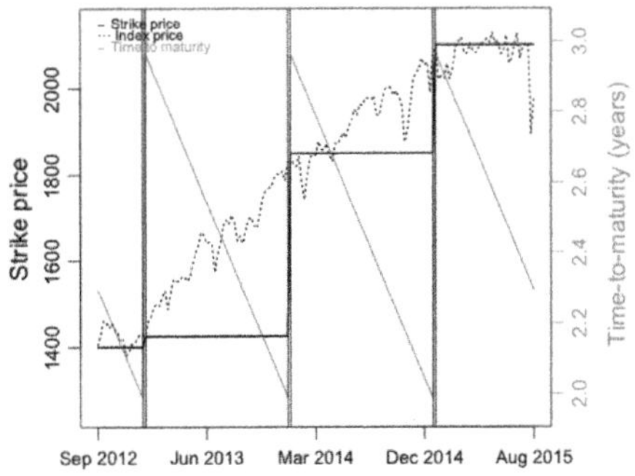


**Fig. 3** **Left figure**: historical market prices of call options on the S&P 500 index. The figure shows 157 bid/offer quotes (converted to zero-dividend prices) from the period August 31st, 2012 to August 31st, 2015. **Right figure**: the strike price and time-to-maturity of the call options

3. For the calculation of pricing bounds for the call option we employ a numerical method described in the Appendix. Here we initially have the following numerical considerations: Firstly, for the parameter estimates in Table 1 (based on weekly data) we have $\sqrt{4\beta} = 0.751 < 0.775 = \sigma$ which implies that the implicit Milstein scheme may fail due to the generation of negative outcomes (see Appendix A.3). To prevent this, we include a truncation step[16] to the simulation according to the suggested method of [4]. Further, we increase the number of time steps to $n = 1{,}000$ for the forward simulation to prevent the generation of negative variance values.[17] We then down-sample the simulated price and variance to the original time-grid of $n = 25$ steps for the backward simulation. Secondly, the driver of the backward process will explode for variance values approaching zero. As the forward simulation may output negative/zero values, we cancel the control: $u(X_t, Z_t, Y_t) \equiv 0$ giving $H(X_t, Z_t, Y_t) = -rY_t$, each time the variance is smaller than a threshold, $V_t < \varepsilon$, and we set $\varepsilon = 0.00041$, which is the minimum value of the S&P 500 variance.
4. We simulate the optimally controlled value-process by the explicit scheme with $Y$-recursion and the MARS method of degree 2 for the regressions, see Appendix A.4. The control variate is included and we simulate $N = 100{,}000$ paths of the forward-backward process over a time grid with $n = 25$ time steps (with $n = 1,000$ down-sampled to $n = 25$ for the forward process). For each option price, we run a separate simulation with the estimated model parameters and corresponding maturity/strike, risk-free rate and initial values for the forward process (S&P 500 index level and variance) from the market data. We simulate the backward process with the minimised driver (for the lower pricing bound)

[16]The time-stepping of the scheme fails whenever $V^{\pi}_{t_i} < 0$ due to the computation of $\sqrt{V^{\pi}_{t_i}}$ and the truncation step is simply to replace with $\sqrt{(V^{\pi}_{t_i})^+}$. Although this prevents the scheme to fail, note that negative values may still be generated and in particular when the time-step $\Delta_i$ is large.

[17]Bookkeeping the sign of the generated variance values yields positive outcomes 99.3% of the time when using $n = 1{,}000$ time steps and 96.7% for $n = 25$.

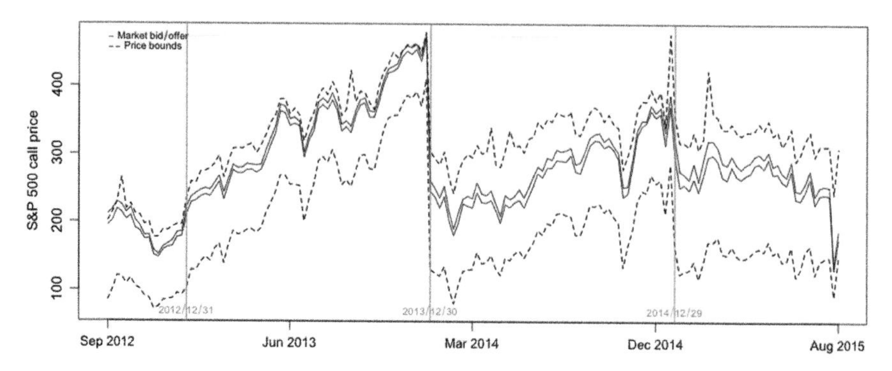


**Fig. 4** European call options on the S&P 500 index: upper and lower pricing bounds (dashed lines) as calculated by simulations of the optimally controlled value process. The graph shows model prices and historical market quotes of bid and offer prices (solid lines) on a weekly basis of options with four different strike/maturity structures (see right pane of Fig. 3)

and the maximised driver (for the upper bound) based on the same forward simulation. Each evaluation of the optimal driver is calculated with the covariance matrix $\Sigma_{r,\kappa,\beta}$ where we use estimated standard errors and correlation for $\kappa$, $\beta$ from Table 1 and a standard deviation of 0.00005 for the interest rate ($r$ uncorrelated with $\kappa$, $\beta$). As before, we use a confidence level of 95% for the uncertainty region.

5. Finally, we also calculate the corresponding minimum/maximum prices for each considered call option by numerical optimisation of Heston's pricing function over the same 95% uncertainty region—see Appendix A.3 for details. Henceforth, we refer to these as formula-optimal prices.

The resulting upper and lower pricing bounds from the simulations are presented in Fig. 4 together with the corresponding market quotes of bid and offer prices. The formula-optimal prices are depicted in Fig. 5. The dates at which there is a common change in strike price and maturity (see right pane of Fig. 3) have been marked in the figures to separate the different options. We label these options (I)–(IV) and give some additional results in Table 2.

A note on the interpretation of these results are in order here.

- First, we see that the market bid/offer quotes fall inside the model bounds for almost all considered call prices (154 out of 157) and in particular, for all prices when looking at the latest two options (III)–(IV), see Table 2. The lower bound is always covering the bid quote and the price interval of the bounds is fairly symmetrical around the mid-market price for options (III)–(IV).
- For option (I), the offer prices are close to the upper bound (occasionally above) and the same holds for option (II). This option's moneyness is strongly increasing with time (see right pane of Fig. 3) while the distance from the upper bound to the offer quote is shrinking. A possible explanation may be that the model is unable to capture the slope and skew of market prices/implied volatilities for the parameters we use, in particular since these are estimated from historical index prices and not from option prices.

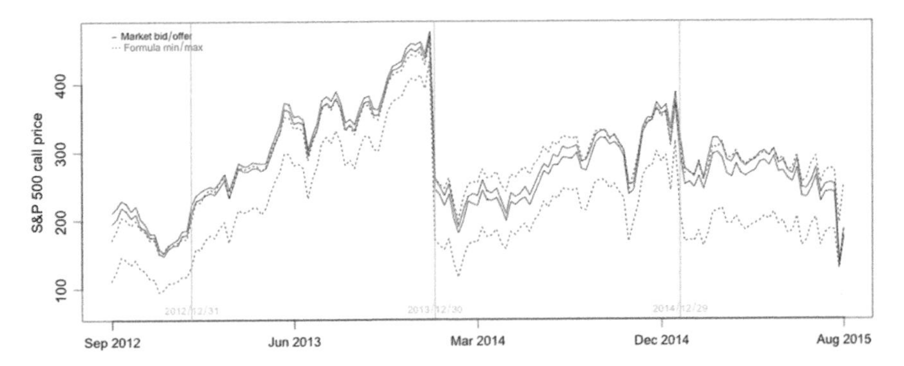


**Fig. 5** The minimum and maximum price (red dotted lines) as obtained from the optimisation of Heston's pricing formula for call options. The solid black lines show the market bid/offer quotes of the S&P 500 options

**Table 2** Key figures for the model prices of S&P 500 call options from the optimally controlled value process (the pricing bounds) and for the formula-optimal price interval. The spread-to-price ratio gives the average size of the market spread as a percentage of the average mid-market price of each option. Similarly, the bounds-to-spread ratio compares the range of the pricing bounds to that of the market spread. The in-bounds figures give the proportion of market quotes that fall inside the model bounds. Corresponding figures are calculated for the intervals of the formula-optimal prices

| European call options on the S&P 500 index | | | | |
|---|---|---|---|---|
| Option | (I) | (II) | (III) | (IV) |
| Duration | 12/09/04–<br>12/12/24 | 12/12/31–<br>13/12/23 | 13/12/30–<br>14/12/22 | 14/12/29–<br>15/08/31 |
| Maturity | 14/12/20 | 15/12/19 | 16/12/16 | 17/12/15 |
| Spread:price (%) | 5.6 | 3.4 | 5.6 | 7.1 |
| Bounds:spread | 14.9 | 12.4 | 13.9 | 12.8 |
| In bounds (%) | 88.2 | 98.1 | 100 | 100 |
| Optim:spread | 7.8 | 6.1 | 7.2 | 6.6 |
| In interval (%) | 11.8 | 0 | 76.9 | 72.2 |

For an investigation of this point, we take a look at the market prices and model boundaries for a range of strikes at the first and last date of option (II), plotted in terms of implied volatilities in Fig. 6.

- The left pane of Fig. 6, which shows prices form the starting date, shows a strongly skewed volatility curve for the market prices, while the prices from Heston's optimised formula yield much flatter, less sloped, curves. This indicates that a higher level of skewness is required to fit the curvature of market volatilities (roughly speaking, a stronger negative correlation to increase the slope and a higher level of "vol-of-vol" to increase the skew), and a higher volatility level overall for the ATM formula-price to fit with the market (a higher mean-reversion level and lower reversion speed).

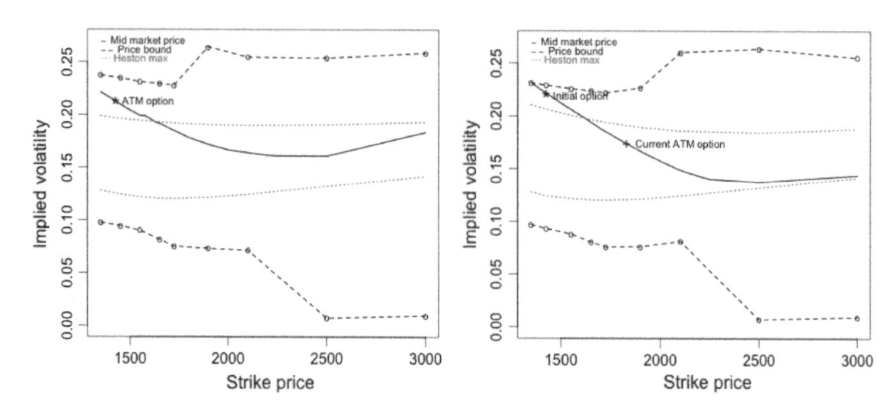


**Fig. 6** **Left figure**: mid-market implied volatility of the S&P 500 call option for different strikes as recorded on 2012-12-31. Corresponding model-boundaries (dashed lines) and formula-optimal prices (red dotted lines), both in terms of implied volatilities. The volatility of the ATM option (II) is marked with a star. **Right figure**: implied volatilities from the mid-market price, model-boundaries and formula-optimum, as recorded on 2013-12-23 (the last date of the considered period for option (II))

- On the other hand, the corresponding pricing-boundaries are wide enough to cover all strikes at both dates, even if the upper boundary is close to the market volatility at the later date where the option is deeply in-the-money—see right pane of Fig. 6. As in the case with formula-optimal prices, we note that the bounds do not exhibit any curvature in line with the market volatilities, supposedly because of the low level of negative correlation and vol-of-vol.

Returning to the prices of all options (I)–(IV), we have in total that 98% of the market quotes are within the pricing bounds whilst the formula optimal prices cover 40% of the quotes. In particular, the market quotes of option (II) are outside the formula-optimal prediction throughout the period, and option (I) has only 11.8% of its quotes covered. Generally, the upper price is too low and the coverage of the formula-price is to small: the (average) ranges of the optimised prices are $\sim$7 times the sizes of market spreads (see Table 2), and thus almost halved compared to the ranges of the model bounds ($\sim$13 times the market spreads).

The optimisation of Heston's formula based on statistical inference is not sufficient to cover for the market quotes of the considered data: the price-interval is too tight, and options in-the-money generally fall outside the model predictions, which indicated that the volatility smile is not captured. This should perhaps be expected since parameters are estimated to fit the underlying index—diffusion parameters in particular—and not to fit the actual options we are trying to price. Further, we use a constant set of parameters to predict option prices over the whole 3-year period, while in practice one would typically update the estimates on regular basis. We have thus faced Heston's model with a challenging task: to price a dynamical set of market options over a long time period while taking in information from the underlying asset alone when estimating the model to data. In return, we allow drift parameters to vary within their 95% confidence region as a representation of the incompleteness of the market model which, after all, gives an optimised price range that cover option quotes to some extend. Only when we generalise the model we obtain conservative pricing bounds wide enough to cover most prices, even if some deep in-the-money options still fall outside. We assume the same 95% confidence region as a representative for the incompleteness, but allow for a much more general view on the uncertainty of the parameters which span the space of risk-neutral measures: not only are they uncertain within the confidence region, the parameters also change dynamically with time. In the end, the former method corresponds to an optimally controlled value process with parameter processes being constants, and we simply have to allow for these parameters to vary in order to cover the market pricing of options in a satisfactory way.

## 4 Stochastic Volatility Models with Jumps

For the purpose of completion, we generalise our modelling framework in this final section to a multi-asset setting under a Markovian stochastic volatility model with jumps. The intension is to give a brief presentation of how the uncertainty pricing transfers to a general model and we deliberately avoid going too deep into details and technical assumptions.

### 4.1 A Generic Markovian Model

We consider a financial market model on a filtered probability space $(\Omega, \mathcal{F}, \{\mathcal{F}_t\}_{t\geq 0}, P)$ that consists of a money account $B$, paying a risk-free interest of deterministic rate $r$, and a $\mathbb{R}^d$-valued stochastic process $S = (S^1, \ldots, S^d)^\top$ representing the price processes of $d$ risky assets. Furthermore, we have $d'$ non-negative stochastic processes, $V$ taking values in $\mathbb{R}^{d'}$, that represents the instantaneous variances. Typically, the two are of equal dimension such that each asset price is diffused by an individual volatility and we also assume that $d' = d$ is the case here. The statistical $P$-dynamics of the $m = 2d$ column-vector $X = (S; V)$ of state variables are assumed to be of the form

$$dX_t = \mu^P(X_t)dt + \sigma(X_t)dW_t + \int_{\mathcal{Z}} h(\xi, X_{t-})\tilde{\mu}(d\xi, dt)$$

where $\mu^P(\cdot)$ is the $m$-dimensional drift-function under the statistical measure, $\sigma(\cdot)$ the $m \times m$-valued diffusion matrix, and $W$ is a $\mathbb{R}^m$-valued Wiener process. The jump part is driven by $\tilde{\mu}$, a compensated Poisson random measure on a Blackwell space $\mathcal{Z}$ with deterministic compensator $\mu_p(d\xi, dt) = \nu(d\xi)dt$, and $h(\cdot)$ is a state-dependent function valued in $\mathbb{R}^m$ that governs the jump sizes of $X$. Since we are working with a $\mathbb{R}^m$-dimensional state processes, we take $\mathcal{Z} = \mathbb{R}^m$. We assume that $\{\mathcal{F}_t\}_{t\geq 0}$ is generated by $W$ and $\tilde{\mu}$ jointly, and augmented to satisfy the usual conditions. The functions $\mu^P$, $\sigma$, $h$ are assumed to be defined such that the SDE admits a unique solution up to a fixed deterministic time $T$ (for instance of linear growth and locally Lipschitz continuous), and $V$ being non-negative almost surely. Further, we assume sufficient integrability conditions such that the market model admits no arbitrage: there exists an equivalent martingale measure $Q$ under which $S$ and $V$ follows

$$\begin{aligned} dS_t &= rS_t dt + \sigma_S(S_t, V_t)d\tilde{W}_t + \int_{\mathbb{R}^d} h(\xi, S_{t-}, V_t)\tilde{\mu}(d\xi, dt) \\ dV_t &= \mu_V(V_t, \Gamma)dt + \sigma_V(V_t)d\tilde{W}_t \end{aligned}$$

where $\sigma_S(\cdot)$ and $\sigma_V(\cdot)$, both with values in $\mathbb{R}^{d\times m}$, are the first and last $d$ rows of $\sigma$. The $\mathbb{R}^d$-valued function $\mu_V(\cdot, \Gamma)$ is the $Q$-drift of the variance with parameters $\Gamma$, and $\tilde{W}$ is an $m$-dimensional Wiener process under $Q$. For convenience, we have assumed that jumps affect the asset prices[18] only; $\mathcal{Z} = \mathbb{R}^d$ and $h(\cdot)$ is an $\mathbb{R}^d$-valued function while the compensator measure $\mu_p(d\xi, dt) = \nu(\gamma, d\xi)dt$ is dependent on $Q$-parameters $\gamma$ (we use the same notation for $\mu_p$ here even if the compensator may be different under $P$ and $Q$). Furthermore, we assume that the continuous variance has drift and diffusion functions dependent on the state of the variance alone. The risky assets are not assumed to carry any dividend payments, although the generalisation to non-zero dividends (as well as jumps and $S$-dependent coefficients for the variance) is straightforward. As under $P$, we assume all coefficients under $Q$ to sufficiently well behaved for an appropriate solution to exists.

The market model $(S, B)$ is free of arbitrage but incomplete as it has more random sources $(2 \times d)$ than traded risky assets $(d)$, and since the asset prices exhibit jumps (i.e. the risk-neutral measure $Q$ is not unique). For a Markovian pricing rule

$$\begin{aligned} D_t &= D(t, S_t, V_t), \quad t \in [0, T], \\ D &: [0, T] \times \mathbb{R}^d \times \mathbb{R}^d \to \mathbb{R}, \quad D \in \mathcal{C}^{1,2}, \end{aligned}$$

of a European option with terminal payoff $g(S_T)$, we have a pricing equation corresponding to (3) as given by

$$\begin{aligned} \frac{\partial D}{\partial t} + \mathcal{L}D - rD = 0 \\ D(T, s, v) = g(s) \end{aligned}$$

[18] As the jumps are generated by a Poisson random measure, $S$ will have jumps given by

$$\Delta S_t = \int_{z\in\mathbb{R}^d} h(z, S_{t-}, V_t)\tilde{\mu}(dz, \{t\}) = h(z_t, S_{t-}, V_t)\mathbf{1}_{\{\Delta S_t \neq 0\}}$$

where $z_t \in \mathbb{R}^d$ is a (unique) point in the set where $\mu(\{z_t\}) = 1$.

where $\mathcal{L}$ is the (time independent) inegro-differential operator that generates $(S; V)$ under $Q$. For a function $f(s,v) \in \mathcal{C}^2$ the operator is defined as

$$\mathcal{L}f(s,v) = \mu^Q(s,v)^\top \nabla_x f(s,v) + \frac{1}{2}\text{tr}\left[\sigma^\top(s,v)\nabla^2_{xx} f(s,v)\sigma(s,v)\right] + \int_{\xi\in\mathbb{R}^d} \left(f(s+h(\xi,s,v),v) - f(s,v) - h(\xi,s,v)^\top \nabla_x f(s,v)\right)\nu(d\xi)$$

where $\nabla_x$, $\nabla^2_{xx}$ with $x=(s,v)$ are the gradient and Hessian operators respectively, and $\text{tr}[\cdot]$ the matrix trace. By the Feynman-Kac representation formula, this is equivalent to the risk-neutral valuation formula $D(t,s,v) = \mathbb{E}^Q\big[\, e^{-r(T-t)}g(S_T)\big|\,(S_t,V_t) = (s,v)\big]$.

## 4.2 *Pricing Under Parameter Uncertainty*

Here we introduce the controlled measure $Q^u$ that represents the parameter uncertainty in our model. Hence, with $\{u_t\}_{t\geq 0}$ being a $\mathcal{F}_t$-predictable control process that takes its values in a compact uncertainty set $U \subset \mathbb{R}^k$, we let the controlled dynamics be

$$dS_t = r_t S_t dt + \sigma_S(S_t,V_t)dW^u_t + \int_{\mathbb{R}^d} h(\xi,S_{t-},V_t)\tilde{\mu}^u(d\xi,dt)$$
$$dV_t = \mu_V(V_t,u_t)dt + \sigma_V(V_t)dW^u_t.$$

where we assume that the controlled drift of the asset price and variance, $r_t S_t$ and $\mu_V(V_t,u_t) = \mu_V(V_t,\Gamma_t)$, to be of the same functional form under $Q$ and $Q^u$. Hence, the control has components $u_t = (r_t,\Gamma_t,\gamma_t)$ where $\Gamma$ are the parameters of $\mu_V$, while $\gamma$ are parameters to the controlled (form invariant) compensator measure

$$\mu^u_p(d\xi,dt) = \nu(\gamma_t,d\xi)dt$$

with Radon–Nikodym density $\beta(\xi,t,u_t) \equiv d\mu^u_p/d\mu_p$ with respect to $\mu_p$. We let $\mu^{Q^u}(S_t,V_t,u_t) \equiv (r_t S_t;\mu_V(V_t,u_t))$ denote the common drift of $(S;V)$ under $Q^u$ (and similarly for the common $Q$-drift). The effect of the control is then defined by the $\mathbb{R}^{1\times m}$-valued process

$$\alpha(S_t,V_t,u_t) = \sigma^{-1}(S_t,V_t)\left(\mu^{Q^u}(S_t,V_t,u_t) - \mu^Q(S_t,V_t) - \int_{\mathbb{R}^d} h(\xi,S_{t-},V_t)(\beta(\xi,t,u_t)-1)\nu(d\xi)\right)$$

to give the linear driver function $f(s,v,y,z,\theta,u) = -ry + z\alpha(s,v,u) + \int_\xi \theta(\xi)(\beta(\xi,t,u)-1)\nu(d\xi)$ where the second last argument is a function $\theta : \mathbb{R}^d \mapsto \mathbb{R}$. Modulo sufficient integrability and Lipschitz conditions, we have that the value function for a fixed admissible control $J_t(u) = \mathbb{E}_u[e^{-\int_t^T r_u du}g(S_T)|\mathcal{F}_t]$, $t\in[0,T]$, is given as part of the solution $(J(u), Z, \Theta)$ to the linear BSDE

$$
\begin{aligned}
dJ_t(u) &= -f(S_t, V_t, J_t(u), Z_t, \Theta_t, u_t)dt + Z_t d\tilde{W}_t + \int_{\xi\in\mathbb{R}^d} \Theta_t(\xi)\tilde{\mu}(d\xi, dt) \\
J_T(u) &= g(S_T)
\end{aligned}
$$

where $Z$ is $\mathbb{R}^{1\times m}$-valued while $\Theta$ is a process taking its values in the space of functions $\theta : \mathbb{R}^d \mapsto \mathbb{R}$. The result follows similarly as in the case of Heston's model: apply Itô's product rule to $\mathcal{E}(\Lambda)J(u)$, where $\Lambda = -\int_0^\cdot r_t dt + \alpha(S_t, V_t, u_t) \bullet \tilde{W} + \int_0^\cdot \int (\beta(\xi, t, u_t) - 1)\tilde{\mu}(d\xi, dt)$, to see that $\mathcal{E}(\Lambda)J(u)$ is a martingale, and use that $\mathcal{E}(\Lambda + \int_0^\cdot r_t dt)_T = dQ^u/dQ$ for the measure change of $\mathcal{E}(\Lambda)_t J_t(u) = \mathbb{E}[\mathcal{E}(\Lambda)_T J_T(u)|\mathcal{F}_t]$ to obtain the original expression for the value process after rearrangement. Further, defining the pointwise optimised driver functions over the compact uncertainty set

$$
H^{\pm}(s, v, y, z, \theta) = \left| \operatorname*{ess\,sup}_{u\in U} \pm f(s, v, y, z, \theta, u) \right|,
$$

we have by the comparison theorem that the the optimally controlled value processes (the upper/lower pricing boundaries) $\{D_t^{\pm}\}_{t\in[0,T]} = \{|\operatorname{ess\,sup}_{\{u_t\}} \pm J_t(u)|\}_{t\in[0,T]}$ are solutions to the BSDEs

$$
\begin{aligned}
dD_t^{\pm} &= -H^{\pm}(S_t, V_t, D_t^{\pm}, Z_t, \Theta_t)dt + Z_t d\tilde{W}_t + \int_{\xi\in\mathbb{R}^d} \Theta_t(\xi)\tilde{\mu}(d\xi, dt) \\
D_T^{\pm} &= g(S_T).
\end{aligned}
$$

Here as well, this is a consequence of the fact that we have a linear driver in $y$, $z$ and $\theta$, and from the comparison theorem for BSDEs. The proof, which we omit for brevity, follows in the same fashion as in the previous case with Heston's model. As well, since we work in a Markovian setting, we have that the solution can be written with a deterministic function $D_t = D(t, S_t, V_t)$, and the same holds for the optimal control: there exists a function $u^*(t, s, v)$ such that the feedback control $u_t^* = u^*(t, S_t, V_t)$ is the optimal control among all admissible controls.

Finally, as in the case with Heston's model, we have by the semilinear Feynman–Kac formula that $D(t, s, v)$ satisfies a semilinear partial differential equation

$$
\frac{\partial D}{\partial t} + \frac{1}{2}\mathrm{tr}[\sigma^\top \nabla_{xx}^2 D\sigma] + \operatorname*{ess\,inf}_{(r,\Gamma,\gamma)\in U} \left\{ -rD + (\mu^{Q^u})^\top \nabla_x D + \int_\xi (\Delta D - h^\top \nabla_x D)\nu(\gamma, d\xi) \right\} = 0
$$

with terminal condition $D(T, s, v) = g(s)$, and where $\Delta D$ is shorthand notation for $D(t, s + h(\xi, s, v), v) - D(t, s, v)$. Although many numerical methods exist for a PIDE of this type, one may opt for simulating the BSDE solution instead (see e.g. [16]), especially when the dimensional of the problem is high.

## 5 Conclusion

Model uncertainty, here represented by parameter uncertainty, is an acknowledged concept formalised by [45] and its importance has been studied in the financial context at least since [26]. The focus of this paper has been to investigate how parameter uncertainty could be incorporated into a stochastic volatility model, and how it affects prices of European option. The considered uncertainty was fairly general: interest rate and volatility drift parameters where allowed to change over time (constant-parameters being a special case) within a pre-described uncertainty region inferred from statistical estimation. The effect on pricing was then studied from a worst-case perspective with boundaries for the option price that could be embedded into a control problem, with the control playing the role of uncertain parameters.

With Heston's model as a working example, the control problem–BSDE duality was exploited and an explicit equation for the pricing boundary (the optimal value process) was derived in the form of a Markovian linear BSDE. A numerical scheme with several suggested modifications was considered for the solution of this BSDE, and an evaluation of the schemes was made in a known-outcome setting analogous to the dynamic-parameter setting. Based on bias/variance (and computational) considerations, a scheme was proposed for an empirical study of the methodology applied to real-world market data. Studying a set of bid/offer market quotes of European call options on the S&P 500 index and their corresponding model-price bounds, it was found that even if the model (and uncertainty set) was estimated from historical prices of the underlying, 98% of the market option prices was within the model-prescribed bounds. In contrast, ∼40% of the market quotes was within the maximum/minimum model-price interval when constant parameters where used.

In both the dynamic and constant parameter setting, it was seen that the model implied volatilities did not follow the curvature of the market implied volatilities. A natural explanation for this observation is that the diffusion parameters, which effectively decide the slope and skew of the implied volatility curve, were estimated from asset-price data, and not from option-price data. An interesting empirical sequel would therefore be to study how the shape and coverage of model-price bounds change when parameters are calibrated from market option prices instead. We leave this for further investigation.

Finally, we note that prior beliefs and preferences about the uncertainty are not taken into consideration by the conservative approach with pricing boundaries. However, with $L(u_t, u'_t)$ being some function that assigns a loss when $u_t$ is used instead of the true parameters $u'_t$, we could incorporate beliefs with a value function of the form

$$J_t(u) = \mathbb{E}_u \left[ e^{\int_t^T r_s ds} G + \int_t^T L(u_s, u'_s) ds \,\middle|\, \mathcal{F}_t \right]$$

that would lead to a similar linear BSDE. The loss could be based on the (approximative) normality of estimated parameters or some quantity related to an economic value, for instance a hedging error. In both cases, the value of the loss must be related to the value of the option payoff, an intricate task that we leave for further research along with this approach.

**Acknowledgements** We thank two anonymous referees for detailed and helpful recommendations on this paper. Tegnér gratefully acknowledges support from the Wallenberg Foundations and the Oxford-Man Institute for Quantitative Finance. Cohen gratefully acknowledges support from Oxford-Nie Financial Big Data Laboratory and the Oxford-Man Institute for Quantitative Finance.

# A Numerical Methods for BSDEs

The optimally controlled value process (or the value process for a fixed feedback control, i.e. $u_t = u(t, S_t, V_t)$ for a deterministic function $u$) is given by the solution to the decoupled forward-backward stochastic differential equation (2)–(12). In general, there is not much hope of finding closed-form solutions to neither forward nor backward SDEs and one typically has to consider numerical methods. For our purposes, we consider the simulation technique by [17].

## *A.1 The Simulation Scheme by Bouchard and Touzi*

For a time gird $\pi : 0 = t_0 < \cdots < t_n = T$, [17] propose a method to generate a discrete-time approximation $(X^\pi, Y^\pi)$ of the solution to a decoupled equation with forward component $X$ and backward component $Y$. In the first part of the scheme, the forward component $X^\pi$ is simulated over the time grid $\pi$ with a standard Euler-Maruyama approximation to generate $N$ paths of $X^\pi$ (see e.g. [44]). The component $Y^\pi$ is then generated by the backward induction

$$\begin{aligned}
Y^\pi_{t_n} &= g(X^\pi_{t_n}) \\
Z^\pi_{t_{i-1}} &= \frac{1}{\Delta_i}\mathbb{E}\left[ Y^\pi_{t_i}\Delta W_{t_i} \,\middle|\, X^\pi_{t_{i-1}}\right] \\
Y^\pi_{t_{i-1}} &= \mathbb{E}\left[ Y^\pi_{t_i} + f(X^\pi_{t_{i-1}}, Y^\pi_{t_{i-1}}, Z^\pi_{t_{i-1}})\Delta_i \,\middle|\, X^\pi_{t_{i-1}}\right]
\end{aligned} \tag{28}$$

where $\Delta_i \equiv t_i - t_{i-1}$ and $\Delta W_{t_i} \equiv W_{t_i} - W_{t_{i-1}}$ are the $i$th time- and Wiener increments from the generation of $X^\pi$. The last equation in (28) is obtained by applying $\mathbb{E}[\cdot|\mathcal{F}_{t_{i-1}}]$ to the following simple discretization of the BSDE

$$Y^\pi_{t_i} - Y^\pi_{t_{i-1}} = -f(X^\pi_{t_{i-1}}, Y^\pi_{t_{i-1}}, Z^\pi_{t_{i-1}})\Delta_i + Z^\pi_{t_{i-1}}\Delta W_{t_i} \tag{29}$$

and using the Markov property of $X$ and the fact that $Y_t$ and $Z_t$ are both deterministic functions of $X_t$ for all $t \in [0, T]$. The second equation for $Z$ is obtained similarly by multiplying (29) with $\Delta W_{t_i}$ and taking conditional expectations.

For the backward induction (28) one has to compute the conditional expectations and to make the scheme operational, this is made with an approximation $\hat{E}[\cdot|X^{\pi}_{t_{i-1}}]$ of the regression function $\mathbb{E}[\cdot|X^{\pi}_{t_{i-1}}]$ based on simulated training data. That is, the data

$$\{Y^{\pi(j)}_{t_i}, \Delta W^{(j)}_{t_i}, X^{\pi(j)}_{t_{i-1}}\}_{1\le j\le N}$$

is used for the first regression in (28) where $X^{\pi(j)}$ is the $j$th simulated path of $X^{\pi}$ and $Y^{\pi(j)}$ is the corresponding value from the induction of the previous time step. For the second regression, $\{Y^{\pi(j)}_{t_i}, Z^{(j)}_{t_{i-1}}, X^{\pi(j)}_{t_{i-1}}\}_{1\le j\le N}$ is used accordingly.

As an example of a non-parametric regression estimator, it is suggested to use the Nadaraya-Watson weighted average for a kernel estimator. We conveniently employ the $k$-nearest neighbour kernel for this purpose: for $X^{\pi}_{t_{i-1}}$ and a generic $\xi \in \mathcal{F}_{t_i}$, each with simulated outcomes $\{X^{\pi(j)}_{t_{i-1}}, \xi^{(j)}\}_{1\le j\le N}$, we approximate $\mathbb{E}[\xi|X^{\pi}_{t_{i-1}} = X^{\pi(j)}_{t_{i-1}}]$, $j = 1, \ldots, N$, with

$$\hat{E}\left[\xi\left|X^{\pi(j)}_{t_{i-1}}\right.\right] = \frac{\sum_{l=1}^{N} \xi^{(l)} \mathbf{1}\left(||X^{\pi(l)}_{t_{i-1}} - X^{\pi(j)}_{t_{i-1}}|| \le d^{(j)}_k\right)}{k+1} \tag{30}$$

where $d^{(j)}_k$ is the distance between $X^{\pi(j)}_{t_{i-1}}$ and its $k$th nearest neighbour, and $\mathbf{1}(\cdot)$ is the indicator function. The regression (30) together with (28) yields an implicit simulation method and as a last step of the scheme, it is suggested to truncate $Z^{\pi}_{t_{i-1}}$ and $Y^{\pi}_{t_{i-1}}$ if one has appropriate (possibly $t$ and $X^{\pi}_{t_{i-1}}$ dependent) bounds for $\mathbb{E}[Y^{\pi}_{t_i}\Delta W_{t_i}|X^{\pi}_{t_{i-1}}]$, $\mathbb{E}[Y^{\pi}_{t_i}|X^{\pi}_{t_{i-1}}]$ and $Y_{t_{i-1}}$.

The $k$-nearest neighbours estimator (30) approximates the regression function in a local neighbourhood with a constant. As such, it has low bias but high variance, and it suffers from the curse of dimensionality at the boundaries. For an alternative regression estimator, we consider the MARS method[19] (multivariate adaptive regression spines) which uses piecewise linear basis functions in an adaptive manner to approximate the regression function. The model has the linear form

$$\hat{E}\left[\xi\left|X^{\pi}_{t_{i-1}}\right.\right] = \beta_0 + \sum_{m=1}^{M} \beta_m h_m(X^{\pi}_{t_{i-1}}) \tag{31}$$

where each basis function $h_m(X)$ is constructed from a collection of paired piecewise-linear splines

$$\mathcal{C} = \left\{(X^k - \eta)^+, (\eta - X^k)^+ : \eta \in \{X^{k,\pi(j)}_{t_{i-1}}\}_{j=1}^{N}, k = 1, 2\right\}.$$

[19] We are using the R package "earth" by [49].

The *knots* $\eta$ are hence placed at any value in the set of $X$-observations (with superscript $k$ referring to the components of $X$). A basis function $h_m(X)$ is also allowed to be a $d$-times product of the functions in $\mathcal{C}$, where $d$ denotes a chosen model order.

The model (31) is built up sequentially by multiplying a current $h_m$ with a new function from $\mathcal{C}$ to form candidates: all pairs in $\mathcal{C}$ are tested, and only the term which yields the largest reduction of residual error is added to (31). Here, all coefficients $\beta_0, \beta_1, \ldots$ are estimated in each step by least squares. The model building then continues until there is a prescribed maximum number of terms $M$. Finally, the model is "pruned" with a deleting procedure (again based on the *smallest* increase of error) where the optimal number of terms is estimated by cross-validation (for details, see [37]).

## A.2 Modified Simulation Schemes

As a first modification of the Bouchard-Touzi method, we consider an explicit version of the implicit scheme by replacing the second regression in (28):

$$\begin{aligned} Y_{t_n}^\pi &= g(X_{t_n}^\pi) \\ Z_{t_{i-1}}^\pi &= \frac{1}{\Delta_i}\mathbb{E}\left[ Y_{t_i}^\pi \Delta W_{t_i} \,\middle|\, X_{t_{i-1}}^\pi \right] \\ Y_{t_{i-1}}^\pi &= \mathbb{E}\left[ Y_{t_i}^\pi + f(X_{t_{i-1}}^\pi, Y_{t_i}^\pi, Z_{t_{i-1}}^\pi)\Delta_i \,\middle|\, X_{t_{i-1}}^\pi \right]. \end{aligned} \tag{32}$$

This comes from a discretization of the BSDE at the right time-point $Y_{t_i}$ instead of $Y_{t_{i-1}}$ and since $Y$ is a continuous process, the effect of using the value at the right time-point is vanishing as the time-grid becomes tighter. The discretization is used by [32], and the benefit is that this allows for an explicit calculation of $Y_{t_{i-1}}^\pi$ in the second regression step of each iteration.

As an additional step, to obtain an implicit method with a fixed point procedure, we may employ (32) to get a first candidate $\tilde{Y}_{t_{i-1}}^\pi$ and supplement each step in the backward induction with a small number of implicit iterations of

$$\tilde{Y}_{t_{i-1}}^\pi = \mathbb{E}\left[ Y_{t_i}^\pi + f(X_{t_{i-1}}^\pi, \tilde{Y}_{t_{i-1}}^\pi, Z_{t_{i-1}}^\pi)\Delta_i \,\middle|\, X_{t_{i-1}}^\pi \right], \tag{33}$$

and keeping $Y_{t_{i-1}}^\pi = \tilde{Y}_{t_{i-1}}^\pi$ as our final value for the next backward step—see [33].

Secondly, to improve the stability of the scheme, we consider a modification of (32) based on the following recursion for the backward component

$$\begin{aligned} Y_{t_{i-1}}^\pi &= Y_{t_i}^\pi + f(X_{t_{i-1}}^\pi, Y_{t_i}^\pi, Z_{t_{i-1}}^\pi)\Delta_i - Z_{t_{i-1}}^\pi \Delta W_{t_i} \\ &= Y_{t_{i+1}}^\pi + f(X_{t_{i-1}}^\pi, Y_{t_i}^\pi, Z_{t_{i-1}}^\pi)\Delta_i + f(X_{t_i}^\pi, Y_{t_{i+1}}^\pi, Z_{t_i}^\pi)\Delta_{i+1} - Z_{t_{i-1}}^\pi \Delta W_{t_i} - Z_{t_i}^\pi \Delta W_{t_{i+1}} \\ &= Y_{t_n}^\pi + \sum_{k=i}^{n} f(X_{t_{k-1}}^\pi, Y_{t_k}^\pi, Z_{t_{k-1}}^\pi)\Delta_k - Z_{t_{k-1}}^\pi \Delta W_{t_k} \end{aligned}$$

such that we may write the explicit backward induction (32) as

$$
\begin{aligned}
Y^{\pi}_{t_n} &= g(X^{\pi}_{t_n}) \\
Z^{\pi}_{t_{i-1}} &= \frac{1}{\Delta_i}\mathbb{E}\left[\left(Y^{\pi}_{t_n} + \sum_{k=i+1}^{n} f(X^{\pi}_{t_{k-1}}, Y^{\pi}_{t_k}, Z^{\pi}_{t_{k-1}})\Delta_k\right)\Delta W_{t_i}\middle| X^{\pi}_{t_{i-1}}\right] \\
Y^{\pi}_{t_{i-1}} &= \mathbb{E}\left[Y^{\pi}_{t_n} + \sum_{k=i}^{n} f(X^{\pi}_{t_{k-1}}, Y^{\pi}_{t_k}, Z^{\pi}_{t_{k-1}})\Delta_k\middle| X^{\pi}_{t_{i-1}}\right].
\end{aligned}
\tag{34}
$$

This idea appears in [11] and is explored further in [34]. The benefit is that errors due to approximating the conditional expectation do not accumulate at the same rate. As in the previous modification, we may complement (34) with a small number of iterations

$$
\tilde{Y}^{\pi}_{t_{i-1}} = \mathbb{E}\left[Y^{\pi}_{t_n} + \sum_{k=i}^{n} f(X^{\pi}_{t_{k-1}}, \tilde{Y}^{\pi}_{t_{k-1}}, Z^{\pi}_{t_{k-1}})\Delta_k\middle| X^{\pi}_{t_{i-1}}\right] \tag{35}
$$

for an implicit method.

For an alternative type of simulation schemes, recall that for Markovian forward-backward equations, both $Y_t$ and $Z_t$ may be written as functions of the current forward state $(t, X_t)$. Hence, we use the regression estimator of (32) to write

$$
\begin{aligned}
Y^{\pi}_{t_{i-1}} &= \hat{E}\left[Y^{\pi}_{t_i} + f(X^{\pi}_{t_{i-1}}, Y^{\pi}_{t_i}, Z^{\pi}_{t_{i-1}})\Delta_i \middle| X^{\pi}_{t_{i-1}}\right] \\
&\equiv \hat{y}_{i-1}(X^{\pi}_{t_{i-1}})
\end{aligned}
\tag{36}
$$

that is, the function $y(t, x)$ that generates $Y_t = y(t, X_t)$ is approximated with $\hat{y}(\cdot)$. Further, if we use $Z^{\pi}_{t_i}$ in the driver of (36) ($Z^{\pi}_{t_i}$ from the previous time-step) to obtain $\hat{y}_{i-1}(\cdot)$, we get the following scheme

$$
\begin{aligned}
Y^{\pi}_{t_n} &= g(X^{\pi}_{t_n}), \quad Z^{\pi}_{t_n} = \partial_x g(X^{\pi}_{t_n})\sigma(X^{\pi}_{t_n}), \\
Y^{\pi}_{t_{i-1}} &= \hat{E}\left[Y^{\pi}_{t_i} + f(X^{\pi}_{t_{i-1}}, Y^{\pi}_{t_i}, Z^{\pi}_{t_i})\Delta_i \middle| X^{\pi}_{t_{i-1}}\right] \implies \hat{y}_{i-1}(\cdot), \\
Z^{\pi}_{t_{i-1}} &= \partial_x \hat{y}_{i-1}(X^{\pi}_{t_{i-1}})\sigma(X^{\pi}_{t_{i-1}}),
\end{aligned}
\tag{37}
$$

since the function that generates $Z_t$ is given by $Z_t = \partial_x y(t, X_t)\sigma(X_t)$, see footnote 8. In particular, if we employ the MARS regression, $\hat{y}(\cdot)$ will be a sum of piecewise linear splines and products thereof up to the specified degree. Hence, the partial derivatives $(\partial_s \hat{y}(\cdot), \partial_v \hat{y}(\cdot))$ are easily calculated analytically. Further, the last two calculations of (37) may be iterated with $Y^{\pi}_{t_{i-1}}, Z^{\pi}_{t_{i-1}}$ for an implicit version of the scheme.

For a second type of modifications, we may include additional predictors for the regression functions. As an example, let $C^{\text{He}}(t, x)$ denote the pricing function of an option with terminal payoff $g(X_T)$ calculated under Heston's model. As the pricing bound $Y_t$ lies in a neighbourhood of the price $C^{\text{He}}(t, X_t)$, we may add this as a predictor to our regression estimator

$$Y^\pi_{t_{i-1}} = \hat{E}\left[ Y^\pi_{t_i} + f(X^\pi_{t_{i-1}}, Y^\pi_{t_i}, Z^\pi_{t_{i-1}})\Delta_i \,\middle|\, X^\pi_{t_{i-1}}, C^{\text{He}}(t_{i-1}, X^\pi_{t_{i-1}})\right]. \tag{38}$$

Finally, we mention a modification of the first regression in the standard scheme (28), as proposed by [2]

$$Z^\pi_{t_{i-1}} = \frac{1}{\Delta_i}\mathbb{E}\left[ \left(Y^\pi_{t_i} - \mathbb{E}[Y^\pi_{t_i}|X^\pi_{t_{i-1}}]\right) \Delta W_{t_i} \,\middle|\, X^\pi_{t_{i-1}}\right] \tag{39}$$

with the purpose being a variance reduction of the regression estimate. The motivation is the following: since $Y^\pi_{t_i} = y(t_i, X^\pi_{t_i})$ for some continuous function $y(t,x)$, we have $Z^\pi_{t_{i-1}} = \mathbb{E}[y(t_{i-1} + \Delta_i, X^\pi_{t_{i-1}} + \Delta X^\pi_{t_i})\Delta W_{t_i}/\Delta_i | X^\pi_{t_{i-1}}]$ and the estimator thereof

$$\frac{1}{N}\sum_{j=1}^{N} y(t_{i-1} + \Delta_i, X^\pi_{t_{i-1}} + \Delta X^{\pi(j)}_{t_i}) \frac{\sqrt{\Delta_i} z^{(j)}}{\Delta_i} \tag{40}$$

where $z^{(j)}$ are independent standard normal random variables. As $\Delta X^{\pi(j)}_{t_i} = \text{drift} \times \Delta_i + \text{diff} \times \sqrt{\Delta_i} z^{(j)}$, we have that the variance of the estimate (40) is approximately $y(t_{i-1}, X^\pi_{t_{i-1}})^2/(N\Delta_i)$ for small $\Delta_i$ and hence, it blows up as $\Delta_i \to 0$. In return if we use

$$\frac{1}{N}\sum_{j=1}^{N} \left(y(t_i, X^\pi_{t_i}) - y(t_{i-1}, X^\pi_{t_{i-1}}) + f_{i-1}\Delta_i\right) \frac{\sqrt{\Delta_i} z^{(j)}}{\Delta_i} \tag{41}$$

where $f_{i-1} \equiv f(X^\pi_{t_{i-1}}, Y^\pi_{t_{i-1}}, Z^\pi_{t_{i-1}})$ and $y(t_{i-1}, X^\pi_{t_{i-1}}) - f_{i-1}\Delta_i = \mathbb{E}[Y^\pi_{t_i}|X^\pi_{t_{i-1}}]$ from (28), we have that the estimator (41) of (39) will have approximate variance $2y_x(t_{i-1}, X^\pi_{t_{i-1}})^2/N + \Delta_i f_{i-1}/N$ which do not depend on $\Delta_i$ as this goes to zero.

We end this section with a demonstration of the simulation schemes based on (32) and (34) in the following example.

*Example 1* For the forward process, we simulate $N = 100{,}000$ paths of Heston's model (2) with parameters $(r, \kappa, \theta, \sigma, \rho) = (0, 5.07, 0.0457, 0.48, -0.767)$, initial value $(S^\pi_0, V^\pi_0) = (100, \theta)$ over an equidistant time grid with $n = 25$ points and terminal time $T = 1$. For the backward process, we consider the trivial driver $f(X_t, Y_t, Z_t) = 0$, i.e. $dY_t = Z_t d\tilde{W}_t$, together with the terminal condition $Y_T = S_T$. Hence, $Y$ is a martingale and we have

$$Y_t = \mathbb{E}^Q\left[S_T|\mathcal{F}_t\right] = S_t$$

since for a zero interest rate, $S$ is a $Q$-martingale as well. As there is no dependency of $Z$ in the driver, the backward induction simplifies to the regression $Y^\pi_{t_{i-1}} = \hat{E}[Y^\pi_{t_i}|X^\pi_{t_{i-1}}]$ repeated for $i = n, \ldots, 1$ and a starting value $Y^\pi_{t_n} = S^\pi_{t_n}$. With $k = 5$ nearest neighbours of the regression estimator (30), the left pane of Fig. 7 shows five simulated paths of the backward process $Y^\pi$ with the explicit scheme

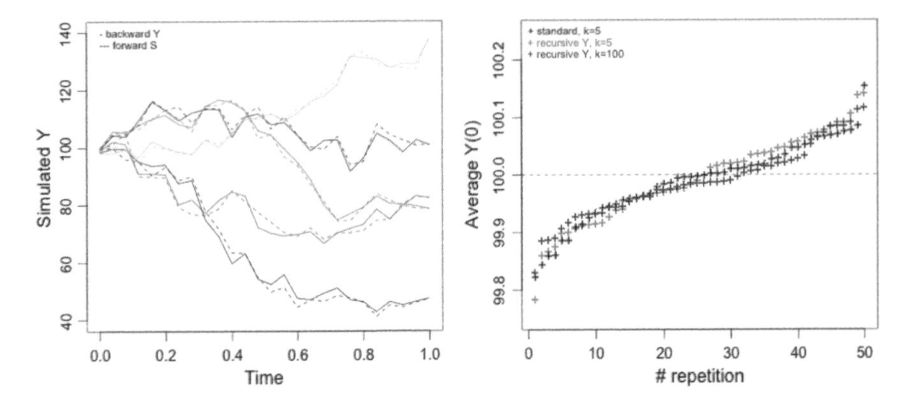


**Fig. 7** **Left figure**: five simulated paths of $Y^\pi$ (solid lines) with the zero-driver of Example 1. The explicit scheme (30)–(32) is employed with $k = 5$ nearest neighbours. The forward component $X^\pi = (S^\pi, V^\pi)$ is simulated from Heston's model and the dashed lines show the corresponding paths of $S^\pi$. **Right figure**: the $N$-sample average of $Y_0^\pi$ (in increasing order) from 50 repetitions of the simulation with the $k = 5$ explicit scheme (32) (black crosses), the recursive-based scheme (34) with $k = 5$ (red crosses) and $k = 100$ (blue crosses)

(32) along with the corresponding paths of $S^\pi$ and it can be seen that the components follow each other quite closely. Looking at the initial time value, the $N$-sample of $Y_0^\pi$ has an average 98.532 to be compared with the true value $Y_0 = \mathbb{E}^Q[S_T] = S_0 = 100$, while the sample of $Y_{t_n}^\pi = S_{t_n}^\pi$ averages to 99.998.

If we repeat the simulation 50 times and calculate the average of $Y_0^\pi$ for each repetition, we obtain the result in the right pane of Fig. 7. The first explicit scheme based on (32) yields sample averages quite close to the true value and if we repeat the simulations with the $Y^\pi$-recursion scheme (34) instead, we obtain similar results. For comparison, we have included the recursive scheme with $k = 100$ nearest neighbours as well.

Finally, notice that this example corresponds to $g(x) = x$ and an effect $\alpha(S_t, V_t, u_t) = (0, 0)^\top$ such that $Q^u \equiv Q$. Hence, with $\varrho = 0$ for the driver (10), we have that the value process $J_t(u) = \mathbb{E}_u[g(S_T)|\mathcal{F}_t] = \mathbb{E}^Q[S_T|\mathcal{F}_t]$ is the solution to (11).

## A.3 Simulation Results for European Options

For numerical calculation of the pricing bounds for European options, we consider the parameter setting given in Table 5 and a set of call options with strike-maturity as given in Table 3. The call prices are calculated from the semi-closed pricing formula of Heston's model, i.e. by numerical integration of the inverse Fourier transform of the price (see e.g. [31]). The corresponding implied volatilities are obtained from Black-Scholes formula by numerical optimisation.

**Table 3** Prices and implied volatilities (in parenthesis) of European call options calculated by the semi-closed pricing formula of Heston's model with parameters from Table 5

| European call prices | | | |
|---|---|---|---|
| Strike/expiry | 75 | 100 | 125 |
| 4m | 26.0044 (0.2823) | 4.8239 (0.2106) | 0.0070 (0.1518) |
| 1y | 29.4915 (0.2482) | 10.9174 (0.2124) | 1.8403 (0.1832) |
| 10y | 57.4959 (0.2220) | 46.4060 (0.2174) | 37.1943 (0.2138) |

**Table 4** Prices and implied volatilities of European call options, calculated by numerical minimisation/maximisation of the Heston pricing formula over the parameters $(r, \kappa, \theta)$ constrained by the parameter uncertainty region

| Optimised Heston pricing function | | | |
|---|---|---|---|
| Strike/expiry | 75 | 100 | 125 |
| 4m | [25.9316, 26.2591]<br>(0.0520, 0.3651) | [4.5758, 5.0572]<br>(0.1980, 0.2225) | [0.0040, 0.0124]<br>(0.1441, 0.1610) |
| 1y | [28.6578, 30.4061]<br>(0.0303, 0.3060) | [9.9716, 11.8229]<br>(0.1872, 0.2364) | [1.3840, 2.4824]<br>(0.1659, 0.2053) |
| 10y | [54.5102, 62.3675]<br>(0.0195, 0.3190) | [40.2004, 51.9955]<br>(0.1085, 0.2925) | [30.7291, 43.0811]<br>(0.1444, 0.2754) |

Prior to considering the pricing bounds as obtained from the optimally controlled value process, we take a look at prices by minimising/maximising Heston's pricing formula $C_{\text{He}}(\cdot)$ over the parameter uncertainty set $U$ represented by the elliptic constraint in (19) with a 95% confidence level. That is

$$C^{\pm}_{\text{He}} = \left| \min_{(r,\kappa,\theta)\in U} \pm C^{\text{He}}(S, V; \tau, K, \Theta) \right| \tag{42}$$

where $\Theta$ is the vector of model parameters including $(r, \kappa, \theta)$ while $K$ is the strike and $\tau$ the time to maturity. From numerical optimisation of (42) with parameters and elliptic uncertainty region based on Table 5, we get the results in Table 4. We use these as a reference point for our forthcoming simulation study.

**Simulation of the Forward Component**

The forward component $X = (S, V)$ of the SDE (2) governing the asset price and variance is simulated in the first stage of the simulation scheme for the forward-backward equation. We employ the standard Euler-Maruyama scheme for the log-price and an implicit Milstein scheme to generate the variance

**Table 5** Parameter setting and covariance matrix used for the numerical calculation of pricing bounds for European options

**Model parameters**

| $S_0$ | $V_0$ | $r$ | $\kappa$ | $\theta$ | $\sigma$ | $\rho$ |
|---|---|---|---|---|---|---|
| 100 | 0.0457 | 0.05 | 5.070 | 0.0457 | 0.4800 | -0.767 |

| | $r$ | $\kappa$ | $\beta$ |
|---|---|---|---|
| $r$ | 2.5e-05 | 0 | 0 |
| $\kappa$ | 0 | 0.25 | 0 |
| $\beta$ | 0 | 0 | 1e-04 |

$$\log S^{\pi}_{t_i} = \log S^{\pi}_{t_{i-1}} + \left(\mu - \frac{1}{2}V^{\pi}_{t_{i-1}}\right)\Delta_i + \sqrt{V^{\pi}_{t_{i-1}}}(\rho\Delta W^1_{t_i} + \sqrt{1-\rho^2}\Delta W^2_{t_i})$$

$$V^{\pi}_{t_i} = \frac{V^{\pi}_{t_{i-1}} + \kappa\theta\Delta_i + \sigma\sqrt{V^{\pi}_{t_{i-1}}}\Delta W^2_{t_i} + \frac{1}{4}\sigma^2((\Delta W^2_{t_i})^2 - \Delta_i)}{1 + \tilde{\kappa}\Delta_i}$$

where $\Delta W^1_{t_i}$, $\Delta W^2_{t_i}$ are independent variables generated from the zero-mean normal distribution with variance $\Delta_i$. If the parameters satisfy $4\kappa\theta > \sigma^2$ this discretization scheme generates positive variance paths and we do not have to impose any truncation as in the case with the standard Euler-Maruyama scheme, see [4] or [3]. We simulate $N = 100{,}000$ paths over an equidistant time gird with $n = 25$ knots.

**The Optimised Heston Formula by Backward Simulation**

As a first simulation example of the backward component $Y$, we consider the formula-optimal price of the at-the-money call with maturity 1 year (prices given in Table 4). Hence, we simulate the backward component with the non-optimised driver $f(X_t, Y_t, Z_t, u_t)$ of Eq. (10) with a constant $u_t$ based on the resulting parameters from the price-optimisation (42) of the considered call option. This allows us to evaluate the accuracy of our simulations schemes in a situation where we know the true values. The reason for simulating the optimised price of (42) instead of the null-controlled plain price of the call (given in Table 3) is that the optimised-price simulation relies on a $Z$-dependent driver, while the $Q$-price has an effect being zero in (10) such that the $Z$-regression step of the simulation scheme expires for the plain price (cf. Example 1).

For starters, we consider the following four variations of the simulation schemes from the previous section:

1. the explicit scheme (32) with $k = 5$ nearest neighbours regression (30)
2. the explicit scheme with MARS regression (31) of degree 2
3. the explicit-recursive scheme (34) with $k = 5$ nearest neighbours regression
4. the explicit-recursive scheme with MARS regression of degree 2.

**Table 6** Accuracy of the simulated formula-optimised price of an at-the-money call option with maturity 1 year (true values in Table 4) for $N = 100{,}000$ and $n = 25$. Sample-average, bias and root mean square error calculated from 100 repetitions of each simulation

| Backward simulated optimised Heston price | | | |
|---|---|---|---|
| Scheme | Ave. $\mathbb{E}(\hat{\pi})$ | Bias: $\mathbb{E}(\hat{\pi}) - \pi$ | RMSE: $\sqrt{\mathbb{E}[(\hat{\pi} - \pi)^2]}$ |
| Explicit knn | 11.6552 | −0.1677 | 0.1783 |
| Explicit MARS | 11.7378 | −0.0851 | 0.0987 |
| Recursive knn | 11.7968 | −0.0261 | 0.0608 |
| Recursive MARS | 11.8164 | −0.0065 | 0.0534 |
| Forward MC | 11.8041 | −0.0188 | 0.0508 |
| Explicit knn | 9.9960 | 0.0244 | 0.0511 |
| Explicit MARS | 10.4993 | 0.5277 | 0.5292 |
| Recursive knn | 10.0351 | 0.0635 | 0.0766 |
| Recursive MARS | 10.0004 | 0.0288 | 0.0509 |
| Forward MC | 9.9719 | 0.0003 | 0.0380 |

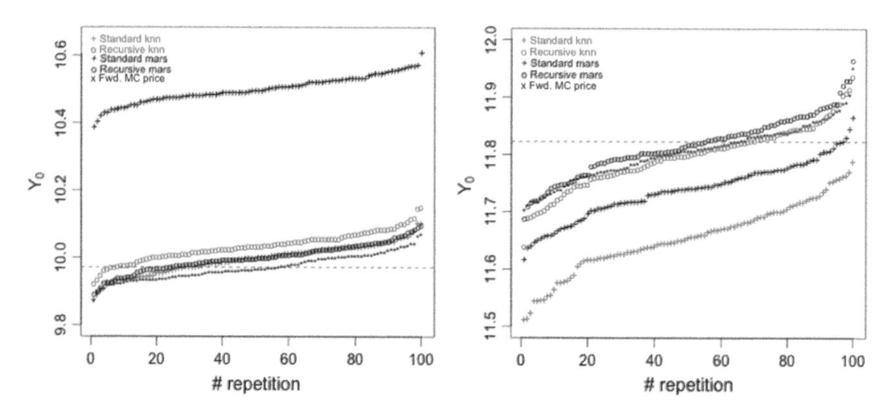


**Fig. 8** Numerical calculation of the formula-optimal price of the 1-year at-the-money call (Table 4). **Left figure**: the $N$-sample average of the minimised price $Y_0^{\pi}$ from 100 repetitions of the simulation (in increasing order). We use a equidistant time-grid with $n = 25$ time-points and generate $N = 100{,}000$ paths of $Y^{\pi}$ in every simulation. **Right figure**: the corresponding maximised price. The figures show the results from four explicit schemes based on the $k = 5$ nearest neighbours estimator (red marks) and the MARS estimator of degree 2 (black marks). The dashed lines indicate the true call price as calculated by the (optimised) Heston's formula while the blue stars show the Monte-Carlo price as calculated from the $N$ simulated paths of $X^{\pi} = (S^{\pi}, V^{\pi})$ for each repetition

For each of the schemes **1–4**, we repeatedly simulate the formula-optimal price 100 times and calculate sample- bias and root mean square errors. The results are given in Table 6, while Fig. 8 shows the prices from all repetitions of the simulation.

From Table 6 we see that the explicit-recursive-MARS scheme performs best in terms of low bias and low RMSE although the simple explicit-knn scheme performs well for the lower price. Comparing the backward simulation with the Monte Carlo price calculated directly from forward simulation we have close to equal performance

**Table 7** Accuracy of the simulated formula-optimised price of an at-the-money call option with maturity 1 year (true values in Table 4) for $N = 100{,}000$ and $n = 25$. Sample-average, bias and root mean square error calculated from 100 repetitions of each simulation

Backward simulated optimised Heston price II

| Scheme | $\mathbb{E}(\hat{\pi})$ | $\mathbb{E}(\hat{\pi}) - \pi$ | $\sqrt{\mathbb{E}[(\hat{\pi} - \pi)^2]}$ |
|---|---|---|---|
| Forward MC | 11.8094 | −0.0135 | 0.0411 |
| Rec. MARS, var. reduction | 11.8169 | −0.0060 | 0.0433 |
| Rec. MARS, two implicit | 11.8196 | −0.0033 | 0.0468 |
| Rec. MARS, var. red. & two imp. | 11.8164 | −0.0065 | 0.0433 |
| Rec. MARS, call-predictor | 11.8166 | −0.0063 | 0.0435 |
| Rec. MARS, $Z$-function | 11.6868 | −0.1361 | 0.1489 |
| Rec. MARS, $Z$-fun. & three imp. | 11.6794 | −0.1435 | 0.1558 |
| Forward MC | 9.9719 | 0.0003 | 0.0380 |
| Rec. MARS, var. reduction | 10.0075 | 0.0359 | 0.0501 |
| Rec. MARS, two implicit | 10.0027 | 0.0311 | 0.0495 |
| Rec. MARS, var. red. & two imp. | 10.0094 | 0.0378 | 0.0515 |
| Rec. MARS, call-predictor | 10.0082 | 0.0366 | 0.0507 |
| Rec. MARS, $Z$-function | 10.1096 | 0.1380 | 0.1502 |
| Rec. MARS, $Z$-fun. & three imp. | 10.1661 | 0.1945 | 0.2034 |

for the higher price. Since the backward simulation step is dependent of the forward step, we can not expect any improvement in accuracy beyond that of the forward simulation.

Next, we continue with the following modifications of the simulation schemes:

**5**. explicit-recursive-MARS with variance reduction (39)
**6**. explicit-recursive-MARS with two implicit iterations (35)
**7**. a combination of **5** and **6**
**8**. explicit-recursive-MARS with call-price predictor[20] (38).

The results are recorded in Table 7 and if we compare these with the result for the plain explicit-recursive-MARS scheme **4**, we observe similar accuracies for all of them. However, as both implicit schemes **6** and **7** add $N$ regressions and $N$ evaluations of the driver to the computational cost for each implicit iteration, we opt for the schemes **4** or **5**.

At last, we consider two schemes based on the MARS derivative: (**9**) explicit-recursive-MARS with $Z$-function (37), (**10**) explicit-recursive-MARS with $Z$-function and three implicit iterations. Both these modifications yield poor accuracy, see Table 7.

[20]The calculation of the pricing-formula for the call relies on numerical integration and we need $N = 100{,}000$ such evaluations for each of $n = 25$ time-step which makes the scheme very computer intensive. For this reason, we calculate a subset of 500 call prices and use a polynomial regression to predict the remaining call prices. As the pricing formula is a "nice" function of $S$ and $V$, this approximation only has a limited impact.

### *A.4 The Optimally Controlled Value-Process of a European Call Option*

Here we simulate the backward component of Eq. (12) that governs the optimally controlled upper/lower pricing bound of the European call option with strike-maturity structure as in Table 3 based on parameters in Table 5. Hence, we simulate $Y$ with an initial (terminal) condition $Y^{\pi}_{t_n} = (S^{\pi}_{t_n} - K)^{+}$ and an optimised driver $H(X_t, Y_t, Z_t)$ as of Eq. (20) with a confidence level of 95% for the parameter uncertainty region based on the covariance matrix in Table 5.

As before, we simulate the forward component $X^{\pi} = (S^{\pi}, V^{\pi})$ with the Euler-Maruyama implicit-Milstein scheme and for a start, we use $N = 100{,}000$ paths over an equidistant time-gird with $n = 25$ points. Note that for each backwards time-step, we perform $3 \times N$ regressions to obtain the one-step recursion of $Z^{1\pi}$, $Z^{2\pi}$, $Y^{\pi}$ and $N$ evaluations of the matrix multiplication in (20) for the optimal driver. For the implicit versions of the schemes, we iterate two (or three) times which adds $2 \times N$ regressions and $2 \times N$ matrix multiplications to each time-step.

For a demonstrative example, we again consider the 1-year the at-the-money call option and run 100 repetitions of the following simulation schemes:

1. explicit-recursive-MARS of degree 2
2. explicit-recursive-MARS of degree 2 with variance reduction
3. explicit-recursive-MARS of degree 2 with $Z$ calculated from the MARS derivative
4. two implicit fixed-point iterations added to scheme number **1**
5. two implicit fixed-point iterations added to scheme number **2**
6. explicit-recursive-MARS of degree 2 with call-price predictor and variance reduction.

The resulting pricing bounds are shown in Fig. 9 where, for clarity, we have plotted only the results from scheme number **1**, **2** and **3**.

From Fig. 9 we see that if we add variance reduction to the explicit-recursive-MARS we obtain slightly higher (lower) prices for the lower (upper) boundary and a somewhat lower variance. Further, if we consider the two-step implicit versions of these schemes, we have that **1** and **4** coincide almost perfectly, and also **2** and **5**, for both the upper and lower bounds (these schemes are excluded from Fig. 9 only for clarity). The same holds if we add the call-price predictor: **2** and **6** coincide for both the upper and lower bounds. As in the case for the formula-optimised price, the $Z$-function scheme yields a high lower bound (similar to the formula minimised price) and a upper bound similar to the other schemes. Both bounds also have a very high variance, and for these reasons we henceforth omit the $Z$-function schemes.

Based on the previous results for the 1-year ATM call, we choose to employ the explicit-recursive-MARS with variance reduction as our working scheme for pricing-boundary calculations. The simulation based results for the considered call options of Table 3 are given in Table 8 and if we compare these with the formula-optimal prices of Table 4, we generally see wider pricing intervals for the optimally controlled value process. This is what we should expect: the formula-optimal prices

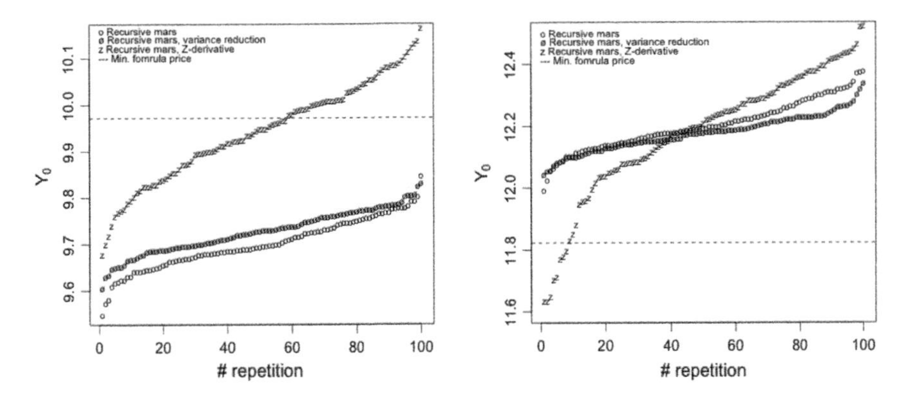


**Fig. 9** Numerical calculation for the pricing bounds of the 1-year at-the-money call with $N = 100{,}000$ paths over $n = 25$ time-points. **Left figure**: the $N$-sample average $Y_0^\pi$ of the lower bound for the call price (in increasing order) calculated from each of 100 repetitions of the simulation. **Right figure**: the corresponding upper bound. The dashed lines indicate the call price as calculated by the optimised Heston's formula

**Table 8** Pricing bounds for the European call option and corresponding Black-Scholes implied volatilities. Calculated from numerical simulation schemes for the backward process with $N = 100{,}000$ simulated paths of the forward process following Heston's model over an equidistant time grid with $n = 25$ points

Recursive MARS degree 2 with variance reduction, $n = 25$

| Strike/expiry | 75 | 100 | 125 |
|---|---|---|---|
| 4m | [25.7771, 26.2877] | [4.5005, 5.1597] | [0.0016, 0.0175] |
| | (0.0585, 0.3714) | (0.1942, 0.2277) | (0.1335, 0.1672) |
| 1y | [28.5910, 30.5482] | [9.7418, 12.1603] | [ 1.2374, 2.6306] |
| | (0.0329, 0.3138) | (0.1811, 0.2454) | (0.1600, 0.2101) |
| 10y | [52.7314, 64.9297] | [40.5299, 54.2037] | [30.7684, 45.1219] |
| | (0.0319, 0.3645) | ( 0.1176, 0.3214) | ( 0.1448, 0.2970) |

correspond to a controlled value-process with parameters held constant throughout the lifetime of the option, while in the former case, the parameters are allowed to vary in an optimal way. An illustration of this point is given in Fig. 10 where it is shown how the parameters vary for the optimally controlled 1-year at-the-money call option.

The previous pricing bounds where obtained from a simulation of $N = 100{,}000$ paths over a regular time-grid of $n = 25$ points. While $N$ is chosen to be a high number for the gain of a low error of the simulation-based regression estimator, the discretization time-step $\Delta = T/n$ is relatively large (for the 1-year option, $n = 25$ corresponds to having a time-step in the size of 2 weeks while for practical Monte Carlo pricing, one typically uses daily or even finer time-steps). For this reason we repeat the calculation of Table 8 with a finer time-step of $n = 100$. The results given

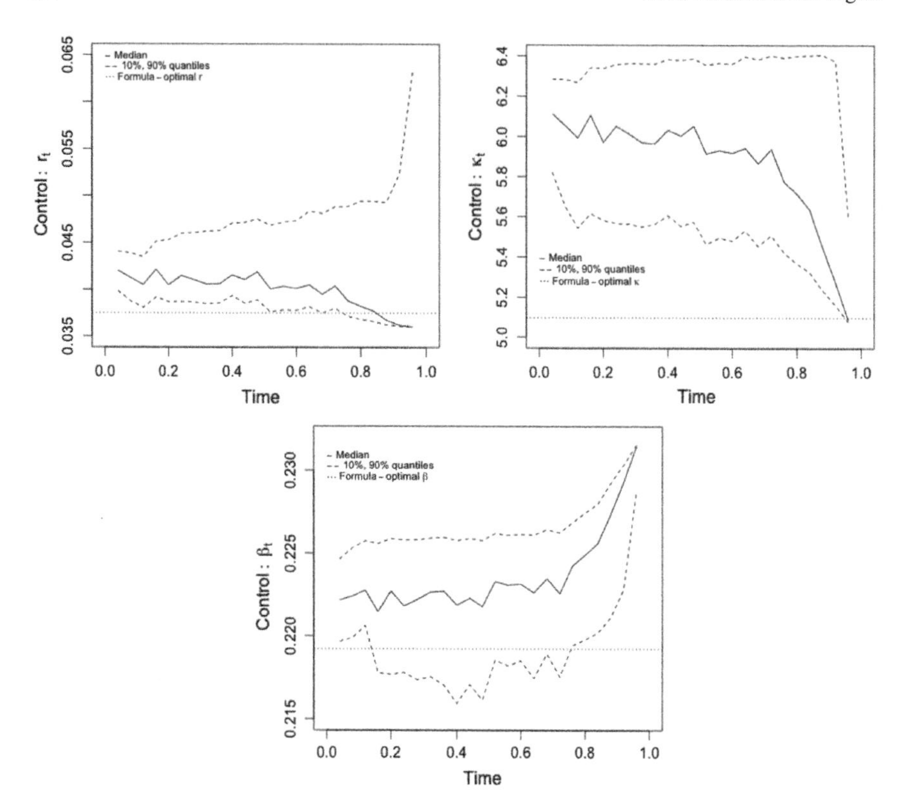


**Fig. 10** The optimal controls $u_t^* = (r_t^*, \kappa_t^*, \beta_t^*)$ as outputted from the optimisation of the driver $H^-$ for the 1-year ATM call option. Plotted median and quantiles of $N = 100{,}000$ simulation paths. The dotted lines show the corresponding constant parameter choice from the optimised Heston formula

**Table 9** Pricing bounds for the European call option and corresponding Black-Scholes implied volatilities. Calculated from numerical simulation schemes for the backward process with $N = 100{,}000$ simulated paths of the forward process following Heston's model over an equidistant time grid with $n = 100$ points

| Recursive MARS degree 2 with variance reduction, $n = 100$ | | | |
|---|---|---|---|
| Strike/expiry | 75 | 100 | 125 |
| 4m | [25.7349, 26.3130]<br>(0.05824, 0.3767) | [4.4748, 5.1885]<br>(0.1929, 0.2292) | [0.0005054, 0.02127]<br>(0.1225, 0.1710) |
| 1y | [28.3640, 30.6353]<br>(0.0330, 0.3184) | [9.6158, 12.2530]<br>(0.1777, 0.2478) | [1.1033, 2.6726]<br>(0.1543, 0.2115) |
| 10y | [48.5895, 67.3999]<br>(0.0217, 0.4076) | [36.9068, 57.1310]<br>(0.0199, 0.3598) | [27.4504, 48.1860]<br>(0.1053, 0.3298) |

in Table 9 show wider pricing bounds for all strikes/maturities when comparing to Table 8 and the difference between the two step sizes increases with the maturity. A natural explanation for this is that with a higher number of $n$ we also have a higher number of time-steps at which we optimise the driver $H^{\pm}$, and this should lead to a value process $Y^{\pi}$ optimised to a higher degree. This effect is obvious for the long-maturity options while it is less apparent for the four-month option (the implied volatilities agrees down to $10^{-2}$) which also indicates that the simulation error is not particularly affected by the finer time-discretization.

## References


1. Aït-Sahalia, Y., Kimmel, R.: Maximum likelihood estimation of stochastic volatility models. J. Financ. Econ. **83**(2), 413–452 (2007)
2. Alanko, S., Avellaneda, M.: Reducing variance in the numerical solution of BSDEs. Comptes Rendus Mathematique **351**(3), 135–138 (2013)
3. Alfonsi, A.: Affine Diffusions and Related Processes: Simulation. Springer, Theory and Applications (2016)
4. Andersen, L.B., Jckel, P., Kahl, C.: Simulation of Square-Root Processes. Wiley (2010)
5. Andersen, T., Benzoni, L.: Realized volatility. In: Handbook of Financial Time Series 555–575 (2009)
6. Avellaneda, M., Levy, A., Parás, A.: Pricing and hedging derivative securities in markets with uncertain volatilities. Appl. Math. Financ. **2**(2), 73–88 (1995)
7. Avellaneda, M., Paras, A.: Managing the volatility risk of portfolios of derivative securities: the lagrangian uncertain volatility model. Appl. Math. Financ. **3**(1), 21–52 (1996)
8. Bannör, K.F., Scherer, M.: Model risk and uncertainty—illustrated with examples from mathematical finance. In: Risk—A Multidisciplinary Introduction, pp. 279–306. Springer (2014)
9. Barczy, M., Pap, G.: Asymptotic properties of maximum-likelihood estimators for Heston models based on continuous time observations. Statistics **50**(2), 389–417 (2016)
10. Bates, D.S.: Jumps and stochastic volatility: exchange rate processes implicit in deutsche mark options. Rev. Financ. Stud. **9**(1), 69–107 (1996)
11. Bender, C., Denk, R.: A forward scheme for backward SDEs. Stochast. Process. Appl. **117**(12), 1793–1812 (2007)
12. Beneš, V.: Existence of optimal strategies based on specified information, for a class of stochastic decision problems. SIAM J. Control **8**, 179–188 (1970)
13. Bibby, B.M., Sørensen, M.: Martingale estimation functions for discretely observed diffusion processes. Bernoulli **17–39** (1995)
14. Black, F.: Studies of stock price volatility changes. In: Proceedings of the 1976 Meetings of the American Statistical Association, pp. 171–181 (1976)
15. Black, F., Scholes, M.: The pricing of options and corporate liabilities. J. Polit. Econ. **637–654**, (1973)
16. Bouchard, B., Elie, R.: Discrete-time approximation of decoupled forward-backward SDE with jumps. Stochast. Process. Appl. **118**(1), 53–75 (2008)
17. Bouchard, B., Touzi, N.: Discrete-time approximation and monte-carlo simulation of backward stochastic differential equations. Stochast. Process. Appl. **111**(2), 175–206 (2004)
18. Breeden, D.T.: An intertemporal asset pricing model with stochastic consumption and investment opportunities. J. Financ. Econ. **7**(3), 265–296 (1979)
19. Bunnin, F.O., Guo, Y., Ren, Y.: Option pricing under model and parameter uncertainty using predictive densities. Stat. Comput. **12**(1), 37–44 (2002)
20. Carr, P., Geman, H., Madan, D.B., Yor, M.: Stochastic volatility for lévy processes. Math. Financ. **13**(3), 345–382 (2003)

21. Carr, P., Sun, J.: A new approach for option pricing under stochastic volatility. Rev. Deriv. Res. **10**(2), 87–150 (2007)
22. Cohen, S.N., Elliott, R.J.: Stochastic Calculus and Applications. Springer (2015)
23. Cont, R., Tankov, P.: Financial Modelling with Jump Processes. Chapman & Hall/CRC (2004)
24. Cox, J.C., Ingersoll Jr., J.E., Ross, S.A.: A theory of the term structure of interest rates. Econ.: J. Econ. Soc. **385–407** (1985)
25. de Chaumaray, M.D.R.: Weighted least squares estimation for the subcritical Heston process. arXiv preprint arXiv:1509.09167 (2015)
26. Derman, E.: Model risk. Risk Mag. **9**(5), 34–37 (1996)
27. El Karoui, N., Peng, S., Quenez, M.: Backward stochastic differential equations in finance. Math. Financ. **7**(1), 1–71 (1997)
28. El Karoui, N., Quenez, M.-C.: Dynamic programming and pricing of contingent claims in an incomplete market. SIAM J. Control Optim. **33**(1), 29–66 (1995)
29. Feller, W.: Two singular diffusion problems. Ann. Math. **173–182** (1951)
30. Filippov, A.: On certain questions in the theory of optimal control. Vestnik Moskov. Univ. Ser. Mat. Meh. Astronom. **2**, 25–42 (1959). English trans. J. Soc. Indust. Appl. Math. Ser. A. Control **1**, 76–84 (1962)
31. Gatheral, J.: The Volatility Surface: A Practitioner's Guide, vol. 357. Wiley (2011)
32. Gobet, E.: Numerical simulation of BSDEs using empirical regression methods: theory and practice. In: Proceedings of the Fifth Colloquium on BSDEs (29th May–1st June 2005, Shangai). http://hal.archives-ouvertes.fr/hal00291199/fr (2006)
33. Gobet, E., Lemor, J.-P., Warin, X.: A regression-based monte carlo method to solve backward stochastic differential equations. Ann. Appl. Probab. (2005)
34. Gobet, E., Turkedjiev, P.: Linear regression MDP scheme for discrete backward stochastic differential equations under general conditions. Math. Comput. **85**(299), 1359–1391 (2016)
35. Godambe, V., Heyde, C.: Quasi-likelihood and optimal estimation. Int. Stat. Rev./Revue Internationale de Statistique 231–244 (1987)
36. Gupta, A., Reisinger, C., Whitley, A.: Model uncertainty and its impact on derivative pricing. In: Rethinking Risk Measurement and Reporting, p. 122 (2010)
37. Hastie, T., Tibshirani, R., Friedman, J., Franklin, J.: The elements of statistical learning: data mining, inference and prediction. Math. Intell. **27**(2), 83–85 (2005)
38. Heston, S.L.: A closed-form solution for options with stochastic volatility with applications to bond and currency options. Rev. Financ. Stud. **6**(2), 327–343 (1993)
39. Heston, S.L.: A simple new formula for options with stochastic volatility (1997)
40. Hull, J., White, A.: The pricing of options on assets with stochastic volatilities. J. Financ. **42**(2), 281–300 (1987)
41. Jacquier, E., Jarrow, R.: Bayesian analysis of contingent claim model error. J. Econ. **94**(1), 145–180 (2000)
42. Kessler, M.: Estimation of an ergodic diffusion from discrete observations. Scand. J. Statis. **24**(2), 211–229 (1997)
43. Keynes, J.M.: A treatise on probability (1921)
44. Kloeden, P.E., Platen, E.: Numerical Solution of Stochastic Differential Equations, vol. 23. Springer (1992)
45. Knight, F.H.: Risk, Uncertainty and Profit. Houghton Mifflin, Boston (1921)
46. Lehmann, E.L., Romano, J.P.: Testing Statistical Hypotheses. Springer Science & Business Media (2006)
47. Lyons, T.J.: Uncertain volatility and the risk-free synthesis of derivatives. Appl. Math. Financ. **2**(2), 117–133 (1995)
48. McShane, E., Warfield Jr., R.: On Filippov's implicit functions lemma. Proc. Am. Math. Soc. **18**, 41–47 (1967)
49. Milborrow. Derived from mda:mars by T. Hastie and R. Tibshirani, S.: Earth: Multivariate Adaptive Regression Splines. R package (2011)
50. Peng, S.: A generalized dynamic programming principle and Hamilton-Jacobi-Bellman equation. Stochast. Stochast. Rep. **38**, 119–134 (1992)

51. Prakasa-Rao, B.: Asymptotic theory for non-linear least squares estimator for diffusion processes. Statis.: J. Theor. Appl. Statis. **14**(2), 195–209 (1983)
52. Quenez, M.-C.: Stochastic Control and BSDEs. Addison Wesley Longman, Harlow (1997)
53. Rogers, L.C.G.: The relaxed investor and parameter uncertainty. Financ. Stochast. **5**(2), 131–154 (2001)
54. Stein, E.M., Stein, J.C.: Stock price distributions with stochastic volatility: an analytic approach. Rev. Financ. Stud. **4**(4), 727–752 (1991)
55. Stein, J.: Overreactions in the options market. J. Financ. **44**(4), 1011–1023 (1989)
56. Tegnér, M., Roberts, S.: A probabilistic approach to nonparametric local volatility. arXiv preprint arXiv:1901.06021 (2019)
57. Wong, B., Heyde, C.: On changes of measure in stochastic volatility models. Int. J. Stochast. Anal. (2006)

# Construction of an Aggregate Consistent Utility, Without Pareto Optimality. Application to Long-Term Yield Curve Modeling

**Nicole El Karoui, Caroline Hillairet and Mohamed Mrad**

**Abstract** The aim of this paper is to describe globally the behavior and preferences of heterogeneous agents. Our starting point is the aggregate wealth of a given economy, with a given repartition of the wealth among investors, which is not necessarily Pareto optimal. We propose a construction of an aggregate forward utility, market consistent, that aggregates the marginal utility of the heterogeneous agents. This construction is based on the aggregation of the pricing kernels of each investor. As an application we analyze the impact of the heterogeneity and of the wealth market on the yield curve.

**Keywords** Utility aggregation · Heterogeneous preferences · Market-consistent progressive utility · Yield curve

**MSC 2010** 60H15 · 91B16 · 91B69

## 1 Introduction

Most of general equilibrium macroeconomic models are simplified by assuming that consumers and/or firms could be described as a representative agent. That is agents may differ and act differently, but at equilibrium the sum of their choices is mathematically equivalent to the decision of one individual or many identical individuals. The way that preferences of multiple agents aggregate at equilibrium is a difficult task, and even if each individual preference is modeled by a simple function,

N. El Karoui
LPSM, UMR CNRS 8001, Sorbonne Université, 4 Place Jussieu, 75015 Paris, France

C. Hillairet (✉)
Crest, UMR CNRS 9194, Ensae Paris, 5 Avenue Le Chatelier, 91120 Palaiseau, France
e-mail: Caroline.Hillairet@ensae.fr

M. Mrad
LAGA, UMR CNRS 7539, Université Paris 13, 9 Avenue Jean Baptiste Clément, 93430 Villetaneuse, France



S. N. Cohen et al. (eds.), *Frontiers in Stochastic Analysis - BSDEs, SPDEs and their Applications*, Springer Proceedings in Mathematics & Statistics 289,
https://doi.org/10.1007/978-3-030-22285-7_6

it is unlikely that the aggregate utility could be reduced into a simple expression (unless all agents are identical). Heterogeneity of investors is an unavoidable feature that should be taken into account.

The literature on equilibrium risk sharing in complete markets with heterogeneous risk preferences starts with the seminal paper by Dumas [9], with two agents with heterogeneous risk preferences. Chan and Kogan [5] consider an extension of the Wang [36] model, with a continuum of agents with heterogeneous risk aversions. Yan [37] and Jouini et al. [22] stress the impact of relative wealth fluctuations on the equilibrium characteristics. Cvitanić et al. [7] were the first to propose an equilibrium model dealing with three types of heterogeneity: investors may differ in their beliefs, in their level of risk aversion and in their time-preference rate. They identify the channels through which heterogeneity impacts the different equilibrium characteristics. In their model, the aggregate parameters can be written as a risk tolerance weighted average of the individual parameters.

In the meantime, the existence of an equilibrium is not always satisfied and equilibrium are often stated and studied in a complete market setting. One key point for the existence of equilibrium is that agents agree on the same state price density process (also called pricing kernel), which is the same for all agents. However, if no equilibrium exists, is it still possible to propose a representative utility aggregating the preferences of all investors in a given economy? In this paper, a given economy may be understood for example as a market or an exchange in a given country (France, England, USA ...).

In this paper, we start from the weaker hypothesis of non arbitrage, and we consider an incomplete market, with given exogenous market parameters. Our aim is to propose a way of describing globally the behavior of heterogeneous agents investing in this market, heterogeneous by their preferences, their weights or sizes. To do this we construct a stochastic utility process corresponding to the aggregate wealth of the economy and to the aggregate pricing kernels. We do not deal with agents interactions, nor equilibrium, neither Pareto optimality: the repartition of the wealth among market investors is given. The wealth of the economy is naturally defined by the aggregation of the wealth of all individuals. The problem consists then in deriving a utility process for which this aggregate wealth is optimal. This is related to a calibration approach, and to do this the progressive framework is well adapted (see [12]). Besides, the progressive approach has also many advantages. First of all, it allows to model the change of the preferences of the investors along time. Indeed, in a dynamic and stochastic environment, the standard notion of utility function is not flexible enough to help us to make good choices in the long run. The utility criterion must be adaptative and adjusted to the information flow. Musiela and Zariphopoulou [29, 30] were the first to suggest to use instead of the classic criterion the concept of progressive dynamic utility, that gives an adaptive way to model possible changes over the time of individual preferences of an agent. The particular case of time-decreasing progressive utilities has been studied by Berrier et al. in [2, 3] and Zariphopoulou et al. in [31]. Zitkovic in [38] gives a dual characterization of exponential progressive utilities. Characterization of market-consistent progressive utility in a general setting has been then studied in El Karoui and Mrad [15, 16].

Secondly, the theoretical study of progressive utility emphasizes the dependency of the optimal processes with respect to their initial conditions. This dependency and some non linearity effects are illustrated in the example of the valuation of the discount rates. In the economic modeling, interest rates are determined endogenously at equilibrium, mainly in an economy composed of identical investors (see for example the well known Vasicek [35] or Cox et al. [6] models). In our framework, the market is incomplete and in place of the traditional (complete) pricing rule, we price the zero-coupon bonds using the indifference pricing rule, based on the marginal indifference pricing. A numerical example is proposed based on an extension of the Vasicek model of the yield curve.

The paper is organized as follow. First we define in Sect. 2 the investment universe and we recall the framework and the main properties of market consistent progressive utilities, and the characterization of a consistent utility from its optimal primal and dual processes. Section 3 states the main results about preferences aggregation: from the characteristics of the investors, we construct an aggregate consistent progressive utility process, by aggregating the wealth of each investors and their pricing kernels. To illustrate this theory, we give the example of aggregating power utilities. In particular we show that aggregating power utilities does not lead to a power utility, except if all investors share the same risk aversion. Thus taking a power utility for the representative agent, as it is done in many economic papers, assumes actually a very strong hypothesis of homogeneity of the different investors in the economy. Section 4 studies the impact of the heterogeneity of investors, that induces dependency and non-linearity in the valuation of financial assets. The particular example developed here consists in the valuation of discount rates and the impact of the aggregate wealth on this rates. Some numerics illustrate the impact of the different parameters on the yield curve. Technical regularity conditions are postponed in the Appendix.

## 2 Investment Universe and Consistent Progressive Utility

The economy (sector, class) we are interested in is part of a larger one (for example the world economy). We model below this global market called the "investment universe".

### 2.1 The Investment Universe

Let us consider an incomplete Itô market, defined on a filtered probability space $(\Omega, (\mathcal{F}_t), \mathbb{P})$ (satisfying usual conditions of completion and right continuity) driven by a $n$-standard Brownian motion $W$. As usual, the market is characterized by some exogenous progressive processes: the short rate $(r_t)$ and a $n$-dimensional risk premium $(\eta_t)$, satisfying the integrability condition $\int_0^T (r_s + \|\eta_s\|^2)ds < \infty$ for any $T$. The agent may invest in this financial market and we assume that her strategies do not affect the market prices. To be short, we give the mathematical definition of the class of admissible strategies[1] $(\kappa_t)$, without specifying the risky assets. The incompleteness of the market is expressed by restrictions on the risky portfolios $\kappa_t$ constrained to live in a given progressive vector space $\mathcal{R}_t$. To fix the idea, if the incompleteness follows only from the fact that the number $d$ of assets is less than the dimension $n$ of the Brownian motion, then typically $\mathcal{R}_t = \sigma_t^T(\mathbb{R}^d)$. For an Itô market, good references are Karatzas et al. [24] or the book of Karatzas and Shreve [23], and in a more general context Kramkov and Schachermayer [26].

To avoid technicalities, we assume throughout the paper that all the processes satisfy the necessary (progressive) measurability and integrability conditions such that the following formal manipulations and statements are meaningful. The following short notations will be used extensively. Let $\mathcal{R}$ be a vector subspace of $\mathbb{R}^n$. For any $x \in \mathbb{R}^n$, $x^{\mathcal{R}}$ is the orthogonal projection of the vector $x$ onto $\mathcal{R}$ and $x^{\perp}$ is the orthogonal projection onto $\mathcal{R}^{\perp}$, the orthogonal vector space of $\mathcal{R}$.

**Definition 2.1** (*Admissible portfolio*)

(i) The self-financing dynamics of a wealth process with risky portfolio $\kappa$, starting from the initial wealth $x > 0$, is given by

$$dX_t^{\kappa} = X_t^{\kappa}[r_t dt + \kappa_t.(dW_t + \eta_t dt)], \quad \kappa_t \in \mathcal{R}_t, \quad \text{and} \quad X_0^{\kappa} = x \qquad (2.1)$$

where $\kappa$ is a progressive $n$-dimensional vector measuring the volatility vector of the wealth $X^{\kappa}$, such that $\int_0^T \|\kappa_t\|^2 dt < \infty, a.s.$

(ii) A self-financing strategy $(\kappa_t)$ is admissible if the portfolio $(\kappa_t)$ lives in a given progressive family of vector spaces $(\mathcal{R}_t)$ $a.s.$

(iii) The set of the wealth processes with admissible $(\kappa_t)$ (called admissible wealth processes) starting from the initial wealth $x$ is denoted by $\mathscr{X}(x)$, and $\mathscr{X}$ when the initial wealth is not specified.

Note that the multiplicative form of the wealth dynamics (2.1) implies that the wealth processes are positive. The existence of a risk premium $\eta$ formulates the absence of arbitrage opportunity. Since from (2.1), the impact of the risk premium on the wealth dynamics only appears through the term $\kappa_t.\eta_t$ for $\kappa_t \in \mathcal{R}_t$, there is a "minimal" risk premium $(\eta_t^{\mathcal{R}})$, the projection of $\eta_t$ on the space $\mathcal{R}_t$ $(\kappa_t.\eta_t = \kappa_t.\eta_t^{\mathcal{R}})$, to which we refer in the sequel. In the following definition, we are interested in the class of the so-called state price density processes $Y^{\nu}$ (taking into account the discount factor) which are also called the pricing kernels.

**Definition 2.2** (*State price density process*) A positive Itô semimartingale $Y^{\nu}$ is called an admissible state price density process if for any admissible wealth process $X^{\kappa} \in \mathscr{X}$,

$$X_t^{\kappa} Y_t^{\nu} \quad \text{is a local martingale.} \qquad (2.2)$$

[1] $\kappa_t = \sigma_t.\kappa_t = \sigma_t^T \pi_t$, with $\pi$ being the fraction of wealth invested in the risky assets, and $\sigma$ being the volatility process, and $.^T$ denotes the transpose operator.

The simplest example of such process is the *market state price process* $Y^0$ ($\nu = 0$, $Y^0_0 = 1$). In particular $(X^\kappa_t Y^0_t)$ is a local martingale, whose volatility $(\kappa_t - \eta^{\mathcal{R}}_t)$ belongs to $\mathcal{R}_t$. The martingale property (2.2) can be then expressed in terms of the ratio $(L^\nu_t = Y^\nu_t / Y^0_t)$ as $(X^\kappa_t Y^0_t L^\nu_t)$ is a local martingale or equivalently $(L^\nu_t)$ is a local exponential martingale whose volatility belongs to $\mathcal{R}^\perp_t$.

**Corollary 2.3** *Denote $\mathscr{Y}(y)$ the convex family of all admissible state density processes $Y^\nu(y)$ issued from y, and $\mathscr{Y}$ the set of all $\mathscr{Y}(y)$. Any $Y^\nu(y)$ is the product of the* market state price process $Y^0$ *by an exponential martingale $L^\nu(y)$ whose volatility $\nu$ belongs to $\mathcal{R}^\perp_.$. The differential decomposition of these three processes is*

$$\begin{cases} dY^0_t = Y^0_t[-r_t\, dt - \eta^{\mathcal{R}}_t.dW_t], \quad Y^0_0 = 1 \\ dL^\nu_t = L^\nu_t[\nu_t.dW_t], \quad \nu_t \in \mathcal{R}^\perp_t \;\; L^\nu_0 = y \\ dY^\nu_t = Y^\nu_t[-r_t dt + (\nu_t - \eta^{\mathcal{R}}_t).dW_t], \quad \nu \in \mathcal{R}^\perp_t \;\; Y^\nu_0 = y. \end{cases} \tag{2.3}$$

Interesting discussions on the links between the state price density processes and the admissible market numeraire $1/Y^0_t$, also called GOP (growth optimal portfolio) can be found in El Karoui et al. [17], in Heath and Platen [33], and in Filipovic and Platen [18]. Besides, the state price density processes are also called "pricing kernels" since they are useful for evaluating contingent claims under the historical probability measure $\mathbb{P}$. Not surprisingly, we will focus on them in the application of Sect. 4 about the valuation of zero-coupon bond and the modeling of the yield curve.

## 2.2 *Consistent Progressive Utility and Their Characteristics*

The preferences of the agents investing in the financial market are modeled by consistent progressive utility. The sub-cone of admissible wealth processes $\mathscr{X}$, describing the financial landscape, is considered in this forward setting as a family of *test processes*. As in statistical learning, the utility criteria are dynamically adjusted to this given family of test processes, also called the learning set. The *time-coherence* is then obtained from a dynamic decision criterion adjusted progressively over the time to this set $\mathscr{X}$.

More precisely, a *progressive utility* **U** is defined as a family of càdlàg adapted processes $(U(t, x), x \in \mathbb{R}^+)$ such that $\mathbb{P}.a.s.$, for every $t \geq 0$, the functions $x \in \mathbb{R}^+ \mapsto U(t, x, \omega)$ are standard utility functions. As usual, a *utility function* $u$ is a strictly concave, strictly increasing, and non-negative function defined on $\mathbb{R}^+$, with continuous *marginal utility* the derivative $u_x$, satisfying the Inada conditions $\lim_{x\to\infty} u_x(x) = 0$ and $\lim_{x\to 0} u_x(x) = \infty$. The risk aversion coefficient $\mathrm{R_A}(u)$ is measured by the ratio $\mathrm{R_A}(u)(x) = -u_{xx}(x)/u_x(x)$ and the relative risk aversion by $\mathrm{R^r_A}(u)(x) = x\,\mathrm{R_A}(u)(x)$. The asymptotic elasticity $E_A(u)(x) = \limsup_{x\to\infty} x u_x(x)/u(x)$ is a key parameter in the optimization problem (see Kramkov [25]). As usual, the dual problem is based on the Fenchel-Legendre convex conjugate transformation $\tilde{u}(y)$ of a utility function $u$, where $\tilde{u}$ satisfies $\tilde{u}(y) = \sup_{x>0} \big(u(x) - yx\big)$. In particular,

$\tilde{u}(y) \geq u(x) - yx$ and the maximum is attained at $u_x(x) = y$. Under Inada conditions, $\tilde{u}$ is twice continuously differentiable, strictly convex, strictly decreasing, with $\tilde{u}(0^+) = u(+\infty)$, $\tilde{u}(+\infty) = u(0^+)$, $a.s.$ Moreover, the marginal utility $u_x$ is the inverse of the opposite of the marginal conjugate utility $\tilde{u}_y$; that is $u_x^{-1}(y) = -\tilde{u}_y(y)$; $\tilde{u}(y) = u\big(-\tilde{u}(y)\big) + \tilde{u}_y(y)\,y$, and $u(x) = \tilde{u}\big(u_x(x)\big) + x\,u_x(x)$.

Throughout the paper, we adopt the convention of small letters for deterministic utilities and capital letters for stochastic utilities.

### 2.2.1 Characteristics of the Consistent Progressive Utility

The progressive utilities are adjusted to the learning set $\mathscr{X}$. The satisfaction provided by a test process $X^\kappa \in \mathscr{X}$ is measured by the dynamic criterion $(U(t, X_t^\kappa))$. Since $\mathscr{X}$ is a learning set, there is no satisfaction to invest in the set $\mathscr{X}$, in other words in mean the future is less preferable than the present. From the mathematical point of view, this is equivalent to the *supermartingale property* of the dynamic preference process $(U(t, X_t^\kappa))$. Moreover, to ensure that the stochastic utility $(U(t,x))$ is optimally adjusted, we make the additional assumption that the previous supermartingale constraint is binded by some optimal process $\kappa^*$ whose preference criterion $(U(t, X_t^{\kappa^*}))$ is a martingale. This leads to the following definition of a consistent progressive utility as formulated in the seminal paper [28].

**Definition 2.4** (*Consistent progressive utility*) Let $\mathbf{U}$ be a progressive utility with learning set $\mathscr{X}$.

(i) The utility $\mathbf{U}$ is said to be $\mathscr{X}$-consistent, if for any admissible test process $X^\kappa \in \mathscr{X}$, the preference process $(U(t, X_t^\kappa))$ is a non-negative supermartingale.
(ii) The consistent utility $\mathbf{U}$ is said to be $\mathscr{X}$-strongly consistent if there exists an optimal process $X^* := X^{\kappa^*} \in \mathscr{X}$, with $\kappa_t^* \in \mathcal{R}_t$, binding the constraint, in the sense that the optimal preference process $(U(t, X_t^*))$ is a martingale.

The value function $(\mathcal{U}(t,x))$ of the classical optimization problem with terminal horizon $T_H$ is an example of strongly $\mathscr{X}$-consistent utility, defined from its terminal condition $\mathcal{U}(T_H, x) = u(x)$ (see [12] for a general discussion between the forward and the backward viewpoints of utility functions).

The *consistency property* of the progressive utility $\mathbf{U}$ has a natural equivalent for dual progressive utility, as stated in the following proposition (see [16] for the proof).

**Proposition 2.5** $\mathbf{U}$ *is a consistent progressive utility with the class* $\mathscr{X}$ *if and only if its Fenchel transform* $\widetilde{\mathbf{U}}$ *is consistent with the class* $\mathcal{Y}$ *in the sense that* $\widetilde{U}(t, Y_t)$ *is a submartingale for any* $Y \in \mathcal{Y}$*, and there exists some* $Y^* \in \mathcal{Y}$ *(called dual optimal process) such that* $\widetilde{U}(t, Y_t^*)$ *is a martingale. Moreover, the two optimal processes are related by the main identity* $U_x(t, X_t^*(x)) = Y_t^*(u_x(x))$.

Rogers provides in [34] a unified (and very simple) approach to get very quickly a simple heuristic of the main identity $U_x(t, X_t^*(x)) = Y_t^*(u_x(x))$, that will be at the cornerstone of this paper.

*Local Characteristics of Consistent Forward Utility*
The "global" supermartingale property implied by the consistency condition may be transferred into local conditions on the differential characteristics of the utility process **U**. El Karoui and Mrad [16] obtained a non linear HJB-SPDE under the general assumption that the utility random field **U** is a "regular" Itô random field with differential decomposition,

$$dU(t,x) = \beta(t,x)dt + \gamma(t,x).dW_t, \tag{2.4}$$

where $\beta(t,x)$ is the drift random field and $\gamma(t,x)$ is the multivariate diffusion random field. The regularity assumption recalled in the Appendix, allows in particular to use the *Itô-Ventzel formula* and to show that the marginal utility $(U_x(t,x))$ is also an Itô random field with local characteristics $(\beta_x(t,x), \gamma_x(t,x))$. We give the main result about the consistency characterization through a HJB constraint:

**Theorem 2.6** (Consistency) *Let* **U** *be a "regular utility"*[2] *and* $(\beta,\gamma)$ *its local characteristics. The utility random field* **U** *is strongly consistent with the family of test processes* $\mathscr{X} = \{X^\kappa, |\kappa \in \mathcal{R}\}$ **if and only if** $(a)$ *and* $(b)$ *hold*

*(i)* (a) *The drift random field* $\beta$ *satisfies the HJB-constraint,* $d\mathbb{P} \times dt.a.s.$

$$\begin{cases} \beta(t,x) &= -U_x(t,x)r_t x - \frac{1}{2}\sup_{\sigma\in\mathcal{R}}\left\{U_{xx}(t,x)\Big(\|\sigma_t\|^2 + 2\sigma_t.\big(\frac{U_x(t,x)\eta_t^{\mathcal{R}}+\gamma_x(t,x)}{U_{xx}(t,x)}\big)\Big)\right\}.\\ &= -U_x(t,x)xr_t + \frac{1}{2U_{xx}(t,x)}\|\,U_x(t,x)\eta_t^{\mathcal{R}} + \gamma_x^{\mathcal{R}}(t,x))\|^2.\\ &= -U_x(t,x)xr_t + \frac{1}{2}U_{xx}(t,x)\|\sigma^*(t,x))\|^2. \end{cases} \tag{2.5}$$

*The quantity* $\frac{\gamma_x^{\mathcal{R}}(t,x)}{U_x(t,x)}$ *can be interpreted as an "utility risk premium".*
(b) *The stochastic differential equation* $SDE^{\mathcal{R}}(\sigma^*)$

$$\begin{cases} dX_t^* &= r_t X_t^* dt + \sigma^*(t, X_t^*).(dW_t + \eta_t^{\mathcal{R}}dt),\\ \sigma^*(t,x) &= -\frac{U_x(t,x)}{U_{xx}(t,x)}\big(\eta_t^{\mathcal{R}} + \frac{\gamma_x^{\mathcal{R}}(t,x)}{U_x(t,x)}\big) = x\kappa^*(t,x) \end{cases} \tag{2.6}$$

*admits a strong solution* $X^*$*, which is an optimal portfolio in the preference sense.*

*(ii) In addition, the positive process* $U_x(t, X_t^*(x))$ *is the optimal dual state price process* $Y_t^*(u_x(x))$*, solution of the* $SDE^{\perp}(\vartheta^{*,\perp})$ *issued from* $y = u_x(x)$

$$\begin{cases} dY_t^* = -r_t Y_t^* dt + \big(\vartheta^{*,\perp}(t, Y_t^*) - \eta_t^{\mathcal{R}} Y_t^*\big).dW_t,\\ \vartheta^{*,\perp}(t,y) = \gamma_x^{\perp}(t, U_x^{-1}(t,y)) = y\nu^{*,\perp}(t,y). \end{cases} \tag{2.7}$$

The regularity assumptions on **U** recalled in the Appendix imply that the coefficients of the SDEs (2.6) and (2.7) are regular enough to ensure that $X^*$ and $Y^*$ are monotonic (increasing) with respect to their respective initial condition $x$ and $y$ with range $[0,\infty]$ (see [16]). Note that the case of time-decreasing consistent dynamic utilities studied in [2, 31] corresponds to $\gamma \equiv 0$.

[2] That is **U** is a $\mathcal{K}_{loc}^{1,\delta} \cap \mathcal{C}^2$-semimartingale, see the Appendix and Theorem 5.1$(iv)$.

### 2.2.2 Consistent Power Utility and Separability

Power utilities with constant relative risk aversion $\theta \in ]0,1[$, $u^{(\theta)}(x) = \frac{x^{1-\theta}}{1-\theta}$ are the standard framework in the economic literature, useful for its simplicity and the easy interpretation of the parameters. In particular, the parameter $\theta$ is the relative risk aversion coefficient $\mathrm{R}^{\mathrm{r}}_{\mathrm{A}}(u^{(\theta)})(x) = -xu^{(\theta)}_{xx}(x)/u^{(\theta)}_x(x) = \theta$.

Consistent progressive power utilities $U^{(\theta)}(t,x)$ are the product of their initial condition $u^{(\theta)}(x)$ by a coefficient $Z^{(\theta)}_t$. Despite their stochastic structure, their relative risk aversion coefficients are still constants, $\mathrm{R}^{\mathrm{r}}_{\mathrm{A}}(U^{(\theta)})(t,x) = \mathrm{R}^{\mathrm{r}}_{\mathrm{A}}(u)(x) = \theta$.

The role of the stochastic process $Z^{(\theta)}_t$ is to guarantee the market consistency of dynamics power utility. Since $u^{(\theta)}_x(1) = 1$, we have $Z^{(\theta)}_t = U^{(\theta)}_x(t,1)$. Since $U^{(\theta)} = Z^{(\theta)}_{.}\, u^{(\theta)}$, its local characteristics $(\beta^{(\theta)}, \gamma^{(\theta)})$ are proportional to $u^{(\theta)}$, with $\beta^{(\theta)}(t,x) = \mu^{(\theta)}_t u^{(\theta)}(x)$ and $\gamma^{(\theta)}(t,x) = Z^{(\theta)}_t \delta^{(\theta)}_t u^{(\theta)}(x)$, $\big(\mu^{(\theta)}_t, Z^{(\theta)}_t \delta^{(\theta)}_t\big)$ being the stochastic parameters of the semimartingale $Z^{(\theta)}$. Thanks to Theorem 2.6, one can characterizes the optimal processes of power progressive utilities.

**Proposition 2.7** *Let $\big(U^{(\theta)}(t,x) = Z^{(\theta)}_t\, u^{(\theta)}(x)\big)$ be a power consistent progressive utility, $(Z^{(\theta)}_t)$ being a positive semimartingale with parameters $\big(\mu^{(\theta)}_t, Z^{(\theta)}_t \delta^{(\theta)}_t\big)$.*

*(i) The optimal processes $X^{(*,\theta)}_t(x)$ and $Y^{(*,\theta)}_t(y)$ are linear with respect to their initial conditions, $X^{(*,\theta)}(x) = x\bar{X}^{(*,\theta)}$ and $Y^{(*,\theta)}(x) = y\bar{Y}^{(*,\theta)}$, with dynamics*

$$\begin{cases} d\bar{X}^{(*,\theta)}_t = \bar{X}^{(*,\theta)}_t \big[r_t + \frac{1}{\theta}(\eta^{\mathcal{R}}_t + \delta^{(\theta),\mathcal{R}}_t).(dW_t + \eta^{\mathcal{R}}_t dt)\big], \\ d\bar{Y}^{(*,\theta)}_t = \bar{Y}^{(*,\theta)}_t \big[-r_t dt + (\delta^{(\theta),\perp}_t - \eta^{\mathcal{R}}_t).dW_t\big]. \end{cases} \tag{2.8}$$

*The coefficient $\delta^{(\theta),\mathcal{R}}_t$ describes how the stochasticity of the utility influences the investment strategy $\kappa^{(*,\theta)}_t = \frac{1}{\theta}(\eta^{\mathcal{R}}_t + \delta^{(\theta),\mathcal{R}}_t)$.*

*(ii) The drift of the process $(Z^{(\theta)}_t)$ is not free, since the consistency condition (equivalent to the HJB constraint) implies that*

- $Z^{(\theta)}_t = \big[\bar{X}^{(*,\theta)}_t\big]^{\theta}\, \bar{Y}^{(*,\theta)}_t$, *and*
- $\mu^{(\theta)}_t = -(1-\theta) Z^{(\theta)}_t \big(r_t + \frac{1}{2\theta}\|\eta^{\mathcal{R}}_t + \delta^{(\theta),\mathcal{R}}_t\|^2\big)$.

*The consistent power utilities are completely specified by the volatility $(\delta^{(\theta)}_t)$ of the dynamics coefficient $Z^{(\theta)}_t$.*

*Proof* (i) By Eq. (2.6), the volatility of the optimal process $X^{(*,\theta)}_t(x)$ is linear with respect to the initial wealth $x$, $\sigma^{(*,\theta)}(t,x) = \frac{x}{\theta}(\eta^{\mathcal{R}}_t + \delta^{(\theta),\mathcal{R}}_t)$. Since the drift is also linear, the optimal process is linear with respect to the initial wealth, $X^{(*,\theta)}(x) = x\bar{X}^{(*,\theta)}$ where the dynamics of $\bar{X}^{(*,\theta)} = X^{(*,\theta)}(1)$ is given by Eq. (2.8). The dual process $Y^{(*,\theta)}_t(y)$ is also linear with respect to $y$, and by Eq. (2.7), $\vartheta^{(*,\theta)}(t,y) = y\,\delta^{(\theta),\perp}_t$. Then, the dynamic of $(\bar{Y}^{(*,\theta)}_t)$ is given by Eq. (2.8).

(ii) By the optimality relation of $U^{(\theta)}_x(X^{(*,\theta)}_t(x)) = u_x(x)\bar{Y}^{(*,\theta)}$. This property is equivalent to the HJB constraint on the drift $\beta^{(\theta)}(t,x) = \mu^{(\theta)}_t u^{(\theta)}(x)$ of the power utility. A consequence is that $Z_t^{(\theta)} = \left[\bar{X}_t^{(*,\theta)}\right]^{\theta} \bar{Y}_t^{(*,\theta)}$. The linearity of the different processes yields

$$dZ_t^{(\theta)} = Z_t^{(\theta)}\left[-\left((1-\theta)r_t + \frac{1-\theta}{2\theta}\|\eta_t^{\mathcal{R}} + \delta_t^{(\theta),\mathcal{R}}\|^2\right)dt + \delta_t^{(\theta)}.dW_t\right].$$

The drift of $Z^{(\theta)}$ depends only of the market parameters $(r_t, \eta_t^{\mathcal{R}})$ and its volatility $\delta_t^{(\theta)}$. □

*Remark 2.1* Power utilities have also the remarkable property to be the only consistent separable progressive utilities $U(t,x) = Z_t\, u(x)$. The HJB equation (2.5) leads to a contradiction as soon as the functions $\phi^1 = xu_x/u$ and $\phi^2 = xu_{xx}/u_x$ are not constant, since the HJB constraint on the drift $\beta(t,x)$ implies that the time function $\rho_t^Z$ satisfies $\rho_t^Z = -\phi^1(x)\, r_t + \phi^1(x)/\phi^2(x)\|\eta_t^{\mathcal{R}} + \delta_t^{Z,\mathcal{R}}\|^2$ for any $x$. An exception is given by the case where $\delta_t^{Z,\mathcal{R}} = -\eta_t^{\mathcal{R}}$ and $r_t = 0$. In this case, $Z_t$ is an exponential martingale with volatility $\eta_t^{\mathcal{R}}$ multiplied by an orthogonal exponential martingale with volatility $\delta_t^{Z,\perp}$.

## 2.3 *Reverse Problem*

One remarkable feature proved in [16] is that properties given in Theorem 2.6 are in fact necessary and sufficient conditions to reconstruct a consistent progressive utility from two optimal processes $X^*$ and $Y^*$, when these processes are monotonic with respect to their initial condition. This construction relies on the identity $U_x(t, X_t^*(x)) = Y_t^*(u_x(x))$, using monotonicity and regularity of optimal random fields, and some integrability condition near zero of the initial utility. This is close to the point of view of Dybvig and Rogers [10], where the authors solve the recovery problem from the observation of one trajectory of the observed wealth process in the Merton framework, with the additional assumption that the state price density process at maturity is log-normal.

Let us consider two *increasing* monotonic processes $X^*(x) \in \mathscr{X}(x)$ and $Y^*(y) \in \mathscr{Y}(y)$, strong regular solutions of the two SDEs

$$\begin{cases} dX_t^* = r_t X_t^* dt + \sigma^*(t, X_t^*).(dW_t + \eta_t^{\mathcal{R}})dt, & X_0^* = x, \\ dY_t^* = -r_t Y_t^* dt + \left(\vartheta^{*,\perp}(t, Y_t^*) - \eta_t^{\mathcal{R}} Y_t^*\right).dW_t, & Y_0^* = y. \end{cases} \tag{2.9}$$

The dynamics of $X^* \in \mathscr{X}$ is uniquely determined by its diffusion coefficient $\sigma^* \in \mathcal{R}$; the corresponding SDE is denoted $\mathrm{SDE}^{\mathcal{R}}(\sigma^*)$. Similarly the dynamics of $Y^* \in \mathscr{Y}$ is uniquely determined by its diffusion coefficient $\vartheta^{*,\perp} \in \mathcal{R}^{\perp}$; the corresponding SDE is denoted $\mathrm{SDE}^{\perp}(\vartheta^{*,\perp})$.

We now give sufficient conditions on the coefficients $\sigma^*(t,x)$ and $\vartheta^{*,\perp}(t,y)$ which ensure on the one hand the monotonicity of the solutions of Equations (2.9) and the semimartingale decomposition of the random field $\mathcal{X}^*$ the inverse flow of $X^*$; and on the other hand that the random field $\mathbf{V}$ defined by $V(t,x) := Y_t^*\left(u_x\left(\mathcal{X}^*(t,x)\right)\right)$ is the derivative of a progressive utility $\mathbf{U}$. The sufficient regularity conditions we state below are proved in [16].

### 2.3.1 Technical Results

In this presentation we clearly favor the SDE point of view for the processes $X^*$ and $Y^*$. This allows us to use the existing results in SDE's theory and provide sufficient regularity conditions $(\overline{\mathcal{K}}_{\mathbf{b}}^{0,1} \cap \overline{\mathcal{K}}_{loc}^{3,\delta})$[3] on the coefficients to ensure the existence of regular SDE solutions. Global Lipschitz condition $(\overline{\mathcal{K}}_{\mathbf{b}}^{0,1})$ is enough to obtain strong and monotonic solutions whereas the regularity is ensured by the local conditions $(\overline{\mathcal{K}}_{loc}^{m,\delta})$. But this point of view is not necessary as soon as one starts from non-explosive monotonic solutions $X^*$ and $Y^*$. We first recall the present version of some results in [16].

**Proposition 2.8** (Regularity) *Let us consider the two stochastic equations* $\mathrm{SDE}^{\mathcal{R}}(\sigma^*)$ *and* $\mathrm{SDE}^{\perp}(\vartheta^{*,\perp})$ *defined in* (2.9) *and assume*

$$\sigma^* \in \overline{\mathcal{K}}_{\mathbf{b}}^{0,1} \cap \overline{\mathcal{K}}_{loc}^{3,\delta}, \; and \;\; \vartheta^{*,\perp} \in \overline{\mathcal{K}}_{\mathbf{b}}^{0,1} \cap \overline{\mathcal{K}}_{loc}^{2,\delta} \;\; for\ some\ \delta \in (0,1]. \tag{2.10}$$

*(i) Then, the differential equations* $\mathrm{SDE}^{\perp}(\vartheta^{*,\perp})$ *and* $\mathrm{SDE}^{\mathcal{R}}(\sigma^*)$ *admit two regular monotonic solutions* $Y^*$ *and* $X^*$ *with different regularity.*

- *The solution* $Y^*$ *belongs to* $\mathcal{K}_{loc}^{2,\varepsilon}$*, and its diffusion local characteristic* $\psi(.,y) := \vartheta^{*,\perp}(., Y^*_{.}(y))$ *is in* $\overline{\mathcal{K}}_{loc}^{2,\varepsilon}$ *for all* $\varepsilon \in [0,\delta[$.
- *The solution* $X^*$ *belongs to* $\mathcal{K}_{loc}^{3,\varepsilon}$ *and its diffusion local characteristics* $\phi(.,x) := \sigma^*(., X^*_{.}(x))$ *is in* $\overline{\mathcal{K}}_{loc}^{3,\varepsilon}$ *for all* $\varepsilon \in [0,\delta[$.

*(ii) The range of the maps* $x \mapsto X^*(x)$ *and* $y \mapsto Y^*(y)$ *is* $]0,+\infty[$. *The inverse* $\mathcal{X}^*$ *of* $X^*$ *is a semimartingale, unique monotonic solution of the stochastic PDE,*

$$\begin{cases} d\mathcal{X}^*(t,x) = -\mathcal{X}^*_x(t,x)\big[\sigma^*(t,x).(dW_t + \eta^{\mathcal{R}}_t dt) + r_t\, x dt\big] + \widehat{L}^*_{t,x}(\mathcal{X})dt \\ \widehat{L}^*_{t,x} \quad\quad := \frac{1}{2}\partial_x(\|\sigma^*(t,x)\|^2\partial_x). \end{cases} \tag{2.11}$$

For integrability reasons, we need to control the speed of convergence of $X^*$ and $Y^*$ at 0 and $\infty$. The following results are standard under Lipschitz conditions, satisfied in our setting, see [27].

**Corollary 2.9** *The asymptotic behaviors of* $X^*$ *and* $Y^*$ *are similar and well-controlled in time. The short notation* $\max(Z_T(z)) = \sup_{0\le t\le T} Z(t,z)$ *is used in the sequel. More precisely, if* $Z$ *is one of the two processes* $X^*$ *and* $Y^*$*, for any* $T$ *almost surely, for any* $\varepsilon \in (0,1)$*, uniformly on* $[0,T]$*, the asymptotic limits in* $\infty$ *or* 0 *are,*

[3] See the Appendix for the definition of these classes of regularity.

$$\begin{cases} \lim_{z\to+\infty}\left(z^{-(1+\varepsilon)}\max(Z_T(z))\right)=0 \text{ and } \lim_{z\to+\infty}\left(z^{-\varepsilon}\max(Z_T(z))\right)=\infty, \\ \lim_{z\to 0}\left(z^{-\varepsilon}\max(Z_T(z))\right)=0 \text{ and } \lim_{z\to 0}\left(z^{-(1+\varepsilon)}\max(Z_T(z))\right)=\infty. \end{cases} \quad (2.12)$$

Sometimes, it is more interesting to consider SDE's solutions as random fields $X^*(t,x)$ or $Y^*(t,y)$ with local characteristics $\phi^*(t,x)=\sigma^*(t,X_t^*(x))$ or $\psi^*(t,y)=\vartheta^{*,\perp}(t,Y_t^*(y))$. With the random fields point of view, non negativity and monotonicity are not so easy to prove.

**Corollary 2.10** *Let $(X_t^*(x))$ and $(Y_t^*(y))$ be two monotonic random fields,*

$$\begin{cases} dX_t^*(x)=r_tX_t^*(x)dt+\phi^*(t,x).(dW_t+\eta_t^{\mathcal{R}}), \quad X_0^*(x)=x,\ \phi^*(t,x)\in\mathcal{R} \\ dY_t^*(y)=-r_tY_t^*(y)dt+\left(\psi^*(t,y)-\eta_t^{\mathcal{R}}Y_t^*(y)\right).dW_t, \quad Y_0^*(y)=y,\ \psi^*(t,y)\in\mathcal{R}^{\perp} \end{cases} \quad (2.13)$$

*and assume that $\phi^*\in\overline{\mathcal{K}}_{loc}^{3,\delta}$, and $\psi^*\in\overline{\mathcal{K}}_{loc}^{2,\delta}$ for $\delta\in(0,1]$. Then, the random fields $X^*$ and $Y^*$ have the same properties as the processes of Proposition 2.8.*

*Proof* Using Theorem 5.1 in the Appendix, one deduces that $X^*(x)\in\mathcal{K}_{loc}^{3,\varepsilon}$ and $Y^*(y)\in\mathcal{K}_{loc}^{2,\varepsilon}$. Then one show exactly as in [16] that the inverse flow $\mathcal{X}^*$ is a regular semimartingale. □

### 2.3.2 Main Result Concerning the Reverse Problem

Let us consider two random fields, $X^*$ and $Y^*$, solution of the two SDEs (2.9) with coefficients satisfying the assumptions (2.10) of Proposition 2.8. Their properties are recalled in Proposition 2.8 and in Corollary 2.9. As denoted previously $\mathcal{X}^*$ is the inverse process of $X^*$ and $u$ is the initial utility.

The main result on the construction of consistent forward utility is obtained in two stages: the first concerns the properties of the decreasing random field $(U_x(t,x)=Y_t^*(u_x(\mathcal{X}_t^*(x))))$ and of its primitive as semimartingale; the second concerns the $\mathscr{X}$-consistency of this forward utility and the optimality of the process $X^*$.

**Theorem 2.11** (Utility Characterization) *Let us assume that the given initial utility $u$ is of class $\mathcal{C}^3$ and $u_x(x)\sim x^{-\xi}$ $(\xi<1)$ in the neighborhood of $x=0$. Under the assumptions and notations of Proposition 2.8, $(X_t^*(x))$ and $(Y_t^*(y))$ are the unique monotonic solutions of the SDEs* (2.9). *Then*

*(i) The random field defined by $V^*(t,x)=Y_t^*(u_x(\mathcal{X}_t^*(x)))$ is a semimartingale, integrable in the neighborhood of $x=0$, which is the derivative of a progressive utility $U(t,x)$ $(V^*(t,x)=U_x(t,x))$ with regular local characteristics $(\beta(t,x),\gamma(t,x))$ with*

$$\begin{cases} \gamma_x^{\mathcal{R}}(t,x) &= -\sigma^*(t,x)V_x^*(t,x)-V^*(t,x)\eta_t^{\mathcal{R}} \\ \gamma_x^{\perp}(t,x) &= \vartheta^{*,\perp}(t,V^*(t,x)) \\ \beta(t,x) &= -V^*(t,x)x\,r_t+\frac{1}{2}V_x^*(t,x)\|\sigma^*(t,x)\|^2. \end{cases} \quad (2.14)$$

(ii) *By Theorem 2.6,* **U** *is strongly consistent with the class* $\mathscr{X}$*, that is for any* $X \in \mathscr{X}$*,* $U(t, X_t)$ *is a supermartingale and martingale for* $X_t^*$.

This result is proved in [16] in a SDE point of view, therefore we do not reproduce it here. A similar proof, this time in a random field point of view, is given in the context of aggregation in Theorem 3.3.

The system (2.14) can be inverted to express the characteristics of optimal processes in terms of progressive utility characteristics.

**Corollary 2.12** *Since* $X^*$ *and* $Y^*$ *are optimal, their characteristics* $(\phi^*(t,x))$ *and* $(\psi^*(t,y))$ *are explicit functionals of the progressive utility* **U** *and its derivatives as well as of its volatility vectors* $\gamma_x$ *along the optimal wealth process. So,*

$$\phi^*(t,x) = -\frac{\partial_x[\gamma^{\mathcal{R}}(t, X_t^*(x)) + U(t, X_t^*(x))\eta_t^{\mathcal{R}}]}{\partial_x[U_x(t, X_t^*(x))]} \quad \text{and} \quad \psi^*(t, u_x(x)) = \gamma_x^{\perp}(t, X_t^*(x)). \tag{2.15}$$

## 3 Aggregating Multi-agents Preferences

Consider a group of agents who invest in the financial market according to their own preferences. In the following, our aim is to characterize a representative agent and his representative preference for this group. The main question is: is it possible to describe globally the behavior of all the agents by a single utility stochastic process? How could we define an aggregate utility taking into account the preferences and the sizes/weights of each agent? If all agents have the same characteristics/behaviors, then the answer is obvious. Otherwise, we classify the agents into classes with characteristics represented by the pair $(\mathbf{U}^\theta, m(d\theta))$: a consistent progressive utility and a weight.

This framework can be applied at different granularity levels. For example, one may aggregate each agent individually, that is $(\mathbf{U}^\theta, m(d\theta))$ corresponds to the characteristics of one single agent. Or one can aggregate different classes of agents having the same preferences and the same strategy inside the class (for example $\theta$ may be interpreted as the risk aversion parameter of the class and $m(d\theta)$ the proportion of this class among the whole). One may also aggregate different classes of agents who are in the same sector but who do not necessarily share the same characteristics, and whose individual characteristics are not always observable, so that one can not proceed by aggregation of each agent individually. One alternative is then to rely on a representative utility $\mathbf{U}^\theta$ of the sector, that is computed beforehand, using eventually different aggregation rules. Then the only information at disposal to aggregate the different sectors consists in this representative utility $\mathbf{U}^\theta$ of each sector and $m(d\theta)$ the relative size/weight of the sector in the economy. For example, we aggregate the different sectors (commodities, industrials, financials, etc) of the economy of a given country, whose market trades with the others countries' market or exchanges.

## 3.1 *Aggregation of the Marginal Utilities*

### 3.1.1 Aggregation of the Initial Utilities

The aggregate initial wealth $x$ of the economy is the sum of the individual wealths: for each $\theta$, the $\theta$-agent/class starts (at time 0) with a proportion $\alpha^\theta$ of the initial global wealth $x$ so that $x = \int \alpha^\theta x \, m(d\theta)$. There are several possible choices[4] to aggregate utilities at time 0. In the standard setting, the individual preferences $u^\theta$ are "scaled" into the utilities $\frac{1}{\alpha^\theta} u^\theta(\alpha^\theta x)$, and the global utility is then the function $u(x)$,

$$u(x) = \int \frac{1}{\alpha^\theta} u^\theta(\alpha^\theta x) \, m(d\theta), \quad \int \alpha^\theta m(d\theta) = 1. \tag{3.1}$$

*Technical Remark*. The measure $m(d\theta)$ can be a discrete finite measure, in this case differentiability under the integral sign is straightforward. One may also consider measures with density with respect to the Lebesgue measure. Then to ensure the 3 times-differentiability under the integral sign, locally-domination conditions are necessary: we assume that for any interval $I \subset \mathbb{R}_+^*$ there exist integrable functions $\phi_k^I(x, \theta)$ such that $|\partial_x^k u^\theta(x)| \le \phi_k^I(x, \theta)$, $\forall x \in I$ and for $k = 1, 2, 3$. In all cases, we can pass to the limits and show $\lim_{x \to +\infty} u_x(x) = 0$ and $\lim_{x \to 0} u_x(x) = +\infty$. Note that for $k = 1$, since $u_x^\theta$ is by definition decreasing, it follows that for any $x_0 \in \mathbb{R}_+^*$, $0 \le u_x^\theta(x) \le u_x^\theta(x_0)$, $\forall x \ge x_0$, thus it suffices to take $\phi_k^{[x_0, +\infty[}(\theta) = u_x^\theta(x_0)$ and assume that it is $\theta$-integrable.

Then, from (3.1), the marginal utility $u_x$ of the global utility is the sum of the marginal utilities, in the sense that

$$u_x(x) = \int u_x^\theta(\alpha^\theta x) \, m(d\theta). \tag{3.2}$$

The same kind of representation holds also for the inverse function of $u_x$, $-\tilde{u}_y$, using the correspondence between the derivatives of the utility and its dual

$$y = \int y^\theta(y) \, m(d\theta), \quad y^\theta(y) = u_x^\theta(-\alpha^\theta \tilde{u}_y(y)), \tag{3.3}$$

[4]Actually, one may choose any deterministic initial utility, as soon as it satisfies sufficient integrability conditions, as the ones required in Theorem 2.11.

which leads to the remarkable feature that for any $\theta$, $\tilde{u}_y(y) = \frac{1}{\alpha^\theta}\tilde{u}^\theta_y(y^\theta(y))$; this relation is the dual version of the $\alpha^\theta$-repartition of the initial wealth, $x = \frac{1}{\alpha^\theta}(\alpha^\theta x)$. Observe that the relative risk aversion coefficient $\mathrm{R}^{\mathrm{r}}_{\mathrm{A}}(u)$ is a "probabilistic" mixture of the different risk aversion coefficients,

$$\mathrm{R}^{\mathrm{r}}_{\mathrm{A}}(u)(x) = \frac{-xu_{xx}(x)}{u_x(x)} = \int \mathrm{R}^{\mathrm{r}}_{\mathrm{A}}(u^\theta)(\alpha^\theta x)\frac{u^\theta_x(\alpha^\theta x)}{\int u^\theta_x(\alpha^\theta x)\, m(d\theta)}\, m(d\theta)$$

bounded if the family of individual risk aversion coefficients $\mathrm{R}^{\mathrm{r}}_{\mathrm{A}}(u^\theta)(\alpha^\theta x)$ is uniformly bounded in $\theta$.

### 3.1.2 Aggregation of the Optimal Processes

Up to the time $t$, the individuals invest optimally in a portfolio $X^{*,\theta}(\alpha^\theta x)$ with preferences characterized by their consistent progressive utility $\mathbf{U}^\theta$. It is then natural to define the aggregate wealth in the considered economy at any time $t$, $(X^*_t)$, as the weighted sum of the individual wealths $(X^{*,\theta}_t)$,

$$X^*_t(x) := \int X^{*,\theta}_t(\alpha^\theta x)m(d\theta). \tag{3.4}$$

Remark that the relative weights evolve stochastically in time and are given by

$$\alpha^\theta_t := \frac{X^{*,\theta}_t(\alpha^\theta x)}{\int X^{*,\theta}_t(\alpha^\theta x)m(d\theta)}.$$

Motivated by the construction of the initial utility and of its derivative $u_x(x) = \int u^\theta_x(\alpha^\theta x)\, m(d\theta) = \int y^\theta(u_x(x))\, m(d\theta)$, a natural choice is to define $Y^*_t(u_x(x))$ as a mixture of individual state price processes, which is still an admissible state price process issued from $u_x(x)$

$$Y^*_t(u_x(x)) := \int Y^{*,\theta}_t(u^\theta_x(\alpha^\theta x))m(d\theta) = \int Y^{*,\theta}_t(y^\theta(u_x(x)))m(d\theta). \tag{3.5}$$

Now, the problem is formulated as a reverse problem (Sect. 2.3) based on the increasing aggregate processes, $X^*_t(x)$ and $Y^*_t(y)$. Remark that the consistent utility is unique as soon as the optimal processes $X^*_t(x)$, $Y^*_t(y)$ and the initial utility $u$ are given. The last difficulty is to study the regularity of those aggregate processes $X^*_t(x)$ and $Y^*_t(y)$ from the regularity of the individual processes $X^{*,\theta}_t(x)$ and $Y^{*,\theta}_t(y)$. Notice that the aggregation of processes is easier when they are considered as random fields rather than solutions of SDEs. Also, we use the representation of optimal processes given in Corollary 2.10 for the processes $X^{*,\theta}$, $Y^{*,\theta}$,

$$\begin{cases} dX_t^{*,\theta}(x) = r_t X_t^{*,\theta}(x)dt + \phi^{*,\theta}(t,x).(dW_t + \eta_t^{\mathcal{R}}), & X_0^{*,\theta}(x) = x, \ \phi^*(t,x) \in \mathcal{R} \\ dY_t^{*,\theta}(y) = -r_t Y_t^{*,\theta}(y)dt + \big(\psi^{*,\theta}(t,y) - \eta_t^{\mathcal{R}} Y_t^{*,\theta}(y)\big).dW_t, & Y_0^{*,\theta}(y) = y, \ \psi^*(t,y) \in \mathcal{R}^{\perp}. \end{cases}$$

Any linear combination of portfolios $X_t^{*,\theta}(\alpha^\theta x)$ is an admissible portfolio issued from the linear combination of their initial wealth $\alpha^\theta x$. The same property is still true for continuous combination (under some integrability conditions). Then, the aggregate wealth process $X_\cdot^*(x) = \int X_\cdot^{*,\theta}(\alpha^\theta x)m(d\theta)$ is an admissible portfolio in $\mathscr{X}(x)$ and

$$\begin{cases} dX_t^*(x) = r_t X_t^*(x)dt + \phi^*(t,x).(dW_t + \eta_t^{\mathcal{R}}\, dt) \\ \phi^*(t,x) := \int \phi^{*,\theta}(t, \alpha^\theta x))m(d\theta). \end{cases} \tag{3.6}$$

By similar arguments, the aggregate dual process $Y^*$ is an admissible one, with more complex dynamics, because of its dependence in $u_x(x)$

$$\begin{cases} dY_t^*(u_x(x)) = -r_t Y_t^*(u_x(x))dt + \big(\psi^*(t, u_x(x)) - Y_t^*(u_x(x))\eta_t^{\mathcal{R}}\big).dW_t. \\ \psi^*(t, u_x(x)) := \int \psi^{*,\theta}(t, y_x^\theta(u_x(x)))m(d\theta). \end{cases}$$

Since for any $\theta$, $X^{*,\theta}$ and $Y^{*,\theta}$ are optimal, their characteristics are given in terms of the volatility vectors $\gamma^\theta$ of $\mathbf{U}^{\grave{}}$, which yields

$$\begin{cases} \phi^*(t,x) = \displaystyle\int \Big(\frac{U_x^\theta}{U_{xx}^\theta}\big(\frac{\gamma_x^{\theta,\mathcal{R}}}{U_x^\theta} + \eta_t^{\mathcal{R}}\big)\Big)(t, X_t^{*,\theta}(\alpha^\theta x))m(d\theta) \\ \psi^*(t, u_x(x)) = \displaystyle\int \gamma_x^{\theta,\perp}(t, X_t^{*,\theta}(\alpha^\theta x))m(d\theta). \end{cases}$$

## 3.2 *The Aggregate Utility*

The goal from now is to show the existence of dynamic utility $\mathbf{U}$ generating $X^*$ and $Y^*$ as optimal processes. As in the previous section, if $\mathbf{U}$ exists, then necessarily the master identity $U_x(t, X_t^*(x)) = Y_t^*(u_x(x))$ has to be satisfied. The problem has a simple solution in the case of power utilities.

### 3.2.1 Aggregating Power Utilities

We come back to the standard example of power utilities and their aggregation, detailed in Sect. 2.2.2. We assume in this subsection that not only the initial utility functions but all the progressive utilities to be aggregated are power utilities with different risk aversion coefficient.

By definition the initial utility is a mixture of scaled power utilities

$$u(x) = \int \frac{1}{\alpha^{(\theta)}} \frac{(\alpha^{(\theta)} x)^{1-\theta}}{1-\theta} m(d\theta),$$

which is no longer a power utility. More generally, all utility processes $\mathbf{U}^{(\theta)}$ are power utilities with constant relative risk aversion coefficient $\theta$ ($0 < \theta < 1$). As recalled in Sect. 2.2.2, $U^{(\theta)}(t,x) = Z_t^{(\theta)} \frac{x^{1-\theta}}{1-\theta}$ for some process $Z^{(\theta)}$ and the optimal primal and dual processes are linear with respect to their initial conditions.

$$X_t^{*,(\theta)}(x) = x\bar{X}_t^{*,(\theta)}, \quad Y_t^{*,(\theta)}(y) = y\bar{Y}_t^{*,(\theta)}, \quad Z_t^{(\theta)} = \bar{Y}_t^{*,(\theta)}(\bar{X}_t^{*,(\theta)})^{\theta}.$$

The characterization of the aggregate optimal processes is easy to obtain from the definition,

$$\begin{cases} X_t^*(x) \quad = x\bar{X}_t^*, \quad \bar{X}_t^* = \int \alpha^{(\theta)} \bar{X}_t^{*,(\theta)} m(d\theta) \\ Y_t^*(u_x(x)) = \int (\alpha^{(\theta)} x)^{-\theta} \bar{Y}_t^{*,(\theta)} \, m(d\theta), \quad u_x(x) = \int (\alpha^{(\theta)} x)^{-\theta} \, m(d\theta). \end{cases} \tag{3.7}$$

Remark that whereas the aggregate wealth $X^*$ is a linear process with respect to its initial value $x$, this is not true anymore for the aggregate state price density process $Y^*$.

The construction of a progressive utility with optimal processes $(x\bar{X}_t^*, Y_t^*(y))$, based on the main identity $U_x(t,x) = Y_t^*(u_x(\frac{x}{\bar{X}_t^*}))$, yields easily to the following characterization.

**Proposition 3.1** *The marginal utility $U_x(t,x)$ is the deterministic aggregation of the power marginal progressive utilities with random repartition of the optimal wealth,*

$$U_x(t,x) = \int \bar{Y}_t^{*,(\theta)} \Big(\frac{\alpha^{(\theta)} x}{\bar{X}_t^*}\Big)^{-\theta} m(d\theta) = \int U_x^{(\theta)}\Big(t, \frac{\alpha^{(\theta)} \bar{X}_t^{*,(\theta)}}{\bar{X}_t^*} x\Big) m(d\theta). \tag{3.8}$$

*The ratio $\bar{\alpha}_t^{(\theta)} = \frac{\alpha^{(\theta)} \bar{X}_t^{*,(\theta)}}{\bar{X}_t^*}$ is the stochastic ratio of the optimal wealths at time $t$.*

As for the Pareto utility in [21], aggregating power utilities provides a family of consistent progressive utilities which is more flexible, while benefiting from some interesting features of power utilities (such as tractability).

Aggregating general consistent utilities is not as straightforward as for power utilities, and it involves the "reverse problem" techniques detailed previously in Sect. 2.3.

### 3.2.2 The General Case

The general case will be considered as a reverse problem. Following the results in Sect. 2.3, some preliminary technical results are needed.

**Lemma 3.2** *The optimal processes $X^{*,\theta}$ and $Y^{*,\theta}$ are assumed to satisfy the regularity conditions of Theorem 2.11, with the same $\delta$ for each $\theta$ and with Lipschitz constants $C^{X,\theta}$ and $C^{Y,\theta}$ satisfying $\int C^{X,\theta} m(d\theta), \int C^{Y,\theta} m(d\theta) < \infty$. We also* *assume that for any interval $I \subset \mathbb{R}_+^*$ there exist integrable functions $\phi_k^I(x,\theta)$ such that $|\partial_x^k u^\theta(x)| \le \phi_k^I(x,\theta)$, $\forall x \in I$ and for $k = 1,2,3$. Then,*

*(i) There exists a constant $K$, such that for any $\theta$ and any $x, y > 0$, $\mathbb{E}(X_t^{*,\theta}(x)) \le C^{X,\theta}Ktx$ and $\mathbb{E}(Y_t^{*,\theta}(y)) \le C^{Y,\theta}Kty$. Consequently, the integrals (3.4) and (3.5) are well defined.*

*(ii) The monotonic random fields $X^*$, defined by (3.6), is $\in \mathcal{K}_{loc}^{3,\varepsilon}$ for any $\varepsilon \in [0,\delta[$ and its inverse flow $\mathcal{X}^*$ is a semimartingale. Moreover $Y^* \in \mathcal{K}_{loc}^{2,\varepsilon}$ for any $\varepsilon \in [0,\delta[$.*

*(iii) As $u_x^\theta$ is of class $\mathcal{C}^2(0,\infty)$, $u_x \in \mathcal{C}^2(0,\infty)$ and the marginal utility process $U_x(t,x) = Y_t^*(u_x(\mathcal{X}_t^*(x)))$ is a $\mathcal{K}_{loc}^{2,\varepsilon}$ semimartingale for any $\varepsilon \in [0,\delta[$.*

*Proof* (i) is a standard result, obtained from Burkholder-Davis-Gundy and the Jensen inequalities, see [27], Lemmas 4.5.3 and 4.5.5.

(ii) Combining Assumptions of this result with Theorem 5.3 leads to $X^{*,\theta} \in \mathcal{K}_{loc}^{3,\varepsilon}$ and $Y^{*,\theta} \in \mathcal{K}_{loc}^{2,\varepsilon}$ for any $\varepsilon \in [0,\delta[$. So $\sigma^{*,\theta}(t, X^{*,\theta}(x)) \in \overline{\mathcal{K}}_{loc}^{3,\varepsilon}$ and $\vartheta^{*,\theta}(t, Y^{*,\theta}(y)) \in \overline{\mathcal{K}}_{loc}^{2,\varepsilon}$. It follows that $\phi^*(t,x)(= \sigma^*(t, X_t^*(x)))$ and $\psi^*(t,y)$ $(= \vartheta^*(t, Y_t^*(y))$ are respectively in $\overline{\mathcal{K}}_{loc}^{3,\varepsilon}$ and $\overline{\mathcal{K}}_{loc}^{2,\varepsilon}$. We then conclude as in Corollary 2.10. Statement (iii) becomes obvious. □

These regularity results allow us to consider the problem of consistency of the aggregate utility as a reverse problem as in Theorem 2.11.

**Theorem 3.3** *Under Assumptions of Lemma 3.2,* **U** *defined by $U(t,x) = \int \int_0^x U_x^\theta(t, X_t^{*,\theta}(\alpha^\theta \mathcal{X}_t^*(z)))dz\, m(d\theta)$ is a consistent semimartingale progressive utility. The optimal primal and dual processes are $(X_t^*(x))$ and $\big(Y_t^*(u_x(x)) = U_x(t, X_t^*(x))\big)$ and*

$$\begin{cases} \gamma_x^{\mathcal{R}}(t,x) = -U_x(t,x)\eta_t^{\mathcal{R}} - U_{xx}(t,x)\phi^*(t, \mathcal{X}^*(t,x)). \\ \gamma_x^{\perp}(t,x) = \psi^*(t, u_x(\mathcal{X}^*(t,x))). \\ \beta(t,x) = -r_t x U_x(t,x) + \dfrac{1}{2}U_{xx}(t,x)||\phi^*(t, \mathcal{X}^*(t,x))||^2. \end{cases} \tag{3.9}$$

Since $\phi^*(t,x) = \sigma^*(t, X_t^*(x))$ and $\psi^*(t,y) = \vartheta^*(t, Y_t^*(y))$, it is easy to check the equivalence between the systems (3.9) and (2.14).

*Proof* It is a consequence of Theorem 2.11, since $X^*$ and $Y^*$ satisfies respectively the SDE (2.6) and (2.7) and the regularity conditions are satisfied. We produce here the proof in this context of aggregation; the proof being still valid in the general setting of Theorem 2.11. It relies on the identity $Y_t^*(u_x(x)) = U_x(t, X_t^*(x))$.
By statement (iii) of previous Lemma, the random field $\mathbf{U}_x$ is sufficiently regular to apply Itô Ventzel's formula to compute the dynamics of $U_x(t, X_t^*(x))$:

$$\begin{aligned} dU_x(t, X_t^*(x)) = \Big(&\beta_x(t, X_t^*(x)) + \frac{1}{2}U_{xxx}(t, X_t^*(x))||\phi^*(t,x)||^2) \\ &+ U_{xx}(t, X_t^*(x))(r_t X_t^*(x) + \phi^*(t,x))\eta_t^{\mathcal{R}}) + \gamma_{xx}(t, X_t^*(x))\phi^*(t,x)\Big)dt \\ &+ \big(\gamma_x(t, X_t^*(x)) + U_{xx}(t, X_t^*(x))\phi^*(t,x)\big).dW_t. \end{aligned}$$

(i) By identification of the diffusion term with the one of

$$dY_t^*(u_x(x)) = -r_t Y_t^*(u_x(x))dt + \big(\phi^*(t, u_x(x)) - Y_t^*(u_x(x))\eta_t^{\mathcal{R}}\big).dW_t$$

and by the fact that $\psi^*(t, u_x(x)) = \int \vartheta^{*,\theta}(t, Y_t^{*,\theta}(u_x^\theta(\alpha^\theta x))m(d\theta)$ and $\vartheta^{*,\theta}(t, Y_t^{*,\theta}(u_x^\theta(x)) = \gamma_x^{\theta,\perp}(t, X_t^{*,\theta}(x))$ it follows that

$$\gamma_x(t,x) + U_{xx}(t,x)\phi^*(t, \mathcal{X}^*(t,x)) = \psi^*(t, u_x(\mathcal{X}^*(t,x))) - U_x(t,x)\eta_t^{\mathcal{R}}$$

or equivalently by projecting on $\mathcal{R}$ and $\mathcal{R}^\perp$,

$$\begin{cases} \gamma_x^{\mathcal{R}}(t,x) = -U_x(t,x)\eta_t^{\mathcal{R}} - U_{xx}(t,x)\phi^*(t, \mathcal{X}^*(t,x)) \\ \gamma_x^{\perp}(t,x) = \psi^*(t, u_x(\mathcal{X}^*(t,x))) = \displaystyle\int \gamma_x^{\theta,\perp}(t, X_t^{*,\theta}(\alpha^\theta \mathcal{X}_t^*(x)))m(d\theta). \end{cases}$$

(ii) Identifying the drift term, it is also easy to prove that **U** satisfies the HJB constraint (2.5). Indeed, using the characterization of $\sigma^*$,

$$\begin{aligned} \gamma_{xx}(t,x)\phi^*(t, \mathcal{X}^*(t,x)) &= \gamma_{xx}^{\mathcal{R}}(t,x)\phi^*(t, \mathcal{X}^*(t,x)) \\ &= \partial_x\big(U_{xx}(t,x)\phi^*(t, \mathcal{X}^*(t,x)) + U_x(t,x)\eta_t^{\mathcal{R}}\big)\phi^*(t, \mathcal{X}^*(t,x)). \end{aligned}$$

It follows, after arranging the terms and identifying with the drift term of $dY_t^{*,\theta}(y)$

$$\beta_x(t,x) = \partial_x\Big(-r_t x U_x(t,x) + \frac{1}{2}U_{xx}(t,x)||\phi^*(t, \mathcal{X}^*(t,x))||^2\Big)$$

and integrating with respect to $x$ gives the HJB constraint. □

## 3.3 *Particular Case of a Pareto Optimal Allocation of the Initial Wealth*

In this work, given the wealth of each class $x^\theta$, we get the aggregate wealth of the given economy as $x = \int x^\theta m(d\theta)$. In other words, the $\alpha^\theta$ are imposed and given by $\alpha^\theta := x^\theta/x$. In the literature, the approach is rather the opposite: given the global wealth of the economy $x$, the problem is to find the fair allocation of the wealth $x$ between the different classes such that the allocation is Pareto optimal, that is there are no possible alternative allocations whose realization would cause every class to gain. The Pareto optimal allocation is determined by the initial wealths $x^{*,\theta}$ with $\int x^{*,\theta}m(d\theta) = x$ such that $u(x) = \int u(x^{*,\theta})m(d\theta) = \sup\{\int u(x^\theta)m(d\theta) | x^\theta \geq 0$ and $\int x^\theta m(d\theta) = x\}$. One important consequence of Pareto optimality is that the optimal pricing kernel $Y^{*,\theta}$ is the same for all agents. See for example the paper of Bank and Kramkov [1] that aggregates utilities parameterized by Pareto weights, for a finite number of investors, or Mrad [14] for a continuum of agents and a general mixture framework. In this work, the initial repartition of the wealth is assumed to be given a priori, without reference to any "optimal allocation". The $\alpha^{\theta}$ are fixed (at time 0) and correspond to the initial proportion of the total wealth hold by the $\theta$-class. Therefore the aggregate utility $\mathbf{U}$ and the aggregate pricing kernel $Y^{*}$ are not standard, but they are natural candidate for aggregating different points of view of several agents, in a context without an equilibrium. It thus allows a richer class of pricing kernel that will add flexibility to capture some financial features, such that the impact of the wealth on the valuation of financial assets.

We choose to illustrate this methodology in measuring its impact on the yield curve.

# 4 Application to the Yield Curve

Numerous economic issues involve the optimization of the aggregate utility of the economy. Besides, among this economic literature involving utility optimization, many papers focus on long term issues. Therefore the use of stochastic utility is particularly relevant in such modeling frameworks with long horizon. Besides, since the theoretical study of progressive utility emphasizes the dependency of the processes with respect to their initial conditions, this framework is also well adapted to study the dependency and the non-linearity of macroeconomic processes with respect to the initial value of economic indexes.

One particular example developed here consists in the valuation of long term interest rates of the considered economy (country). Modeling accurately long term interest rates is a crucial challenge in many financial topics, such as the financing of ecological project, or the pricing of longevity-linked securities or any other investment with long term impact. In the economic setting, the interest are determined endogenously at equilibrium, from the equilibrium optimal pricing kernel (see for example Vasicek [35], Cox et al. [6], Björk [4] and Piazzesi [32]). The financial settings only assume no arbitrage and with an exogenous short term interest rate, in a framework of incomplete market.

## *4.1 Yield Curve in Incomplete Market*

In the context of the high illiquidity of the bond market for longer maturities, the financial evaluation we consider is the marginal utility indifference pricing of zero-coupon bond. The link with the economic discount rate given by the Ramsey rule (in an equilibrium setting) is studied in El Karoui et al. [11, 13].

### 4.1.1 Utility Indifference Pricing

In incomplete market, for the pricing of non replicable contingent claims, there are different ways to evaluate the risk coming from the unhedgeable part, yielding to a bid-ask spread. A way is the pricing by indifference (as in Henderson and Hobson [20]).

Utility indifference price of a quantity $q$ of a positive claim $\xi_T$ delivered in $T$ is the cash amount $(p_t(x,q))_{t\in[0,T]}$ for which the investor is indifferent from investing or not in the claim

$$\mathcal{U}^{\xi}(t, x + p_t(x,q), q) = \mathcal{U}(t,x), \quad \text{for all } t \in [0,T]$$

with the two following maximization problems (with and without the claim $\xi_T$):

$$\begin{aligned} \mathcal{U}^{\xi}(t,x,q) &:= \sup\nolimits_{\kappa\in\mathcal{X}(t,x)} \mathbb{E}[U(T, X_T^{\kappa,c} - q\xi_T)|\mathcal{F}_t], \\ \mathcal{U}(t,x) &:= \sup\nolimits_{\kappa\in\mathcal{X}(t,x)} \mathbb{E}[U(T, X_T^{\kappa,c})|\mathcal{F}_t], \quad t \leq T. \end{aligned}$$

When the investors are aware of their sensitivity to the unhedgeable risk, they can try to transact for only a little amount in the risky contract. In this case, the buyer wants to transact at the seller's "fair price" (also called *Davis price* or *marginal utility price*, see Davis [8]), which corresponds to the zero marginal rate of substitution $(\pi^*_{t,T}(\xi_T)(y))_{t\in[0,T]}$ (with $y = \mathcal{U}_x(t,x)$) determined at any time $t$ by the relationship

$$\partial_q \mathcal{U}^{\xi}(t, x + \pi^*_{t,T}(\xi_T)(y), q)|_{q=0} = 0, \quad \text{for all } t \in [0,T].$$

It is given via the dual parametrization

$$\pi^*_{t,T}(\xi_T)(y) = \mathbb{E}\big[\xi_T \frac{Y^*_T(y)}{Y^*_t(y)}|\mathcal{F}_t\big], \quad y = \mathcal{U}_x(t,x). \tag{4.1}$$

The *marginal utility indifference pricing* at time $t$ is not based on the "universal" *market state price density* $Y^0$ (as in complete market), but on the optimal state price density $Y^*(y)$ of the progressive dual utility $\widetilde{\mathbf{U}}$ of $\mathbf{U}$ (Proposition 2.5). With this point of view, the price of some derivative $\xi_T$ is not given by $\pi^0_{0,T}(\xi_T) = \mathbb{E}(Y^0_T\,\xi_T)$, $(Y^0_0 = 1)$, as in a complete market but by $\pi^*_{0,T}(\xi_T)(y) = \frac{1}{y}\mathbb{E}(Y^*_T(y)\,\xi_T)$, making the price depending on the global wealth $x$ of the economy via the correspondence $u_x(x) = y$. The pricing rule $\pi^0_{0,T}$, that is independent of the wealth, will be called *market pricing rule*.

*Dynamic Marginal Utility Indifference Pricing* By definition (see (2.3)), any state price density $(Y^*_t(y))$ can be written as $Y^*_t(y) = Y^0_t L^*_t(y)$ with $L^*_t(y) := L^{\nu^*}_t(y) = e^{\int_0^t \nu^{*,\perp}_s(y)dW_s - \frac{1}{2}\int_0^t \|\nu^{*,\perp}_s(y)\|^2 ds}$. All the dependencies on the wealth $x$ (or $y = u_x(x)$) is supported by the exponential martingale $L^*_t(y)$, normalized by its value at time 0, and denoted $L^*_{0,t}(y) := \frac{1}{y}L^*_t(y)$.

The marginal utility pricing rule becomes the "market pricing" of some modified pay-off $\pi^*_{0,T}(\xi_T)(y) = \pi^0_{0,T}(L^*_{0,T}(y)\xi_T)$. The extension of the pricing rules to any date in the future is straightforward, using the conditional expectation, and the relative state price density $Y^*_{t,T}(y) := \frac{Y^*_T(y)}{Y^*_t(y)}$, so that

$$\pi^0_{t,T}(\xi_T) = \mathbb{E}(Y^0_{t,T}\,\xi_T|\mathcal{F}_t) \text{ and } \pi^*_{t,T}(\xi_T)(y) = \mathbb{E}(Y^*_{t,T}(y)\,\xi_T|\mathcal{F}_t) = \pi^0_{t,T}(L^*_{t,T}(y)\xi_T).$$

*Wealth Sensitivity Analysis* By Corollary 2.2, the volatility of $L^*_{0,t}(y)$ is the regular volatility random field $\upsilon^{*,\perp}_t(y) := \nu^{*,\perp}(t, Y^*_t(y)) = (Y^*_t(y))^{-1}\vartheta^{*,\theta}_t(Y^*_t(y))$ and

$$\ln(L^*_{0,t}(y)) = \int_0^t \upsilon^{*,\perp}_s(y).dW_s - \frac{1}{2}\|\upsilon^{*,\perp}_s(y)\|^2 ds.$$

Its sensitivity in $y$ is given by

$$\frac{\partial_y L^*_{0,t}(y)}{L^*_{0,t}(y)} = \int_0^t \partial_y \upsilon^{*,\perp}_s(y).(dW_s - \upsilon^{*,\perp}_s(y)ds) = \int_0^t \partial_y \upsilon^{*,\perp}_s(y).(dW_s + (\eta^{\mathcal{R}}_s - \upsilon^{*,\perp}_s(y))ds).$$

The second equality uses the orthogonality of the vectors $\upsilon^{*,\perp}_s(y)$ and $\eta^{\mathcal{R}}_s$.

The remarkable property is that $\frac{\partial_y L^*_{0,t}(y)}{L^*_{0,t}(y)}$ is a martingale under the probability measure with density martingale $\Lambda^*_{0,t}(y) = \exp(\int_0^t r_s ds)Y^*_{0,t}(y)$ whose volatility is the $Y^*$-volatility $(-\upsilon^{*,\perp}_s(y) + \eta^{\mathcal{R}}_s)$ .

*Marginal Utility Bond Curve* Applying the previous theory to the zero-coupon bond, that delivers **one** unit of cash at maturity $T$, we get the market bond price $B^0_t(T)$, as well as the indifference bond price $B^*_t(T, y)$ that depends on $x$ by the initial relation $y = u_x(x)$

$$B^0_t(T) := \mathbb{E}\big(Y^0_{t,T}|\mathcal{F}_t\big), \text{ and } B^*_t(T, y) := \mathbb{E}\big(Y^*_{t,T}(y)|\mathcal{F}_t\big) = \mathbb{E}\big(Y^0_{t,T}L^*_{t,T}(y)|\mathcal{F}_t\big). \quad (4.2)$$

(a) The sensitivity of the zero-coupon bonds with respect to their maturity is interpreted in any yield market as a forward rate, that is the instantaneous short rate for an operation starting in the future at time $T$. Then, we make the distinction between *market or indifference forward rate*

$$f^0_t(T) := -\partial_T \ln B^0_t(T), \text{ respectively, } f^*_t(T, y) := -\partial_T \ln B^*_t(T, y).$$

The *yield curve* at current date $t$ is the function,

$$\delta \mapsto R^*_t(\delta, y) := \frac{1}{\delta}\int_t^{t+\delta} f^*_t(u, y)du = -\frac{1}{\delta}\ln B^*_t(t+\delta, y).$$

(b) The sensitivity of the bonds with respect to the initial wealth at the current date $t$ is

$$\partial_y B_t^*(T,y) = \pi_t^0(\partial_y L_{t,T}^*(y)) = \mathbb{E}(Y_{t,T}^*(y)\, \xi_{t,T}(y)|\mathcal{F}_t) \tag{4.3}$$

where $-\xi_{t,T}(y) = \int_t^T \partial_y v_s^{*,\perp}(y).(dW_s + (\eta_s^{\mathcal{R}} - v_s^{*,\perp}(y))ds)$. As it is often useful for financial interpretations (see Geman et al. [19]), relation (4.3) can be reinterpreted by using a change of probability measure, associated to a numeraire change,

$$\partial_y B_t^*(T,y) = \mathbb{E}_{\mathbb{Q}_{(y)}^{*,T}}\left(e^{-\int_t^T r_s ds}\, \xi_{t,T}(y) \mid \mathcal{F}_t\right)$$

where $\mathbb{Q}_{(y)}^{*,T}$ is the probability measure with density $\Lambda_{0,T}^*(y)$ and under which $dW_s^{*,T} = dW_s + (\eta_s^{\mathcal{R}} - v_s^{*,\perp}(y))ds$ is a martingale. Sometimes this probability measure is called *indifference forward neutral probability*.

### 4.1.2 Yield Curve in Aggregate Economy

We come back to the framework of an economy with multi-agents having access to the same market, and so having the same market price density $Y^0$. They have their own progressive utilities $\mathbf{U}^\theta$, and then their own marginal utility pricing rules driven by their own optimal state price density $Y_t^{*,\theta}(y) = Y_t^0 L_t^{*,\theta}(y)$.

Then, each agent gives a different "marginal utility price" for the zero-coupon bonds, $B_t^{*,\theta}(T,y) = \mathbb{E}\big(Y_{t,T}^{*,\theta}(y)|\mathcal{F}_t\big)$. In particular, the bond curves today $B_0^{*,\theta}(T,y^\theta)$ are different and a priori depend on the individual wealth $y^\theta$ of the agent; but a large part of the curve is explained by the common market curve $B_0^0(T)$.

*Aggregate Yield Curves* In the aggregate economy, the initial marginal utility is defined as a mixture of the individual marginal utilities, $u_x(x) = \int u_x^\theta(\alpha^\theta x)\, m(d\theta)$. Similarly, the optimal state price density $Y_t^*(y)$ is a mixture of the individual optimal state prices defined as $Y_t^*(y) = \int Y_t^{*,\theta}(y^\theta) m(d\theta)$ where $y^\theta(u_x(x)) = u_x^\theta(\alpha^\theta x)$. Thanks to Eq. (4.2), the bond curve $B_t^*(T,y)$ in the aggregate market satisfies

$$Y_t^*(y)B_t^*(T,y) = \mathbb{E}\big(Y_T^*(y)|\mathcal{F}_t\big) = \int \mathbb{E}(Y_T^{*,\theta}(y^\theta)|\mathcal{F}_t) m(d\theta) = \int B_t^{*,\theta}(T,y) Y_t^{*,\theta}(y^\theta) m(d\theta).$$

The aggregate bond curve is a mixture of different bond curves with respect to the non normalized densities $Y_t^{*,\theta}(y^\theta)$, whose integral is by definition $Y_t^*(y)$.

It is easy to take the derivative in maturity in the previous equality, and to use intensively that $\partial_T\, B_t^*(T,y) = -f_t^*(T,y) B_t^*(T,y)$ where $f_t^*(T,y)$ is the instantaneous forward rate in the aggregate market. Thus we obtain that

$$f_t^*(T,y) Y_t^*(y) B_t^*(T,y) = \int f_t^{*,\theta}(T,y) Y_t^{*,\theta}(y^\theta) B_t^{*,\theta}(T,y) m(d\theta).$$

The remarkable feature of these two decompositions is that the non normalized mixing processes may be chosen to be martingales:

- It is obvious in the case of spot forward rates where the mixing processes are $Y_t^{*,\theta}(y^\theta)B_t^{*,\theta}(T,y^\theta)$ which are by definition the exponential martingales determining the volatility of the bond.
- For the mixing of the bonds, the non normalized coefficients $Y_t^{*,\theta}(y^\theta)=Y_t^0L_t^{*,\theta}(y^\theta)$ (having the common factor $Y_t^0$) can be replaced by the martingales $L_t^{*,\theta}(y^\theta)$ without change after renormalization.

All these results are gathered in the next proposition:

**Proposition 4.1** *In an aggregate economy,*

*(i) The marginal utility bond curve $B_t^*(T,y)$ is a normalized mixture of individual bond curves, based on the martingales $L_t^{*,\theta}$,*

$$B_t^*(T,y)=\int B_t^{*,\theta}(T,y^\theta)\frac{L_t^{*,\theta}(y^\theta)\,m(d\theta)}{\int L_t^{*,\theta}(y^\theta)m(d\theta)}. \tag{4.4}$$

*(ii) The marginal utility spot forward rates $f_t^*(T,y)$ is a normalized mixture of individual spot forward rates curve based on the martingales $Y_t^{*,\theta}(y^\theta)B_t^{*,\theta}(T,y^\theta)$*

$$f_t^*(T,y)=\int f_t^{*,\theta}(T,y^\theta)\frac{B_t^{*,\theta}(T,y^\theta)L_t^{*,\theta}(y^\theta)}{\int B_t^{*,\theta}(T,y^\theta)L_t^{*,\theta}(y^\theta)m(d\theta)}\,m(d\theta). \tag{4.5}$$

**Indifference Bonds Pricing for Power Utilities** We come back to the framework of aggregate power utilities, that will be used in the forthcoming numerical application in Sect. 4.2. We consider $N$ agents with consistent power utilities characterized by their relative risk aversion parameters $\theta_1<\cdots<\theta_N$, as studied in Sect. 2.2.2. Then, their optimal state prices $Y_t^{*,\theta}(y)$ are linear in $y$ with coefficient $\bar{Y}_t^{*,\theta}$, and the individual price of zero-coupon bonds with maturity $T$ does not depend on $y$ and more generally, $\bar{B}_t^{*,\theta}(T)=\mathbb{E}\big(\bar{Y}_{t,T}^{*,\theta}|\mathcal{F}_t\big)$. The aggregate indifference zero-coupon price $\bar{B}_0^*(T,y)$, computed at time 0 for simplicity, is given by

$$\bar{B}_0^*(T,y)=\frac{\sum_{i=1}^N y^{\theta_i}(y)\bar{B}_0^{*,\theta_i}(T)}{y}=\frac{\sum_{i=1}^N(\alpha_i x)^{-\theta_i}\bar{B}_0^{*,\theta_i}(T)}{\sum_{i=1}^N(\alpha_i x)^{-\theta_i}}$$

with $y=\sum_{i=1}^N y^{\theta_i}(y)=u_x(x)$ and for power utilities, $y^{\theta_i}(y)=u_x^{\theta_i}(\alpha_i x)=(\alpha^i x)^{-\theta_i}$.
*Asymptotic Behavior* Using the linearity of the optimal state prices $Y_t^{*,\theta}(y^{\theta_i})$ in $y^{\theta_i}$ and the form of the marginal initial power utilities $u_x^{\theta_i}$, the asymptotic behavior of the aggregate zero-coupon price, for $y$ around 0 (respectively $\infty$), is straightforward and relies on the convergence of the random measure $\frac{\sum_{i=1}^N y^{\theta_i}(y)\,\delta_{\theta_i}}{y}$ towards a dirac measure that charges the agent with the smallest (respectively the largest) risk aversion $\theta_i$. Indeed, scaling the initial wealth $x$ with a factor $\lambda\in\mathbb{R}^+$, leads to the following asymptotics (for $\lambda\to 0$ or $\infty$)

$$\lim_{y\to 0} B_0^*(T,y) = B_0^{\theta_1}(T) \text{ and } \lim_{y\to +\infty} B_0^*(T,y) = B_0^{\theta_N}(T).$$

This means that, when the wealth tends to infinity, the aggregate zero-coupon price converges to the one priced by the less risk averse agent, whereas when the wealth tends to zero, it converges to the one priced by the more risk averse agent. This is a similar result as the ones stated in Cvitanic et al. [7].

## 4.2 Numerical Results

The numerical illustration is developed in the simple model of an economy where three agents invest in an incomplete market with two independent Brownian motions: the market is characterized by the market state price density $Y^0$ with a constant market risk premium $(\eta, 0)$ and a stochastic interest rate. The volatility vector of admissible portfolios only depends on the first component $\kappa_t = (\kappa_t^1, 0)$; the optimal dual orthogonal volatility $(0, \upsilon^{*,\theta})$ is also assumed to be constant in time and independent of $y$, where $\theta \in ]0,1[$ is the relative risk aversion parameter characterizing the agent starting with a power utility:

$$Y_t^{*,\theta}(y) = y e^{-\int_0^t r_s ds} e^{-\eta W_t^1 + \upsilon^{*,\theta} W_t^2 - \frac{1}{2}(\eta^2 + (\upsilon^{*,\theta})^2)t}. \tag{4.6}$$

We also need to specify a model for the spot rate common for all agents. The simplest and currently used in financial market is the Vasicek model [35].
*Vasicek Model for the Spot Rate* $r_t$: We assume a Vasicek model for the spot rate $r_t$,

$$dr_t = a(b - r_t)dt + \sigma_1 dW_t^1 + \sigma_2 dW_t^2,$$

which is a Gaussian Ornstein-Uhlenbeck process given by

$$r_t = r_0 e^{-at} + b(1 - e^{-at}) + \int_0^t e^{-a(t-s)}(\sigma_1 dW_t^1 + \sigma_2 dW_t^2).$$

The *market* zero-coupon bond is given from the market state price density $Y_t^0$, by $B_t^0(T) = \mathbb{E}(Y_{t,T}^0 \mid \mathcal{F}_t)$. From Gaussian standard calculus, it is well-known that the *market* yield curve $R_t^0(\delta) = -\frac{1}{\delta} \ln B_t^0(t+\delta)$ has the following form

$$\begin{aligned} R_t^0(\delta) &= R_\infty^0 - (R_\infty^0 - r_t)\tfrac{(1-e^{-a\delta})}{a\delta} + \tfrac{\mathfrak{s}^{2,r}}{a^2}\tfrac{(1-e^{-a\delta})^2}{4a\delta} \\ \text{where } R_\infty^0 &= b - \tfrac{1}{2}\left[\tfrac{\mathfrak{s}^{2,r}}{a^2} + 2\tfrac{\sigma_1\eta}{a}\right] \text{ and } \mathfrak{s}^{2,r} = (\sigma_1)^2 + (\sigma_2)^2. \end{aligned}$$

*Indifference Yield Curve* In this example, the indifference yield curve is obtained by multiplication of the market price density $(Y_t^0)$ by the exponential martingale $L_t^{*,\theta} = \exp\left(\upsilon^{*,\theta} W_t^2 - \frac{1}{2}(\upsilon^{*,\theta})^2 t\right)$ which depends on the Brownian motion $W^2$ only.

The bonds are obtained as previously, using the probability measure $\mathbb{Q}^{\perp,\theta} = L_T^{*,\theta}.\mathbb{P}$ in place of $\mathbb{P}$. Under $\mathbb{Q}^{\perp,\theta}$, $W^1$ is still a Brownian motion but $W^2$ becomes $W_t^2 = W_t^{2,\perp,\theta} + \upsilon^{*,\theta} t$ where $W^{2,\perp,\theta}$ is a $\mathbb{Q}^{\perp,\theta}$-Brownian motion. The spot rate $r_t$ remains an Ornstein-Uhlenbeck process under $\mathbb{Q}^{\perp,\theta}$, only the level $b$ is modified into $b^{\perp,\theta} = b + \frac{1}{a}\sigma_2\upsilon^{*,\theta}$. This modification has an impact on the infinite rate, that becomes $R_\infty^{*,\theta} = R_\infty^0 + \frac{1}{a}\sigma_2\upsilon^{*,\theta}$.

The new yield curve is now:

$$\begin{cases} R_t^{*,\theta}(\delta) = R_\infty^{*,\theta} - (R_\infty^{*,\theta} - r_t)\frac{(1-e^{-a\delta})}{a\delta} + \frac{\mathfrak{s}^{2,r}}{a^2}\frac{(1-e^{-a\delta})^2}{4a\delta} \\ R_t^{*,\theta}(\delta) = R_t^0(\delta) - \sigma_2\upsilon^{*,\theta}\big(\frac{1-e^{-a\delta}-a\delta}{a^2\delta}\big). \end{cases}$$

The same kind of equation holds for the different forward rates. In particular the spread between the indifference curve and the market curve is given by:

$$f_t^{*,\theta}(T) - f_t^0(T) = -\sigma_2\upsilon^{*,\theta}\big(\frac{1-e^{-a(T-t)}}{a}\big).$$

*Aggregate Bond Curve* For the aggregate bond curve, we consider the aggregation of three agents with power utility and risk aversion parameter $(\theta_i)$ and with a given initial repartition of the wealth $(\alpha_i)$ (see Sect. 3.2.1). In this case, the zero-coupon bond is evaluated as

$$B_t^*(T, x, \alpha) = \frac{\sum_{i=1}^3 (\alpha_i x)^{-\theta_i} B_t^{*,\theta_i}(T)}{u_x(x)} \tag{4.7}$$

where $x$ and $\alpha$ stand here to remind that the aggregate price depends on the initial wealth and on the initial choice of the parameters $(\alpha_i)$.

The ratio $\frac{B_t^*(T,x,\alpha)}{B_t^0(T)}$ is particularly simple, using the notation $\xi_2(T-t) = \sigma_2\big(\frac{1-e^{-a(T-t)}-a(T-t)}{a^2}\big)$

$$\frac{B_t^*(T,x,\alpha)}{B_t^0(T)} = \frac{\sum_{i=1}^3 (\alpha_i x)^{-\theta_i}\exp(-\xi_2(T-t)\upsilon^{*,\theta_i})}{u_x(x)} = \frac{\sum_{i=1}^3 \exp(-\theta_i \ln(\alpha_i x) - \xi_2(T-t)\upsilon^{*,\theta_i})}{u_x(x)}.$$

**Simulations** The following simulations are provided taking $\upsilon^{*,\theta} = \theta\upsilon^*$ for some given constant $\upsilon^*$. For any $\theta$, the individual yield curve $\delta \mapsto R_0^{*,\theta}(\delta)$ does not depend on the wealth and is a Vasicek curve with infinite rate $R_0^{*,\theta}(\delta) = R_0^0(\delta) - \sigma_2\theta\upsilon^*\big(\frac{1-e^{-a\delta}-a\delta}{a^2\delta}\big)$.

In the figures we choose the following numerical values for the parameters

$$r_0 = 5\%,\ a = 1,\ b = 0.2,\ \sigma_1 = 20\%,\ \sigma_2 = 15\%,\ \eta = 20\%,\ \upsilon^* = 80\%.$$

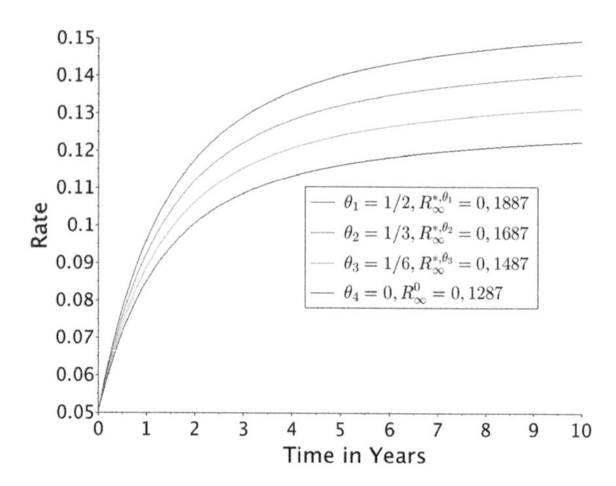


**Fig. 1** Individual yield curve $R_0^{*,\theta}(\delta)$ for different values of $\theta$

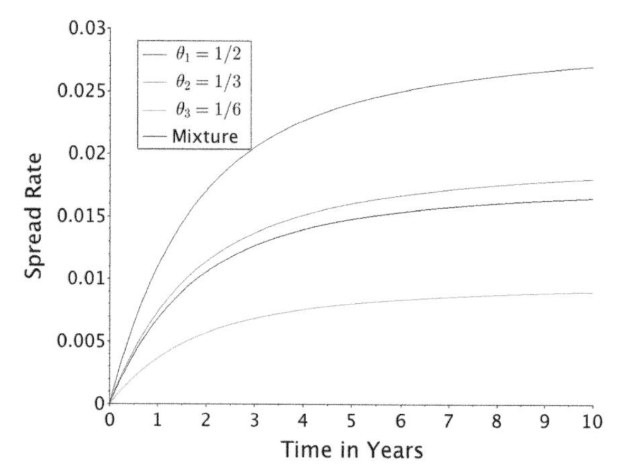


**Fig. 2** Individual and aggregate yield curve spread

It provides a standard increasing yield curve, but Vasicek model can also achieve others forms of curve (not monotonic and with bumps) for other parameters values.

In Fig. 1 we draw the individual yield curve $R_0^{*,\theta}(\delta)$ of each class (Vasicek yield curves), for different values of $\theta$.

From now on, we will represent only the spreads between the different rate curves and the market yield curve $R_0^0(\delta)$, namely $R_0^{*,\theta}(\delta) - R_0^0(\delta)$. Figure 2 represents the spread of the individual curve for three different values of $\theta$ as well as the spread of the aggregate curve. The spread of the aggregate curve depends on $x$ and the $\alpha_i$, we choose here $x = 10$, $\alpha_1 = 1/8$, $\alpha_2 = 1/2$, $\alpha_3 = 3/8$ (unless other numerical values are specified).

Figure 3 (respectively Fig. 4) illustrates the sensitivity of the aggregate yield curve on the initial wealth $x$ available on the market (respectively on the initial proportion parameters $\alpha^\theta$).

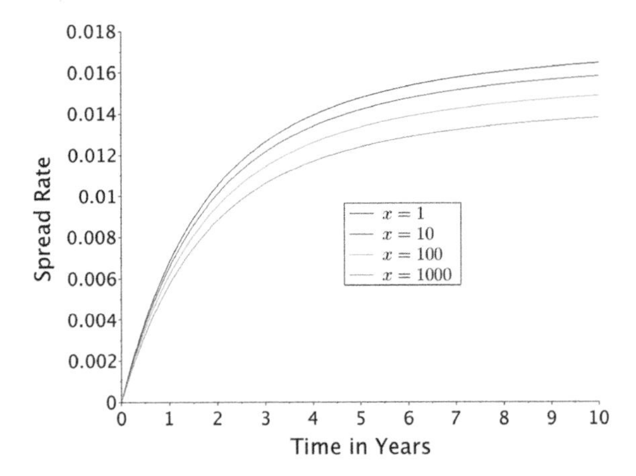


**Fig. 3** Aggregate yield curve spread depending of the wealth $x$

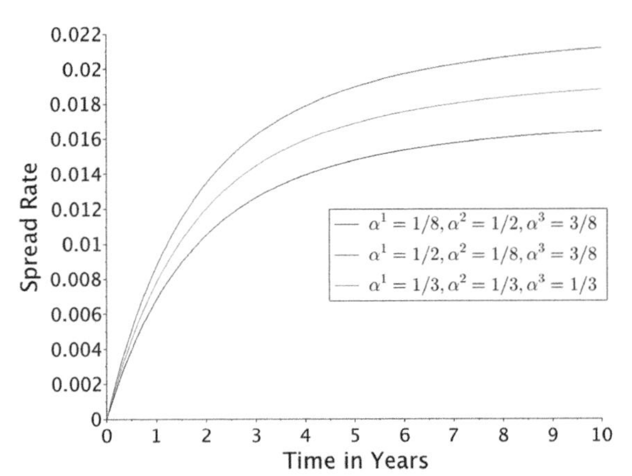


**Fig. 4** Aggregate yield curve spread depending on the initial proportion parameters $\alpha$

# 5 Appendix

**Itô-Ventzel's Formula** In this paper we used the *Itô-Ventzel formula* that gives the decomposition of the compound random field $G(t, X_t)$ for $G(t, x) = G(0, x) + \int_0^t \phi(s, x)ds + \int_0^t \psi(s, x).dW_s$ regular enough (see Theorem 5.1 below) and for any Itô semimartingale $X$. This decomposition is the sum of three terms: the first one is the "differential in $t$" of $G$, the second one is the classic Itô's formula (without differentiation in time) and the third one is the infinitesimal covariation between the martingale part of $G_x$ and the martingale part of $X$, all these terms being taken in $X_t$.

$$\begin{aligned} dG(t, X_t) &= \big(\phi(t, X_t)\,dt + \psi(t, X_t).dW_t\big) \\ &\quad + \Big(G_x(t, X_t)dX_t + \frac{1}{2}G_{xx}(t, X_t)d < X, X >_t\Big) + \big( < \psi_x(t, X_t).dW_t, dX_t > \big). \end{aligned} \tag{5.1}$$

When $G$ has only finite variation, the formula is reduced to a classic Itô's formula, since in this case $\psi(t,x) \equiv 0$, $\phi(t,X_t) = \partial_t G_t(t,X_t)$.

**Different Spaces of Regular Random Fields** We give here the regularity conditions needed in Theorem 2.11 to characterize a consistent progressive utility from its optimal processes. These regularity conditions are related to the SDEs' coefficients.. Let $(\phi,\psi)$ be continuous $\mathbb{R}^k$-valued progressive random fields and let $m$ be a non-negative integer, and $\delta$ a number in $(0,1]$. We need to control the asymptotic behavior in 0 and $\infty$ of $\phi$ and $\psi$, and the regularity of their Hölder derivatives (when they exist). More precisely, let $\phi \in \mathcal{C}^{m,\delta}(]0,+\infty[)$ be $(m,\delta)$-times[5] continuously differentiable in $x$ for any $t$, a.s.

For any subset $K \subset ]0,+\infty[$, we define the family of random (Hölder) $K$-semi-norms

$$\begin{cases} \|\phi\|_{m:K}(t,\omega) = \sup_{x\in K} \frac{\|\phi(t,x,\omega)\|}{x} + \sum_{1\le j\le m} \sup_{x\in K} \|\partial_x^j \phi(t,x,\omega)\| \\ \|\psi\|_{m,\delta:K}(t,\omega) = \|\psi\|_{m:K}(t,\omega) + \sup\limits_{x,y\in K} \dfrac{\|\partial_x^m \psi(t,x,\omega) - \partial_x^m \psi(t,y,\omega)\|}{|x-y|^\delta}. \end{cases} \tag{5.2}$$

When $K$ is all the domain $]0,+\infty[$, we simply write $\|.\|_m(t,\omega)$, or $\|.\|_{m,\delta}(t,\omega)$. Calligraphic notation recalls that these semi-norms are random.

(a) $\mathcal{K}_{loc}^{m,\delta}$ (resp. $\overline{\mathcal{K}}_{loc}^{m,\delta}$) denotes the set of all $\mathcal{C}^{m,\delta}$-random fields such that for any compact $K \subset ]0,+\infty[$, and any $T$, $\int_0^T \|\phi\|_{m,\delta:K}(t,\omega)dt < \infty$, (resp. $\int_0^T \|\psi\|^2_{m,\delta:K}(t,\omega)dt < \infty$ ).
(b) When these different norms are well-defined on the whole space $]0,+\infty[$, we use the notations $\mathcal{K}_b^m, \overline{\mathcal{K}}_b^m$ or $\mathcal{K}_b^{m,\delta}, \overline{\mathcal{K}}_b^{m,\delta}$.

**Differentiability of Itô Random Fields** We discuss the regularity of an Itô semi-martingale random field

$$G(t,x) = G(0,x) + \int_0^t \phi(s,x)ds + \int_0^t \psi(s,x).dW_s \tag{5.3}$$

in connection with the regularity of its local characteristics $(\phi,\psi)$. An Itô random field $\mathbf{G}$ is said to be a $\mathcal{K}_{loc}^{m,\delta}$**-semimartingale**, whenever $G(0,x)$ is of class $\mathcal{C}^{m,\delta}$, $B^G(t,x) = \int_0^t \phi(s,x)ds$ is of class $\mathcal{K}_{loc}^{m,\delta}$, and $M^G(t,x) = \int_0^t \psi(s,x).dW_s$ is of class $\overline{\mathcal{K}}_{loc}^{m,\delta}$. As in Kunita [27], we are concerned with the regularity of $\mathbf{G}$ (the regularity of its local characteristics $(\phi,\psi)$ being given) and conversely with the regularity of $(\phi,\psi)$ (the regularity of $\mathbf{G}$ being given). Theorem 3.1.2, Theorem 3.1.3 and Theorem 3.3.3 in [27] give the differential rules (term by term) of the dynamics of an Itô random field and the minimal condition to apply Itô-Ventzel's formula.

[5]That is $\phi$ is $m$-times continuously differentiable with $\phi^{(m)}$ being $\delta$-Hölder.

**Theorem 5.1** (Differential Rules) *Let $\delta \in (0, 1]$ and $\mathbf{G}$ be an Itô semimartingale random field with local characteristics $(\phi, \psi)$, $G(t, x) = G(0, x) + \int_0^t \phi(s, x)ds + \int_0^t \psi(s, x).dW_s$*

(i) *If $\mathbf{G}$ is a $\mathcal{K}_{loc}^{m,\delta}$-semimartingale for some $m \geq 0$, its local characteristics $(\phi, \psi)$ are of class $\mathcal{K}_{loc}^{m,\varepsilon} \times \overline{\mathcal{K}}_{loc}^{m,\varepsilon}$ for any $\varepsilon < \delta$, and conversely.*

(ii) *Conversely, if the local characteristics $(\phi, \psi)$ are of class $\mathcal{K}_{loc}^{m,\delta} \times \overline{\mathcal{K}}_{loc}^{m,\delta}$, then $\mathbf{F}$ is a $\mathcal{K}_{loc}^{m,\varepsilon}$-semimartingale for any $\varepsilon < \delta$.*

(iii) *For $m \geq 1$, the derivative random field $\mathbf{G_x}$ is an Itô random field with local characteristics $(\phi_x, \psi_x)$, and for $m \geq 2$ the Itô-Ventzel formula is applicable.*

(iv) *Moreover, if $\mathbf{G}$ is a $\mathcal{K}_{loc}^{1,\delta} \cap C^2$-semimartingale, for any Itô process $X$, $G(., X_.)$ is a continuous Itô semimartingale satisfying the Itô-Ventzel formula* (5.1).

**Differentiability of SDEs Solutions** It is well known from the SDE's theory that is sufficient (but not necessary) to take a coefficients $(\mu, \sigma) \in \mathcal{K}_b^{0,1}, \overline{\mathcal{K}}_b^{0,1}$ to ensure the existence of monotonic global (non-explosive) solution of SDE$(\mu, \sigma)$ with range $[0, \infty)$ and a good behavior near to zero and infinity (see the discussion in [16] or Kunita's book [27]). Otherwise, local regularity on SDEs coefficients appears as a kind of minimal assumption to ensure the regularity of a global solution *if there exists*.

**Definition 5.2** A SDE$(\mu, \sigma)$ is said to be of class $\mathcal{S}^{m,\delta}$ if

(a) the coefficients $(\mu, \sigma)$ are in the spaces $(\mathcal{K}_{loc}^{m,\delta}, \overline{\mathcal{K}}_{loc}^{m,\delta})$

(b) the maximal solution $X$ is non explosive.

Classical examples of $\mathcal{S}^{m,\delta}$ SDEs are given by SDE$(\mu, \sigma)$ when $(\mu, \sigma)$ are in the spaces $(\mathcal{K}_b^m, \overline{\mathcal{K}}_b^m)$, or even in $(\mathcal{K}_b^0, \overline{\mathcal{K}}_b^0) \cap (\mathcal{K}_{loc}^{m,\delta}, \overline{\mathcal{K}}_{loc}^{m,\delta})$.

**Theorem 5.3** (Flows property of SDE) *We consider the SDE* $(\mu, \sigma)$

$$dX_t = \mu(t, X_t)dt + \sigma(t, X_t).dW_t, \quad X_0 = x. \tag{5.4}$$

*Let $m \geq 1$, $\delta \in (0, 1]$ and $\varepsilon < \delta$.*

(i) *Assume uniformly Lipschitz coefficients, that is $(\mu, \sigma) \in \mathcal{K}_{\mathbf{b}}^{0,1} \times \overline{\mathcal{K}}_{\mathbf{b}}^{0,1}$. Then, (5.4) admits a unique strong solution $X$ which is strictly monotonic satisfying $X(0) = 0$ and $X(+\infty) := \lim_{x \to +\infty} X(x) = +\infty$.*

(ii) *Assume $\mu \in \mathcal{K}_{\mathbf{b}}^{m,\delta}$ and $\sigma \in \overline{\mathcal{K}}_{\mathbf{b}}^{m,\delta}$.*

 (a) *Then the solution $\mathbf{X} = (X_t(x), x > 0)$ is a $\mathcal{K}_{loc}^{m,\varepsilon}$ semimartingale the derivatives $\mathbf{X_x}$ and $\mathbf{1/X_x}$ are $\mathcal{K}_{loc}^{m-1,\varepsilon}$-semimartingales. Its inverse $\mathbf{X^{-1}}$ is also of class $C^m$.*

 (b) *The local characteristics of $\mathbf{X}$, $\lambda^X(t, x) = \mu(t, X_t(x))$ and $\theta^X(t, x) = \sigma(t, X_t(x))$ have only local properties and belong to $\mathcal{K}_{loc}^{m,\varepsilon} \times \overline{\mathcal{K}}_{loc}^{m,\varepsilon}$.*

**Acknowledgements** The authors thank the financial supports of Chaire "Risques Financiers", of Labex Ecodec and of Labex MME-DII.

## References


1. Bank, P., Kramkov, D.: A model for a large investor trading at market indifference prices. ii: continuous-time case. Ann. Appl. Probab. **25**(5), 2708–2742 (2015)
2. Berrier, F.P., Rogers, L.G.C., Tehranchi, M.R.: A characterization of forward utility functions. Preprint (2009)
3. Berrier, F.P., Tehranchi, M.R.: Forward utility of investment and consumption. Preprint (2011)
4. Bjork, T.: Equilibrium theory in continuous time. In: Lecture Notes (2012)
5. Chan, Y.L., Kogan, L.: Catching up with the joneses: heterogeneous preferences and the dynamics of asset prices. J. Polit. Econ. **110**, 1255–1285 (2002)
6. Cox, J.C., Ingersoll, J.C., Ross, S.A.: A theory of the term structure of interest rates. Econometrica **53**(2), 385–403 (1985)
7. Cvitanić, J., Jouini, E., Malamud, S., Napp, C.: Financial markets equilibrium with heterogeneous agents. Rev. Financ. **16**(1), 285–321 (2011)
8. Davis, M.H.A.: Option pricing in incomplete markets. In Pliska, S.R. (ed.) Mathematics of Derivative Securities, pp. 216–226. M.A.H. Dempster and S.R. Pliska, Cambridge University Press edition (1998)
9. Dumas, B.: Two-person dynamic equilibrium in the capital market. Rev. Financ. Stud. **2**(2), 157–188 (1989)
10. Dybvig, P.H., Rogers, L.C.G.: Recovery of preferences from observed wealth in a single realization. Rev. Financ. Stud. **10**(1), 151–174 (1997)
11. El Karoui, N., Hillairet, C., Mrad, M.: Affine long term yield curves: an application of the Ramsey rule with progressive utility. J. Financ. Eng. **1**(1) (2014)
12. El Karoui, N., Hillairet, C., Mrad, M.: Consistent utility of investment and consumption: a forward/backward spde viewpoint. Stochastics (2017)
13. El Karoui, N., Hillairet, C., Mrad, M.: Ramsey rule with forward/backward utility for long term yield curves modeling. Preprint (2017)
14. El Karoui, N., Mrad, M.: Mixture of consistent stochastic utilities, and a priori randomness. Preprint (2016)
15. El Karoui, N., Mrad, M.: Recover dynamic utility from monotonic characteristic/extremal processes (2019), in revision. https://hal.archives-ouvertes.fr/hal-01966312
16. El Karoui, N., Mrad, M.: An exact connection between two solvable SDEs and a non linear utility stochastic PDEs. SIAM J. Financ. Math. **4**(1), 697–736 (2013)
17. El Karoui, N., Geman, H., Rochet, J.C.: Changes of numeraire, changes of probability measure and option pricing. J. Appl. Probab. **32**(2), 443–458 (1995)
18. Filipovic, D., Platen, E.: Consistent market extensions under the benchmark approach. Math. Financ. **19**(1), 41–52 (2009)
19. Geman, H., El Karoui, N., Rochet, J.-C.: Changes of numeraire, changes of probability measure and option pricing. J. Appl. Probab. **32**(2), 443–458 (1995)
20. Henderson, V., Hobson, D.: Indifference Pricing: Theory and Application (Ed. R. Carmona). Princeton University Press, Princeton (2009)
21. Ikefuji, M., Laeven, R.J.A., Magnus, J.R., Muris, C.: Pareto utility. Theor. Decis. **75**(1), 43–57 (2013)
22. Jouini, E., Napp, C.: Unbiased disagreement in financial markets, waves of pessimism and the risk-return trade-off. Rev. Financ. **15**(3), 575–601 (2010)
23. Karatzas, I., Shreve, S.E.: Methods of Mathematical Finance. Springer (2001)
24. Karatzas, I., Lehoczky, J.P., Shreve, S.E.: Optimal portfolio and consumption decisions for a small investor on a finite horizon. SIAM J. Control Optim. **25**, 1557–1586 (1987)
25. Kramkov, D., Schachermayer, W.: The asymptotic elasticity of utility functions and optimal investment in incomplete markets. Ann. Appl. Probab. **9**(3), 904–950 (1999)
26. Kramkov, D., Schachermayer, W.: Necessary and sufficient conditions in the problem of optimal investment in incomplete markets. Ann. Appl. Probab. **13**(4), 1504–1516 (2003)

27. Kunita, H.: Stochastic flows and stochastic differential equations. In: Cambridge Studies in Advanced Mathematics, vol. 24. Cambridge University Press, Cambridge (1997). Reprint of the 1990 original
28. Musiela, M., Zariphopoulou, T.: Backward and forward utilities and the associated pricing systems: the case study of the binomial model. In: Indifference Pricing: Theory and Application, pp. 3–44. Princeton University Press (2009)
29. Musiela, M., Zariphopoulou, T.: Investment and valuation under backward and forward dynamic exponential utilities in a stochastic factor model. In: Advances in Mathematical Finance, pp. 303–334. Birkhäuser, Boston (2007)
30. Musiela, M., Zariphopoulou, T.: Stochastic partial differential equations in portfolio choice. In: Chiarella, C., Novikov, A. (eds.) Contemporary Quantitative Finance: Essays in Honour of Eckhard Platen, pp. 161–170 (2010)
31. Musiela, M., Zariphopoulou, T.: Portfolio choice under space-time monotone performance criteria. SIAM J. Financ. Math. **1**(1), 326–365 (2010)
32. Piazzesi, M.: Affine term structure models. Handbook Financ. Econ. **1**, 691–766 (2010)
33. Platen, E., Heath, D.: A benchmark approach to quantitative finance. In: Springer Finance. Springer-Verlag, Berlin (2006)
34. Rogers, L.C.G.: Duality in constrained optimal investment and consumption problems: a synthesis. In: Paris-Princeton Lectures on Mathematical Finance 2002, pp. 95–131. Springer (2003)
35. Vasicek, O.: An equilibrium characterization of the term structure. J. Financ. Econ. **5**(2), 177–188 (1977)
36. Wang, J.: The term structure of interest rates in a pure exchange economy with heterogeneous investors. J. Financ. Econ. **41**(1), 75–110 (1996)
37. Yan, H.: Is noise trading cancelled out by aggregation? Manag. Sci. **56**(7), 1047–1059 (2010)
38. Zitkovic, G.: A dual characterization of self-generation and log-affine forward performances. Ann. Appl. Probab. **19**(6), 2176–2210 (2009)

# BSDEs and Enlargement of Filtration

**Monique Jeanblanc and Dongli Wu**

**Abstract** In this paper we study the solution of a BSDE in a large filtration, and we show that the projection (on a smaller filtration) of its semimartingale part has coefficients that can be explicitely given in terms of the coefficients in the large filtration.

**Keywords** BSDE · Enlargement of filtration · Absolute continuity Jacod's hypothesis · Projection · Brownian motion · Poisson random measure

## 1 Introduction

Starting with the seminal paper by Pardoux and Peng [18], the study of Backward Stochastic Differential Equations (BSDE) and their applications in finance (see e.g., El Karoui et al. [6]) is an important area of research. El Karoui and Rouge [21] were the first to apply this methodology to study exponential utility problems.

The filtration enlargement theory was developed by Jacod, Jeulin and Yor in [10, 13–15]. Starting with a reference filtration $\mathbb{F}$, these authors have studied the behavior of $\mathbb{F}$-martingales in a larger filtration $\mathbb{G}$, under specific filtration enlargements. In particular, they gave conditions so that the $(\mathscr{H}')$-hypothesis holds, i.e., any $\mathbb{F}$-martingale is a $\mathbb{G}$-semimartingale for two kinds of enlargement: initial and progressive (see below for a precise definition). Brémaud and Yor [3] have investigated the stronger assumption of immersion, i.e., the case where any $\mathbb{F}$-martingale is a $\mathbb{G}$-martingale. The reader can refer to Aksamit and Jeanblanc [1] for a study of immersion and details and various conditions which imply $(\mathscr{H}')$-hypothesis.

M. Jeanblanc (✉)
LaMME, Univ Evry, Université Paris Saclay, Paris, France
e-mail: monique.jeanblanc@univ-evry.fr

D. Wu
SGCIB R&D, Paris, France
e-mail: dongli.wu-ext@sgcib.com



S. N. Cohen et al. (eds.), *Frontiers in Stochastic Analysis - BSDEs, SPDEs and their Applications*, Springer Proceedings in Mathematics & Statistics 289,
https://doi.org/10.1007/978-3-030-22285-7_7

In the first part of this paper, we study the relation of the solutions of BSDEs in two filtrations $\mathbb{F}$ and $\mathbb{G}$ satisfying $\mathbb{F} \subset \mathbb{G}$. Firstly, we define a BSDE in the filtration $\mathbb{G}$ with terminal condition $\zeta \in \mathscr{G}_T$, and we denote by $Y^{\mathbb{G}}$ the semimartingale part of the solution of this BSDE. We then obtain a BSDE in the smaller filtration $\mathbb{F}$ satisfied by the projection of $Y^{\mathbb{G}}$ on $\mathbb{F}$ and we study the relations between the solutions. We study the particular cases where $\mathbb{G}$ is the initial or progressive enlargement of $\mathbb{F}$. Then, we consider a linear driver with $\mathbb{F}$-adapted coefficients and we give the link between solutions of the BSDEs with same driver and same terminal condition in the filtration $\mathbb{G}$ and in the filtration $\mathbb{F}$. In the second part of the paper, we focus on the utility maximization problems for an exponential utility, by using $\mathbb{F}$ and $\mathbb{G}$-predictable strategies. We solve them by using BSDEs.

## 2 Background

In this first section, we recall some useful facts on enlargement of filtrations and BSDEs.

### 2.1 Facts on Enlargement of Filtration

Let $(\Omega, \mathscr{G}, \mathbb{F}, \mathbb{P})$ be a given filtered probability space and $\mathbb{G}$ a filtration larger than $\mathbb{F}$, i.e., $\mathscr{F}_t \subset \mathscr{G}_t \subset \mathscr{G}$.

The problem of enlargement of filtration is the study of $\mathbb{F}$-martingales considered as $\mathbb{G}$-adapted processes and the questions to answer are

- under which conditions are all the $\mathbb{F}$-martingales $\mathbb{G}$-martingales?
- under which conditions are all the $\mathbb{F}$-martingales $\mathbb{G}$-semimartingales and in that case give their $\mathbb{G}$-semimartingale decomposition.

The answer to the first question is well known: all the $\mathbb{F}$-martingales are $\mathbb{G}$-martingales (in other terms, $\mathbb{F}$ is immersed in $\mathbb{G}$ or immersion holds between $\mathbb{F}$ and $\mathbb{G}$) if and only if $\mathscr{G}_t$ and $\mathscr{F}_\infty$ are conditionally independent w.r.t. $\mathscr{F}_t$ for all $t \geq 0$. This is equivalent to

$$\forall t, \forall X \in L^1(\mathscr{F}_\infty),\ \mathbb{E}(X|\mathscr{F}_t) = \mathbb{E}(X|\mathscr{G}_t)\,. \tag{2.1}$$

Immersion holds between $\mathbb{F}$ and $\mathbb{G}$ if (2.1) is satisfied for any bounded r.v. $X$.

The hypothesis leading to a positive answer to the second question is called hypothesis $(\mathscr{H}')$ in the literature. It is deeply studied in the following two particular cases:

- Initial enlargement, i.e., when one considers the smallest right-continuous filtration $\mathbb{G} = (\mathscr{G}_t, t \geq 0)$ which contains $\mathscr{F}_t \vee \sigma(L)$ (i.e., $\mathscr{F}_t \vee \sigma(L) \subset \mathscr{G}_t$) where $L$ is a given r.v. with law $\nu$. This initially enlarged filtration is denoted by $\mathbb{F}^{\sigma(L)}$. Jacod [10] has introduced the following absolute continuity condition: there exists a family of $\mathbb{F}$-martingales $p(x)$ such that

$$\mathbb{P}(L \in dx|\mathscr{F}_t) = p_t(x)\nu(dx) . \tag{2.2}$$

Then, he has established that under absolute continuity condition,[1] any $\mathbb{F}$-martingale $X$ admits an $\mathbb{F}^{\sigma(L)}$-semimartingale decomposition

$$X_t = X_t^L + \int_0^t \frac{d\langle X, p(x)\rangle_s^{\mathbb{F}}|_{x=L}}{p_{s-}(L)}$$

where $X^L$ is an $\mathbb{F}^{\sigma(L)}$-martingale.
We also recall an important result concerning the predictable representation property (PRP). In the case of initial enlargement, it is proved in Fontana [8] that if $X$ is an $\mathbb{F}$-martingale (possibly multidimensional) which satisfies PRP in $\mathbb{F}$, and if Jacod's absolute continuity condition (2.2) holds, then $X^L$ enjoys PRP in $\mathbb{F}^{\sigma(L)}$.

- Progressive enlargement, i.e., when one considers the smallest right-continuous filtration containing $\mathscr{F}_t \vee \sigma(\tau \wedge t)$ where $\tau$ is a random time, i.e., a nonnegative r.v. with law $\nu$. This progressively enlarged filtration is denoted $\mathbb{F}^\tau$. It is known (see Jeanblanc and Le Cam [11]) that, if Jacod's absolute continuity condition (2.2) holds, then any $\mathbb{F}$-martingale $X$ admits an $\mathbb{F}^\tau$-semimartingale decomposition

$$X_t = X_t^{(\tau)} + \int_0^{t\wedge\tau} \frac{d\langle X, m\rangle_s^{\mathbb{F}}}{Z_{s-}} + \int_{t\wedge\tau}^t \frac{d\langle X, p(x)\rangle_s^{\mathbb{F}}|_{x=\tau}}{p_{s-}(\tau)}$$

where $X^{(\tau)}$ is an $\mathbb{F}^\tau$-martingale, $Z_t = \mathbb{P}(\tau > t|\mathscr{F}_t)$ and (roughly speaking) $m$ is the martingale part of the Doob-Meyer decomposition of the supermartingale $\mathbb{P}(\tau \geq t|\mathscr{F}_t)$. Another case where hypothesis $(\mathscr{H}')$ holds is when the random time $\tau$ is honest (see [1] and the references therein for more information).

## 2.2 Notation

We define some spaces associated to a filtration $\mathbb{K}$, or to a $\sigma$-algebra $\mathscr{K}$, which will be relevant in what follows.

- $S(\mathbb{K}, [0, T])$ is the subset of $\mathbb{R}$-valued, càdlàg, $\mathbb{K}$-adapted processes $(k_t)_{t\in[0,T]}$ such that

$$\mathbb{E}[\sup_{t\in[0,T]} k_t^2] < \infty .$$

[1] The needed regularity of $p$ w.r.t. $x$ is established in Jacod [10].

- For $p \in \{1, 2\}$, $L^p\big(\mathbb{K}, [0, T] \times \Omega\big)$ is the set of $\mathbb{K}$-adapted processes such that

$$\mathbb{E}\left(\int_0^T |k_t|^p dt\right) < \infty .$$

- For $p \in \{1, 2\}$, $L^p(\Omega, \mathscr{K})$ is the set of $\mathscr{K}$-measurable random variables $\zeta$ such that

$$\mathbb{E}\left(|\zeta|^p\right) < \infty .$$

- $\mathscr{M}^2\big(\mathbb{K}, [0, T] \times \Omega\big)$ is the set of square-integrable $\mathbb{K}$-martingales $m$ defined on $[0, T]$, i.e., $\mathbb{K}$-martingales $m$ such that

$$\sup_{s \in [0,T]} \mathbb{E}\big(m_s^2\big) < \infty .$$

## 3 A Toy Model for BSDEs in Two Filtrations

In this section, we consider a simple case where the driver of the BSDE is null and where $\mathbb{F}$ is a Brownian filtration.

### 3.1 *General Framework*

Assume that the filtration $\mathbb{G}$ is larger than $\mathbb{F}$ and that any $\mathbb{F}$-martingale is a $\mathbb{G}$-semimartingale. We also assume that there exists $\mu \in L^1(\mathbb{G}, [0, T] \times \Omega)$ such that $W_t = W_t^{\mathbb{G}} + \int_0^t \mu_s ds$ for all $t \in [0, T]$, where $W^{\mathbb{G}}$ is a $\mathbb{G}$-Brownian motion. Since any $\mathbb{F}$-martingale is a $\mathbb{G}$-semimartingale, this implies that, for any $\theta \in L^2(\mathbb{F}, [0, T] \times \Omega)$, one has $\int_0^T |\theta_s \mu_s| ds < \infty$. Let $\zeta$ be a $\mathscr{G}_T$-measurable bounded random variable. Then, the two following BSDEs have unique solution

$$dY_t^{\mathbb{G}} = Z_t^{\mathbb{G}} dW_t^{\mathbb{G}} + dM_t^{\perp},\; Y_T^{\mathbb{G}} = \zeta , \tag{3.1}$$

$$dY_t = Z_t dW_t,\; Y_T = \mathbb{E}(\zeta|\mathscr{F}_T) , \tag{3.2}$$

where $M^{\perp}$, part of the solution of (3.1), is a square integrable $\mathbb{G}$-martingale orthogonal to $W^{\mathbb{G}}$. The solution of BSDE (3.1) is a triplet $(Y^{\mathbb{G}}, Z^{\mathbb{G}}, M^{\perp})$ taking values in the space $S(\mathbb{G}, [0, T])$, $L^2(\mathbb{G}, [0, T] \times \Omega)$, $\mathscr{M}^2\big(\mathbb{G}, [0, T] \times \Omega\big)$ and the solution of BSDE (3.2) is a pair $(Y, Z)$, taking values in the space $S(\mathbb{F}, [0, T])$, $L^2(\mathbb{F}, [0, T] \times \Omega)$.

Obviously $Y_t^{\mathbb{G}} = \mathbb{E}(\zeta|\mathscr{G}_t)$ and $Y_t = \mathbb{E}(Y_t^{\mathbb{G}}|\mathscr{F}_t) = \mathbb{E}(\zeta|\mathscr{F}_t)$; moreover $Y^{\mathbb{G}}$ and $Y$ are both bounded.

The processes $Y$ and $Y^{\mathbb{G}}$ are equal for any $\zeta \in \mathscr{F}_T$ if and only if $\mathbb{F}$ is immersed in $\mathbb{G}$ up to time $T$ (i.e., $\mathbb{F}$-martingales stopped at time $T$ are $\mathbb{G}$-martingales): indeed,

$Y = Y^{\mathbb{G}}$ is equivalent to, for any $t \in [0, T]$,

$$Y_t = \mathbb{E}(\zeta|\mathscr{F}_t) = \mathbb{E}(\zeta|\mathscr{G}_t) = Y_t^{\mathbb{G}}$$

that is any bounded $\mathbb{F}$-martingale stopped at time $T$ is a $\mathbb{G}$-martingale.

In general, we have the following relationship between the solutions $(Y^{\mathbb{G}}, Z^{\mathbb{G}})$ and $(Y, Z)$.

**Proposition 3.1** *The solutions of (3.1) and (3.2) satisfy, for $t \in [0, T]$, $Y_t = \mathbb{E}(Y_t^{\mathbb{G}}|\mathscr{F}_t)$ and $Z_t = \mathbb{E}(Y_t^{\mathbb{G}}\mu_t + Z_t^{\mathbb{G}}|\mathscr{F}_t)$.*

*Proof* The first equality $Y_t = \mathbb{E}(Y_t^{\mathbb{G}}|\mathscr{F}_t)$ is obvious. The equality $Y_T = \mathbb{E}(Y_T^{\mathbb{G}}|\mathscr{F}_T)$ is equivalent to: for any bounded $\mathscr{F}_T$-measurable random variable $X$, one has $\mathbb{E}(Y_T X) = \mathbb{E}(Y_T^{\mathbb{G}} X)$. Using the predictable representation property of $W$ in $\mathbb{F}$, we can write $X$ as $X = x + \int_0^T \theta_s dW_s$. For any $\theta \in L^2(\mathbb{F}, [0, T] \times \Omega)$ such that $\int_0^T \theta_s dW_s$ is bounded, we have

$$\mathbb{E}(Y_T \int_0^T \theta_s dW_s) = \mathbb{E}(Y_T^{\mathbb{G}} \int_0^T \theta_s dW_s) . \tag{3.3}$$

Let $J$ be the bounded $\mathbb{F}$-martingale $J_t := \int_0^t \theta_s dW_s$. The left-hand side of (3.3) is equal to $\mathbb{E}(\int_0^T Z_s\theta_s ds)$. By integration by parts, and the orthogonality between $M^{\perp}$ and $W^{\mathbb{G}}$, the right-hand side is

$$\mathbb{E}(Y_T^{\mathbb{G}} J_T) = \mathbb{E}(\int_0^T J_s dY_s^{\mathbb{G}} + \int_0^T Y_s^{\mathbb{G}}\theta_s dW_s + \int_0^T Z_s^{\mathbb{G}}\theta_s ds) .$$

Using the fact that $Y^{\mathbb{G}}$ is a martingale, the relation between $W$ and $W^{\mathbb{G}}$, and the boundness property of $J$ and of $Y^{\mathbb{G}}$, the right-hand side is

$$\mathbb{E}(\int_0^T Y_s^{\mathbb{G}}\theta_s dW_s + \int_0^T Z_s^{\mathbb{G}}\theta_s ds) = \mathbb{E}(\int_0^T (Y_s^{\mathbb{G}}\theta_s\mu_s + Z_s^{\mathbb{G}}\theta_s)ds) ,$$

where the integral in the right-hand side is well defined since $\int_0^T |\theta_s\mu_s|ds < \infty$ and the expectation of $\int_0^T Y_s^{\mathbb{G}}\theta_s\mu_s ds$ is finite as the difference of two finite quantities. Finally, for any $\theta \in L^2(\mathbb{F}, [0, T] \times \Omega)$ such that $\int_0^T \theta_s dW_s$ is bounded, one has

$$\mathbb{E}(\int_0^T Z_s\theta_s ds) = \mathbb{E}(\int_0^T \theta_s(Z_s^{\mathbb{G}} + \mu_s Y_s^{\mathbb{G}})ds) .$$

For $\theta \in L^2(\mathbb{F}, [0, T] \times \Omega)$, we consider $U := \int_0^T \theta_s dW_s$ and set $U^n = \sup(-n, \inf(n, U))$. The random variable $U^n$ is bounded and there exists $\theta^n \in L^2(\mathbb{F}, [0, T] \times$

$\Omega$) such that $U^n = \int_0^T \theta_s^n dW_s$. Since $U$ is square integrable, $U^n$ converges to $U$ in $L^2$ therefore $\theta^n$ converges to $\theta$ in $L^2(\mathbb{F}, [0, T] \times \Omega)$ as $\mathbb{E}(\int_0^T (\theta_s^n - \theta_s)^2 ds) = \mathbb{E}((U^n - U)^2)$ converges to 0. Therefore

$$\mathbb{E}(\int_0^T Z_s \theta_s ds) = \mathbb{E}(\int_0^T \theta_s (Z_s^{\mathbb{G}} + \mu_s Y_s^{\mathbb{G}}) ds)$$

holds for any $\theta \in L^2(\mathbb{F}, [0, T] \times \Omega)$, hence $Z_t = \mathbb{E}(Z_t^{\mathbb{G}} + \mu_t Y_t^{\mathbb{G}} | \mathscr{F}_t)$.

□

**Comment 3.2** This result is useful in case of change of probability: if $\ell$ is a positive $\mathbb{G}$-martingale (e.g., in a financial setting, the density of some equivalent martingale measure for a price process which is an $\mathbb{F}$-semimartingale, so that the enlargement of filtration does not induce an arbitrage), the projection of $\ell$ on $\mathbb{F}$ will be the density of an equivalent martingale measure for the price process in the filtration $\mathbb{F}$.

## 3.2 *Examples*

### 3.2.1 Independence

Let $\mathbb{G} = \mathbb{F} \vee \widetilde{\mathbb{F}}$, where $\mathbb{F}$ is a Brownian filtration independent from $\widetilde{\mathbb{F}}$. In that case, $\mathbb{F}$ is immersed in $\mathbb{G}$ and $W^{\mathbb{G}} = W$. Let $\zeta$ be an integrable $\widetilde{\mathscr{F}}_T$-measurable r.v. Then, $Y_t^{\mathbb{G}} = \mathbb{E}(\zeta | \mathscr{G}_t) = \mathbb{E}(\zeta | \widetilde{\mathscr{F}}_t)$ is an $\widetilde{\mathbb{F}}$-martingale orthogonal to any $\mathbb{F}$-martingale which implies that the unique solution of the BSDE in $\mathbb{G}$ is $M_t^{\perp} = \mathbb{E}(\zeta | \widetilde{\mathscr{F}}_t) = Y_t^{\mathbb{G}}$, $Z^{\mathbb{G}} \equiv 0$. Furthermore, $\mu = 0$, $Y_t^{\mathbb{F}} = \mathbb{E}(\zeta)$ and $Z^{\mathbb{F}} \equiv 0$.

### 3.2.2 Equivalence Jacod's Hypothesis

Let $\mathbb{F}$ be the filtration generated by a Brownian motion $W$ and $\mathbb{F}^{\sigma(L)}$ be the initial enlargement of $\mathbb{F}$ by a r.v. $L$ which satisfies Jacod's equivalence hypothesis, i.e. (2.2) is reinforced with the hypothesis that, for any $x$, the martingale $p(x)$ is strictly positive. Then, all $\mathbb{F}$-martingales are $\mathbb{F}^{\sigma(L)}$-semimartingales and $W_t = W_t^L + \int_0^t \mu_s ds$, where $W^L$ is an $\mathbb{F}^{\sigma(L)}$-Brownian motion and

$$\mu_t dt = \frac{d\langle W, p(x)\rangle_t^{\mathbb{F}}|_{x=L}}{p_t(L)} . \tag{3.4}$$

Since $W$ enjoys PRP in $\mathbb{F}$ and that, for any $x \in \mathbb{R}$, the process $p(x)$ is an $\mathbb{F}$-martingale, there exists a predictable process $\sigma(x)$ such that $dp_t(x) = p_t(x)\sigma_t(x)dW_t$, hence $\mu_t = \sigma_t(L)$. Moreover, $W^L$ enjoys PRP in $\mathbb{F}^{\sigma(L)}$. Let $\zeta$ be a bounded r.v. in $\mathscr{F}_T^{\sigma(L)}$. The two following BSDEs have unique solutions in the spaces $S(\mathbb{G}, [0, T])$, $L^2(\mathbb{G}, [0, T] \times \Omega)$ (resp. $S(\mathbb{F}, [0, T])$, $L^2(\mathbb{F}, [0, T] \times \Omega)$)

$$dY_t^L = Z_t^L dW_t^L, \ Y_T^L = \zeta, \tag{3.5}$$
$$dY_t = Z_t dW_t, \ Y_T = \mathbb{E}(\zeta|\mathscr{F}_T). \tag{3.6}$$

From results on enlargement of filtrations (see e.g., Corollary 4.21 in [1])

$$\mathbb{E}(\zeta|\mathscr{F}_t^{\sigma(L)}) = \frac{1}{p_t(L)}\mathbb{E}(\zeta p_T(x)|\mathscr{F}_t)|_{x=L}.$$

For any $x \in \mathbb{R}$, setting $\Theta_t(x) = \mathbb{E}(\zeta p_T(x)|\mathscr{F}_t)$, we see that there exists an $\mathbb{F}$-predictable process $\theta(x)$ such that

$$\Theta_t(x) = \mathbb{E}(\zeta p_T(x)) + \int_0^t \theta_s(x) dW_s$$

and the solution of the BSDE

$$dY_t^L = Z_t^L dW_t^L, \ Y_T^L = \zeta$$

satisfies $Y_t^L = \mathbb{E}(\zeta|\mathscr{F}_t^{\sigma(L)}) = \frac{\Theta_t(L)}{p_t(L)}$, hence, by integration by parts and using the fact that if $dK_t(x) = k_t(x)dW_t$, then $dK_t(L) = k_t(L)(dW_t^L + \mu_t^L dt)$, one gets

$$d(Y_t^L p_t(L)) = (Y_t^L p_t(L)\sigma_t(L) + p_t(L)Z_t^L)dW_t^L + (\cdots)dt = d\Theta_t^L = \theta_t(L)dW_t^L + (\cdots)dt,$$

hence, by identification of the coefficients of $W^L$,

$$Z_t^L = \frac{1}{p_t(L)}\Big(\theta_t(L) - Y_t^L p_t(L)\sigma_t(L)\Big) = \frac{\theta_t(L)}{p_t(L)} - Y_t^L \mu_t^L.$$

The process $Z$, part of solution of the $\mathbb{F}$-BSDE

$$dY_t = Z_t dW_t, \ Y_T = \mathbb{E}(\zeta|\mathscr{F}_T)$$

is given by $Z_t dt = d\langle Y, W\rangle_t^{\mathbb{F}}$. Note that, since $\int_{\mathbb{R}} p_T(x)\nu(dx) = 1$, one obtains

$$\begin{aligned} Y_t = \mathbb{E}(\zeta|\mathscr{F}_t) &= \mathbb{E}(\zeta \int_{\mathbb{R}} p_T(x)\nu(dx)|\mathscr{F}_t) = \int_{\mathbb{R}} \Theta_t(x)\nu(dx) \\ &= \mathbb{E}(\zeta) + \int_0^t \Big(\int_{\mathbb{R}} \theta_s(x)\nu(dx)\Big) dW_s, \end{aligned}$$

therefore $Z_t = \int_{\mathbb{R}} \theta_t(x)\nu(dx)$.

We now check that our results are coherent: in the previous section, we have obtained $Z_t = \mathbb{E}(Z_t^L + \mu_t(L)Y_t^L|\mathscr{F}_t)$. Note that, from the above computations, $Z_t^L + \mu_t(L)Y_t^L = \theta_t(L)/p_t(L)$, hence we get

$$\mathbb{E}(Z_t^L + \mu_t(L)Y_t^L|\mathscr{F}_t) = \mathbb{E}(\frac{\theta_t(L)}{p_t(L)}|\mathscr{F}_t) = \int_{\mathbb{R}} \theta_t(x)\nu(dx)\,.$$

*Example 3.3* Let $T = 1/2$, $\zeta = W_{1/2}$, then $Y = W$ and $Z \equiv 1$. Take $L = W_1$. From classical results on Brownian bridge (see, e.g., [1, Chap. 4]),

$$W_t^L = W_t - \int_0^t \frac{W_1 - W_s}{1-s} ds = W_t - \int_0^t \mu_s ds$$

and for $t \le 1/2$

$$Y_t^L = \mathbb{E}(W_{1/2}|\mathscr{F}_t^{\sigma(W_1)}) = W_t + \frac{1-2t}{2(1-t)}(W_1 - W_t)$$

and

$$dY_t^L = \frac{1}{2(1-t)} dW_t^L\,.$$

Hence $Z_t^L = \frac{1}{2(1-t)}$ and one can check easily that $Z_t = \mathbb{E}(Z_t^L + \mu_t Y_t^L|\mathscr{F}_t)$. Computing

$$\mathbb{E}(f(W_1)|\mathscr{F}_t) = \int_{\mathbb{R}} f(x)p_t(x)\nu(dx) = \Phi(t, W_t)\,,$$

where $\Phi(t,x) = \mathbb{E}(f(W_{1-t}+x))$, one obtains that

$$p_t(x)\frac{1}{\sqrt{2\pi}}\exp(-\frac{x^2}{2}) = \frac{1}{\sqrt{2\pi(1-t)}}\exp\Big(-\frac{(x-W_t)^2}{2(1-t)}\Big)\,.$$

Setting $Y_t^x := W_t + \frac{1-2t}{2(1-t)}(x - W_t)$, it is easy, even if tedious, to show

$$\theta_t(x) = p_t(x)\Big(\frac{x - W_t}{1-t}Y_t^x + \frac{1}{2(1-t)}\Big)$$

and $\int_{\mathbb{R}} \theta_s(x)\frac{1}{\sqrt{2\pi}}\exp(-\frac{x^2}{2})dx = 1$.

**Comment 3.4** Here, we comment the results of Eyraud-Loisel [7]. In that paper, the author works under equivalent density hypothesis, and introduces the change of probability $\mathbb{Q}$ defined on $\mathbb{F}^{\sigma(L)}$ as

$$d\mathbb{Q}|_{\mathscr{F}_t^{\sigma(L)}} = \frac{1}{p_t(L)} d\mathbb{P}|_{\mathscr{F}_t^{\sigma(L)}}\,.$$

It is known that under $\mathbb{Q}$, $L$ is independent from $\mathscr{F}_t$ for any $t$ and the restriction of $\mathbb{Q}$ to $\mathbb{F}$ is $\mathbb{P}$, hence the $(\mathbb{P},\mathbb{F})$ Brownian motion $W$ is a $(\mathbb{Q},\mathbb{F}^{\sigma(L)})$ Brownian motion. Then, as mentioned in [7], the equation $dY_t = Z_t dW_t$ reads, under $(\mathbb{Q},\mathbb{F}^{\sigma(L)})$ as $dY_t = Z_t dW_t$, since $W$ is a $(\mathbb{Q},\mathbb{F}^{\sigma(L)})$ Brownian motion. Immersion holds under $\mathbb{Q}$, and, for $\zeta \in \mathscr{F}_T$, one has $Y_t = \mathbb{E}_{\mathbb{Q}}(\zeta|\mathscr{F}_t) = \mathbb{E}_{\mathbb{Q}}(\zeta|\mathscr{F}_t^{\sigma(L)})$. Moreover the coincidence of $\mathbb{Q}$ and $\mathbb{P}$ on $\mathbb{F}$ implies that $\mathbb{E}_{\mathbb{Q}}(\zeta|\mathscr{F}_t) = \mathbb{E}_{\mathbb{P}}(\zeta|\mathscr{F}_t)$.

Note that the equation $dY_t = Z_t dW_t$ under $(\mathbb{Q}, \mathbb{F}^{\sigma(L)})$ is not the one we are studying, since we are working only under $\mathbb{P}$. The equality $\mathbb{E}_{\mathbb{Q}}(\zeta|\mathscr{F}_t) = \mathbb{E}_{\mathbb{P}}(\zeta|\mathscr{F}_t)$ does not imply that $\mathbb{E}_{\mathbb{Q}}(\zeta|\mathscr{F}_t^{\sigma(L)}) = \mathbb{E}_{\mathbb{P}}(\zeta|\mathscr{F}_t^{\sigma(L)})$. In fact

$$\mathbb{E}_{\mathbb{Q}}(\zeta|\mathscr{F}_t^{\sigma(L)}) = p_t(L)\mathbb{E}_{\mathbb{P}}(\frac{\zeta}{p_T(L)}|\mathscr{F}_t^{\sigma(L)}),$$

and, since

$$\mathbb{E}_{\mathbb{P}}(\frac{\zeta}{p_T(L)}|\mathscr{F}_t^{\sigma(L)}) = \frac{1}{p_t(L)}\mathbb{E}_{\mathbb{P}}(\frac{\zeta}{p_T(u)}p_T(u)|\mathscr{F}_t)|_{u=\zeta}$$

we obtain, as expected $\mathbb{E}_{\mathbb{Q}}(\zeta|\mathscr{F}_t^{\sigma(L)}) = \mathbb{E}_{\mathbb{P}}(\zeta|\mathscr{F}_t)$.

As mentioned in [7], if $f$ is an $\mathbb{F}$-adapted driver, and $\zeta \in \mathscr{F}_T$, the solution $(Y, Z)$ of the BSDE $dY_t = f(t, Y_t, Z_t)dt + Z_t dW_t,\ Y_T = \zeta$ computed under $\mathbb{P}$ and the solution $(Y^L, Z^L)$ of $dY_t^L = f(t, Y_t^L, Z_t^L)dt + Z_t^L dW_t,\ Y_T^L = \zeta$ computed under $\mathbb{Q}$ (where $W$ is a $\mathbb{P}$ and a $\mathbb{Q}$-Brownian motion) are equal.

### 3.2.3 Progressive Enlargement

Let $\mathbb{F}$ be the filtration generated by a Brownian motion $W$ and $\mathbb{G} := \mathbb{F}^\tau$ be the progressive enlargement of $\mathbb{F}$ by the positive random variable $\tau$. We assume that any $\mathbb{F}$-martingale is a $\mathbb{G}$-semimartingale and that there exists a process $\mu \in L^1(\mathbb{G}, [0, T] \times \Omega)$ such that $W_t = W_t^{(\tau)} + \int_0^t \mu_s ds$ where $W^{(\tau)}$ is an $\mathbb{F}^\tau$-Brownian motion. These hypotheses hold if $\tau$ satisfies Jacod's absolute continuity hypothesis or if $\tau$ is honest. In that setting, $W^{(\tau)}$ does not satisfy PRP in $\mathbb{F}^\tau$. We then consider the following BSDEs

$$dY_t^{(\tau)} = Z_t^{(\tau)} dW_t^{(\tau)} + dM_t^\perp,\ Y_T^{(\tau)} = \zeta \tag{3.7}$$

$$dY_t = Z_t dW_t,\ Y_T = \mathbb{E}(\zeta|\mathscr{F}_T) \tag{3.8}$$

where, for the first equation, the solution is a triple $(Y^{(\tau)}, Z^{(\tau)}, M^\perp)$ with $M^\perp$ a square integrable $\mathbb{F}^\tau$-martingale orthogonal to $W^{(\tau)}$.

Under additional hypothesis on $\tau$, the pair $(W^{(\tau)}, M)$ enjoys PRP, where $M$ is the compensated $\mathbb{F}^\tau$-martingale associated with $\tau$. We refer the reader to Jeanblanc and Song [12] for details. In that case, (3.7) can be written

$$dY_t^{(\tau)} = Z_t^{(\tau)} dW_t^{(\tau)} + U_t^{(\tau)} dM_t,\ Y_T^{(\tau)} = \zeta,$$

where the solution is a triple $(Y^{(\tau)}, Z^{(\tau)}, U^{(\tau)})$. The following proposition is a consequence of Proposition 3.1.

**Proposition 3.5** *The solutions* $(Y^{(\tau)}, Z^{(\tau)})$ *of (3.7) and* $(Y, Z)$ *of (3.8) satisfy*

$$Y_t = \mathbb{E}(Y_t^{(\tau)}|\mathscr{F}_t), \quad \text{and } Z_t = \mathbb{E}(Y_t^{(\tau)}\mu_t + Z_t^{(\tau)}|\mathscr{F}_t)\,.$$

# 4 Enlargement of a Brownian-Poisson Filtration and BSDEs

We have seen that, in a Brownian filtration, under absolute continuity Jacod's hypothesis, PRP holds in $\mathbb{F}^{\sigma(L)}$ w.r.t. $W^L$. However, even if we assume that $f(t, y, z)$ is $\mathbb{F}$-adapted, and $\zeta \in \mathscr{F}_T$, the link between the well defined BSDEs

$$dY_t = f(t, Y_t, Z_t)dt + Z_t dW_t,\ Y_T = \zeta$$

and

$$dY_t^L = f(t, Y_t^L, Z_t^L)dt + Z_t^L dW_t^L,\ Y_T^L = \zeta \tag{4.1}$$

is still unknown.

## 4.1 BSDE Satisfied by the Projection of $Y^L$

We can characterize the BSDE satisfied by the projection $Y^{\mathbb{F}}$ of $Y^L$ defined in (4.1) on $\mathbb{F}$.

**Proposition 4.1** *If* $Y_t^{\mathbb{F}} := \mathbb{E}(Y_t^L|\mathscr{F}_t)$, *then*

$$dY_t^{\mathbb{F}} = \mathbb{E}(f(t, Y_t^L, Z_t^L)|\mathscr{F}_t)dt + Z_t^{\mathbb{F}} dW_t$$

*where* $Z_t^{\mathbb{F}} = \mathbb{E}(Z_t^L + \mu_t Y_t^L|\mathscr{F}_t)$.

*Proof* We shall study a more general situation in the following Proposition. □

Let $\mathbb{F}$ be the filtration generated by $\mathbf{N}$ and $W$ where

- $\mathbf{N}$ is a Poisson random measure on $([0, T], \mathscr{E})$ where $\mathscr{E}$ is the $\sigma$-field of Borel sets of $E := \mathbb{R} - \{0\}$, with compensator $\nu(dx, dt) = \lambda(dx)dt$ so that

$$\mathbf{M}([0, t] \times A) := (\mathbf{N} - \nu)([0, t] \times A)\,, \quad t \geq 0$$

  is a martingale for all $A \in \mathscr{E}$ satisfying $\lambda(A) < \infty$. Here $\lambda$ is a $\sigma$-finite and positive measure on $(E, \mathscr{E})$ such that $\int_E (1 \wedge x^2)\lambda(dx) < \infty$,
- $W$ is a Brownian motion independent of $\mathbf{N}$.

Assume that the filtration $\mathbb{G}$ is larger than $\mathbb{F}$ and that

1. Every $\mathbb{F}$-martingale is a $\mathbb{G}$-semimartingale. In other terms, the $(\mathscr{H}')$-hypothesis is satisfied.
2. There exists $\mu \in L^1(\mathbb{G}, [0,T] \times \Omega)$ such that $W_t = W_t^{\mathbb{G}} + \int_0^t \mu_s ds$ for all $t \in [0,T]$, where $W^{\mathbb{G}}$ is a $\mathbb{G}$-Brownian motion.
3. There exists a positive $\sigma$-finite measure $\lambda^{\mathbb{G}}$ on $(E, \mathscr{E})$ satisfying $\int_E (1 \wedge x^2)\lambda^{\mathbb{G}}(dx) < \infty$ such that
$$\mathbf{M}^{\mathbb{G}}([0,t] \times A) = (\mathbf{N} - \nu^{\mathbb{G}})([0,t] \times A)\,, \quad t \geq 0$$
is a $\mathbb{G}$-martingale for all $A \in \mathscr{E}$ satisfying $\lambda^{\mathbb{G}}(A) < \infty$ where $\nu^{\mathbb{G}}(dx, dt) := \lambda^{\mathbb{G}}(dx)dt$. Moreover, we assume that $\lambda^{\mathbb{G}}(dx) = \kappa(x)\lambda(dx)$ where $\kappa$ is a square integrable deterministic function.

We denote by
- $\mathscr{P}$ (resp. $\mathscr{P}^{\mathbb{G}}$) the $\sigma$-field of $\mathbb{F}$-predictable (resp. $\mathbb{G}$-predictable) sets on $[0,T] \times \Omega$
- $L^2(\mathbb{F}, [0,T] \times \Omega, \mathbf{M})$ (resp. $L^2(\mathbb{G}, [0,T] \times \Omega, \mathbf{M}^{\mathbb{G}})$ the set of $\mathscr{P} \otimes \mathscr{E}$-measurable (resp. $\mathscr{P}^{\mathbb{G}} \otimes \mathscr{E}$-measurable) processes $u$ such that $\mathbb{E}(\int_0^T \int_E |u_s(x)|^2 \lambda(dx)ds) < \infty$ (resp. $\mathbb{E}(\int_0^T \int_E |u_s(x)|^2 \kappa(x)\lambda(dx)ds) < \infty$)

For a bounded random variable $\zeta \in \mathscr{G}_T$, we consider the following $\mathbb{G}$-BSDE

$$\begin{aligned} dY_t^{\mathbb{G}} &= f\big(t, Y_t^{\mathbb{G}}, Z_t^{\mathbb{G}}, U_t^{\mathbb{G}}\big)dt + Z_t^{\mathbb{G}} dW_t^{\mathbb{G}} + \int_E U_t^{\mathbb{G}}(x)\mathbf{M}^{\mathbb{G}}(dt, dx) + dM_t^{\perp} \\ Y_T^{\mathbb{G}} &= \zeta \end{aligned} \tag{4.2}$$

where the solution, if it exists, is a quadruplet $(Y^{\mathbb{G}}, Z^{\mathbb{G}}, U^{\mathbb{G}}, M^{\perp})$, with $Z^{\mathbb{G}}$ a $\mathbb{G}$-predictable process, $M^{\perp}$ a $\mathbb{G}$-martingale orthogonal to $W^{\mathbb{G}}$ and $\mathbf{M}^{\mathbb{G}}$. The process $Y_t^{\mathbb{G}} - \int_0^t f\big(s, Y_s^{\mathbb{G}}, Z_s^{\mathbb{G}}, U_s^{\mathbb{G}}\big)ds$ being a $\mathbb{G}$-martingale, the process $Y_t^{\mathbb{F}} - \int_0^t \widehat{f}_s ds$, where $Y_t^{\mathbb{F}} = \mathbb{E}(Y_t^{\mathbb{G}} | \mathscr{F}_t)$ and $\widehat{f}_t := \mathbb{E}(f(t, Y_t^{\mathbb{G}}, Z_t^{\mathbb{G}}, U_t^{\mathbb{G}}) | \mathscr{F}_t)$, is an $\mathbb{F}$-martingale, hence there exist $Z^{\mathbb{F}}$ and $U^{\mathbb{F}}$ so that

$$dY_t^{\mathbb{F}} = \widehat{f}_t dt + Z_t^{\mathbb{F}} dW_t + \int_E U_t^{\mathbb{F}}(x)\mathbf{M}(dt, dx)\,. \tag{4.3}$$

We assume the standard hypothesis on $f$ (see e.g., Royer [22] or Kruse and Popier [16]) such that there exists a unique solution for the BSDE (4.2) and that $Y$ is bounded. The goal of this section is to exhibit a relationship between $(Y^{\mathbb{G}}, Z^{\mathbb{G}}, U^{\mathbb{G}})$ and $(Y^{\mathbb{F}}, Z^{\mathbb{F}}, U^{\mathbb{F}})$.

**Proposition 4.2** *If the BSDE (4.2) with a $\mathbb{G}$-adapted bounded process $f$ and a bounded random variable $\zeta \in \mathscr{G}_T$, has a solution $(Y^{\mathbb{G}}, Z^{\mathbb{G}}, U^{\mathbb{G}})$, then the processes $Z^{\mathbb{F}}$ and $U^{\mathbb{F}}$ representing the martingale parts of $Y_t^{\mathbb{F}} = \mathbb{E}(Y_t^{\mathbb{G}} | \mathscr{F}_t)$ are given by*

- $Z_t^{\mathbb{F}} = \mathbb{E}\Big(Z_t^{\mathbb{G}} + \mu_t Y_{t-}^{\mathbb{G}}\Big|\mathscr{F}_t\Big)$,
- $U_t^{\mathbb{F}}(x) = \mathbb{E}\Big(U_t^{\mathbb{G}}(x)\kappa(x) + Y_{t-}^{\mathbb{G}}(\kappa(x)-1)|\mathscr{F}_t\Big)$.

*Proof* Note that any bounded $\mathscr{F}_T$-measurable random variable $F_T$ can be written as $F_T = \phi + \int_0^T \psi_t dW_t + \int_0^T \int_E \rho_t(x)\mathbf{M}(dt, dx)$ where $\psi \in L^2(\mathbb{F}, [0, T] \times \Omega)$ and $\rho \in L^2(\mathbb{F}, [0, T] \times \Omega, \mathbf{M})$ for any $x \in E$ are $\mathbb{F}$-predictable processes and $\phi$ a real constant. We define the bounded $\mathbb{F}$-martingale $J$ by $J_t := \int_0^t \psi_s dW_s + \int_0^t \int_E \rho_s(x)\mathbf{M}(ds, dx)$ for all $t \in [0, T]$. In the proof, all processes are considered on the time interval $[0, T]$. As in the previous section, we start with the equality $\mathbb{E}(Y_T^{\mathbb{F}} J_T) = \mathbb{E}(Y_T^{\mathbb{G}} J_T)$. We divide the proof of the Proposition into four steps:
- In a first step we compute $\mathbb{E}(Y_T^{\mathbb{F}} J_T)$. By integration by parts, we get

$$\mathbb{E}(Y_T^{\mathbb{F}} J_T) = \mathbb{E}\Big(\int_0^T Y_{t-}^{\mathbb{F}}\psi_t dW_t + \int_0^T \int_E Y_{t-}^{\mathbb{F}}\rho_t(x)\mathbf{M}(dt, dx) + \int_0^T J_{t-}dY_t^{\mathbb{F}}\Big)$$
$$+\mathbb{E}\Big(\int_0^T Z_t^{\mathbb{F}}\psi_t dt + \int_0^T \int_E U_t^{\mathbb{F}}(x)\rho_t(x)\mathbf{N}(dt, dx)\Big).$$

By using (4.3) we get

$$\mathbb{E}\Big(\int_0^T J_{t-}dY_t^{\mathbb{F}}\Big) = \mathbb{E}\Big(\int_0^T J_{t-}\widehat{f}_t dt + \int_0^T J_{t-}Z_t^{\mathbb{F}} dW_t + \int_0^T \int_E J_{t-}U_t^{\mathbb{F}}(x)\mathbf{M}(dt, dx)\Big),$$

where $\widehat{f}_t := \mathbb{E}(f(t, Y_t^{\mathbb{G}}, Z_t^{\mathbb{G}}, U_t^{\mathbb{G}})|\mathscr{F}_t)$ for all $t \in [0, T]$.

Since $J$ (resp. $Y^{\mathbb{F}}$) is bounded, the predictable process $J\_Z^{\mathbb{F}}$ (resp.$Y^{\mathbb{F}}\psi$) belongs to $L^2(\mathbb{F}, [0, T] \times \Omega)$ and the predictable process $J\_U^{\mathbb{F}}$ (resp. $Y^{\mathbb{F}}\rho$) belongs to $L^2(\mathbb{F}, [0, T] \times \Omega, \mathbf{M})$. Hence, the various stochastic integrals which appear in the two last equalities are martingales and

$$\mathbb{E}(Y_T^{\mathbb{F}} J_T) = \mathbb{E}\Big(\int_0^T J_{t-}\widehat{f}_t dt\Big) + \mathbb{E}\Big(\int_0^T Z_t^{\mathbb{F}}\psi_t dt + \int_0^T \int_E U_t^{\mathbb{F}}(x)\rho_t(x)\mathbf{N}(dt, dx)\Big).$$

Using that $\Big(\int_0^t \int_E U_s^{\mathbb{F}}(x)\rho_s(x)\mathbf{M}(ds, dx)\Big)_{t\in[0,T]}$ is an $\mathbb{F}$-martingale since the jump coefficient $U^{\mathbb{F}}$ is bounded, we obtain

$$\mathbb{E}(Y_T^{\mathbb{F}} J_T) = \mathbb{E}\Big(\int_0^T J_{t-}\widehat{f}_t dt\Big) + \mathbb{E}\Big(\int_0^T Z_t^{\mathbb{F}}\psi_t dt + \int_0^T \int_E U_t^{\mathbb{F}}(x)\rho_t(x)\nu(dt, dx)\Big).$$

- In a second step we compute $\mathbb{E}(Y_T^{\mathbb{G}} J_T)$. Recalling that $dW_t = dW_t^{\mathbb{G}} + \mu_t dt$ for all $t \in [0, T]$ and $\mathbf{M}(dt, dx) = \mathbf{M}^{\mathbb{G}}(dt, dx) + (\nu^{\mathbb{G}} - \nu)(dt, dx)$, and using integration by parts, we get

$$\mathbb{E}(Y_T^{\mathbb{G}} J_T) = \mathbb{E}\Big(\int_0^T Y_{t-}^{\mathbb{G}}\psi_t dW_t^{\mathbb{G}} + \int_0^T \int_E Y_{t-}^{\mathbb{G}}\rho_t(x)\mathbf{M}^{\mathbb{G}}(dt,dx)\Big)$$
$$+\mathbb{E}\Big(\int_0^T Y_{t-}^{\mathbb{G}}\mu_t\psi_t dt + \int_0^T \int_E Y_{t-}^{\mathbb{G}}\rho_t(x)(\nu^{\mathbb{G}}-\nu)(dt,dx)\Big)$$
$$+\mathbb{E}\Big(\int_0^T J_{t-}dY_t^{\mathbb{G}}\Big) + \mathbb{E}\Big(\int_0^T Z_t^{\mathbb{G}}\psi_t dt + \int_0^T \int_E U_t^{\mathbb{G}}(x)\rho_t(x)\mathbf{N}(dt,dx)\Big).$$

From the form of the BSDE satisfied by $Y^{\mathbb{G}}$, we obtain

$$\mathbb{E}\Big(\int_0^T J_{t-}dY_t^{\mathbb{G}}\Big) = \mathbb{E}\Big(\int_0^T J_{t-}f(t,Y_t^{\mathbb{G}},Z_t^{\mathbb{G}},U_t^{\mathbb{G}})dt + \int_0^T J_{t-}Z_t^{\mathbb{G}}dW_t^{\mathbb{G}}\Big)$$
$$+\mathbb{E}\Big(\int_0^T \int_E J_{t-}U_t^{\mathbb{G}}(x)\mathbf{M}^{\mathbb{G}}(dt,dx) + \int_0^T J_{t-}dM_t^{\perp}\Big).$$

Note that, as in the previous section $V := \int_0^T |Y\psi\mu|dt < \infty$ and, by difference $\mathbb{E}(\int_0^T Y\psi\mu dt)$ exists (we use again that $Y$ is bounded). By construction $Y_-^{\mathbb{G}}\phi$ and $J_-Z^{\mathbb{G}}$, (resp. $Y_-^{\mathbb{G}}\rho(x)$ and $(J_-U^{\mathbb{G}}(x))$ are $\mathbb{G}$-predictable and belong to $L^2(\mathbb{G},[0,T]\times\Omega)$ (resp. $L^2(\mathbb{G},[0,T]\times\Omega,\mathbf{M}^{\mathbb{G}})$). The various stochastic integrals are associated with martingales, therefore

$$\mathbb{E}(Y_T^{\mathbb{G}} J_T) = \mathbb{E}\Big(\int_0^T Y_{t-}^{\mathbb{G}}\psi_t\mu_t dt + \int_0^T \int_E Y_{t-}^{\mathbb{G}}\rho_t(x)(\nu^{\mathbb{G}}-\nu)(dt,dx)\Big)$$
$$+\mathbb{E}\Big(\int_0^T J_{t-}f_t dt + \int_0^T Z_t^{\mathbb{G}}\psi_t dt + \int_0^T \int_E U_t^{\mathbb{G}}(x)\rho_t(x)\mathbf{N}(dt,dx)\Big)$$
$$= \mathbb{E}\Big(\int_0^T Y_{t-}^{\mathbb{G}}\psi_t\mu_t dt + \int_0^T \int_E Y_{t-}^{\mathbb{G}}\rho_t(x)(\nu^{\mathbb{G}}-\nu)(dt,dx)\Big)$$
$$+\mathbb{E}\Big(\int_0^T J_{t-}f_t dt + \int_0^T Z_t^{\mathbb{G}}\psi_t dt + \int_0^T \int_E U_t^{\mathbb{G}}(x)\rho_t(x)\nu^{\mathbb{G}}(dt,dx)\Big),$$

where $f_t := f(t,Y_t^{\mathbb{G}},Z_t^{\mathbb{G}},U_t^{\mathbb{G}})$ for all $t\in[0,T]$.

• In a third step we show that $\mathbb{E}\Big(\int_0^T J_{t-}\widehat{f}_t dt\Big) = \mathbb{E}\Big(\int_0^T J_{t-}f_t dt\Big)$. By definition of $\widehat{f}$ and by Fubini's Theorem,

$$\mathbb{E}\Big(\int_0^T J_{t-}\widehat{f}_t dt\Big) = \int_0^T \mathbb{E}\Big(J_{t-}f(t,Y_t^{\mathbb{G}},Z_t^{\mathbb{G}},U_t^{\mathbb{G}})\Big)dt$$
$$= \mathbb{E}\Big(\int_0^T J_{t-}f_t dt\Big).$$

• The equality $\mathbb{E}(Y_T^{\mathbb{F}} J_T) = \mathbb{E}\big(Y_T^{\mathbb{G}} J_T\big)$ implies

$$\begin{aligned}\mathbb{E}\Big(\int_0^T Y_{t-}^{\mathbb{G}}\mu_t\psi_t dt + \int_0^T\int_E Y_{t-}^{\mathbb{G}}\rho_t(x)(\nu^{\mathbb{G}}-\nu)(dt,dx)\Big)\\ +\mathbb{E}\Big(\int_0^T Z_t^{\mathbb{G}}\psi_t dt + \int_0^T\int_E U_t^{\mathbb{G}}(x)\rho_t(x)\nu^{\mathbb{G}}(dt,dx)\Big)\\ =\mathbb{E}\Big(\int_0^T Z_t^{\mathbb{F}}\psi_t dt + \int_0^T\int_E U_t^{\mathbb{F}}(x)\rho_t(x)\nu(dt,dx)\Big).\end{aligned}$$

Using that $\lambda^{\mathbb{G}}(dx) = \kappa(x)\lambda(dx)$, we get

$$\begin{aligned}\mathbb{E}\Big(\int_0^T (Z_t^{\mathbb{G}} + \mu_t Y_{t-}^{\mathbb{G}})\psi_t dt\Big) + \mathbb{E}\Big(\int_0^T\int_E \big(U_t^{\mathbb{G}}(x)\kappa(x) + Y_{t-}^{\mathbb{G}}(\kappa(x)-1)\big)\rho_t(x)\lambda(dx)dt\Big)\\ = \mathbb{E}\Big(\int_0^T Z_t^{\mathbb{F}}\psi_t dt + \int_0^T\int_E U_t^{\mathbb{F}}(x)\rho_t(x)\lambda(dx)dt\Big).\end{aligned}$$

Finally, Fubini's Theorem and the tower property lead to

$$\begin{aligned}\mathbb{E}\Big(\int_0^T \big(Z_t^{\mathbb{F}} - \mathbb{E}(Z_t^{\mathbb{G}} + \mu_t Y_{t-}^{\mathbb{G}}|\mathscr{F}_t)\big)\psi_t dt\Big)\\ + \mathbb{E}\Big(\int_E\int_0^T \big(U_t^{\mathbb{F}}(x) - \mathbb{E}\big(U_t^{\mathbb{G}}(x)\kappa(x) + Y_{t-}^{\mathbb{G}}(\kappa(x)-1)\big|\mathscr{F}_t\big)\big)\rho_t(x)\lambda(dx)dt\Big) = 0\,.\end{aligned} \tag{4.4}$$

Equation (4.4) is true for any $\mathbb{F}$-predictable processes $(\psi,\rho)$ in $L^2(\mathbb{F},[0,T]\times\Omega)\times L^2(\mathbb{F},[0,T]\times\Omega,\mathbf{M})$, such that the associated process $J$ is bounded. The same raisonning as in the first section allows to extend the validity of (4.4) to any square integrable processes and we deduce

$$Z_t^{\mathbb{F}} = \mathbb{E}\big(Z_t^{\mathbb{G}} + \mu_t Y_{t-}^{\mathbb{G}}|\mathscr{F}_t\big)\,,\quad \forall t\in[0,T]\,,$$

and

$$U_t^{\mathbb{F}}(x) = \mathbb{E}\Big(U_t^{\mathbb{G}}(x)\kappa(x) + Y_{t-}^{\mathbb{G}}(\kappa(x)-1)\big|\mathscr{F}_t\Big)\,,\quad \forall t\in[0,T] \text{ and } \forall x\in E\,.$$

□

### 4.2 Particular Case of Linear BSDE

Let $\alpha$, $\beta$ and $\delta$ be bounded $\mathbb{F}$-adapted processes and $\gamma(x)$ a family of bounded $\mathbb{F}$-adapted processes and introduce the driver

$$f(t,y,z,u) = \alpha_t y + \beta_t z + \int_E \gamma_t(x)u(x)\lambda^{\mathbb{G}}(dx) + \delta_t\,.$$

We consider the two BSDEs

$$\begin{aligned} dY_t^{\mathbb{G}} &= f(t, Y_t^{\mathbb{G}}, Z_t^{\mathbb{G}}, U_t^{\mathbb{G}}(x))dt + Z_t^{\mathbb{G}} dW_t^{\mathbb{G}} + \int_{\mathbb{R}} U_t^{\mathbb{G}}(x)\mathbf{M}^{\mathbb{G}}(dt, dx) + dM_t^{\perp}, \\ Y_T^{\mathbb{G}} &= \zeta , \end{aligned} \tag{4.5}$$

and

$$\begin{aligned} dY_t &= f(t, Y_t, Z_t, U_t(x))dt + Z_t dW_t + \int_{\mathbb{E}} U_t(x)\mathbf{M}(dt, dx) , \\ Y_T &= \mathbb{E}(\zeta|\mathscr{F}_T) , \end{aligned} \tag{4.6}$$

where $\zeta \in \mathscr{G}_T$ is bounded.

**Proposition 4.3** *The solution of the BSDEs (4.5) and (4.6) are related by*

$$Y_t = Y_t^{\mathbb{F}} + \frac{1}{L_t}\mathbb{E}\Big(\int_t^T L_s Y_s^{\mathbb{G}}\big(\beta_s\mu_s - \int_{\mathbb{R}} \gamma_s(x)(\kappa(x) - 1)\lambda(dx)\big)ds\Big|\mathscr{F}_t\Big) ,$$

*where $Y_t^{\mathbb{F}} = \mathbb{E}(Y_t^{\mathbb{G}}|\mathscr{F}_t)$ and the process $L$ is the unique solution of the stochastic differential equation*

$$dL_t = L_{t-}(\alpha_t dt + \beta_t dW_t + \int_{\mathbb{R}} \gamma_t(x)\mathbf{M}(dt, dx))$$

*and $L_0 = 1$.*

*Proof* Let $t \in [0, T]$ be fixed. We recall that

$$Y_t^{\mathbb{F}} = \mathbb{E}(\xi|\mathscr{F}_T) + \int_t^T \widehat{f}_s ds - \int_t^T Z_s^{\mathbb{F}} dW_s - \int_t^T \int_{\mathbb{R}} U_s^{\mathbb{F}}(x)\mathbf{M}(ds, dx) ,$$

where

$$\begin{aligned} \widehat{f}_s &:= \mathbb{E}\big[\alpha_s Y_s^{\mathbb{G}} + \beta_s Z_s^{\mathbb{G}} + \int_{\mathbb{R}} \gamma_s(x) U_s^{\mathbb{G}}(x)\lambda^{\mathbb{G}}(dx) + \delta_s|\mathscr{F}_s\big] \\ &= \alpha_s\mathbb{E}[Y_s^{\mathbb{G}}|\mathscr{F}_s] + \beta_s\mathbb{E}[Z_s^{\mathbb{G}}|\mathscr{F}_s] + \int_E \gamma_s(x)\kappa(x)\mathbb{E}[U_s^{\mathbb{G}}(x)|\mathscr{F}_s]\lambda(dx) + \delta_s \end{aligned}$$

for all $s \in [0, T]$. Applying Proposition 4.2, we have that $\mathbb{E}(Z_s^{\mathbb{G}}|\mathscr{F}_s) = Z_s^{\mathbb{F}} - \mathbb{E}(\mu_s Y_{s-}^{\mathbb{G}}|\mathscr{F}_s)$ and $\mathbb{E}\Big(U_s^{\mathbb{G}}(x)\kappa(x)|\mathscr{F}_s\Big) = U_s^{\mathbb{F}}(x) - \mathbb{E}\Big(Y_{s-}^{\mathbb{G}}(\kappa(x) - 1)|\mathscr{F}_s\Big)$, for all $x \in E$ and $s \in [t, T]$, we obtain

$$\begin{aligned}\widehat{f}_s &= \alpha_s Y_s^{\mathbb{F}} + \beta_s Z_s^{\mathbb{F}} + \int_{\mathbb{R}} \gamma_s(x) U_s^{\mathbb{F}}(x)\lambda(dx) + \delta_s + \beta_s \mathbb{E}(\mu_s Y_{s-}^{\mathbb{G}} | \mathcal{F}_s) \\ &\quad - \mathbb{E}\Big[\int_{\mathbb{R}} \gamma_s(x) Y_{s-}^{\mathbb{G}}(\kappa(x)-1)\lambda(dx) \big| \mathcal{F}_s\Big], \quad \forall s \in [t,T]\,. \end{aligned} \tag{4.7}$$

Consider $\overline{Y} := Y^{\mathbb{F}} - Y$, $\overline{Z} := Z^{\mathbb{F}} - Z$ and $\overline{U} := U^{\mathbb{F}} - U$, then from (4.7) and (4.6)

$$\begin{aligned}\overline{Y}_t &= \int_t^T \Big(\alpha_s \overline{Y}_s + \beta_s \overline{Z}_s + \int_{\mathbb{R}} \gamma_s(x) \overline{U}_s(x)\lambda(dx) + \beta_s \mathbb{E}[\mu_s Y_{s-}^{\mathbb{G}} | \mathcal{F}_s] \\ &\quad - \gamma_s \mathbb{E}\Big[\int_{\mathbb{R}} Y_{s-}^{\mathbb{G}}(\kappa(x)-1)\lambda(dx) \big| \mathcal{F}_s\Big]\Big) ds - \int_t^T \overline{Z}_s dW_s - \int_t^T \int_{\mathbb{R}} \overline{U}_s(x) \mathbf{M}(ds,dx)\,,\end{aligned}$$

since this is a linear BSDE with terminal solution $\overline{Y}_T = 0$, the solution is unique and explicit (see [19]), given by

$$\overline{Y}_t = -\frac{1}{L_t}\mathbb{E}\Big[\int_t^T L_s\Big(-\beta_s \mathbb{E}[\mu_s Y_{s-}^{\mathbb{G}} | \mathcal{F}_s] + \mathbb{E}\Big[\int_{\mathbb{R}} \gamma_s(x) Y_{s-}^{\mathbb{G}}(\kappa(x)-1)\lambda(dx) \big| \mathcal{F}_s\Big]\Big) ds \big| \mathcal{F}_t\Big]\,.$$

□

## 5 Optimisation

In this section, we apply BSDEs methodology to solve some optimisation problem. This part is not related with the first part of the paper. Our goal is to compare the optimal portfolio in various filtrations.

We consider a financial market on the time interval $[0, T]$ composed by a riskless bond with an interest rate $r = 0$ and a risky asset. The price process $S$ of the risky asset, is assumed to be the unique solution of the following stochastic differential equation

$$dS_t = S_t(\alpha_t dt + \sigma_t dW_t)\,, \quad \forall t \in [0,T]\,, \quad S_0 = s > 0\,,$$

where $W$ is a Brownian motion with natural filtration $\mathbb{F}$, and $\alpha$ and $\sigma$ are $\mathbb{F}$-predictable processes such that $S_t = S_0 \exp(\int_0^t (\alpha_s - \frac{1}{2}\sigma_s^2) ds + \int_0^t \sigma_s^2 dW_s)$ is well defined on $[0, T]$. The risk premium is $\theta_t = \frac{\alpha_t}{\sigma_t}$.

The wealth $X^{x,\pi}$ associated with strategy $\pi$ is given by

$$X_0^{x,\pi} = x,\ dX_t^{x,\pi} = \pi_t \sigma_t (dW_t + \theta_t dt)\,.$$

For a given $\pi$, we denote by $X^{t,x,\pi}$ the solution of $X_t^{t,x,\pi} = x,\ dX_s^{t,x,\pi} = \pi_s \sigma_s (dW_s + \theta_s ds)$, for $s > t$. We consider a filtration $\mathbb{G}$ larger than $\mathbb{F}$ such that $dW_t = dW_t^{\mathbb{G}} +$

$\mu_t dt$. Denoting by $\mathscr{A}^{\mathbb{F}}[t,T]$ (resp. $\mathscr{A}^{\mathbb{G}}[t,T]$) the set of admissible $\mathbb{F}$ (resp. $\mathbb{G}$) strategies, we are interested in the two following problems

$$V_t^{\mathbb{F}}(x) := \sup_{\pi \in \mathscr{A}^{\mathbb{F}}[t,T]} \mathbb{E}\Big[U\big(X_T^{t,x,\pi}\big)\Big|\mathscr{F}_t\Big],$$

$$V_t^{\mathbb{G}}(x) := \sup_{\pi \in \mathscr{A}^{\mathbb{G}}[t,T]} \mathbb{E}\Big[U\big(X_T^{t,x,\pi}\big)\Big|\mathscr{G}_t\Big],$$

for an exponential utility function $U(x) = -e^{-\gamma x}$ with $\gamma > 0$. The following results are not new (see Delong [4, Chap. 11] and Hu et al. [9] for the proofs), our goal is only to solve the optimisation problem in various filtrations and to emphazise the integrability condition we shall require on the information drift. The case of power utility function can be solved using the same arguments and the results presented in Delong [4, Chap. 11] and Hu et al. [9].

**Comment 5.1** A study of indifference price, i.e., the study of $\sup_{\pi \in \mathscr{A}^{\mathbb{K}}[t,T]} \mathbb{E}\Big[U\big(X_T^{t,x,\pi} - \zeta\big)\Big|\mathscr{K}_t\Big]$, for $\mathbb{K} = \mathbb{F}$ or $\mathbb{K} = \mathbb{G}$ and $\zeta \in \mathscr{G}_T$ is done in Romo Romero [20].

## *5.1 Optimisation in the Reference Filtration* $\mathbb{F}$

In a first step, we recall some classical results for optimisation in the filtration $\mathbb{F}$. The value function $V^{\mathbb{F}}$ is given by $V_t^{\mathbb{F}}(x) = -\exp(-\gamma(x - Y_t^{\mathbb{F}}))$, where $(Y^{\mathbb{F}}, Z^{\mathbb{F}})$ is the solution of the following BSDE

$$\begin{cases} dY_t^{\mathbb{F}} &= \Big(\frac{\theta_t^2}{2\gamma} + \theta_t Z_t^{\mathbb{F}}\Big)dt + Z_t^{\mathbb{F}} dW_t\,, \\ Y_T^{\mathbb{F}} &= 0\,. \end{cases} \tag{5.1}$$

We assume that the process $\eta$ defined as

$$d\eta_t = -\eta_t \theta_t dW_t,\ \eta_0 = 1$$

is a martingale, and that $\mathbb{E}(\eta_T \int_0^T \theta_s^2 ds) < \infty$, in particular the market satisfies NFLVR (there exists a probability measure $\mathbb{Q}$ such that $S$ is a $\mathbb{Q}$-local martingale). We do not require that $\theta$ (the coefficient of $Z^{\mathbb{F}}$ in the drift part) is bounded, the above hypothesis on $\eta$ leads to the existence of solution of the BSDE (5.1), given by $Y_t^{\mathbb{F}}\eta_t = -\frac{1}{2\gamma}\mathbb{E}(\eta_T \int_t^T \theta_s^2 ds|\mathscr{F}_t)$. The optimal strategy associated to this problem is defined by

$$\pi_t^* := \frac{1}{\sigma_t}\Big(\frac{\theta_t}{\gamma} + Z_t^{\mathbb{F}}\Big)\,.$$

**Comment 5.2** See El Karoui and Huang [5] and Mastrolia [17] for conditions on $\theta$ for existence of a solution to (5.1).
(b) In the case where $\int_0^T \theta_s^2 ds = \infty$ with a positive probability, taking $\pi = \frac{\alpha}{\sigma^2}$, we obtain that $X_T^\pi = \exp(\frac{1}{2}\int_0^T \theta_s^2 ds + \int_0^T \theta_s dW_s)$ equals $+\infty$ with positive probability, which leads to an arbitrage, and the value function is null.
An open problem is to detect arbitrages from the no-existence of solution in adequate spaces.

## 5.2 *Initial Enlargement*

Here we set $\mathbb{G} = \mathbb{F}^{\sigma(L)}$. We assume that $L$ satisfies Jacod's absolute continuity assumption, so that $W^L$ enjoys PRP in $\mathbb{F}^{\sigma(L)}$ and

$$dS_t = S_t(\alpha_t + \sigma_t dW_t) = S_t((\alpha_t + \mu_t^L \sigma_t)dt + \sigma_t dW_t^L),$$

where

$$\mu_t^L = \frac{d\langle W, p(x)\rangle_t^{\mathbb{F}}|_{x=L}}{p_t(L)}.$$

The proof of the previous section applies, changing $\theta_t$ into $\theta_t^L = \frac{\alpha_t + \mu_t^L \sigma_t}{\sigma_t}$. In particular, square integrability conditions on $\mu^L$ are required (see Ankirchner et al. [2] for the study of such a condition).

## 5.3 *Progressive Enlargement*

We consider the case where $\mathbb{G} = \mathbb{F}^\tau$ is the progressive enlargement of $\mathbb{F}$ with $\tau$. We assume that
(a) the random time $\tau$ satisfies Jacod's absolute continuity assumption
(b) there exists an $\mathbb{F}$-adapted positive process such that $M_t := H_t - \int_0^{t\wedge\tau} \lambda_s ds$ is a $\mathbb{G}$-martingale where $H_t = \mathbb{1}_{\{\tau \le t\}}$
(c) the pair $(W^{(\tau)}, M)$ enjoys PRP in $\mathbb{F}^\tau$.

Then, we have

$$dS_t = S_t((\alpha_t + \mu_t^{(\tau)}\sigma_t)dt + \sigma_t dW_t^{(\tau)})$$

where

$$\mu_t^{(\tau)} = \mathbb{1}_{\{t<\tau\}} \frac{d\langle W, m\rangle_s^{\mathbb{F}}}{Z_{s-}} + \mathbb{1}_{\{\tau \le t\}} \frac{d\langle W, p(x)\rangle_s^{\mathbb{F}}|_{x=\tau}}{p_{s-}(\tau)}$$

where $m$ is the $\mathbb{F}$-martingale part of the supermartingale $Z$. We assume that the process $\eta^{(\tau)}$, solution of where $\theta_t^{(\tau)} = \frac{\alpha_t+\mu_t^{(\tau)}\sigma_t}{\sigma_t}$ is a martingale.

The value function $V_t^{(\tau)}(x)$ is given by

$$V_t^{(\tau)}(x) = -\exp\big(-\gamma(x - Y_t^{(\tau)})\big)\,,$$

where $(Y^{(\tau)}, Z^{(\tau)}, U^{(\tau)})$ is the solution of the following BSDE

$$\begin{cases} dY_t^{(\tau)} = & \Big(\frac{(\alpha_t+\mu_t^{(\tau)}\sigma_t)^2}{2\gamma\sigma_t^2} + \frac{\alpha_t+\mu_t^{(\tau)}\sigma_t}{\sigma_t} Z_t^{(\tau)} - \lambda_t(1-H_t)\frac{e^{\gamma U_t^{(\tau)}}-1}{\gamma}\Big)dt \\ & +Z_t^{(\tau)} dW_t^{(\tau)} + U_t^{(\tau)} dH_t, \\ Y_T^{(\tau)} = & 0. \end{cases}$$

Moreover, the optimal strategy associated to this problem is defined by

$$\pi_t^* := \frac{\alpha_t + \mu_t^{(\tau)}\sigma_t}{\gamma\sigma_t^2} + \frac{Z_t^{(\tau)}}{\sigma_t}\,.$$

Note that the optimal strategy does not depend explicitly of $U^{(\tau)}$. This is due to the fact that the dynamics of $S$ does not involve any jump part.

## References


1. Aksamit, A., Jeanblanc, M.: Enlargement of Filtration with Finance in View. Springer (2017)
2. Ankirchner, S., Dereich, S., Imkeller, P.: The Shannon information of filtrations and the additional logarithmic utility of insiders. Ann. Probab. **34**(2), 743–778 (2006)
3. Brémaud, P., Yor, M.: Changes of filtrations and of probability measures. Zeitschrift für Wahrscheinlichkeitstheorie und Verwandte Gebiete **45**(4), 269–295 (1978)
4. Delong, Ł.: Backward Stochastic Differential Equations with Jumps and Their Actuarial and Financial Applications: BSDEs with Jumps. EAA Series. Springer, London (2013)
5. El Karoui, N., Huang, S.: A general result of existence and uniqueness of backward stochastic differential equations. In: El Karoui, N., Mazliak, L. (eds.) Backward Stochastic Differential Equations. Pitman Research Notes in Mathematics Series 364, Longman, pp. 27–38 (1997)
6. El Karoui, N., Peng, S., Quenez, M.-C.: Backward stochastic differential equations in finance. Math. Financ. **7**(1), 1–71 (1997)
7. Eyraud-Loisel, A.: Backward stochastic differential equations with enlarged filtration: option hedging of an insider trader in a financial market with jumps. Stochas. Process. Appl. **115**(11), 1745–1763 (2005)
8. Fontana, C.: The strong predictable representation property in initially enlarged filtrations under the density hypothesis. Stochast. Process. Appl. **128**(3), 1007–1033 (2018)
9. Hu, Y., Imkeller, P., Müller, M.: Utility maximization in incomplete markets. Ann. Appl. Probab. **15**(3), 1691–1712 (2005)
10. Jacod, J.: Grossissement initial, hypothèse $(H')$ et théorème de Girsanov. In: Jeulin, T., Yor, M. (eds.) Grossissements de filtrations: exemples et applications. Lecture Notes in Mathematics, vol. 1118, pp. 15–35. Springer, Berlin, Heidelberg (1985)

11. Jeanblanc, M., Le Cam, Y.: Progressive enlargement of filtration with initial times. Stochast. Process. Appl. **119**, 2523–2543 (2009)
12. Jeanblanc, M., Song, S.: Martingale representation property in progressively enlarged filtrations. Stochast. Process. Appl. **125**, 4242–4271 (2015)
13. Jeulin, T.: Grossissement d'une filtration et applications. Séminaire de probabilités de Strasbourg **13**, 574–609 (1979)
14. Jeulin, T.: Semi-martingales et grossissement d'une filtration. In: Lecture Notes in Mathematics, vol. 833. Springer, Berlin (1980)
15. Jeulin, T., Yor, M.: Nouveaux résultats sur le grossissement des tribus. Annales scientifiques de l'École Normale Supérieure **11**(3), 429–443 (1978)
16. Kruse, T., Popier, A.: BSDEs with monotone generator driven by Brownian and Poisson noises in a general filtration. Stochastics **88**(4), 491–539 (2016)
17. Mastrolia, T.: Density analysis of non-markovian BSDEs and applications to biology and finance. Stochast. Process. Appl. **128**(3), 897–938 (2018)
18. Pardoux, É., Peng, S.: Adapted solution of a backward stochastic differential equation. Syst. Control Lett. **14**(1), 55–61 (1990)
19. Quenez, M.-C., Sulem, A.: Bsdes with jumps, optimization and applications to dynamic risk measures. Stochast. Process. Appl. **123**(8), 3328–3357 (2013)
20. Romo Romero, R.: Grossissement de filtrations et applications à la finance. Ph.D. Thesis, Evry University (2016)
21. Rouge, R., El Karoui, N.: Pricing via utility maximization and entropy. Math. Financ. **10**(2), 259–276 (2000)
22. Royer, M.: Backward stochastic differential equations with jumps and related non-linear expectations. Stochast. Process. Appl. **116**(10), 1358–1376 (2006)

# An Unbiased Itô Type Stochastic Representation for Transport PDEs: A Toy Example

**Gonçalo dos Reis and Greig Smith**

**Abstract** We propose a stochastic representation for a simple class of transport PDEs based on Itô representations. We detail an algorithm using an estimator stemming for the representation that, unlike regularization by noise estimators, is unbiased. We rely on recent developments on branching diffusions, regime switching processes and their representations of PDEs. There is a loose relation between our technique and regularization by noise, but contrary to the latter, we add a perturbation and immediately its correction. The method is only possible through a judicious choice of the diffusion coefficient $\sigma$. A key feature is that our approach does not rely on the smallness of $\sigma$, in fact, our $\sigma$ is strictly bounded from below which is in stark contrast with standard perturbation techniques. This is critical for extending this method to non-toy PDEs which have nonlinear terms in the first derivative where the usual perturbation technique breaks down. The examples presented show the algorithm outperforming alternative approaches. Moreover, the examples point toward a potential algorithm for the fully nonlinear case where the method of characteristics break down.

G. dos Reis acknowledges support from the *Fundação para a Ciência e a Tecnologia* (Portuguese Foundation for Science and Technology) through the project UID/MAT/00297/2013 (Centro de Matemática e Aplicações CMA/FCT/UNL, Portugal).
G. Smith was supported by The Maxwell Institute Graduate School in Analysis and its Applications, a Centre for Doctoral Training funded by the UK Engineering and Physical Sciences Research Council (grant EP/L016508/01), the Scottish Funding Council, the University of Edinburgh and Heriot-Watt University.

G. dos Reis (✉)
Maxwell Institute for Mathematical Sciences, School of Mathematics, University of Edinburgh, Edinburgh, UK
e-mail: G.dosReis@ed.ac.uk

Centro de Matemática e Aplicações CMA/FCT/UNL, Caparica, Portugal

G. Smith
Maxwell Institute for Mathematical Sciences, School of Mathematics,
University of Edinburgh, Edinburgh, UK
e-mail: G.Smith-13@sms.ed.ac.uk



S. N. Cohen et al. (eds.), *Frontiers in Stochastic Analysis - BSDEs, SPDEs and their Applications*, Springer Proceedings in Mathematics & Statistics 289,
https://doi.org/10.1007/978-3-030-22285-7_8

**Keywords** Monte Carlo methods · Regime switching diffusion · Probabilistic methods for PDEs

**MSC2010** Primary 65C05 · 65N75; Secondary 60J60

## 1 Introduction

Stochastic techniques to solve PDEs have become increasing popular in recent times with advances in computing power and numerical techniques allowing for solutions of PDEs to be calculated to high precision. Advances in BSDEs (Backward Stochastic Differential Equations) and so-called branching diffusions also allow one to tackle nonlinear PDEs (see [3] and references therein). Stochastic representations for PDEs are useful as they give access to probabilistic Monte Carlo methods, in turn yielding strong numerical gains over deterministic based solvers, especially in high dimensional problems, see [3, 8, 10]. Unlike their deterministic counterparts, stochastic based PDE solvers are less prone to the curse of dimensionality. In [3] the authors used hybrid Monte Carlo & PDE solvers to split the domain of the non-linear PDEs into multiple (independent) parts which allowed one to achieve perfect parallelization drastically reducing the time taken to numerically approximate such equations; a general discussion on such techniques is given there.

In this work we focus on transport PDEs. One of the main limitations when using Itô based stochastic techniques to represent PDEs is the requirement that the PDE is of second order in space (i.e. a "Laplacian" must be present). Thus PDEs with only one spatial and one time derivative (transport PDEs) have been, until now beyond the scope of stochastic techniques. An idea to navigate around this is to perturb the PDE by a "small" Laplacian, then one can use stochastic techniques on the perturbed PDE. Although this does provide a way to approximate the solution, it is very dependent on the perturbation being small enough so that the solution of the perturbed PDE is close to the first order PDE. Of course introducing a perturbation will lead to an error (bias) in the estimation, but more problematic is that the inverse of the perturbation coefficient will appear in the nonlinearities containing derivatives, thus the small perturbation makes the numerical scheme unstable. We discuss this point further in Sect. 5. Let us note that stochastic representations are only important for transport PDEs with nonlinearities in the derivative of the solution, see Remark 1.1.

A string of related literature based on numerical approximations via branching processes has re-emerged due to to recent developments. We do not carry out a review of these developments here but refer to [3] for a review on the state of the art. Branching algorithms offer a useful approach to solve non-linear PDEs and also for unbiased simulation of SDEs (see [6, 12]). However, in order to apply Monte Carlo methods one requires estimators to be square-integrable and of finite computational complexity. For square integrability several works have fine tuned previous results to allow for increasing general cases: [15] introduced a control variate on the final step, which allowed for an unbiased simulation of an SDE with constant diffusion; later,

[12] changed the time stepping scheme from an Exponential to a Gamma random variable, this allowed for the simulation of semilinear PDEs; most recently, [6] used antithetic variables as well as control variates to obtain an unbiased algorithm for an SDE with non constant diffusion.

The material we present requires all of the above mentioned improvements along with new ideas in order to ensure the estimator to be square-integrable. Taking the long view, we believe these techniques to be crucial in extending this type of stochastic representations to the fully non-linear case. The second order parabolic fully non-linear case has been considered in [18] and [19], but the theoretical basis for that case is to the best of our knowledge open. There are also several works looking at branching style algorithms but to tackle different types of PDEs, see [1, 5, 13] for further results.

The *contributions* of this paper are two-fold. Firstly we show how one can take the ideas of branching diffusions and regime switching to construct an unbiased stochastic representation for transport PDE. To the best of our knowledge this is the first result of its kind. Secondly, we improve upon the techniques currently presented in the literature [6, 12, 15] in order to show our representation is square integrable and of finite computational complexity and thus can be used in Monte Carlo simulation. For better readability we also provide a heuristic description of our ideas.

From a *methodological* point of view, the approach in this paper is related to the regime switching algorithms presented in [6, 15], where one adds and subtracts terms in the PDE to change the "driving SDE" defined by the Dynkin operator. Such algorithms were inspired by branching diffusion algorithms as developed in [17] and [11], although there is also a connection with parametrix approach (see [2]) where measure changes are used with corresponding weights to yield an unbiased representation. Here we add and subtract the second order derivative, which leaves us with a nonlinear PDE that can then be solved using regime switching (essentially we perturb the PDE then correct for the perturbation). Crucially this does not require $\sigma$ to be small. Although the transport PDE we consider is simple, one of the main challenges is to keep the representation square integrable, which comes from the added second order term. The general case will be addressed in future work, nonetheless we give numerical examples showing that the general case is within (numerical) reach.

*Remark 1.1* Basic first order PDEs can easily be made to have a stochastic like representation using branching type arguments, for example a PDE of the type,

$$\partial_t u(t,x) + b(t,x)\partial_x u(t,x) + u(t,x)^2 = 0, \quad u(T,x) = g(x).$$

It is possible to write the solution to this as,

$$u(t,x) = g(X_T) + \int_t^T u(s, X_s)^2 \mathrm{d}s,$$

where $X$ is the deterministic process satisfying the ODE $\mathrm{d}X_s = b(t, X_s)\mathrm{d}s$, $X_t = x$. Introducing random times into the solution of $u$ as is done in standard branching we can obtain a solution to $u$ as the expected product of particles at time $T$. A similar argument can also be made for nonlinear ODEs.

What is crucial here though is that this argument only holds when we do not have nonlinearities in the first derivative of the process, since we require Malliavin integration by parts tricks to deal with those. This is also the case when we want to apply the unbiased trick to $b$.

This work is organized as follows. In Sect. 2 we present our notation, the problem and give a heuristic description of our ideas. In Sect. 3 we present and prove our main results. Following that in Sect. 4 we discuss the open problems left by this work. Finally Sect. 5 illustrates numerically our findings to show our method is indeed unbiased. Moreover, we show the capability of our method to tackle problems in the nonlinear setting where the perturbation technique performs poorly.

# 2 Regime Switching Diffusion Representation in General

## 2.1 Notation and Recap of Stochastic Representations

Following the standard notation in stochastic analysis let $C_b^{1,n}([0,T]\times\mathbb{R}^d,\mathbb{R})$ be the set of functions $v:[0,T]\times\mathbb{R}^d\to\mathbb{R}$ with one bounded time derivative and order $n$ bounded spatial derivatives. Further, let $d\geq 1$ and $W$ be a $d$-dimensional Brownian motion, defined on the probability space $(\Omega,\mathcal{F},\mathbb{P},(\mathcal{F}_t)_{t\geq 0})$, with $\mathcal{F}_t$ the filtration of a multidimensional Brownian motion augmented with the null sets (satisfying the usual conditions).

Consider a multidimensional stochastic differential equation (SDE) $X$ starting at time point $t$, $0\leq t\leq T$ of the form,

$$\mathrm{d}X_s=b(s,X_s)\mathrm{d}s+\sigma(s,X_s)\mathrm{d}W_s\,,\quad \text{for } s\in[t,T] \text{ and } X_t=x\,,$$

where the drift $b:[0,T]\times\mathbb{R}^d\to\mathbb{R}^d$ and diffusion $\sigma:[0,T]\times\mathbb{R}^d\to\mathbb{S}^d$ satisfy the usual Lipschitz conditions so that the above SDE has a unique strong solution, $\mathbb{S}^d$ denotes the set of $d$-by-$d$ dimensional real valued matrices.

We associate with the SDE the infinitesimal generator $\mathcal{L}$, which when applied to any function $\phi\in C_b^{1,2}([0,T]\times\mathbb{R}^d,\mathbb{R})$ in the domain of $\mathcal{L}$ is,

$$(\mathcal{L}\phi)(t,x)=b(t,x)\cdot D\phi(t,x)+\frac{1}{2}a(t,x):D^2\phi(t,x)\,,\quad \text{for all } (t,x)\in[0,T]\times\mathbb{R}^d\,,$$

where we define $a(t,x)=\sigma(t,x)\sigma(t,x)^\intercal$, $A:B:=\mathrm{trace}(AB^\intercal)$, $^\intercal$ is the transpose of a matrix and $D$, $D^2$ denotes the usual multi-dimensional spatial differential operators of order one and two (see [7]).

It well known by the Feynman-Kac formula that if a unique classical solution $v\in C_b^{1,2}$ exists to the following PDE,

$$\begin{cases} \partial_t v(t,x) + \mathcal{L}v(t,x) = 0\,, \\ v(T,x) = g(x)\,, \end{cases}$$

for $g$ a Lipschitz continuous function, then the solution of this PDE admits a stochastic representation, $v(t,x) = \mathbb{E}[g(X_T)|X_t = x]$. Further, by the use of branching diffusions (see [12]) or BSDEs (see [4]), one is able to obtain a stochastic representation for semi-linear PDEs of the form

$$\begin{cases} \partial_t v(t,x) + \mathcal{L}v(t,x) = f(t,x,v,Dv)\,, \\ v(T,x) = g(x)\,, \end{cases}$$

for $f$ and $g$ nice enough.

## 2.2 *Heuristic Derivation of the Idea of Our Work*

Much of the current literature on branching diffusions and regime switching is technical and complex, to aid the presentation of this paper we give an introductory outline of our work. The ultimate goal of our paper is to construct a stochastic representation of PDEs with only first order spatial derivatives and develop a way to deal with the corresponding 2nd order nonlinearity. We consider PDEs of the form

$$\begin{cases} \partial_t v(t,x) + b(t,x)\cdot Dv(t,x) = 0\,, \\ v(T,x) = g(x)\,, \end{cases} \tag{2.1}$$

for notational convenience we will work in one spatial dimension here (hence $D = \partial_x$). The problem with constructing a stochastic representation involving the use of Itô's formula is that we automatically obtain a second order derivative. However, it is known that arguments from branching diffusion can be used to deal with higher order derivatives through the Bismut-Elworthy-Li formula (automatic differentiation as developed in [9]). Let us assume that $v$ solving (2.1) is a unique classical solution which is $C_b^{1,2}$ (i.e. we can apply Itô's formula to $v$), then we can consider the following equivalent PDE

$$\begin{cases} \partial_t v(t,x) + b(t,x)\partial_x v(t,x) + \frac{1}{2}\sigma_0^2\partial_{xx}v(t,x) - \frac{1}{2}\sigma_0^2\partial_{xx}v(t,x) = 0\,, \\ v(T,x) = g(x)\,, \end{cases}$$

where $\sigma_0$ is some constant. In fact, as considered in [15], we can consider the equivalent PDE,

$$\begin{cases} \partial_t v(t,x) + b_0 \partial_x v(t,x) + \frac{1}{2}\sigma_0^2 \partial_{xx} v(t,x) + \big(b(t,x) - b_0\big)\partial_x v(t,x) - \frac{1}{2}\sigma_0^2 \partial_{xx} v(t,x) = 0\,, \\ v(T,x) = g(x)\,, \end{cases} \tag{2.2}$$

where $b_0$ is also some constant.

**Stochastic Representation**. Using the Feynman-Kac formula one can easily obtain the following stochastic representation of the solution to (2.2),

$$v(t,x) = \mathbb{E}\Big[g(\bar{X}_T) + \int_t^T \Big(\big(b(s,\bar{X}_s) - b_0\big)\partial_x v(s,\bar{X}_s) - \frac{1}{2}\sigma_0^2 \partial_{xx} v(s,\bar{X}_s)\Big)\mathrm{d}s \,\Big|\, \bar{X}_t = x\Big]\,, \tag{2.3}$$

where the driving SDE satisfies

$$\mathrm{d}\bar{X}_s = b_0 \mathrm{d}s + \sigma_0 \mathrm{d}W_s\,, \quad \bar{X}_t = x \quad s \in [t,T]. \tag{2.4}$$

One can observe that such a representation holds provided our constants are $\mathcal{F}_t$ measurable.

**Introduce a New Random Variable**. Following a standard branching diffusion style argument, alongside the Brownian motion, $W$, we also consider an independent random variable $\tau$ with density $f > 0$ on $[0, T - t + \epsilon]$ for $\epsilon > 0$ and denote by $\overline{F}$ the corresponding survival function, namely for $s \in \mathbb{R}_+$ $\overline{F}(s) := \int_s^\infty f(r)\mathrm{d}r$. Consider some nice functions $\psi$ and $\phi$, then following representation holds

$$\begin{aligned} \psi(T) + \int_t^T \phi(s)\mathrm{d}s &= \frac{\psi(T)\overline{F}(T-t)}{\overline{F}(T-t)} + \int_t^T \frac{1}{f(s-t)}\phi(s) f(s-t)\mathrm{d}s \\ &= \mathbb{E}_f\Big[\mathbb{1}_{\{\tau \geq T-t\}} \frac{\psi(T)}{\overline{F}(T-t)} + \mathbb{1}_{\{\tau < T-t\}} \frac{1}{f(\tau)}\phi(t+\tau)\Big]\,, \end{aligned}$$

where $\mathbb{E}_f$ denotes the expectation for the random variable $\tau$.

**Rewriting the Stochastic Representation** (2.3). Applying this to the Feynman-Kac representation (2.3) yields,

$$\begin{aligned} v(t,x) =& \mathbb{E}\Big[\frac{g(\bar{X}_T)}{\overline{F}(T-t)}\mathbb{1}_{\{t+\tau \geq T\}} + \mathbb{1}_{\{t+\tau < T\}} \frac{1}{f(\tau)}\Big[-\frac{1}{2}\sigma_0^2 \partial_{xx} v(t+\tau, \bar{X}_{t+\tau}) \\ &+ \big(b(t+\tau, \bar{X}_{t+\tau}) - b_0\big)\partial_x v(t+\tau, \bar{X}_{t+\tau})\Big] \,\Big|\, \bar{X}_t = x\Big]. \end{aligned} \tag{2.5}$$

One may note the abuse of notation here, the original Feynman-Kac representation expectation was only w.r.t. the Brownian motion, while (2.5) is w.r.t. both $\tau$ and the Brownian motion. To make the notation easier we now introduce the following stochastic sequence of times (stochastic mesh on the interval $[t,T]$), $t =: T_0 < T_1 < \cdots < T_{N_T} < T_{N_T+1} := T$ constructed as follows, take a sequence of i.i.d. copies of $\tau$, then set $T_{k+1} = (T_k + \tau^{(k)}) \wedge T$ for $k \in \Lambda \subset \mathbb{N}$, where $\Lambda$ is the set of integers (of stochastic length) $\{1, \ldots, N_T + 1\}$. Using this mesh we then define $\Delta T_{k+1} = T_{k+1} - T_k$ and $\Delta W_{T_{k+1}} = W_{T_{k+1}} - W_{T_k}$.

**Choosing the SDE's coefficients**. Let us now consider a good choice of constant for $b_0$ (we define $\sigma_0$ later). As discussed in [6, 15], one can use the so called *frozen coefficient* function which defines the Euler scheme. That is, we may define the SDE $\bar{X}$ recursively over the random mesh by

$$\bar{X}_{T_k} = \bar{X}_{T_{k-1}} + b(T_{k-1}, \bar{X}_{T_{k-1}})\Delta T_k + \sigma_{k-1}\Delta W_{T_k}, \quad \bar{X}_0 = x, \tag{2.6}$$

for $k \in \Lambda$. Define $\theta_{k-1}$ as the times in the mesh and position of the SDE up to time $T_{k-1}$ i.e. $\theta_{k-1} := (T_1, \dots, T_{k-1}, x, \bar{X}_{T_1}, \dots, \bar{X}_{T_{k-1}})$. Furthermore define the functions $\bar{b}(\theta_{k-1}, s, \bar{X}_s) = b(T_{k-1}, \bar{X}_{T_{k-1}})$ and $\sigma(\theta_{k-1}, s) = \sigma_{k-1}$ for $T_{k-1} < s$. Then the SDE defined recursively by,

$$\bar{X}_{T_k} = \bar{X}_{T_{k-1}} + \int_{T_{k-1}}^{T_k} \bar{b}(\theta_{k-1}, s, \bar{X}_s)\mathrm{d}s + \int_{T_{k-1}}^{T_k} \sigma(\theta_{k-1}, s)\mathrm{d}W_s, \tag{2.7}$$

is the Euler scheme in (2.6). Moreover, it is clear that the coefficients $\bar{b}(\theta_k, \cdot)$ and $\sigma(\theta_k, \cdot)$ are $\mathcal{F}_{T_k}$-adapted, hence can be used in (2.5). Using the coefficients coming from the Euler scheme is key here since we can simulate an Euler scheme exactly and hence the SDE appearing in (2.5) can be simulated exactly (which leads to the unbiased representation).

*Remark 2.1* We draw attention to a subtlety in the notation, we will define $\sigma$ on intervals of the form $(\cdot, \cdot]$, thus $\sigma$ is constant over each interval in the time mesh (as is the case in the Euler scheme).

**Obtaining a Representation for the Derivatives**. The only terms left to consider in (2.5) are the derivatives of $v$. We will formulate rigorous results in Sect. 3, for now let us assume that all functions are sufficiently smooth and with good properties. We construct the Bismut-Elworthy-Li formula (automatic differentiation) w.r.t. the SDE (2.7). From [9, Assumption 3.1] the following integration by parts relation holds for any square integrable function $\phi$,

$$\partial_x \mathbb{E}[\phi(X_s)|X_t = x] = \mathbb{E}\left[\phi(X_s)\int_t^s \sigma(u)^{-1}Y(u)\mu(u)\mathrm{d}W_u \,\Big|\, X_t = x\right],$$

where $Y$ is the first variation process of the SDE $X$ and $\mu$ is any function such that $\int_t^s \mu(u)\mathrm{d}u = 1$. In the case of the SDE being (2.7), it is clear that the first variation process is constant equal to one (note $\sigma$ does not have a space dependence). Typically one takes constant $\mu = 1/(s-t)$, thus for (2.7) we obtain,

$$\partial_x \mathbb{E}[\phi(X_{T_1})|X_t = x] = \mathbb{E}\left[\phi(X_{T_1})\frac{1}{\Delta T_1}\int_t^{T_1} \sigma(\theta_0, u)^{-1}\mathrm{d}W_u \,\Big|\, X_t = x\right],$$

The same method yields a similar expression for the second derivative

$$\partial_{xx}\mathbb{E}[\phi(X_{T_1})|X_t=x]=\mathbb{E}\left[\frac{\phi(X_{T_1})}{\Delta T_1^2}\left(\left(\int_t^{T_1}\sigma(\theta_0,u)^{-1}\mathrm{d}W_u\right)^2-\int_t^{T_1}(\sigma(\theta_0,u)^{-1})^2\mathrm{d}u\right)\Big|\,X_t=x\right],$$

From this result and using the fact that $\sigma$ is constant between mesh points we obtain for the second derivative

$$\begin{aligned}\partial_{xx}v(t,x)=&\mathbb{E}\Big[\frac{g(X_T)}{\bar{F}(\Delta T_1)}\mathbb{1}_{\{T_1\geq T\}}\frac{1}{\Delta T_1^2}\left(\left(\sigma(\theta_0,T_1)^{-1}\Delta W_{T_1}\right)^2-(\sigma(\theta_0,T_1)^{-1})^2\Delta T_1\right)\\&+\frac{\mathbb{1}_{\{T_1<T\}}}{f(\Delta T_1)}\left((b(T_1,\bar{X}_{T_1})-\bar{b}(\theta_0,T_1,\bar{X}_{T_1}))\partial_x v(T_2,\bar{X}_{T_2})-\frac{1}{2}\sigma(\theta_0,T_1)^2\partial_{xx}v(T_1,X_{T_1})\right)\\&\times\frac{(\sigma(\theta_0,T_1)^{-1}\Delta W_{T_1})^2-(\sigma(\theta_0,T_1)^{-1})^2\Delta T_1}{\Delta T_1^2}\,\Big|\,X_t=x\Big],\end{aligned}$$

the $\partial_x v$ term is similar. The idea of branching diffusion style algorithms is to continuously substitute in terms involving the solution until we remove the dependence on it. Of course, $v(t,x)$ does not appear inside the expectation, however, by using the tower property and flow property of the SDE we are able to derive the corresponding representations for $\partial_x v(T_k,X_{T_k})$ and $\partial_{xx}v(T_k,X_{T_k})$.

**Rewriting the Stochastic Representation** (2.5). Substituting in the expressions for $\partial_x v(T_1,X_{T_1})$ and $\partial_{xx}v(T_1,X_{T_1})$ into (2.5) yields,

$$\begin{aligned}&v(t,x)\\&=\mathbb{E}\Big[\frac{g(\bar{X}_T)}{\bar{F}(\Delta T_1)}\mathbb{1}_{\{T_1\geq T\}}+\mathbb{1}_{\{T_1<T\}}\frac{1}{f(\Delta T_1)}\mathbb{E}\Big[\overline{\mathcal{W}}_2\Big\{\frac{g(\bar{X}_T)}{\bar{F}(\Delta T_2)}\mathbb{1}_{\{T_2\geq T\}}+\mathbb{1}_{\{T_2<T\}}\frac{1}{f(\Delta T_2)}\\&\times\Big[(b(T_2,\bar{X}_{T_2})-\bar{b}(\theta_1,T_2,\bar{X}_{T_2}))\partial_x v(T_2,\bar{X}_{T_2})-\frac{1}{2}\sigma(\theta_1,T_2)^2\partial_{xx}v(T_2,\bar{X}_{T_2})\Big]\Big\}\Big|\,\bar{X}_{T_1}\Big]\Big|\,\bar{X}_t=x\Big],\end{aligned}$$

where $\overline{\mathcal{W}}_k$ is the so-called Malliavin weight stemming from the automatic differentiation,

$$\overline{\mathcal{W}}_k:=\frac{b(T_{k-1},\bar{X}_{T_{k-1}})-\bar{b}(\theta_{k-2},T_{k-1},\bar{X}_{T_{k-1}})}{\sigma(\theta_{k-1},T_k)}\frac{\Delta W_{T_k}}{\Delta T_k}-\frac{1}{2}\frac{\sigma(\theta_{k-2},T_{k-1})^2}{\sigma(\theta_{k-1},T_k)^2}\left(\frac{\Delta W_{T_k}^2-\Delta T_k}{\Delta T_k^2}\right).$$

One observes that this Feynman-Kac representation now only depends on the solution $v$ if $T_2<T$.

**Taking the Limit**. Following the standard procedure in branching diffusions (see [11, 12, 14]), executing the same argument multiple times removes the dependence on $v$ on the right hand side. Following [6] we introduce the following notation,

$$M_{k+1}=\Delta b_k\sigma(\theta_k,T_{k+1})^{-1}\frac{\Delta W_{T_{k+1}}}{\Delta T_{k+1}}\quad\text{and}\quad V_{k+1}=-\frac{1}{2}\frac{\sigma(\theta_{k-1},T_k)^2}{\sigma(\theta_k,T_{k+1})^2}\left(\frac{\Delta W_{T_{k+1}}^2-\Delta T_{k+1}}{\Delta T_{k+1}^2}\right).\tag{2.8}$$

where $\Delta b_k=b(T_k,\bar{X}_{T_k})-\bar{b}(\theta_{k-1},T_k,\bar{X}_{T_k})=b(T_k,\bar{X}_{T_k})-b(T_{k-1},\bar{X}_{T_{k-1}})$. Further define the terms

$$P_{k+1} := \frac{M_{k+1} + \frac{1}{2} V_{k+1}}{f(\Delta T_k)} \quad \text{for } k \in \Lambda. \tag{2.9}$$

It is then clear that the solution to the PDE can be written as follows,

$$v(t,x) = \mathbb{E}\left[ \frac{g(\bar{X}_{T_{N_T+1}})}{\bar{F}(\Delta T_{N_T+1})} \prod_{k=2}^{N_T+1} P_k \,\middle|\, \bar{X}_t = x \right]. \tag{2.10}$$

Although this relation is useful for us, in its current form it is not square integrable, thus we need to use some variance reduction techniques in order to use Monte Carlo. Moreover, many of the operations above require some form of integrability, these points will be the main focus of the next section.

## 3 Stochastic Representation for a Toy Transport PDE

The goal of the paper is to derive a square-integrable representation that solves a PDE of the form,

$$\begin{cases} \partial_t v(t,x) + b(t) \cdot Dv(t,x) = 0 \quad \text{for all } (t,x) \in [0,T) \times \mathbb{R}^d, \\ v(T,x) = g(x). \end{cases} \tag{3.1}$$

*Remark 3.1* We show the representation in the case $b : [0,T] \to \mathbb{R}$ is independent of space. This ensures finite variance, we shall return to the case of space dependency later.

We wish to consider SDEs of the form (2.7), in $d$-dimensions this is,

$$\mathrm{d}\bar{X}_s = \bar{b}(\theta, s)\mathrm{d}s + \sigma(\theta, s)\mathbb{I}_d \mathrm{d}W_s, \quad \text{for } s \in [t,T] \text{ and } \bar{X}_t = x,$$

where $\mathbb{I}_d$ is the $d$-dimensional identity matrix. Unlike typical stochastic representations, $\sigma$ is not fixed by the PDE, thus we have the freedom to choose $\sigma$. Although, the representation is somewhat independent of the precise choice of $\sigma$, the variance of the estimate (and hence the usefulness) heavily depends on $\sigma$.

In order to keep our representation and in particular our proofs as readable as possible, we consider only the one dimensional case. As one can clearly see though, due to fact that $\sigma$ is a scalar multiplied by the identity, all our arguments generalise to the higher dimensional case. Of course as $\sigma$ is not fixed in this case, it may be that other representations especially in high dimension may yield superior results. However, our goal here is purely to obtain a representation with finite variance.

The previous section outlined how one builds the stochastic representation without going into detail about when the various steps are applicable. We now want to show that this representation holds under some integrability and regularity assumptions. In the previous section we required two types of random variable, namely a driving

Brownian motion and an i.i.d. sequence of random times $\tau^{(k)}$ with density $f$, independent of the Brownian motion and $k \in \Lambda$ as before. Thus consider the probability space $(\Omega, \mathcal{F}, \mathbb{P})$ generated by these random variables, we also denote by $\mathbb{P}_W$ and $\mathbb{P}_f$ the probability measure ($\mathbb{E}_W$ and $\mathbb{E}_f$ the corresponding expectation) restricted to the Brownian motion and random times respectively. With this notation, one may think of $\mathbb{P}$ as the product measure $\mathbb{P}_W \otimes \mathbb{P}_f$. The corresponding filtration $\mathcal{F}_t$ is the sigma-algebra generated by the set of random times up to $t$ i.e. $\max\{k : T_k \leq t\}$ and the Brownian motion up to $t$, hence, $\mathcal{F}_t := \sigma(T_1, \ldots, T_k, (W_s)_{s \leq t})$.

Let us first state the assumptions we will use.

**Assumption 3.2** We assume the drift, $b$ is uniformly Lipschitz in time.

The analysis we carry out using regime switching techniques is sufficiently difficult to present that we assume the existence of a good enough solution to the transport PDE, as opposed to assuming sufficient conditions that would allow us to derive the said solution. Waiving the next assumption is left for future work.

**Assumption 3.3** Firstly we assume that there exists a unique solution $v \in C_b^{1,3}$ $([0, T], \mathbb{R}^d)$ to (3.1). In particular, we have that the terminal condition function $g$ of the PDE satisfies $g \in C_b^2$.

The assumption on $g$ is not necessary since it follows from $v \in C_b^{1,3}$, however, we make this explicit since it is all we require for our estimator to be of finite variance. It is possible to put some conditions on $b$ and $g$ leading to a unique solution for general transport PDEs see [16] for example. We do not go into detail here as this will again be the subject of future work.

We consider the particles to have a life time given by Gamma distributed random variables, i.e. $\tau$ has density,

$$f(s) := f_\Gamma^{\kappa,\eta}(s) = \frac{s^{\kappa-1}\exp(-s/\eta)}{\Gamma(\kappa)\eta^\kappa}, \quad \text{for all } s > 0 \text{ where } \kappa,\ \eta > 0, \tag{3.2}$$

where $\Gamma$ is the Euler function $\Gamma(y) = \int_0^\infty x^{y-1}\exp(-x)\mathrm{d}x$.

We will use a *mesh dependent* coefficient for $\sigma$ relying on the times at which the regime switching occurs,

$$\sigma(\theta_{k-1}, s) := \sigma_0 \prod_{i=1}^{k-1} \Delta T_i^n \quad \text{for } s \in (T_{k-1}, T_k]\,,\ k = 1, \ldots, N_T + 1\,,\ \ n \in \mathbb{R} \text{ and } \sigma_0 \in \mathbb{R}_+\,, \tag{3.3}$$

hence $\sigma(\theta_{k-1}, T_k) = \sigma_0 \prod_{i=1}^{k-1} \Delta T_i^n$, with the convention $\prod_{i=1}^{0} \cdot = 1$.

*Remark 3.4* (Adaptedness of $\sigma$) Even though our $\sigma$ depends on the stochastic mesh, it is $\mathcal{F}_t$-adapted. This is of fundamental importance to show that the estimator in (3.5) solves the PDE (3.1).

We make an assumption on the parameters of $\sigma$ and $f$.

**Assumption 3.5** The power exponent $n$ in the diffusion coefficient (3.3) satisfies $n \leq -1$. The shape parameter of the Gamma random variable, (3.2), is $\kappa = 1/2$.

*Remark 3.6* Under Assumption 3.5, $\sigma$ is a positive function bounded from below away from zero.[1] The bounds on $n$ and $\kappa$ are mainly for convenience in order for the proof of Proposition 3.8 to follow.

As was alluded to in Sect. 2, (2.10) was not useful since it did not have finite second moment. To solve this problem we employ variance reduction techniques, namely antithetic variables and control variates. Consider the following auxiliary random variables, $\beta := (\beta_1 + \beta_2)/2$ with

$$\begin{cases} \beta_1 := \dfrac{g(\bar{X}_{T_{N_T+1}}) - g(\bar{X}_{T_{N_T}} + b(T_{N_T})\Delta T_{N_T+1})}{\overline{F}(\Delta T_{N_T+1})} \dfrac{M_{N_T+1} + \frac{1}{2}V_{N_T+1}}{f(\Delta T_{N_T})}, \\ \beta_2 := \dfrac{g(\hat{X}_{T_{N_T+1}}) - g(\bar{X}_{T_{N_T}} + b(T_{N_T})\Delta T_{N_T+1})}{\overline{F}(\Delta T_{N_T+1})} \dfrac{-M_{N_T+1} + \frac{1}{2}V_{N_T+1}}{f(\Delta T_{N_T})}, \end{cases} \tag{3.4}$$

where $\hat{X}$ is the antithetic random variable associated to $\bar{X}$ i.e. the Euler scheme defined by, $\hat{X}_{T_k} = \bar{X}_{T_{k-1}} + b(T_{k-1})\Delta T_k - \sigma(\theta_{k-1}, T_k)\Delta W_{T_k}$ and $V$ and $M$ as defined in (2.8). It is straightforward to see that the additional $g$ term is a control variate since its input is independent of Brownian motion $\Delta W_{T_{N_T+1}}$. One can further understand $(\beta_1, \beta_2)$ as an antithetic pair.

We now state our main result of the paper.

**Theorem 3.7** (Representation Solves the PDE) *Let Assumptions 3.2, 3.3 and 3.5 hold, and let us denote by* $\hat{v} : [0, T] \times \mathbb{R} \to \mathbb{R}$ *the following function,*

$$\hat{v}(t, x) := \mathbb{E}\left[\beta \prod_{k=2}^{N_T} P_k \mathbb{1}_{\{N_T \geq 1\}} \,\middle|\, \sigma(\theta_0, t),\ X_t = x\right] + \mathbb{E}\left[\frac{g(\bar{X}_{T_1})}{\overline{F}(\Delta T_1)} \mathbb{1}_{\{N_T = 0\}} \,\middle|\, \sigma(\theta_0, t),\ X_t = x\right], \tag{3.5}$$

*with* $\{P_k\}_k$ *as defined in* (2.9). *Then* $\hat{v}$ *solves the PDE* (3.1), *namely* $\hat{v} = v$ *(hence* $\hat{v}$ *is an unbiased estimator of* $v$*). Moreover, the stochastic process generating* $\hat{v}$ *is square integrable and hence of finite variance.*

**Outline of proof** The proof of Theorem 3.7 requires several steps which we show in the following order.

1. Take $\hat{v}$ in (3.5), which is the expected value of a stochastic process (estimator).
2. Show that the estimator is square integrable, Proposition 3.8.
3. Show that under enough integrability a stochastic representation to (3.1) exists when a solution in $C_b^{1,3}([0, T], \mathbb{R})$ exists, Theorem 3.10.
4. Show that (3.5), satisfies the integrability conditions in Theorem 3.10 and thus solves (3.1), Theorem 3.13.

[1] To see this, note that $n < 0$, hence for $\sigma$ to be zero, we require a set of $\Delta T_k \geq 0$ for $k = 1, \ldots, N_{T+1}$, such that $\sum_{k=1}^{N_T+1} \Delta T_k = T$ and $\prod_{k=1}^{N_T+1} \Delta T_k = \infty$. Which clearly does not exist.

## 3.1 Variance Analysis for a Specific Diffusion Coefficient

Since our regime switching algorithm does not create new particles, our computational complexity for any Monte Carlo realisation is only $O(C(N_T+1))$, since $T<\infty$, it is clear we have finite computational complexity. We therefore only need to consider the variance of the estimator. We obtain the following.

**Proposition 3.8** *Let Assumptions 3.2, 3.3 and 3.5 hold. Then the random variable appearing in* (3.5),

$$\beta\prod_{k=2}^{N_T} P_k \mathbb{1}_{\{N_T\geq 1\}} + \frac{g(\bar{X}_{T_1})}{\overline{F}(\Delta T_1)}\mathbb{1}_{\{N_T=0\}}$$

*has finite variance.*

Although this proof is argued in a similar style to the proof of Proposition 4.1 in [6], there are many subtle differences and we overall require a more refined analysis of the various terms to ensure our estimator has finite second moment. We point in particular to the "Interval splitting" argument in order to deal with instability in the last time point of the random mesh. This is essential to deal with the second order term that appears.

*Proof* (*Finite variance of the estimator*) Consider $\overline{\mathcal{F}}_k$ the sigma-algebra generated by the set of random times up to $T_{k+1}$ and the Brownian motion up to $T_k$, hence,[2] $\overline{\mathcal{F}}_k := \sigma(T_1,\ldots,T_{k+1},(W_s)_{s\leq T\wedge T_k})$.

Throughout the proof, for ease of writing we suppress the condition in the expectation of the process starting at $x$ at time $t$.

In order to show finite variance we only need to show finite second moment (the dominant term), further note that due to the indicators we obtain no cross term. Looking first at the second term of (3.5), by the bounds on the coefficients on the SDE and the Lipschitz property of $g$ we have $\mathbb{E}[g(\bar{X}_{T_1})^2]<\infty$, and $\overline{F}(T-t)>0$, thus we have finite variance on the second term. For the first term in (3.5), we can rewrite the second moment as,

$$\mathbb{E}\left[\left(\beta\prod_{k=2}^{N_T}P_k\right)^2\mathbb{1}_{\{N_T\geq 1\}}\right] = \sum_{\ell=1}^{\infty}\mathbb{E}\left[\left(\beta\prod_{k=2}^{N_T}P_k\right)^2\middle| N_T=\ell\right]\times\mathbb{P}[N_T=\ell].$$

In order to tackle this term we split the proof into several steps by bounding various quantities then combining them together to show the sum is bounded. We also note that we often work with conditional expectations, hence statements involving them are to be understood in the $\mathbb{P}$-a.s. sense.

*Step 1: Bounding* $\mathbb{E}[\beta^2|\overline{\mathcal{F}}_{N_T}, N_T=\ell]$, *for* $\beta$ *from* (3.4). As is standard practice when we only care about showing an estimate to be finite we use $C$ to denote some

[2]One should note the small but critical distinction between $\mathcal{F}_t$ and $\overline{\mathcal{F}}_k$.

finite constant which can change over inequalities but crucially can only depend on "known" constants such as $T$ etc. By the tower property we can rewrite any term in the sum as,

$$\mathbb{E}\left[\left(\beta\prod_{k=2}^{N_T}P_k\right)^2\middle|N_T=\ell\right]=\mathbb{E}\left[\mathbb{E}[\beta^2|\overline{\mathcal{F}}_{N_T},N_T=\ell]\prod_{k=2}^{N_T}P_k^2\middle|N_T=\ell\right].$$

Rewriting $\beta$ with $M_{N_T+1}$ and $V_{N_T+1}$ as common factors then using Young's inequality we obtain,

$$\begin{aligned}
&\mathbb{E}[\beta^2|\bar{\mathcal{F}}_{N_T},N_T=\ell]\\
&\le C\mathbb{E}\left[\frac{\left(g(\bar{X}_{T_{N_T+1}})-g(\hat{X}_{T_{N_T+1}})\right)^2}{\overline{F}(\Delta T_{N_T+1})^2}\frac{M_{N_T+1}^2}{f(\Delta T_{N_T})^2}\middle|\overline{\mathcal{F}}_{N_T},N_T=\ell\right]\\
&+C\mathbb{E}\left[\frac{\left(g(\bar{X}_{T_{N_T+1}})+g(\hat{X}_{N_T+1})-2g(\bar{X}_{T_{N_T}}+b(T_{N_T})\Delta T_{N_T+1})\right)^2}{\overline{F}(\Delta T_{N_T+1})^2}\frac{\frac{1}{2}V_{N_T+1}^2}{f(\Delta T_{N_T})^2}\middle|\overline{\mathcal{F}}_{N_T},N_T=\ell\right].
\end{aligned}$$

Considering the first term on the RHS, we note by the Lipschitz property of $g$ that,

$$|g(\bar{X}_{T_{N_T+1}})-g(\hat{X}_{T_{N_T+1}})|\le L|\bar{X}_{T_{N_T+1}}-\hat{X}_{T_{N_T+1}}|\le C|\sigma(\theta_{N_T},T_{N_T+1})\Delta W_{T_{N_T+1}}|.$$

Hence using this bound and the representation for $M_{N_T+1}$ (see (2.8)),

$$\begin{aligned}
&\mathbb{E}\left[\frac{\left(g(\bar{X}_{T_{N_T+1}})-g(\hat{X}_{T_{N_T+1}})\right)^2}{\overline{F}(\Delta T_{N_T+1})^2}\frac{M_{N_T+1}^2}{f(\Delta T_{N_T})^2}\middle|\overline{\mathcal{F}}_{N_T},N_T=\ell\right]\\
&\le C\frac{\Delta b_{N_T}^2}{f(\Delta T_{N_T})^2}\mathbb{E}\left[\left(\Delta W_{T_{N_T+1}}\sigma(\theta_{N_T},T_{N_T+1})\right)^2\left(\frac{\Delta W_{T_{N_T+1}}}{\Delta T_{N_T+1}}\sigma(\theta_{N_T},T_{N_T+1})^{-1}\right)^2\middle|\overline{\mathcal{F}}_{N_T},N_T=\ell\right]\\
&=C\frac{\Delta b_{N_T}^2}{f(\Delta T_{N_T})^2},
\end{aligned}$$

where we used $1/\overline{F}(\Delta T_{N_T+1})^2\le C$ in the inequality. For the second term on the RHS, it is more complex, let us first split the terms using Cauchy-Schwarz,

$$\begin{aligned}
&\mathbb{E}\left[\left(g(\bar{X}_{T_{N_T+1}})+g(\hat{X}_{N_T+1})-2g(\bar{X}_{T_{N_T}}+b(T_{N_T})\Delta T_{N_T+1})\right)^2V_{N_T+1}^2\middle|\overline{\mathcal{F}}_{N_T},N_T=\ell\right]\\
&\le\mathbb{E}\left[\left(g(\bar{X}_{T_{N_T+1}})+g(\hat{X}_{N_T+1})-2g(\bar{X}_{T_{N_T}}+b(T_{N_T})\Delta T_{N_T+1})\right)^4\middle|\overline{\mathcal{F}}_{N_T},N_T=\ell\right]^{1/2}\\
&\times\mathbb{E}\left[V_{N_T+1}^4\middle|\overline{\mathcal{F}}_{N_T},N_T=\ell\right]^{1/2}.
\end{aligned}$$

Let us firstly focus on the $g$ term. Consider the ODE on the interval $s \in [T_{N_T}, T_{N_T+1}]$,

$$\frac{\mathrm{d}Y_s}{\mathrm{d}s} = b(T_{N_T}), \qquad Y_{T_{N_T}} = \bar{X}_{T_{N_T}}.$$

Then, the solution is $Y_{T_{N_T+1}} = \bar{X}_{T_{N_T}} + b(T_{N_T})\Delta T_{N_T+1}$. Consequently,

$$\begin{aligned} &g\Big(\bar{X}_{T_{N_T}} + b(T_{N_T})\Delta T_{N_T+1}\Big) - g(\bar{X}_{T_{N_T}}) \\ &\quad = \int_{T_{N_T}}^{T_{N_T+1}} g'(Y_s)\mathrm{d}Y_s = \int_{T_{N_T}}^{T_{N_T+1}} g'\Big(\bar{X}_{T_{N_T}} + b(T_{N_T})(s - T_{N_T})\Big)b\big(T_{N_T}\big)\mathrm{d}s. \end{aligned} \tag{3.6}$$

By applying Itô's formula to $g(\bar{X}_{T_{N_T+1}})$ and $g(\hat{X}_{T_{N_T+1}})$ (recall $g \in C_b^2$), and using (3.6) we obtain,

$$\begin{aligned} &g(\bar{X}_{T_{N_T+1}}) + g(\hat{X}_{T_{N_T+1}}) - 2g(\bar{X}_{T_{N_T}} + b(T_{N_T})\Delta T_{N_T+1}) \\ &= \frac{1}{2}\sigma(\theta_{N_T}, T_{N_T+1})^2 \int_{T_{N_T}}^{T_{N_T+1}} (g''(\bar{X}_s) + g''(\hat{X}_s))\mathrm{d}s + \sigma(\theta_{N_T}, T_{N_T+1}) \int_{T_{N_T}}^{T_{N_T+1}} (g'(\bar{X}_s) - g'(\hat{X}_s))\mathrm{d}W_s \\ &\quad + \int_{T_{N_T}}^{T_{N_T+1}} \Big(g'(\bar{X}_s) + g'(\hat{X}_s) - 2g'\Big(\bar{X}_{T_{N_T}} + b(T_{N_T})(s - T_{N_T})\Big)\Big)b(T_{N_T})\mathrm{d}s. \end{aligned} \tag{3.7}$$

Since $g'$ is Lipschitz, we obtain,

$$\begin{aligned} \big|g'(\bar{X}_s) - g'\Big(\bar{X}_{T_{N_T}} + b(T_{N_T})(s - T_{N_T})\Big)\big| &\le C|\bar{X}_s - \bar{X}_{T_{N_T}} + b(T_{N_T})(s - T_{N_T})| \\ &\le C\sigma(\theta_{N_T}, T_{N_T+1})|W_s - W_{T_{N_T}}|, \end{aligned}$$

the same bound holds for the $g(\hat{X}_s)$ term. Thus the following bound can be obtained for the final integral in (3.7)

$$\begin{aligned} &\int_{T_{N_T}}^{T_{N_T+1}} \Big(g'(\bar{X}_s) + g'(\hat{X}_s) - 2g'\Big(\bar{X}_{T_{N_T}} + b(T_{N_T})(s - T_{N_T})\Big)\Big)b(T_{N_T})\mathrm{d}s \\ &\qquad \le C|b(T_{N_T})|\sigma(\theta_{N_T}, T_{N_T+1}) \int_{T_{N_T}}^{T_{N_T+1}} |W_s - W_{T_{N_T}}|\mathrm{d}s. \end{aligned}$$

Recalling that we are interested in the fourth moment, using Doob's maximal inequality,

$$\begin{aligned} &\mathbb{E}\left[\left(\int_{T_{N_T}}^{T_{N_T+1}} |W_s - W_{T_{N_T}}|\mathrm{d}s\right)^4 \,\Big|\, \overline{\mathcal{F}}_{N_T}, N_T = \ell\right] \\ &\quad \le C\Delta T_{N_T+1}^4 \mathbb{E}\left[\sup_{T_{N_T} \le s \le T_{N_T+1}} |W_s - W_{T_{N_T}}|^4 \,\Big|\, \overline{\mathcal{F}}_{N_T}, N_T = \ell\right] \le C\Delta T_{N_T+1}^6. \end{aligned}$$

For the stochastic integral in (3.7), again taking the fourth moment we obtain,

$$\sigma(\theta_{N_T}, T_{N_T+1})^4 \mathbb{E}\left[\left(\int_{T_{N_T}}^{T_{N_T+1}} (g'(\bar{X}_s) - g'(\hat{X}_s))\mathrm{d}W_s\right)^4 \middle| \mathcal{F}_{N_T}, N_T = \ell\right]$$
$$= 3\sigma(\theta_{N_T}, T_{N_T+1})^4 \mathbb{E}\left[\left(\int_{T_{N_T}}^{T_{N_T+1}} (g'(\bar{X}_s) - g'(\hat{X}_s))^2 \mathrm{d}s\right)^2 \middle| \mathcal{F}_{N_T}, N_T = \ell\right].$$

Using that $g'$ is Lipschitz and the difference is given by

$$|g'(\bar{X}_s) - g'(\hat{X}_s)| \leq C|\sigma(\theta_{N_T}, T_{N_T+1})(W_s - W_{T_{N_T}}) + \sigma(\theta_{N_T}, T_{N_T+1})(W_s - W_{T_{N_T}})|.$$

This along with a similar Doob's maximal inequality implies that we can bound the stochastic integral by,

$$\sigma(\theta_{N_T}, T_{N_T+1})^4 \mathbb{E}\left[\left(\int_{T_{N_T}}^{T_{N_T+1}} (g'(\bar{X}_s) - g'(\hat{X}_s))\mathrm{d}W_s\right)^4 \middle| \mathcal{F}_{N_T}, N_T = \ell\right] \leq C\sigma(\theta_{N_T}, T_{N_T+1})^8 \Delta T_{N_T+1}^4.$$

Recalling that $g''$ is bounded, we can bound the remaining term in (3.7) by a similar term to the stochastic integral to obtain,

$$\mathbb{E}\left[(g(\bar{X}_{T_{N_T+1}}) + g(\hat{X}_{T_{N_T+1}}) - 2g(\bar{X}_{T_{N_T}} + b(T_{N_T})\Delta T_{N_T+1}))^4 | \mathcal{F}_{N_T}, N_T = \ell\right] \tag{3.8}$$
$$\leq C\sigma(\theta_{N_T}, T_{N_T+1})^8 \Delta T_{N_T+1}^4.$$

The above bound was obtained using differentiability and Itô's formula, however, it will also be useful for us to note that just using the Lipschitz property yields,

$$\mathbb{E}\left[(g(\bar{X}_{T_{N_T+1}}) + g(\hat{X}_{T_{N_T+1}}) - 2g(\bar{X}_{T_{N_T}} + b(T_{N_T})\Delta T_{N_T+1}))^4 | \mathcal{F}_{N_T}, N_T = \ell\right]$$
$$\leq C\sigma(\theta_{N_T}, T_{N_T+1})^4 \Delta T_{N_T+1}^2.$$

Hence we obtain the following stronger bound for the $g$ terms

$$\mathbb{E}\left[(g(\bar{X}_{T_{N_T+1}}) + g(\hat{X}_{T_{N_T+1}}) - 2g(\bar{X}_{T_{N_T}} + b(T_{N_T})\Delta T_{N_T+1}))^4 | \mathcal{F}_{N_T}, N_T = \ell\right]$$
$$\leq C \min\left[\sigma(\theta_{N_T}, T_{N_T+1})^4 \Delta T_{N_T+1}^2, \sigma(\theta_{N_T}, T_{N_T+1})^8 \Delta T_{N_T+1}^4\right].$$

For the $V$ term,

$$\mathbb{E}\left[V_{N_T+1}^4 | \mathcal{F}_{N_T}, N_T = \ell\right] \leq C \frac{\sigma(\theta_{N_T-1}, T_{N_T})^8}{\sigma(\theta_{N_T}, T_{N_T+1})^8} \frac{1}{\Delta T_{N_T+1}^8} \mathbb{E}\left[\left(\Delta W_{T_{N_T+1}}^2 - \Delta T_{N_T+1}\right)^4 \middle| \mathcal{F}_{N_T}, N_T = \ell\right]$$
$$\leq C \frac{\sigma(\theta_{N_T-1}, T_{N_T})^8}{\sigma(\theta_{N_T}, T_{N_T+1})^8} \frac{1}{\Delta T_{N_T+1}^4}.$$

Hence using Cauchy-Schwarz we obtain,

$$\mathbb{E}\left[\frac{\left(g(\bar{X}_{T_{N_T+1}})+g(\hat{X}_{T_{N_T+1}})-2g(\bar{X}_{T_{N_T}}+b(T_{N_T})\Delta T_{N_T+1})\right)^2}{\overline{F}(\Delta T_{N_T+1})^2}\frac{\frac{1}{2}V_{N_T+1}^2}{f(\Delta T_{N_T})^2}\middle|\mathcal{F}_{N_T},N_T=\ell\right]$$

$$\leq C\frac{\sigma(\theta_{N_T-1},T_{N_T})^4}{\sigma(\theta_{N_T},T_{N_T+1})^4}\frac{1}{\Delta T_{N_T+1}^2}\min\left[\sigma(\theta_{N_T},T_{N_T+1})^2\Delta T_{N_T+1},\sigma(\theta_{N_T},T_{N_T+1})^4\Delta T_{N_T+1}^2\right]\frac{1}{f(\Delta T_{N_T})^2}.$$

Therefore, the conditional expectation of $\beta^2$ can be bounded by,

$$\mathbb{E}[\beta^2|\mathcal{F}_{N_T},N_T=\ell]\leq\frac{C}{f(\Delta T_{N_T})^2}\left(\Delta b_{N_T}^2+\frac{\sigma(\theta_{N_T-1},T_{N_T})^4}{\sigma(\theta_{N_T},T_{N_T+1})^2}\frac{\min\left[1,\sigma(\theta_{N_T},T_{N_T+1})^2\Delta T_{N_T+1}\right]}{\Delta T_{N_T+1}}\right).$$

*Step 2: Bounding* $\mathbb{E}[P_{k+1}^4|\mathcal{F}_k,N_T=\ell]$. Let $k\in\Lambda$ and note by Assumption 3.2 we obtain,

$$\mathbb{E}\big[\Delta b_k^4|\mathcal{F}_{k-1},N_T=\ell\big]\leq C\Delta T_k^4.$$

From (2.8) we observe the following,

$$\mathbb{E}[M_{k+1}^4|\mathcal{F}_k,N_T=\ell]\leq C\frac{\Delta b_k^4}{\Delta T_{k+1}^2}\frac{1}{\sigma(\theta_k,T_{k+1})^4}\leq C\frac{\Delta T_k^4}{\Delta T_{k+1}^2}\frac{1}{\sigma(\theta_k,T_{k+1})^4},$$

$$\mathbb{E}[V_{k+1}^4|\mathcal{F}_k,N_T=\ell]\leq C\frac{\sigma(\theta_{k-1},T_k)^8}{\sigma(\theta_k,T_{k+1})^8}\frac{1}{\Delta T_{k+1}^4}.$$

By Assumption 3.5 and the fact that $\sigma$ is bounded from below implies that the $V$ term dominates the $M$ term, hence, we obtain,

$$\mathbb{E}[P_{k+1}^4|\mathcal{F}_k,N_T=\ell]\leq C\frac{1}{f(\Delta T_k)^4}\frac{\sigma(\theta_{k-1},T_k)^8}{\sigma(\theta_k,T_{k+1})^8}\frac{1}{\Delta T_{k+1}^4}.\tag{3.9}$$

We are now able to consider bounding the term we originally set out to. Using the bound we obtained for $\beta^2$,

$$\mathbb{E}\left[\beta^2\prod_{k=2}^{N_T}P_k^2\middle|N_T=\ell\right]$$

$$\leq\mathbb{E}\left[\frac{C}{f(\Delta T_{N_T})^2}\left(\Delta b_{N_T}^2+\frac{\sigma(\theta_{N_T-1},T_{N_T})^4}{\sigma(\theta_{N_T},T_{N_T+1})^2}\frac{\min\left[1,\sigma(\theta_{N_T},T_{N_T+1})^2\Delta T_{N_T+1}\right]}{\Delta T_{N_T+1}}\right)\prod_{k=2}^{N_T}P_k^2\middle|N_T=\ell\right].\tag{3.10}$$

One can view this product as having two components, one which does not depend on $\Delta T_{N_T+1}$ which comes from the $\Delta b_{N_T}$ and a component that does depend on $\Delta T_{N_T+1}$. In order to show that the second moment is finite we split these two components and show each of them is finite.

*Step 3: Bounding each product in* (3.10). Let us start by considering the product from the $\Delta b_{N_T}$ term

$$\mathbb{E}\left[\frac{\Delta b_{N_T}^2}{f(\Delta T_{N_T})^2}\prod_{k=2}^{N_T} P_k^2 \middle| N_T=\ell\right] = \mathbb{E}\left[\frac{1}{f(\Delta T_{N_T})^2}\mathbb{E}[\Delta b_{N_T}^2 P_{N_T}^2 | \overline{\mathcal{F}}_{N_T-1}, N_T=\ell] \prod_{k=2}^{N_T-1} P_k^2 \middle| N_T=\ell\right].$$

Applying Cauchy-Schwarz to the internal expectation and using the previous bounds we obtain,

$$\mathbb{E}[\Delta b_{N_T}^2 P_{N_T}^2 | \overline{\mathcal{F}}_{N_T-1}, N_T=\ell] \le C\frac{1}{f(\Delta T_{N_T-1})^2}\frac{\sigma(\theta_{N_T-2}, T_{N_T-1})^4}{\sigma(\theta_{N_T-1}, T_{N_T})^4}. \tag{3.11}$$

Note that this bound and (3.9) have no dependence on the Brownian motion, therefore we can isolate each $P_k$ by recursively conditioning, i.e.

$$\begin{aligned}
&\mathbb{E}\left[\frac{C}{f(\Delta T_{N_T})^2}\mathbb{E}[\Delta b_{N_T}^2 P_{N_T}^2 | \overline{\mathcal{F}}_{N_T-1}, N_T=\ell] \prod_{k=2}^{N_T-1} P_k^2 \middle| N_T=\ell\right] \\
&\le \mathbb{E}\left[\frac{C}{f(\Delta T_{N_T})^2}\frac{1}{f(\Delta T_{N_T-1})^2}\frac{\sigma(\theta_{N_T-2}, T_{N_T-1})^4}{\sigma(\theta_{N_T-1}, T_{N_T})^4}\prod_{k=2}^{N_T-1} P_k^2 \middle| N_T=\ell\right] \\
&= \mathbb{E}\left[\frac{C}{f(\Delta T_{N_T})^2}\frac{1}{f(\Delta T_{N_T-1})^2}\frac{\sigma(\theta_{N_T-2}, T_{N_T-1})^4}{\sigma(\theta_{N_T-1}, T_{N_T})^4}\mathbb{E}[P_{N_T-1}^2 | \overline{\mathcal{F}}_{N_T-2}, \Delta T_{N_T}, N_T=\ell] \prod_{k=2}^{N_T-2} P_k^2 \middle| N_T=\ell\right].
\end{aligned}$$

Using our results and noting that most of the $\sigma$ terms cancel yields the following bound,

$$\begin{aligned}
&\mathbb{E}\left[\frac{\Delta b_{N_T}^2}{f(\Delta T_{N_T})^2}\prod_{k=2}^{N_T} P_k^2 \middle| N_T=\ell\right] \\
&\le \mathbb{E}\left[\frac{C^{N_T}}{f(\Delta T_{N_T})^2}\frac{1}{f(\Delta T_{N_T-1})^2}\frac{\sigma(\theta_{N_T-2}, T_{N_T-1})^4}{\sigma(\theta_{N_T-1}, T_{N_T})^4}\prod_{k=2}^{N_T-1}\frac{1}{f(\Delta T_{k-1})^2}\frac{\sigma(\theta_{k-2}, T_{k-1})^4}{\sigma(\theta_{k-1}, T_k)^4}\frac{1}{\Delta T_k^2} \middle| N_T=\ell\right] \\
&= \mathbb{E}\left[\frac{C^{N_T}}{f(\Delta T_{N_T})^2}\frac{1}{f(\Delta T_1)^2}\frac{\sigma(\theta_0, T_1)^4}{\sigma(\theta_{N_T-1}, T_{N_T})^4}\prod_{k=2}^{N_T-1}\frac{1}{f(\Delta T_k)^2}\frac{1}{\Delta T_k^2} \middle| N_T=\ell\right].
\end{aligned}$$

Recall the goal here is to ultimately bound this by a term of the form $C^{N_T}$, which holds provided all $\Delta T_k$ dependence is to a positive power. Recall that since $f$ is the density for the Gamma distribution with shape $\kappa$, we have that,

$$f(\Delta T_k) \ge C\Delta T_k^{\kappa-1} \quad\Longrightarrow\quad \frac{1}{f(\Delta T_{N_T})^2} \le C\Delta T_{N_T}^{2-2\kappa}.$$

Using the representation for $\sigma$ we obtain terms of the form $\Delta T_k^{2-2\kappa-2-4n}$, hence we require $2\kappa - 4n \ge 0$, which suggests $n \le -\kappa/2$. Since Assumption 3.5 implies these conditions on $n$ and $\kappa$ hold,[3] one obtains

[3]Note that $\kappa = 1/2$ also implies $1/f(\Delta T_1) \le C$.

$$\mathbb{E}\left[\frac{\Delta b_{N_T}^2}{f(\Delta T_{N_T})^2}\prod_{k=2}^{N_T} P_k^2 \,\middle|\, N_T=\ell\right] \le \mathbb{E}\left[C^{N_T}\,\middle|\, N_T=\ell\right]. \tag{3.12}$$

Showing this is finite is done in [6]. As it turns out the other term in (3.10) also dominates this term, hence we do not discuss it further.

For the second term in (3.10) we note that the $\sigma$ terms do not depend on the Brownian motion, hence we can again condition to isolate the various $P_k$ terms, hence,

$$\begin{aligned}
&\mathbb{E}\left[\frac{C}{f(\Delta T_{N_T})^2}\frac{\sigma(\theta_{N_T-1},T_{N_T})^4}{\sigma(\theta_{N_T},T_{N_T+1})^2}\frac{\min\left[1,\sigma(\theta_{N_T},T_{N_T+1})^2\Delta T_{N_T+1}\right]}{\Delta T_{N_T+1}}\prod_{k=2}^{N_T} P_k^2\,\middle|\,N_T=\ell\right]\\
&\le \mathbb{E}\left[\frac{C}{f(\Delta T_{N_T})^2}\frac{\sigma(\theta_{N_T-1},T_{N_T})^4}{\sigma(\theta_{N_T},T_{N_T+1})^2}\frac{\min\left[1,\sigma(\theta_{N_T},T_{N_T+1})^2\Delta T_{N_T+1}\right]}{\Delta T_{N_T+1}}\right.\\
&\qquad\left.\times\prod_{k=2}^{N_T}\frac{1}{f(\Delta T_{k-1})^2}\frac{\sigma(\theta_{k-2},T_{k-1})^4}{\sigma(\theta_{k-1},T_k)^4}\frac{1}{\Delta T_k^2}\,\middle|\,N_T=\ell\right].
\end{aligned} \tag{3.13}$$

By cancelling repeating $\sigma$ terms in the product and again using $1/f(\Delta T_1)\le C$, we obtain the following simpler result,

$$(3.13) \le C\mathbb{E}\left[\frac{\sigma(\theta_0,T_1)^4}{\sigma(\theta_{N_T},T_{N_T+1})^2}\frac{\min\left[1,\sigma(\theta_{N_T},T_{N_T+1})^2\Delta T_{N_T+1}\right]}{\Delta T_{N_T+1}}\prod_{k=2}^{N_T}\frac{1}{f(\Delta T_k)^2}\frac{1}{\Delta T_k^2}\,\middle|\,N_T=\ell\right]. \tag{3.14}$$

Using the fact that $\sigma(\theta_0,T_1)=\sigma_0$ and $f$ is the density for the Gamma distribution we can bound (3.14) by,

$$\begin{aligned}
&\mathbb{E}\left[C^{N_T}\frac{\min\left[1,\sigma(\theta_{N_T},T_{N_T+1})^2\Delta T_{N_T+1}\right]}{\sigma(\theta_{N_T},T_{N_T+1})^2}\Delta T_{N_T+1}^{-1}\prod_{k=2}^{N_T}\Delta T_k^{-2\kappa}\,\middle|\,N_T=\ell\right]\\
&\le \mathbb{E}\left[C^{N_T}\frac{\sigma(\theta_{N_T},T_{N_T+1})^{\nu}\Delta T_{N_T+1}^{\nu/2}}{\sigma(\theta_{N_T},T_{N_T+1})^2}\Delta T_{N_T+1}^{-1}\prod_{k=2}^{N_T}\Delta T_k^{-2\kappa}\,\middle|\,N_T=\ell\right] \quad \text{for } \nu\in[0,2],
\end{aligned} \tag{3.15}$$

where the inequality comes from the observation that,

$$\min\left[1,\sigma(\theta_{N_T},T_{N_T+1})^2\Delta T_{N_T+1}\right]\le \sigma(\theta_{N_T},T_{N_T+1})^{\nu}\Delta T_{N_T+1}^{\nu/2} \quad \text{for any } \nu\in[0,2].$$

The presence of $\Delta T_{N_T+1}^{-1}$ makes (3.15) more challenging. Of course, one could take $\nu=2$ to remove $\Delta T_{N_T+1}^{-1}$, however, this also removes $\sigma$ and since $\kappa>0$ we are still left with an unbounded product. Therefore we must chose $\nu$ carefully and apply a delicate argument to appropriately bound (3.15).

One can note the similarity between (3.15) and (3.12). However, (3.15) is more complex and as it turns out, the bound we eventually achieve for it dominates (3.12). We therefore complete the proof showing (3.15) is bounded, since this implies (3.12) is bounded.

*Step 4: Interval splitting*. Recall we are interested in proving convergence of the sum

$$\sum_{\ell=1}^{\infty} \mathbb{E}\left[\left(\beta \prod_{k=2}^{\ell} P_k\right)^2 \middle| N_T = \ell\right] \mathbb{P}[N_T = \ell] .$$

Let us split this into two components, $\ell = 1$ and $\ell \geq 2$. When $\ell = 1$ we obtain nothing from the product and are thus only showing that $\beta$ is square integrable, such is obvious from our previous calculations. We now concentrate on the case $\ell \geq 2$. Recall that for $i = 1, \ldots, M$, if $Y_i \sim \Gamma(a, b)$ i.i.d. then $\sum_{i=1}^{M} Y_i \sim \Gamma(aM, b)$ and fix $\ell \geq 2$, we can then partition the expectation as follows,

$$\mathbb{E}\left[\left(\beta \prod_{k=2}^{\ell} P_k\right)^2 \middle| N_T = \ell\right] = \mathbb{E}\left[\left(\beta \prod_{k=2}^{\ell} P_k\right)^2 \middle| N_T = \ell, \Delta T_{N_T+1} \geq \frac{T}{\ell}\right] \mathbb{P}\left[\Delta T_{N_T+1} \geq \frac{T}{\ell} \middle| N_T = \ell\right]$$
$$+ \sum_{m=1}^{\infty} \mathbb{E}\left[\left(\beta \prod_{k=2}^{\ell} P_k\right)^2 \middle| N_T = \ell, \frac{T}{\ell^{m+1}} \leq \Delta T_{N_T+1} < \frac{T}{\ell^m}\right] \mathbb{P}\left[\frac{T}{\ell^{m+1}} \leq \Delta T_{N_T+1} < \frac{T}{\ell^m} \middle| N_T = \ell\right] .$$

Firstly, we note that when $\Delta T_{N_T+1} \geq T/\ell$, the expectation is simple to bound since we can take the minimum as 1 (the $\nu = 0$ case in (3.15)) then use the fact $\sigma(\theta_{N_T}, T_{N_T+1})^{-2} = \sigma_0^{-2} \prod_{i=1}^{\ell} \Delta T_i^{2n}$ and $\kappa < -n$ by Assumption 3.5. Hence the following bound holds,

$$\mathbb{E}\left[\left(\beta \prod_{k=2}^{\ell} P_k\right)^2 \middle| N_T = \ell, \Delta T_{N_T+1} \geq \frac{T}{\ell}\right] \mathbb{P}\left[\Delta T_{N_T+1} \geq \frac{T}{\ell} \middle| N_T = \ell\right] \leq \ell C^{\ell} .$$

For the case $m \geq 1$, we have that

$$\mathbb{P}\left[\frac{T}{\ell^{m+1}} \leq \Delta T_{N_T+1} < \frac{T}{\ell^m} \middle| N_T = \ell\right] = \mathbb{P}\left[T - \frac{T}{\ell^m} \leq \sum_{i=1}^{\ell} \Delta T_i < T - \frac{T}{\ell^{m+1}} \middle| N_T = \ell\right] .$$

Due to the fact $\kappa = 1/2$ by Assumption 3.5, the distribution of $\sum_{i=1}^{\ell} \Delta T_i$ is Gamma with shape parameter at least 1, therefore the density has a finite maximum, unfortunately the conditioning makes this probability difficult to deal with. We therefore expand,

$$\mathbb{P}\left[T - \frac{T}{\ell^m} \leq \sum_{i=1}^{\ell} \Delta T_i < T - \frac{T}{\ell^{m+1}} \middle| N_T = \ell\right]$$
$$= \frac{1}{\mathbb{P}[N_T = \ell]} \mathbb{P}\left[T - \frac{T}{\ell^m} \leq \sum_{i=1}^{\ell} \Delta T_i < T - \frac{T}{\ell^{m+1}}, \sum_{i=1}^{\ell} \Delta T_i < T, \sum_{i=1}^{\ell+1} \Delta T_i \geq T\right]$$
$$\leq \frac{1}{\mathbb{P}[N_T = \ell]} \mathbb{P}\left[T - \frac{T}{\ell^m} \leq \sum_{i=1}^{\ell} \Delta T_i < T - \frac{T}{\ell^{m+1}}\right].$$

Using this form we have removed the conditional dependence on the number of jumps and therefore we can use the distribution of $\sum_{i=1}^{\ell} \Delta T_i$. We note that for $\ell$ large the density of the distribution at point $T$ will be larger than values less than $T$, further, since the density has a finite maximum, for $\ell$ smaller we can bound by some constant multiplied by the value at point $T$, thus,

$$\mathbb{P}\left[T - \frac{T}{\ell^m} \leq \sum_{i=1}^{\ell} \Delta T_i < T - \frac{T}{\ell^{m+1}}\right] \leq C\ell^{-m} f(T) \leq C\ell^{-m} \frac{T^{\ell\kappa-1} e^{-T/\eta}}{\eta^{\ell\kappa}\Gamma(\ell\kappa)},$$

where we have used the p.d.f. of a Gamma random variable to obtain the last inequality. Similar to the case $\ell = 1$ we can bound the expectation by

$$\mathbb{E}\left[\left(\beta \prod_{k=2}^{N_T} P_k\right)^2 \middle| N_T = \ell, \frac{T}{\ell^{m+1}} \leq \Delta T_{N_T+1} < \frac{T}{\ell^m}\right]$$
$$\leq \mathbb{E}\left[C^{N_T} \Delta T_{N_T+1}^{-1+\nu/2} \prod_{k=2}^{N_T} \Delta T_k^{-(2-\nu)n-2\kappa} \middle| N_T = \ell, \frac{T}{\ell^{m+1}} \leq \Delta T_{N_T+1} < \frac{T}{\ell^m}\right].$$

A simple requirement for the product to be bounded is $-(2-\nu)n - 2\kappa \geq 0$, by Assumption 3.5 $\kappa = 1/2$, hence $-n \geq 1/(2-\nu)$. As it turns out, taking $\nu = 1$ is useful to complete the proof, therefore we require $n \leq -1$, which holds by Assumption 3.5. This set of $\kappa$, $\nu$ and $n$ also allow us to bound (3.12), hence we only considered (3.15).

The only term we have to consider in the expectation is $\Delta T_{N_T+1}^{-1+\nu/2}$, but by our conditioning this is bounded by $T\ell^{(1-\nu/2)(m+1)}$, hence for fixed $\ell \geq 2$ and letting $\nu = 1$ we obtain the following,

$$\mathbb{E}\left[\left(\beta \prod_{k=2}^{N_T} P_k\right)^2 \middle| N_T = \ell\right] \leq C^\ell \ell + \frac{1}{\mathbb{P}[N_T = \ell]} \sum_{m=1}^{\infty} C^\ell \ell^{(1/2)(m+1)} \ell^{-m} \frac{T^{\ell\kappa-1} e^{-T/\eta}}{\eta^{\ell\kappa}\Gamma(\ell\kappa)}.$$

One can easily see that the sum in $m$ converges since $(1/2)(m+1) - m \leq 0$ for $m \geq 1$ and $\ell \geq 2$, the sum can be easily bounded by $\sum_{m=1}^{\infty} 2^{-(1/2)m+1/2} = C$ for any

$\ell \geq 2$. One can compare this to the result in [6, Proposition 4.1] where the authors obtain a bound of the form $C^{\ell}$, hence our bound is not as strong but it is still good enough to ensure convergence.

*Step 5: The sum over $N_T$ converges.* The final step of the proof is to show that the overall sum converges. We proceed by observing the following (see [6, Proposition 4.1]),

$$\mathbb{P}[N_T = \ell] \leq \frac{C^{\ell\kappa}}{\ell\kappa\Gamma(\ell\kappa)}.$$

Using a generalisation of Stirling's formula one can approximate $\Gamma(z) \sim z^{z-1/2}e^{-z}\sqrt{2\pi}$. Hence we can bound

$$\begin{aligned}\mathbb{E}\left[\beta \prod_{k=2}^{N_T} P_k \mathbb{1}_{\{N_T \geq 1\}}\right] &\leq \sum_{\ell=1}^{\infty} C^{\ell}\ell \frac{C^{\ell\kappa}}{\ell\kappa\Gamma(\ell\kappa)} + \frac{\mathbb{P}[N_T = \ell]}{\mathbb{P}[N_T = \ell]} \sum_{m=1}^{\infty} C^{\ell}\ell^{(1/2)(m+1)}\ell^{-m} \frac{T^{\ell\kappa-1}e^{-T/\eta}}{\eta^{\ell\kappa}\Gamma(\ell\kappa)} \\ &\leq \sum_{\ell=1}^{\infty} C^{\ell} \frac{C^{\ell\kappa}}{\kappa\Gamma(\ell\kappa)},\end{aligned}$$

and using Stirling's formula,

$$C^{\ell}\frac{C^{\ell\kappa}}{\kappa\Gamma(\ell\kappa)} \sim C^{\ell}\frac{C^{\ell\kappa}e^{\ell\kappa}}{\kappa(\ell\kappa)^{\ell\kappa-1/2}\sqrt{2\pi}} \leq \left(\frac{C^{1/\kappa}e^{1}}{\ell\kappa}\right)^{\ell\kappa-1/2} C^{1/(2\kappa)}e^{1/2},$$

since $\kappa = 1/2$ this gives a sequence that converges under summation. □

*Remark 3.9* (Optimal $\sigma_0$) One can see from the variance calculations that the $\frac{\sigma(\theta,T_0)^4}{\sigma(\theta,T_{N_T})^2}$ will leave a $\sigma_0^2$ term behind. Thus as one would expect the variance will be minimised by taking $\sigma_0$ smaller, however, to deal with terms involving nonlinearities in $\partial_x v$ one obtains terms of the form $\frac{1}{\sigma}$ thus an optimisation needs to be performed in order to set $\sigma_0$ at the correct level. Crucially however, the expected value (bias) is not effected by this choice.

## 3.2 *Estimator Solves the PDE Under Enough Integrability*

At this point we have only proved that the estimator can be approximated via Monte Carlo. We now show that given some extra integrability conditions the estimator solves PDE (2.1). The final step is to show the said integrability conditions hold.

Theorem 3.10 is the analogous result to Theorem 3.5 in [12], however, the representation we derive below is more complex. The reason for the added complexity is the antithetic as well as the control variate on the final jump. Where as the control variate keeps the final Malliavin weight the same, the antithetic changes the weight, this then requires us to have extra terms that [12] does not have.

**Theorem 3.10** *Let Assumptions 3.2, 3.3 and 3.5 hold. Define the following random variables,*

$$\begin{aligned}\tilde{\psi}^{t,x} :=& \Bigg( \frac{\Delta g_{T_{N_T+1}}}{2\overline{F}(\Delta T_{N_T+1})} \frac{\Delta b_{N_T} \mathcal{W}^1_{N_T+1} - \frac{1}{2}\sigma(\theta_{N_T-1}, T_{N_T})^2 \mathcal{W}^2_{N_T+1}}{f(\Delta T_{N_T})} \\ &+ \frac{\Delta \hat{g}_{T_{N_T+1}}}{2\overline{F}(\Delta T_{N_T+1})} \frac{-\Delta b_{N_T} \mathcal{W}^1_{N_T+1} - \frac{1}{2}\sigma(\theta_{N_T-1}, T_{N_T})^2 \mathcal{W}^2_{N_T+1}}{f(\Delta T_{N_T})} \Bigg) \prod_{k=2}^{N_T} \frac{\Delta b_{k-1} \mathcal{W}^1_k - \frac{1}{2}\sigma(\theta_{k-2}, T_{k-1})^2 \mathcal{W}^2_k}{f(\Delta T_{k-1})}, \\ \psi^{t,x} :=& \mathbb{1}_{\{N_T=0\}} \frac{g(\bar{X}_{T_{N_T+1}})}{\overline{F}(\Delta T_{N_T+1})} + \mathbb{1}_{\{N_T \geq 1\}} \tilde{\psi}^{t,x},\end{aligned}$$

*and*

$$\Phi_1^{T_{N_T}, \bar{X}_{T_{N_T}}} = \frac{\Delta g_{T_{N_T+1}} - \Delta \hat{g}_{T_{N_T+1}}}{2\overline{F}(\Delta T_{N_T+1})} \quad \text{and} \quad \Phi_2^{T_{N_T}, \bar{X}_{T_{N_T}}} = \frac{\Delta g_{T_{N_T+1}} + \Delta \hat{g}_{T_{N_T+1}}}{2\overline{F}(\Delta T_{N_T+1})}$$

*where*

$$\begin{aligned}\Delta g_{T_{N_T+1}} &:= g(\bar{X}_{T_{N_T+1}}) - g(\bar{X}_{T_{N_T}} + b(T_{N_T})\Delta T_{N_T+1}), \\ \Delta \hat{g}_{T_{N_T+1}} &:= g(\hat{X}_{T_{N_T+1}}) - g(\bar{X}_{T_{N_T}} + b(T_{N_T})\Delta T_{N_T+1}),\end{aligned}$$

*the first and second order Malliavin weights are given by,*

$$\mathcal{W}^1_{k+1} = \sigma(\theta_k, T_{k+1})^{-1} \frac{\Delta W_{T_{k+1}}}{\Delta T_{k+1}} \quad \text{and} \quad \mathcal{W}^2_{k+1} = \sigma(\theta_k, T_{k+1})^{-2} \left( \frac{\Delta W^2_{T_{k+1}} - \Delta T_{k+1}}{\Delta T^2_{k+1}} \right). \tag{3.16}$$

*The superscript in $\psi$, $\tilde{\psi}$, $\Phi_1$ and $\Phi_2$ denotes the initial condition for the SDE, $\bar{X}$. Further assume that,*

$$\psi^{t,x}, \; \tilde{\psi}^{t,x}\mathcal{W}^1_1, \; \tilde{\psi}^{t,x}\mathcal{W}^2_1, \; f(\Delta T_1)^{-1} \Delta b_1 \tilde{\psi}^{T_1, \bar{X}_{T_1}} \mathcal{W}^1_2, \; f(\Delta T_1)^{-1} \sigma(\theta_0, T_1)^2 \tilde{\psi}^{T_1, \bar{X}_{T_1}} \mathcal{W}^2_2,$$
$$\Phi_1^{T_{N_T}, \bar{X}_{T_{N_T}}} \mathcal{W}^1_{N_T+1}, \; \Phi_2^{T_{N_T}, \bar{X}_{T_{N_T}}} \mathcal{W}^2_{N_T+1},$$

*are uniformly integrable and that $\psi^{T_1, \bar{X}_{T_1}}$, $\Delta b_2 \tilde{\psi}^{T_2, \bar{X}_{T_2}} \mathcal{W}^1_3$, $\sigma(\theta_1, T_2)^2 \tilde{\psi}^{T_2, \bar{X}_{T_2}} \mathcal{W}^2_3$ are $\mathbb{P}$-a.s. uniformly integrable and $\tilde{\psi}^{T_1, \bar{X}_{T_1}} \mathcal{W}^1_2$ and $\tilde{\psi}^{T_1, \bar{X}_{T_1}} \mathcal{W}^2_2$ are $\mathbb{P}$-a.s. integrable.*

*Then, the function $\hat{v}(t, x) := \mathbb{E}[\psi^{t,x} | \mathcal{F}_t]$ solves the PDE* (3.1).

*Remark 3.11* ($\mathbb{P}$-a.s. (uniformly) integrable) Note that some of the processes stated in the theorem, for example $\psi^{T_1, \bar{X}_{T_1}}$ and $\tilde{\psi}^{T_1, \bar{X}_{T_1}} \mathcal{W}^2_2$ depend on random "initial conditions". Hence some of these processes are unbounded, but are finite up to a null set. For example, when we state $\tilde{\psi}^{T_1, \bar{X}_{T_1}} \mathcal{W}^2_2$ is $\mathbb{P}$-a.s. integrable, we mean that, $\mathbb{E}[|\tilde{\psi}^{T_1, \bar{X}_{T_1}} \mathcal{W}^2_2| \, | \mathcal{F}_{T_1}] < \infty$ $\mathbb{P}$-a.s. and similar for the uniform integrability condition. Recall that $\mathbb{P}$ is the product measure $\mathbb{P}_W \otimes \mathbb{P}_f$.

This theorem only shows that the estimator gives rise to the solution of the PDE under certain integrability assumptions. In order to finish our proof we need to show that such integrability conditions hold (Theorem 3.13). Although it is $\psi$ that solves the PDE, our proof relies on various intermediary steps requiring additional integrability on $\psi\mathcal{W}$. Since one does not have this in general, we introduce the seemingly arbitrary $\tilde{\psi}$ and $\Phi$ which have the required integrability. Therefore, throughout the proof we show that one can view these additional processes as $\psi\mathcal{W}$ with a control variate and perform the various steps on $\tilde{\psi}$ and $\Phi$.

*Remark 3.12* The Malliavin weights are given by (2.8) since our unbiased estimation puts us in the simple setting where the SDE has constant coefficients (see [9]).

*Proof* The main idea of this proof is to first show a stochastic representation for the PDE, then show that this representation and $\mathbb{E}[\psi^{t,x}|\mathcal{F}_t]$ are equivalent. Following Sect. 2.2, since a $C_b^{1,3}$ solution is assumed to exist, one can take constants $b_0$ and $\sigma_0$ and define the following PDE (equivalent to (3.1)),

$$\begin{cases} \partial_t v(t,x) + b_0 \partial_x v(t,x) + \frac{1}{2}\sigma_0^2 \partial_{xx} v(t,x) + (b(t)-b_0)\partial_x v(t,x) - \frac{1}{2}\sigma_0^2 \partial_{xx} v(t,x) = 0\,, \\ v(T,x) = g(x)\,. \end{cases}$$

Assume that these constants $b_0$ and $\sigma_0$ are adapted to the filtration $\mathcal{F}_t$ (as defined at the start of Sect. 3). Define $\tilde{X}$ as the solution to the SDE on $s \in [t,T]$

$$\mathrm{d}\tilde{X}_s = b_0 \mathrm{d}s + \sigma_0 \mathrm{d}W_s\,, \quad \tilde{X}_t = x\,.$$

again since $v \in C_b^{1,3}$, one obtains from the Feynman-Kac formula,

$$v(t,x) = \mathbb{E}_W\left[ g(\tilde{X}_T) + \int_t^T (b(s)-b_0)\partial_x v(s,\tilde{X}_s) - \frac{1}{2}\sigma_0^2 \partial_{xx} v(s,\tilde{X}_s)\mathrm{d}s \,\middle|\, \mathcal{F}_t \right].$$

It is important to note that we have not assigned values to the constants $b_0$ and $\sigma_0$ here, only that they are adapted to the initial filtration. Using standard branching arguments, we introduce a random variable independent of Brownian motion, corresponding to the life of the particle which allows us to rewrite the previous expression as,[4]

$$v(t,x) = \mathbb{E}\left[ \frac{g(\tilde{X}_T)}{\overline{F}(\Delta T_1)} \mathbb{1}_{\{T_1=T\}} + \frac{\mathbb{1}_{\{T_1<T\}}}{f(\Delta T_1)} \left\{ (b(T_1)-b_0)\partial_x v(T_1,\tilde{X}_{T_1}) - \frac{1}{2}\sigma_0^2 \partial_{xx} v(T_1,\tilde{X}_{T_1}) \right\} \,\middle|\, \mathcal{F}_t \right]. \tag{3.17}$$

As before, the representation does not depend on the value of the constants, therefore let us take $b_0 := b(t)$ and $\sigma_0 := \sigma_0$ (in the sense of (3.3)), thus $\tilde{X}$ is equivalent to $\bar{X}$.

This can be thought of as the forward representation, the goal now is to reach the same representation going backwards. Namely, starting from the estimator $\psi^{t,x}$, we want to remove the Malliavin weights and obtain the same relationship. We break the remainder of the proof into several steps.

[4] Where $\mathbb{E}$ is the expectation in the product space of the two random variables.

*Step 1: Continuity of the functions.* We start by noting that between any two mesh points, the SDE is continuous w.r.t. its initial condition $(T_k, \bar{X}_{T_k})$, which is clear from the fact that it is just an SDE with constant coefficients. This along with the uniform integrability assumption of $\psi$ implies that the function $\hat{v}$ is jointly continuous. This stems from the fact that we can define $\psi_n^{t,x}$ as $\psi^{t,x}$ but with the $N_T$ replaced by $N_T \wedge n$, hence $\psi^{t,x} = \lim_{n\to\infty} \psi_n^{t,x}$. Then for each $n$ we have a finite product of jointly continuous functions, which is therefore jointly continuous. Then uniform integrability allows us to take the limit as $n \to \infty$ inside to conclude that $(t, x) \to \mathbb{E}[\psi^{t,x}|\mathcal{F}_t]$ must also be a jointly continuous function.

The weights $\mathcal{W}^i$ for $i = 1, 2$ are also continuous w.r.t. the initial condition. Thus by arguing in a similar way to above we have $\mathbb{E}[\tilde{\psi}^{t,x}\mathcal{W}_1^i|\mathcal{F}_t]$ and $\mathbb{E}[\Phi_i^{T_{N_T}, \bar{X}_{T_{N_T}}} \mathcal{W}_{N_T+1}^i|\mathcal{F}_t]$ are jointly continuous by the uniform integrability assumption.

*Step 2: Rewriting the representation.* By construction of $\psi$, there are two main cases, either the particle goes through a regime switch, which implies $\{N_T \geq 1\}$ or it "survives" until the end, $\{N_T = 0\}$. The key difference to the representation is the introduction of the variance reduction techniques when $\{N_T \geq 1\}$, this is also the distinction between $\psi$ and $\tilde{\psi}$. Hence the representation is,

$$
\begin{aligned}
\hat{v}(t,x) = \mathbb{E}\Bigg[ & \mathbb{1}_{\{N_T=0\}} \frac{g(\bar{X}_{T_{N_T+1}})}{\overline{F}(\Delta T_{N_T+1})} \\
& + \mathbb{1}_{\{N_T \geq 1\}} \Bigg( \frac{\Delta g_{T_{N_T+1}}}{2\overline{F}(\Delta T_{N_T+1})} \frac{\Delta b_{N_T} \mathcal{W}_{N_T+1}^1 - \frac{1}{2}\sigma(\theta_{N_T-1}, T_{N_T})^2 \mathcal{W}_{N_T+1}^2}{f(\Delta T_{N_T})} \\
& \qquad + \frac{\Delta \hat{g}_{T_{N_T+1}}}{2\overline{F}(\Delta T_{N_T+1})} \frac{-\Delta b_{N_T} \mathcal{W}_{N_T+1}^1 - \frac{1}{2}\sigma(\theta_{N_T-1}, T_{N_T})^2 \mathcal{W}_{N_T+1}^2}{f(\Delta T_{N_T})} \Bigg) \\
& \times \prod_{k=2}^{N_T} \frac{\Delta b_{k-1} \mathcal{W}_k^1 - \frac{1}{2}\sigma(\theta_{k-2}, T_{k-1})^2 \mathcal{W}_k^2}{f(\Delta T_{k-1})} \Bigg| X_t = x,\ \sigma(\theta_0, t) \Bigg],
\end{aligned}
$$

where we are using conditioning to state the initial condition of the SDE. In order to save space in the future we will stick to conditioning $\mathcal{F}_t$. Concentrating on the case $\{N_T \geq 1\}$, then the random variable $\Delta T_1$ exists and satisfies $t < T_1 < T$. Hence we can consider the filtration up to that point and by the tower property rewrite the $\{N_T \geq 1\}$ term in the expectation as,

$$\mathbb{E}\Bigg[\mathbb{1}_{\{N_T\geq 1\}}\frac{1}{f(\Delta T_1)}\Bigg\{$$

$$\Delta b_1 \mathbb{E}\Bigg[\mathbb{1}_{\{N_T=1\}}\frac{\Delta g_{T_{N_T+1}}-\Delta\hat{g}_{T_{N_T+1}}}{2\overline{F}(\Delta T_{N_T+1})}\mathcal{W}^1_{N_T+1}$$

$$+\mathbb{1}_{\{N_T>1\}}\Bigg(\frac{\Delta g_{T_{N_T+1}}}{2\overline{F}(\Delta T_{N_T+1})}\frac{\Delta b_{N_T}\mathcal{W}^1_{N_T+1}-\frac{1}{2}\sigma(\theta_{N_T-1},T_{N_T})^2\mathcal{W}^2_{N_T+1}}{f(\Delta T_{N_T})}$$

$$+\frac{\Delta\hat{g}_{T_{N_T+1}}}{2\overline{F}(\Delta T_{N_T+1})}\frac{-\Delta b_{N_T}\mathcal{W}^1_{N_T+1}-\frac{1}{2}\sigma(\theta_{N_T-1},T_{N_T})^2\mathcal{W}^2_{N_T+1}}{f(\Delta T_{N_T})}\Bigg)$$

$$\times\prod_{k=3}^{N_T}\frac{\Delta b_{k-1}\mathcal{W}^1_k-\frac{1}{2}\sigma(\theta_{k-2},T_{k-1})^2\mathcal{W}^2_k}{f(\Delta T_{k-1})}\mathcal{W}^1_2\,\Bigg|\,\mathcal{F}_{T_1}\Bigg]$$

$$-\frac{1}{2}\sigma(\theta_0,T_1)^2\mathbb{E}\Bigg[\mathbb{1}_{\{N_T=1\}}\frac{\Delta g_{T_{N_T+1}}+\Delta\hat{g}_{T_{N_T+1}}}{2\overline{F}(\Delta T_{N_T+1})}\mathcal{W}^2_{N_T+1}$$

$$+\mathbb{1}_{\{N_T>1\}}\Bigg(\frac{\Delta g_{T_{N_T+1}}}{2\overline{F}(\Delta T_{N_T+1})}\frac{\Delta b_{N_T}\mathcal{W}^1_{N_T+1}-\frac{1}{2}\sigma(\theta_{N_T-1},T_{N_T})^2\mathcal{W}^2_{N_T+1}}{f(\Delta T_{N_T})}$$

$$+\frac{\Delta\hat{g}_{T_{N_T+1}}}{2\overline{F}(\Delta T_{N_T+1})}\frac{-\Delta b_{N_T}\mathcal{W}^1_{N_T+1}-\frac{1}{2}\sigma(\theta_{N_T-1},T_{N_T})^2\mathcal{W}^2_{N_T+1}}{f(\Delta T_{N_T})}\Bigg)$$

$$\times\prod_{k=3}^{N_T}\frac{\Delta b_{k-1}\mathcal{W}^1_k-\frac{1}{2}\sigma(\theta_{k-2},T_{k-1})^2\mathcal{W}^2_k}{f(\Delta T_{k-1})}\mathcal{W}^2_2\Bigg|\,\mathcal{F}_{T_1}\Bigg]\;\Bigg\}\Bigg|\,\mathcal{F}_t\Bigg], \tag{3.18}$$

where we have used that $\Delta b_1$ and $\sigma(\theta_0, T_1)$ are bounded and our integrability assumptions on $\Phi$ and $\tilde{\psi}^{T_1,\bar{X}_{T_1}}$ to apply the tower property. We see here that the antithetic variable is causing extra difficultly since we need to treat the case $N_T = 1$ separately.

*Step 3: Existence and continuity of derivatives.* In order to obtain the required expression we must also understand the derivatives of the function, hence we must show these derivatives exist and obtain a representation for them. One can identify the terms inside the conditional expectations as $\Phi_i^{T_{N_T},\bar{X}_{T_{N_T}}}\mathcal{W}^i_{N_T+1}$ and $\tilde{\psi}^{T_1,\bar{X}_{T_1}}\mathcal{W}^i_2$ for $i \in \{1, 2\}$.

Let us denote by $\eta(T_1, \bar{X}_{T_1}) := \mathbb{E}[\psi^{t,x}|\mathcal{F}_{T_1}]$, notice that for the same reasons $\psi^{t,x}$ is a continuous function of $x$, $\eta(T_1, \bar{X}_{T_1})$ is continuous w.r.t. $\bar{X}_{T_1}$ (which is in turn continuous w.r.t. $x$). Let us now consider derivatives of this function w.r.t. $x$. However, one should note that this expectation is on the product space of random variables $T_i$ and $W$. While the Malliavin automatic differentiation results only hold differentiating $\mathbb{E}_W[\cdot]$. Therefore we must swap the derivative with the expectation $\mathbb{E}_f$, which we have proved to be valid (actually shown a more general case) in Lemma 7.1 under the assumed integrability. Hence since we have a continuous function over a bounded interval, one can conclude via Lemma 7.1 and automatic differentiation,

$$\partial^i_x\hat{v}(t,x)=\partial^i_x\mathbb{E}\Big[\eta(T_1,\bar{X}_{T_1})\big|\mathcal{F}_t\Big]=\mathbb{E}\left[\eta(T_1,\bar{X}_{T_1})\mathcal{W}^i_1\big|\mathcal{F}_t\right]=\mathbb{E}\left[\psi^{t,x}\mathcal{W}^i_1\big|\mathcal{F}_t\right].$$

Technically we have again used the Tower property to remove the final conditional expectation which requires integrability. We now show this is valid and due to the form of $\psi$ we split into two terms,

$$\mathbb{E}\left[\psi^{t,x}\mathcal{W}_1^i\middle|\mathcal{F}_t\right] = \mathbb{E}\left[\mathbb{1}_{\{N_T=0\}}\psi^{t,x}\mathcal{W}_1^i + \mathbb{1}_{\{N_T\geq 1\}}\psi^{t,x}\mathcal{W}_1^i\middle|\mathcal{F}_t\right].$$

One can automatically see that if $N_T \geq 1$ then $\psi = \tilde{\psi}$, for the case $N_T = 0$, we need to show equivalence between $\psi$ and the corresponding $\Phi$. Firstly let us show,

$$\mathbb{E}\left[\mathbb{1}_{\{N_T=0\}}\psi^{t,x}\mathcal{W}_1^1\middle|\mathcal{F}_t\right] = \mathbb{E}\left[\mathbb{1}_{\{N_T=0\}}\Phi_1^{t,x}\mathcal{W}_1^1\middle|\mathcal{F}_t\right].$$

Expanding out $\Phi_1$ we obtain,

$$\mathbb{E}\left[\mathbb{1}_{\{N_T=0\}}\Phi_1^{t,x}\mathcal{W}_1^1\middle|\mathcal{F}_t\right] = \mathbb{E}\left[\mathbb{1}_{\{N_T=0\}}\frac{g(\bar{X}_{T_{N_T+1}}) - g(\hat{X}_{T_{N_T+1}})}{2\overline{F}(\Delta T_{N_T+1})}\mathcal{W}_1^1\middle|\mathcal{F}_t\right].$$

Using that $W$ and $-W$ have the same distribution and $\mathcal{W}^1$ is an odd function of the Brownian increment $\Delta W$ (see (3.16)) we obtain,

$$\mathbb{E}\left[\mathbb{1}_{\{N_T=0\}}\Phi_1^{t,x}\mathcal{W}_1^1\middle|\mathcal{F}_t\right] = \mathbb{E}\left[2\mathbb{1}_{\{N_T=0\}}\frac{g(\bar{X}_{T_{N_T+1}})}{2\overline{F}(\Delta T_{N_T+1})}\mathcal{W}_1^1\middle|\mathcal{F}_t\right],$$

which shows the required result. Equivalently, we now show the equality

$$\mathbb{E}\left[\mathbb{1}_{\{N_T=0\}}\psi^{t,x}\mathcal{W}_1^2\middle|\mathcal{F}_t\right] = \mathbb{E}\left[\mathbb{1}_{\{N_T=0\}}\Phi_2^{t,x}\mathcal{W}_1^2\middle|\mathcal{F}_t\right].$$

By a similar argument to above,

$$\begin{aligned}&\mathbb{E}\left[\mathbb{1}_{\{N_T=0\}}\Phi_2^{t,x}\mathcal{W}_1^2\middle|\mathcal{F}_t\right]\\ &= \mathbb{E}\left[\mathbb{1}_{\{N_T=0\}}\frac{g(\bar{X}_{T_{N_T+1}}) + g(\hat{X}_{T_{N_T+1}}) - 2g(\bar{X}_{T_{N_T}} + b(T_{N_T})\Delta T_{N_T+1})}{2\overline{F}(\Delta T_{N_T+1})}\mathcal{W}_1^2\middle|\mathcal{F}_t\right].\end{aligned}$$

By the fact that $g(\bar{X}_{T_{N_T}} + b(T_{N_T})\Delta T_{N_T+1})$ is $\overline{\mathcal{F}}_{N_T}$-adapted, and the weight has zero expectation we can remove this term from the expectation. Again, since $W$ and $-W$ have the same distribution, and $\mathcal{W}^2$ is even we obtain,

$$\mathbb{E}\left[\mathbb{1}_{\{N_T=0\}}\Phi_2^{t,x}\mathcal{W}_1^2\middle|\mathcal{F}_t\right] = \mathbb{E}\left[2\mathbb{1}_{\{N_T=0\}}\frac{g(\bar{X}_{T_{N_T+1}})}{2\overline{F}(\Delta T_{N_T+1})}\mathcal{W}_1^2\middle|\mathcal{F}_t\right],$$

again, this yields the required result. Thus the spatial derivatives of $\hat{v}$ satisfy,

$$\partial_x^i \hat{v}(t,x) = \mathbb{E}\left[\mathbb{1}_{\{N_T=0\}}\Phi_i^{t,x}\mathcal{W}_1^i + \mathbb{1}_{\{N_T\geq 1\}}\tilde{\psi}^{t,x}\mathcal{W}_1^i \middle| \mathcal{F}_t\right].$$

Uniform integrability of $\tilde{\psi}\mathcal{W}^i$ and $\Phi_i\mathcal{W}^i$ then implies $\partial_x^i \hat{v}(t,x)$ is a continuous function and one can use this integrability to also conclude $\partial_x^i \hat{v}(t,x) = \mathbb{E}\left[\psi^{t,x}\mathcal{W}_1^i \middle| \mathcal{F}_t\right]$. Thus existence of the first and second spatial derivatives are assured.

*Step 4: Representations match.* Introducing the following notation, $N_T(s) := N_T - N_s$, i.e. the number of regime switches that occur between time $s$ and $T$, with the obvious relation $N_T(t) = N_T$.

To show that the two representations are the same, we need to consider the terms $\partial_x^i \hat{v}(T_1, \bar{X}_{T_1})$ for $t \leq T_1 < T$. One has that,

$$\hat{v}(T_1, \bar{X}_{T_1}) = \mathbb{E}[\psi^{T_1,\bar{X}_{T_1}}|\mathcal{F}_{T_1}].$$

To apply derivatives we again introduce the function $\eta(T_2, \bar{X}_{T_2}) = \mathbb{E}[\psi^{T_1,\bar{X}_{T_1}}|\mathcal{F}_{T_2}]$ and then Lemma 7.1 and Malliavin automatic differentiation implies,

$$\partial_x^i \hat{v}(T_1, \bar{X}_{T_1}) = \mathbb{E}[\psi^{T_1,\bar{X}_{T_1}}\mathcal{W}_2^i|\mathcal{F}_{T_1}] \quad \mathbb{P}\text{-a.s.}$$

Using the same arguments as before we can rewrite this as,

$$\partial_x^i \hat{v}(T_1, \bar{X}_{T_1}) = \mathbb{E}\left[\mathbb{1}_{\{N_T(T_1)=0\}}\Phi_i^{T_1,\bar{X}_{T_1}}\mathcal{W}_2^i + \mathbb{1}_{\{N_T(T_1)\geq 1\}}\tilde{\psi}^{T_1,\bar{X}_{T_1}}\mathcal{W}_2^i \middle| \mathcal{F}_{T_1}\right] \quad \mathbb{P}\text{-a.s.}$$

One then recognises the internal conditional expectations in (3.18) as the derivatives of $\hat{v}$ starting at time $(T_1, \bar{X}_{T_1})$. Thus, by integrability, (3.18) can be simply written as,

$$\mathbb{E}\left[\mathbb{1}_{\{N_T\geq 1\}}\frac{1}{f(\Delta T_1)}\left(\Delta b_1 \partial_x \hat{v}(T_1, \bar{X}_{T_1}) - \frac{1}{2}\sigma(\theta_0, T_1)^2 \partial_{xx}\hat{v}(T_1, \bar{X}_{T_1})\right) \middle| \mathcal{F}_t\right].$$

This leads us to the following nonlinear relation for $\hat{v}$,

$$\hat{v}(t,x) = \mathbb{E}\left[\frac{g(\bar{X}_{T_1})}{\bar{F}(\Delta T_1)}\mathbb{1}_{\{N_T=0\}} + \mathbb{1}_{\{N_T\geq 1\}}\frac{\Delta b_1 \partial_x \hat{v}(T_1, \bar{X}_{T_1}) - \frac{1}{2}\sigma(\theta_0, T_1)^2 \partial_{xx}\hat{v}(T_1, \bar{X}_{T_1})}{f(\Delta T_1)} \middle| \mathcal{F}_t\right].$$

Since this representation and (3.17) are equal we have $v(t,x) = \hat{v}(t,x)$ hence our representation solves the PDE. □

## 3.3 Verifying the Integrability Assumptions

Theorem 3.10 relied on various integrability assumptions and our final result is to show that these assumptions hold.

**Theorem 3.13** *Let Assumptions 3.2, 3.3 and 3.5 hold. Then the integrability conditions in Theorem 3.10 hold.*

*Proof* We start by showing the uniform integrability conditions, recall that for uniform integrability to hold it is sufficient to show the stochastic process is in $L^p$ for $p > 1$ (see [20, Chap. 13] for results on uniform integrability).

Firstly, by Proposition 3.8, one can conclude that $\psi^{t,x} \in L^2$, thus we have the required uniform integrability. Let us now consider $\tilde{\psi}^{t,x}\mathcal{W}_1^1$ and $\tilde{\psi}^{t,x}\mathcal{W}_1^2$. Due to both quantities having very similar forms we consider $\tilde{\psi}^{t,x}\mathcal{W}_1^i$ for $i \in \{1, 2\}$, hence we want to show,

$$\mathbb{E}[|\tilde{\psi}^{t,x}\mathcal{W}_1^i|^p|\mathcal{F}_t] < \infty, \quad \text{for some } p > 1.$$

We show this by borrowing many of the arguments in the proof of Proposition 3.8, hence we take $p = 2$. Using the representation for $\tilde{\psi}^{t,x}$ and taking common factors we obtain,

$$\begin{aligned}\mathbb{E}[|\tilde{\psi}^{t,x}\mathcal{W}_1^i|^2|\mathcal{F}_t] \leq& \mathbb{E}\Bigg[\left(\frac{\Delta g_{T_{N_T+1}} - \Delta \hat{g}_{T_{N_T+1}}}{2\overline{F}(\Delta T_{N_T+1})} \frac{\Delta b_{N_T}\mathcal{W}^1_{N_T+1}}{f(\Delta T_{N_T})}\right)^2 \prod_{k=2}^{N_T} P_k^2 \left(\mathcal{W}_1^i\right)^2 \Big| \mathcal{F}_t\Bigg] \\ &+ \mathbb{E}\Bigg[\left(\frac{\Delta g_{T_{N_T+1}} + \Delta \hat{g}_{T_{N_T+1}}}{2\overline{F}(\Delta T_{N_T+1})} \frac{\frac{1}{2}\sigma(\theta_{N_T-1}, T_{N_T})^2\mathcal{W}^2_{N_T+1}}{f(\Delta T_{N_T})}\right)^2 \prod_{k=2}^{N_T} P_k^2 \left(\mathcal{W}_1^i\right)^2 \Big| \mathcal{F}_t\Bigg].\end{aligned}$$

We now use the same techniques from the proof of Proposition 3.8, firstly, we can condition on $N_T = \ell$ and multiply by the corresponding probability. Then by conditioning on $\overline{\mathcal{F}}_{N_T}$ (see proof of Proposition 3.8) we obtain the following,

$$\mathbb{E}\Bigg[\left(\frac{\Delta g_{T_{N_T+1}} - \Delta \hat{g}_{T_{N_T+1}}}{2\overline{F}(\Delta T_{N_T+1})} \frac{\Delta b_{N_T}\mathcal{W}^1_{N_T+1}}{f(\Delta T_{N_T})}\right)^2 \Big| \overline{\mathcal{F}}_{N_T}, \, N_T = \ell\Bigg] \leq C \frac{\Delta b^2_{N_T}}{f(\Delta T_{N_T})^2},$$

and

$$\begin{aligned}&\mathbb{E}\Bigg[\left(\frac{\Delta g_{T_{N_T+1}} + \Delta \hat{g}_{T_{N_T+1}}}{2\overline{F}(\Delta T_{N_T+1})} \frac{\frac{1}{2}\sigma(\theta_{N_T-1}, T_{N_T})^2\mathcal{W}^2_{N_T+1}}{f(\Delta T_{N_T})}\right)^2 \Big| \overline{\mathcal{F}}_{N_T}, \, N_T = \ell\Bigg] \\ &\leq \frac{C}{f(\Delta T_{N_T})^2} \frac{\sigma(\theta_{N_T-1}, T_{N_T})^4}{\sigma(\theta_{N_T}, T_{N_T+1})^2} \frac{\min\left[1, \sigma(\theta_{N_T}, T_{N_T+1})^2 \Delta T_{N_T+1}\right]}{\Delta T_{N_T+1}}.\end{aligned}$$

We now use these bounds to bound $\tilde{\psi}\mathcal{W}$. Concentrating on the $\Delta b_{N_T}$ term, we follow the finite variance proof and condition out $\Delta b^2_{N_T} P^2_{N_T}$, then use (3.11), namely,

$$\mathbb{E}\Big[\Big(\frac{\Delta g_{T_{N_T+1}} - \Delta \hat{g}_{T_{N_T+1}}}{2\overline{F}(\Delta T_{N_T+1})}\frac{\Delta b_{N_T}\mathcal{W}^1_{N_T+1}}{f(\Delta T_{N_T})}\Big)^2\prod_{k=2}^{N_T}P_k^2\left(\mathcal{W}_1^i\right)^2\Big|\,\mathcal{F}_t,\ N_T=\ell\Big]$$
$$\leq \mathbb{E}\Big[\frac{C}{f(\Delta T_{N_T})^2}\frac{1}{f(\Delta T_{N_T-1})^2}\frac{\sigma(\theta_{N_T-2},T_{N_T-1})^4}{\sigma(\theta_{N_T-1},T_{N_T})^4}\prod_{k=2}^{N_T-1}P_k^2\left(\mathcal{W}_1^i\right)^2\Big|\,\mathcal{F}_t,\ N_T=\ell\Big].$$

By continuing to follow the argument we can bound the above quantity by,

$$\mathbb{E}\Big[\frac{C^{N_T}}{f(\Delta T_{N_T})^2}\frac{\sigma(\theta_0,T_1)^4}{\sigma(\theta_{N_T-1},T_{N_T})^4}\prod_{k=2}^{N_T-1}\frac{1}{f(\Delta T_k)^2\Delta T_k^2}\frac{1}{f(\Delta T_1)^2}\left(\mathcal{W}_1^i\right)^2\Big|\,\mathcal{F}_t,\ N_T=\ell\Big]. \tag{3.19}$$

Since $\sigma_0 > 0$ is constant it is clear that,

$$\mathbb{E}[\left(\mathcal{W}_1^1\right)^2|\overline{\mathcal{F}}_0] \leq C\mathbb{E}[\left(\mathcal{W}_1^2\right)^2|\overline{\mathcal{F}}_0] \leq C\frac{1}{\Delta T_1^2}.$$

Hence we can bound (3.19),

$$\mathbb{E}\Big[\frac{C^{N_T}}{f(\Delta T_{N_T})^2}\frac{\sigma(\theta_0,T_1)^4}{\sigma(\theta_{N_T-1},T_{N_T})^4}\prod_{k=1}^{N_T-1}\frac{1}{f(\Delta T_k)^2\Delta T_k^2}\Big|\,\mathcal{F}_t,\ N_T=\ell\Big] \leq \mathbb{E}\Big[C^{N_T}\Big|\,\mathcal{F}_t,\ N_T=\ell\Big],$$

where the inequality follows from our assumptions on $f$ and $\sigma$.

Using this argument to deal with the extra Malliavin weight and the arguments in Proposition 3.8, we also obtain,

$$\mathbb{E}\Big[\Big(\frac{\Delta g_{T_{N_T+1}} + \Delta \hat{g}_{T_{N_T+1}}}{2\overline{F}(\Delta T_{N_T+1})}\frac{\frac{1}{2}\sigma(\theta_{N_T-1},T_{N_T})^2\mathcal{W}^2_{N_T+1}}{f(\Delta T_{N_T})}\Big)^2\prod_{k=2}^{N_T}P_k^2\left(\mathcal{W}_1^i\right)^2\Big|\,\mathcal{F}_t\Big]$$
$$\leq \mathbb{E}\left[C^{N_T}\frac{\sigma(\theta_{N_T},T_{N_T+1})^\nu\Delta T_{N_T+1}^{\nu/2}}{\sigma(\theta_{N_T},T_{N_T+1})^2}\Delta T_{N_T+1}^{-1}\prod_{k=1}^{N_T}\Delta T_k^{-2\kappa}\Big|N_T=\ell\right],\quad \text{for } \nu\in[0,2].$$

The finiteness of these bounds follows directly from Proposition 3.8.

For the $f(\Delta T_1)^{-1}\Delta b_1\tilde{\psi}^{T_1,\bar{X}_{T_1}}\mathcal{W}_2^1$ and $f(\Delta T_1)^{-1}\sigma(\theta_0,T_1)^2\tilde{\psi}^{T_1,\bar{X}_{T_1}}\mathcal{W}_2^2$ terms, these follow automatically from Proposition 3.8.

For uniform integrability of $\Phi_1\mathcal{W}^1$, take $p=2$ as above. Then use Cauchy-Schwarz and the Lipschitz property of $g$, which yields $|\Delta g_{T_{N_T+1}} - \Delta\hat{g}_{T_{N_T+1}}| \leq C\sigma(\theta_{N_T},T_{N_T+1})|\Delta W_{T_{N_T+1}}|$. One notes that the $\sigma$ and $\Delta T$ terms cancel and hence finite.

Similarly, for $\Phi_2\mathcal{W}^2$, again take $p=2$ and use Cauchy-Schwarz along with (3.8). Again all terms cancel which implies this is also finite and hence uniformly integrable.

The final integrability results we require are all $\mathbb{P}$-a.s. results. We have $\psi^{T_1,\bar{X}_{T_1}}$, $\Delta b_2\tilde{\psi}^{T_2,\bar{X}_{T_2}}\mathcal{W}_3^1$ and $\sigma(\theta_1,T_2)^2\tilde{\psi}^{T_2,\bar{X}_{T_2}}\mathcal{W}_3^2$ are $\mathbb{P}$-a.s. uniformly integrable, and

$\tilde{\psi}^{T_1,\bar{X}_{T_1}}\mathcal{W}_2^1$ and $\tilde{\psi}^{T_1,\bar{X}_{T_1}}\mathcal{W}_2^2$ are $\mathbb{P}$-a.s. integrable. However, these follow from the arguments above along with the fact that $t < T_1 < T_2$ $\mathbb{P}$-a.s. hence $\sigma(\theta_1, T_2) < \infty$ $\mathbb{P}$-a.s. Hence we have shown all the required integrability conditions to use Theorem 3.10. □

The proof of Theorem 3.7 follows in a straightforward way by combining these results.

*Proof of Theorem* 3.7. By letting Assumptions 3.2, 3.3 and 3.5 hold, then Theorems 3.10 and 3.13 imply that our estimator $\tilde{v}$ given in (3.5) solves the PDE (3.1).

Moreover, Proposition 3.8, implies that $\psi$ is square integrable and hence of finite variance. □

# 4 Towards the General Case and Future Work

The methodology presented in this work can be extended to accommodate PDEs of the form,

$$\begin{cases} \partial_t v(t,x) + b(t)\cdot Dv(t,x) + h(t,x) = 0 \quad \text{for all } (t,x) \in [0,T) \times \mathbb{R}^d\,, \\ v(T,x) = g(x)\,, \end{cases} \tag{4.1}$$

where $h$ is a nice function and we still have $v \in C_b^{1,3}$. As in the case of standard branching representations one introduces a further probability measure $\mathbb{P}_B$ on the space $\{0, 1\}$, where 0 signifies the case the particles dies (this can be thought of as a $v^0$ term) at position $(T_k, \bar{X}_{T_k})$ and we evaluate $h$ at this position.

## *4.1 Allowing the Drift to Have a Spatial Dependence*

Throughout this chapter we have made the assumption that the drift $b$ does not depend on space. The main reason for this is to ensure finite variance. One can consider replacing Assumption 3.2, with $b : [0, T] \times \mathbb{R} \to \mathbb{R}$, satisfying 1/2-Hölder in time, Lipschitz in space and uniformly bounded and most of the arguments presented still hold. The bound that changes and makes the arguments more difficult is (3.9), to see this let us observe how $\Delta b$ is bounded under these new assumptions,

$$\begin{aligned} &\mathbb{E}\big[\Delta b_k^4 | \overline{\mathcal{F}}_{k-1}, N_T = \ell\big] \\ &\quad \le C\mathbb{E}\Big[(b(T_k, \bar{X}_{T_k}) - b(T_k, \bar{X}_{T_{k-1}}))^4 + (b(T_k, \bar{X}_{T_{k-1}}) - b(T_{k-1}, \bar{X}_{T_{k-1}}))^4 | \overline{\mathcal{F}}_{k-1}, N_T = \ell\Big]. \end{aligned}$$

For the second term we can use 1/2-Hölder continuity in time of $b$, for the first term we can Lipschitz continuity in space to obtain,

$$\begin{aligned}\mathbb{E}[(b(T_k, \bar{X}_{T_k}) - b(T_k, \bar{X}_{T_{k-1}}))^4 | \overline{\mathcal{F}}_{k-1}, N_T = \ell] \leq & C\mathbb{E}[(\bar{X}_{T_k} - \bar{X}_{T_{k-1}})^4 | \overline{\mathcal{F}}_{k-1}, N_T = \ell] \\ \leq & C\mathbb{E}[(\Delta T_k + \sigma(\theta_{k-1}, T_k)\Delta W_{T_k})^4 | \overline{\mathcal{F}}_{k-1}, N_T = \ell] \\ \leq & C\sigma(\theta_{k-1}, T_k)^4 \Delta T_k^2 .\end{aligned}$$

Since $\sigma$ is bounded from below we can conclude,

$$\mathbb{E}[\Delta b_k^4 | \overline{\mathcal{F}}_{k-1}, N_T = \ell] \leq C\sigma(\theta_{k-1}, T_k)^4 \Delta T_k^2 .$$

It is also straightforward to see the same bound applies if we take $b$ Lipschitz in time. The bounds on $M$ and $V$ still have the form

$$\begin{aligned}\mathbb{E}[M_{k+1}^4 | \overline{\mathcal{F}}_k, N_T = \ell] &\leq C\frac{\Delta b_k^4}{\Delta T_{k+1}^2} \frac{1}{\sigma(\theta_k, T_{k+1})^4} , \\ \mathbb{E}[V_{k+1}^4 | \overline{\mathcal{F}}_k, N_T = \ell] &\leq C\frac{\sigma(\theta_{k-1}, T_k)^8}{\sigma(\theta_k, T_{k+1})^8} \frac{1}{\Delta T_{k+1}^4} ,\end{aligned}$$

although one should note that we cannot use the $\Delta b$ bound above in the $M$ term since they are w.r.t. different conditional expectations. That being said though one can still observe where a problem arises by considering,

$$\begin{aligned}&\mathbb{E}\big[\mathbb{E}[P_{k+1}^4 | \overline{\mathcal{F}}_k, N_T = \ell] \big| \overline{\mathcal{F}}_{k-1}, N_T = \ell\big] \\ &\leq C\mathbb{E}\Big[\frac{1}{f(\Delta T_k)^4}\Big(\frac{\Delta T_k^2}{\Delta T_{k+1}^2} \frac{\sigma(\theta_{k-1}, T_k)^4}{\sigma(\theta_k, T_{k+1})^4} + \frac{\sigma(\theta_{k-1}, T_k)^8}{\sigma(\theta_k, T_{k+1})^8} \frac{1}{\Delta T_{k+1}^4}\Big)\Big| \overline{\mathcal{F}}_{k-1}, N_T = \ell\Big].\end{aligned}$$

Whereas in the proof we can bound (3.9) by the term arising from the $V$ (i.e. the $V$ bound dominates the $M$ bound), that is not the case here. To see this take $n = -1$ for the coefficient in the $\sigma$, we then obtain,

$$C\mathbb{E}\Big[\frac{1}{f(\Delta T_k)^4} \frac{\Delta T_k^6}{\Delta T_{k+1}^2}\Big(1 + \frac{\Delta T_k^2}{\Delta T_{k+1}^2}\Big)\Big| \overline{\mathcal{F}}_{k-1}, N_T = \ell\Big].$$

Therefore the 1 (term arising from the $M$) is larger if $\Delta T_k < \Delta T_{k+1}$, hence we cannot dominate in the same way. As it turns out this a not a problem for obtaining (3.12), however, it does become an issue for obtaining (3.15). This appears because (3.15) relies on a cancelling argument, while this extra term changes the original bound from,

$$\mathbb{E}\Big[\frac{1}{f(\Delta T_k)^4} \frac{\Delta T_k^8}{\Delta T_{k+1}^4}\Big| \overline{\mathcal{F}}_{k-1}, N_T = \ell\Big] \quad \text{to} \quad \mathbb{E}\Big[\frac{1}{f(\Delta T_k)^4} \frac{\Delta T_k^8}{\Delta T_{k+1}^4 \Delta T_k^2}\Big| \overline{\mathcal{F}}_{k-1}, N_T = \ell\Big].$$

This extra $\Delta T_k$ dependency makes the bound far weaker and consequently proving finite variance becomes more difficult. Of course the new bound we have obtained is not sharp, for example in the case $\Delta T_k \geq \Delta T_{k+1}$ we can return the original bound.

If we wish to argue the proof in a similar way one must either look to obtain a stronger bound on $\Delta b$ (this is essentially why $b$ in Assumption 3.2 worked), or one can find a way to make the $V$ term dominate without increasing its size so much to break the remainder of the proof. For example, an interesting route to explore is to add an event probability distribution to the $M$ and $V$ term (similar to other branching diffusion algorithms) applying a judicious choice of probability distribution may give us the means to bound the $M$ term by $V$ again.

There are of course many different approaches one can take to solve this problem and as described, the remaining arguments in Theorems 3.10 and 3.13 follow with a more general $b$. But proving finite variance of this representation remains an open question.

### 4.2 Fully Nonlinear First Order Case

Of course the true end goal of this work is to handle nonlinearities, for example, Burger's type $vDv$, which arise in many applications and for which numerical methods like characteristics cannot apply. Therefore future work will be on addressing explicit conditions under which this method provides solutions to transport PDEs of the form,

$$\begin{cases} \partial_t v(t,x) + b(t,x) \cdot Dv(t,x) = f(t,x,v,Dv)\,, \\ v(T,x) = g(x)\,, \end{cases}$$

where $f$ is polynomial in $v$ and $Dv$.

Handling such general first order PDEs will require additional arguments to what we have presented here. However, ideas from the case $b(t,x)$ along with the (purely numerical) technique presented in [18] may yield the necessary tools to overcome such equations.

*Remark 4.1* [Requirement for Smooth Solutions] In theory this technique should be able to extend to the general, fully nonlinear case, one will still require a sufficiently smooth classical solution to the underlying PDE. The reason for this is due to the fact we assign a representation to $\partial_{xx} v$, thus we automatically require existence of this quantity.

This implies that if we argue that the representation solves the PDE via viscosity solutions then we in fact show a classical solution. Of course this implies the method is not suitable for PDEs with "shocks".

## 5 Examples

We show the potential of this method on two examples to compare this technique against the standard perturbation technique. The first example is a simple linear PDE which satisfies all of our assumptions and hence is only an example to show that our algorithm converges to the true, while the perturbation converges to a different value. The second is a nonlinear first order PDE, this is the more interesting case and we still observe our method giving reasonable results.

## 5.1 *Simple First Order PDE*

Let us consider the following linear PDE,

$$\begin{cases} \partial_t v(t,x) + \partial_x v(t,x) = 0 \quad \text{for all } (t,x) \in [0,1) \times \mathbb{R}, \\ v(1,x) = 10\cos(x-1-5). \end{cases} \tag{5.1}$$

It is then clear to see that $v(t,x) = 10\cos(x-t-5)$ satisfies this PDE. Although such a PDE is easy to solve it serves as a good example to show the issue using a perturbation. We want to solve this PDE at the point (0, 10), where the true solution is $\approx 2.84$. By considering the case where we perturb by $\sigma = 0.1$, and then estimate the expectation using varying amounts of Monte Carlo simulations, see Fig. 1. To get a handle on the variance (error) we ran the simulation 50 times, plotted the average and the approximate 90% confidence interval. That is we view the largest and smallest value as a proxy for convergence of the algorithm. For the unbiased algorithm we also took, $n = -1$ and for the Gamma parameters $\kappa = 1/2$ and $\eta = 2$.

What is clear from Fig. 1 is, as the number of Monte Carlo simulations increase, both algorithms are converging. However the perturbed case stays at a constant level

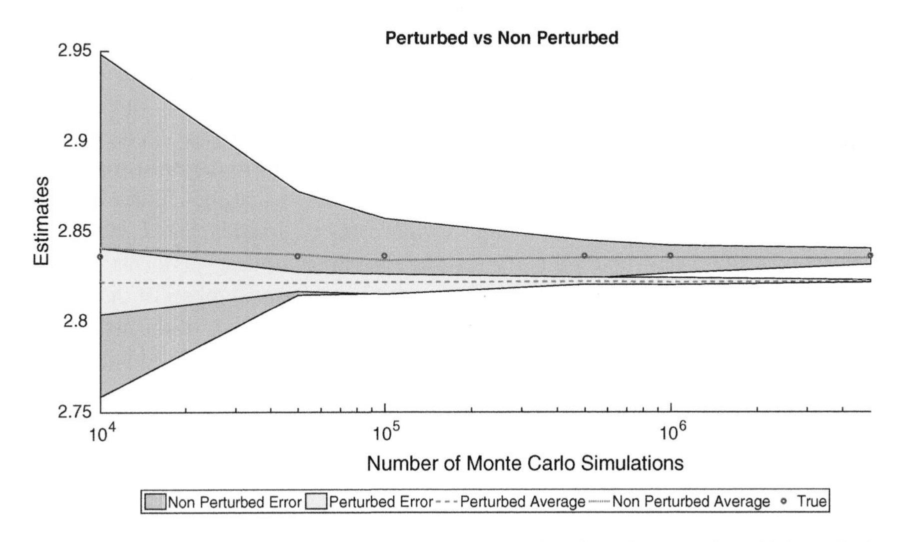


**Fig. 1** Shows the error and estimates of the solution as a function of the number of Monte Carlo simulations. The error corresponds to the approximate 90% confidence interval

away from the true, which implies that the estimate is biased (as was expected). Therefore no amount of Monte Carlo simulations will yield the true solution. For the unbiased algorithm, although having a higher variance, we see that the average hovers around the true value and moreover we observe convergence towards this point.

Hence the stochastic representation we derive indeed yields the true solution of the PDE, what is more fascinating and important about this result though is $\sigma$ is not tending to zero, in fact we can bound it from below, this is the key step when it comes to more complex PDEs.

Moreover, this calculation was carried out using a basic Monte Carlo algorithm, one could look to more sophisticated techniques as appearing in [6] where the authors apply particle methods for an improved convergence.

## 5.2 *Nonlinear PDE*

Let us now generalise to the nonlinear setting and consider the following PDE,

$$\begin{cases} \partial_t v(t,x) + \partial_x v(t,x) + \frac{1}{10}\big((\partial_x v(t,x))^2 + v(t,x)^2 - 1\big) = 0 & \text{for all } (t,x) \in [0,1) \times \mathbb{R}\,, \\ v(1,x) = \cos(1-x)\,. \end{cases} \tag{5.2}$$

We have taken this PDE since it is simple to observe that $v(t,x) = \cos(t-x)$ is the solution. It also is nice enough that one would expect our unbiased algorithm and the perturbation algorithm to work reasonably well. We want to solve this at the point $(0, 1)$.

▷ *Convergence issue for the perturbation algorithm* One can note that, applying the perturbation technique implies that the resulting PDE is a second order semilinear PDE, and hence the corresponding branching algorithm is given in [12]. This creates a problem for the convergence of the algorithm, Assumption 3.10 and Theorem 3.12 of [12] give minimum bounds on the relative size of the drift to the diffusion, even for (5.2) which has a extremely nice solution, we observe that the algorithm fails to converge for $\sigma_0 = 0.5$ and has a large variance for $\sigma_0$ smaller than 1. Needless to say this is not a desirable property for the algorithm to have; perturbation can only work as a method if the perturbation is small and here we observe that there is a lower bound on the size of the perturbation and hence the bias of the estimator. Furthermore, as it turns out, there is no such problem with our unbiased algorithm and one can observe convergence for $\sigma_0 < 0.5$.

With the above in mind, in order to make the two algorithms comparable we set the perturbed algorithm as $\sigma_0 = 1$, but the remaining parameters are as above. Because the variance here is larger than the linear PDE we consider 100 realisations for each Monte Carlo level then take the approximate 80% confidence intervals and the average is then based on these 80 realisations. Furthermore, because we are dealing with nonlinear terms we have a more complex representation and need to

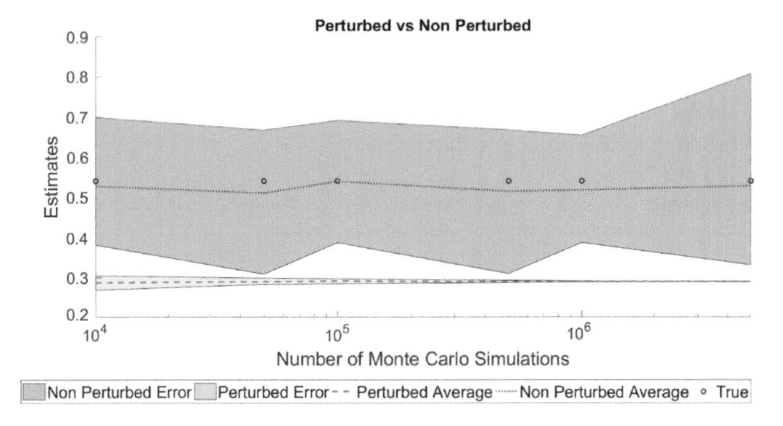


**Fig. 2** Shows the error and estimates of the solution as a function of the number of Monte Carlo simulations. The error corresponds to the approximate 80% confidence interval

establish a probability distribution for the type of event i.e. $v^2$, $(\partial_x v)^2$ etc. This is well understood in the case of the perturbation algorithm (see [12]), however, the variance of our unbiased algorithm seems to be highly dependent on how one chooses this probability distribution.

Figure 2 shows that yet again our unbiased algorithm provides a correction for the second order term. While the perturbation algorithm converges to a different value. However, it is clear that the variance in our algorithm is much higher. One of the reasons for this is because of the uncertainty in what events will be used for each realisation. Namely, for the linear PDE case, there was no probability distribution over events and this allowed us to bound the variance. In this more general case, more work would have to be done in order to bound the variance, and from our numerical example the choice of probability distribution has a role to play here.

## 6 Conclusions and Outlook

We have demonstrated a stochastic algorithm capable of dealing with first order PDEs, where originally such PDEs seemed beyond the reach of stochastic methods without approximation. This has potentially large implications for numerics of such PDEs since stochastic algorithms can easily be parallelised and scale favourable with dimension as argued in [3].

Due to the added difficulty in considering more general transport PDEs we have taken a simple case here. As a consequence we have left some open problems to be addressed, namely:

1. Finite variance estimator when the drift component also depends on space.
2. Dropping the assumption on the initial PDE having a classical solution.

3. Extending to the case of nonlinear terms in both the solution of the PDE and its first spatial derivative.

Our hope is that with the continued research and innovation into branching diffusions that such results will be within reach.

**Acknowledgements** The authors would like to thank Nizar Touzi (École Polytechnique Paris) and Christa Cuchiero (Vienna University) for the helpful discussions.

## 7 Technical Result: Swapping Differentiation with Integration

When deriving the PDE we swapped the operators $\partial_x$ with $\mathbb{E}_f$. This essentially requires taking a limit inside an integral, hence we show this is valid in this setting. A similar result was tackled in [15, Lemma A2], although our proof follows similar ideas to the one presented there, our version relaxes some of the conditions on the second derivative.

**Lemma 7.1** *Let Assumptions 3.2, 3.3 and 3.5 hold. Let* $\psi^{T_1,\bar{X}_{T_1}}$, $\Delta b_2 \tilde{\psi}^{T_2,\bar{X}_{T_2}} \mathcal{W}_3^1$ *and* $\sigma(\theta_1, T_2)^2 \tilde{\psi}^{T_2,\bar{X}_{T_2}} \mathcal{W}_3^2$ *be* $\mathbb{P}$*-a.s. uniformly integrable, let* $\tilde{\psi}^{T_1,\bar{X}_{T_1}} \mathcal{W}_2^1$ *and* $\tilde{\psi}^{T_1,\bar{X}_{T_1}} \mathcal{W}_2^2$ *be* $\mathbb{P}$*-a.s. integrable (as defined in Theorem 3.7), and define the function*

$$\hat{v}(T_1, \bar{X}_{T_1}) := \mathbb{E}_f[\mathbb{E}_W[\psi^{T_1,\bar{X}_{T_1}} | \mathcal{F}_{T_1}] | \mathcal{F}_{T_1}] .$$

*Then for* $i \in \{1, 2\}$,

$$\partial_x^i \hat{v}(T_1, \bar{X}_{T_1}) = \mathbb{E}_f[\partial_x^i \mathbb{E}_W[\psi^{T_1,\bar{X}_{T_1}} | \mathcal{F}_{T_1}] | \mathcal{F}_{T_1}] \quad \mathbb{P}\text{-a.s.}$$

*Proof* Technically, the results below are for random variables and hence should be viewed in the a.s. sense, however, for ease of presentation we suppress writing a.s. at the end of each equation. Let us start by noting that,

$$\psi^{T_1,\bar{X}_{T_1}} = \mathbb{1}_{\{N_T(T_1)=0\}} \frac{g(\bar{X}_{T_{N_T+1}})}{\overline{F}(\Delta T_{N_T+1})} + \mathbb{1}_{\{N_T(T_1)\geq 1\}} \beta \prod_{k=3}^{N_T} P_k ,$$

where $N_T(T_1) = N_T - N_{T_1}$. Observing that we can remove the time integral for the case $N_T(T_1) = 0$, that is,

$$\hat{v}(T_1, \bar{X}_{T_1}) = \mathbb{E}_f[\mathbb{E}_W[\mathbb{1}_{\{N_T(T_1)=0\}} \psi^{T_1,\bar{X}_{T_1}} + \mathbb{1}_{\{N_T(T_1)\geq 1\}} \psi^{T_1,\bar{X}_{T_1}} | \mathcal{F}_{T_1}] | \mathcal{F}_{T_1}] ,$$

and by integrability we have

$$\begin{aligned}\mathbb{E}_f[\mathbb{E}_W[\mathbb{1}_{\{N_T(T_1)=0\}}\psi^{T_1,\bar{X}_{T_1}}|\mathcal{F}_{T_1}]|\mathcal{F}_{T_1}] &= \mathbb{E}_W[\mathbb{E}_f[\mathbb{1}_{\{N_T(T_1)=0\}}\psi^{T_1,\bar{X}_{T_1}}|\mathcal{F}_{T_1}]|\mathcal{F}_{T_1}]\\ &= \mathbb{E}_W[g(\bar{X}_{T_{N_T+1}})|\mathcal{F}_{T_1}].\end{aligned}$$

Hence we only need to consider the case $N_T(T_1) \geq 1$ hence $T_2 < T$. To make the proof easier we define the function $\varphi$ for $T_1 < T_2 < T$ and $\bar{X}_{T_2} \in \mathbb{R}$ as follows,

$$\frac{1}{f(\Delta T_2)}\varphi^{T_1,\bar{X}_{T_1}}(T_2, \bar{X}_{T_2}) = \mathbb{E}[\mathbb{1}_{\{N_T(T_1)\geq 1\}}\psi^{T_1,\bar{X}_{T_1}}|\mathcal{F}_{T_2}].$$

Following the argument as in Theorem 3.10 one can conclude from our uniform integrability assumption that for any $T_1 < T_2 < T$, $\varphi^{T_1,\bar{X}_{T_1}}(T_2, \bar{X}_{T_2})$ is $\mathbb{P}$-a.s. continuous in space i.e. w.r.t. $\bar{X}_{T_2}$. Further for any fixed $t < T_1 < T_2$, $\varphi$ is bounded in space. To see this one can observe for $T_2 < T$,

$$\begin{aligned}|\varphi^{T_1,\bar{X}_{T_1}}(T_2, \bar{X}_{T_2})| =& |f(\Delta T_2)\mathbb{E}[\mathbb{1}_{\{N_T(T_1)\geq 1\}}\psi^{T_1,\bar{X}_{T_1}}|\mathcal{F}_{T_2}]|\\ =& \Bigg|f(\Delta T_2)\mathbb{E}\Bigg[\Bigg(\frac{\Delta g_{T_{N_T+1}}}{2\overline{F}(\Delta T_{N_T+1})}\frac{\Delta b_{N_T}\mathcal{W}^1_{N_T+1} - \frac{1}{2}\sigma(\theta_{N_T-1}, T_{N_T})^2\mathcal{W}^2_{N_T+1}}{f(\Delta T_{N_T})}\\ &+ \frac{\Delta\hat{g}_{T_{N_T+1}}}{2\overline{F}(\Delta T_{N_T+1})}\frac{-\Delta b_{N_T}\mathcal{W}^1_{N_T+1} - \frac{1}{2}\sigma(\theta_{N_T-1}, T_{N_T})^2\mathcal{W}^2_{N_T+1}}{f(\Delta T_{N_T})}\Bigg)\\ &\times\prod_{k=3}^{N_T}\frac{\Delta b_{k-1}\mathcal{W}^1_k - \frac{1}{2}\sigma(\theta_{k-2}, T_{k-1})^2\mathcal{W}^2_k}{f(\Delta T_{k-1})}\Bigg|\mathcal{F}_{T_2}\Bigg]\Bigg|.\end{aligned}$$

Removing $\mathcal{F}_{T_2}$-measurable terms and noticing that the remaining terms are integrable and $\Delta b_{k-1} < C$ independent of $\bar{X}_{T_2}$, we have $\varphi^{T_1,\bar{X}_{T_1}}(T_2, \cdot)$ is bounded in space, as required. Hence we can consider the following bounded Lipschitz approximation to $\varphi$,

$$\varphi_n^{T_1,\bar{X}_{T_1}}(T_2, x) := \inf_{y\in\mathbb{R}}\Big\{\varphi^{T_1,\bar{X}_{T_1}}(T_2, y) + n|x - y|\Big\}.$$

One can observe this approximation is both pointwise convergent and increasing in $n$. We therefore work with this approximation and take the limit to complete the proof.

Let us consider differentiating w.r.t. $x$, and in order to make all steps clear let us explicitly write each expectation. Using the tower property to write $\hat{v}$ in terms of $\varphi$ then making the approximation we obtain,

$$\begin{aligned}&\partial_x\mathbb{E}_f\left[\mathbb{E}_W\left[\mathbb{1}_{\{N_T(T_1)\geq 1\}}\frac{1}{f(\Delta T_2)}\varphi_n^{T_1,\bar{X}_{T_1}}(T_2, \bar{X}_{T_2})\Big|\mathcal{F}_{T_1}\right]\Big|\mathcal{F}_{T_1}\right]\\ &= \lim_{\epsilon\to 0}\mathbb{E}_f\left[\frac{1}{\epsilon}\mathbb{E}_W\left[\mathbb{1}_{\{N_T(T_1)\geq 1\}}\frac{1}{f(\Delta T_2)}(\varphi_n^{T_1,\bar{X}_{T_1}+\epsilon}(T_2, \bar{X}^\epsilon_{T_2}) - \varphi_n^{T_1,\bar{X}_{T_1}}(T_2, \bar{X}_{T_2}))\Big|\mathcal{F}_{T_1}\right]\Big|\mathcal{F}_{T_1}\right],\end{aligned}$$

where we are using the notation $\bar{X}^{\epsilon}_{T_2}$ to denote the SDE with initial condition perturbed by $\epsilon$. Dominated convergence theorem implies we can take the limit inside the expectation if we show the "integrand" to be bounded. Using the Lipschitz assumption on $\varphi_n$, one has that,

$$|\varphi_n^{T_1,\bar{X}_{T_1}+\epsilon}(T_2,\bar{X}^{\epsilon}_{T_2}) - \varphi_n^{T_1,\bar{X}_{T_1}}(T_2,\bar{X}_{T_2})| \leq C|\bar{X}^{\epsilon}_{T_2} - \bar{X}_{T_2}| .$$

As stated in [15, Lemma A2], since $\bar{X}$ has constant coefficients the following bound holds,

$$\mathbb{E}\left[\left|\frac{\bar{X}^{\epsilon}_{T_2} - \bar{X}_{T_2}}{\epsilon}\right|^2 \Big|\mathcal{F}_{T_1}\right] \leq C , \tag{7.1}$$

further, since $1/f(\Delta T_2) \leq C$ by dominated convergence theorem we can take the limit inside $\mathbb{E}_f$ to conclude,

$$\begin{aligned}
&\partial_x \mathbb{E}_f\left[\mathbb{E}_W\left[\mathbb{1}_{\{N_T(T_1)\geq 1\}}\frac{\varphi_n^{T_1,\bar{X}_{T_1}}(T_2,\bar{X}_{T_2})}{f(\Delta T_2)}\Big|\mathcal{F}_{T_1}\right]\Big|\mathcal{F}_{T_1}\right] \\
&\quad= \mathbb{E}_f\left[\partial_x \mathbb{E}_W\left[\mathbb{1}_{\{N_T(T_1)\geq 1\}}\frac{\varphi_n^{T_1,\bar{X}_{T_1}}(T_2,\bar{X}_{T_2})}{f(\Delta T_2)}\Big|\mathcal{F}_{T_1}\right]\Big|\mathcal{F}_{T_1}\right] .
\end{aligned}$$

Completing the proof for the first derivative requires showing one can take the $\lim_{n\to\infty}$, however, we suppress this here and concentrate on the second derivative. One can check this holds by following the arguments presented in the case of the second derivative.

Again using the sequence of bounded Lipschitz functions we consider,

$$\begin{aligned}
&\partial_x^2 \mathbb{E}_f\left[\mathbb{E}_W\left[\mathbb{1}_{\{N_T(T_1)\geq 1\}}\frac{\varphi_n^{T_1,\bar{X}_{T_1}}(T_2,\bar{X}_{T_2})}{f(\Delta T_2)}\Big|\mathcal{F}_{T_1}\right]\Big|\mathcal{F}_{T_1}\right] \\
&= \lim_{\epsilon\to 0}\mathbb{E}_f\left[\frac{1}{\epsilon}\mathbb{E}_W\left[\mathbb{1}_{\{N_T(T_1)\geq 1\}}\frac{\varphi_n^{T_1,\bar{X}_{T_1}+\epsilon}(T_2,\bar{X}^{\epsilon}_{T_2}) - \varphi_n^{T_1,\bar{X}_{T_1}}(T_2,\bar{X}_{T_2})}{f(\Delta T_2)}\mathcal{W}_2^1\Big|\mathcal{F}_{T_1}\right]\Big|\mathcal{F}_{T_1}\right] ,
\end{aligned}$$

where we have used our first derivative result and the fact that $\varphi_n$ is a bounded Lipschitz function to rewrite this derivative with a Malliavin weight. To bound this term one can apply Cauchy-Schwarz, use (7.1) and,

$$\mathbb{E}_W\left[\left(\frac{\mathcal{W}_2^1}{f(\Delta T_2)}\right)^2\Big|\mathcal{F}_{T_1}\right] \leq C .$$

Hence we can again apply dominated convergence theorem to obtain,

$$
\begin{aligned}
&\partial_x^2 \mathbb{E}_f \left[ \mathbb{E}_W \left[ \mathbb{1}_{\{N_T(T_1)\geq 1\}} \frac{\varphi_n^{T_1,\bar{X}_{T_1}}(T_2,\bar{X}_{T_2})}{f(\Delta T_2)} \Big| \mathcal{F}_{T_1} \right] \Big| \mathcal{F}_{T_1} \right] \\
&= \mathbb{E}_f \left[ \partial_x^2 \mathbb{E}_W \left[ \mathbb{1}_{\{N_T(T_1)\geq 1\}} \frac{\varphi_n^{T_1,\bar{X}_{T_1}}(T_2,\bar{X}_{T_2})}{f(\Delta T_2)} \Big| \mathcal{F}_{T_1} \right] \Big| \mathcal{F}_{T_1} \right].
\end{aligned}
$$

To complete the proof we need to also take the $\lim_{n\to\infty}$, and have the expected values the same. Firstly recall that $\varphi$ is an upperbound for $\varphi_n$, hence the result follows from the monotone convergence theorem (see [20, Sect. 5.3]). Alternatively, one can use the upper bound and uniform integrability results in Theorem 3.13 to take the $\lim_{n\to\infty}$. □

## References


1. Agarwal, A., Claisse, J.: Branching diffusion representation of quasi-linear elliptic PDEs and estimation using Monte Carlo method. arXiv preprint arXiv:1704.00328 (2017)
2. Andersson, P., Kohatsu-Higa, A.: Unbiased simulation of stochastic differential equations using parametrix expansions. Bernoulli **23**(3), 2028–2057 (2017)
3. Bernal, F., dos Reis, G., Smith, G.: Hybrid PDE solver for data-driven problems and modern branching. Eur. J. Appl. Math. 1–24 (2017)
4. Crisan, D., Manolarakis, K.: Probabilistic methods for semilinear partial differential equations. Applications to finance. Math. Model. Numer. Anal. **44**(5), 1107 (2010)
5. Cuchiero, C., Teichmann, J.: Stochastic representations of ordinary differential equations via affine processes. Working paper (2017)
6. Doumbia, M., Oudjane, N., Warin, X.: Unbiased monte carlo estimate of stochastic differential equations expectations. ESAIM: Probab. Statis. **21**, 56–87 (2017)
7. Evans, L.C.: Partial Differential Equations. American Mathematical Society, Providence, R.I. (1998)
8. Fahim, A., Touzi, N., Warin, X.: A probabilistic numerical method for fully nonlinear parabolic PDEs. Ann. Appl. Probab. 1322–1364 (2011)
9. Fournié, E., Lasry, J.-M., Lebuchoux, J., Lions, P.-L., Touzi, N.: Applications of malliavin calculus to monte carlo methods in finance. Financ. Stochast. **3**(4), 391–412 (1999)
10. Han, J., Jentzen, A., Weinan, E.: Overcoming the curse of dimensionality: solving high-dimensional partial differential equations using deep learning. arXiv preprint arXiv:1707.02568 (2017)
11. Henry-Labordere, P.: Counterparty risk valuation: a marked branching diffusion approach. SSRN **1995503** (2012)
12. Henry-Labordere, P., Oudjane, N., Tan, X., Touzi, N., Warin, X., et al.: Branching diffusion representation of semilinear PDEs and Monte Carlo approximation. In: Annales de l'Institut Henri Poincaré, Probabilités et Statistiques, vol. 55(1), pp. 184–210, Institut Henri Poincaré (2019)
13. Henry-Labordere, P., Touzi, N.: Branching diffusion representation for nonlinear Cauchy problems and Monte Carlo approximation. arXiv preprint arXiv:1801.08794 (2018)
14. Henry-Labordere, P., Tan, X., Touzi, N.: A numerical algorithm for a class of BSDEs via the branching process. Stochast. Process. Appl. **124**(2), 1112–1140 (2014)
15. Henry-Labordere, P., Tan, X., Touzi, N.: Unbiased simulation of stochastic differential equations. Ann. Appl. Probab. **27**(6), 3305–3341 (2017)

16. Kato, T.: The Cauchy problem for quasi-linear symmetric hyperbolic systems. Arch. Rational Mech. Anal. **58**(3), 181–205 (1975)
17. Rasulov, A., Raimova, G., Mascagni, M.: Monte Carlo solution of Cauchy problem for a nonlinear parabolic equation. Math. Comput. Simul. **80**(6), 1118–1123 (2010)
18. Warin, X.: Variations on branching methods for nonlinear PDEs. arXiv:1701.07660 (2017)
19. Warin, X.: Monte Carlo for high-dimensional degenerated semi linear and full non linear PDEs. arXiv preprint arXiv:1805.05078 (2018)
20. Williams, D.: Probability with martingales. Cambridge University Press (1991)

# Path-Dependent SDEs in Hilbert Spaces

**Mauro Rosestolato**

**Abstract** We study path-dependent SDEs in Hilbert spaces. By using methods based on contractions in Banach spaces, we prove the Gâteaux differentiability of generic order $n$ of mild solutions with respect to the starting point and the continuity of the Gâteaux derivatives with respect to all the data.

**Keywords** Stochastic functional differential equations in Hilbert spaces · Gâteaux differentiability · Contraction mapping theorem

**AMS 2010 Subject Classification** 37C25 · 34K50 · 37C05 · 47H10 · 47J35 · 58C20, 58D25, 60G99, 60H10.

## 1 Introduction

In this paper we deal with mild solutions to path-dependent SDEs evolving in a separable Hilbert space $H$, of the form

$$\begin{cases} dX_s = (AX_s + b((\cdot, s), X))ds + \sigma((\cdot, s), X)dW_s & \forall s \in (t, T] \\ X_s = Y_s & s \in [0, t], \end{cases} \tag{1.1}$$

where $t \in [0, T)$, $Y$ is a $H$-valued adapted process defined on a filtered probability space $(\Omega, \mathscr{F}, \{\mathscr{F}_t\}_{t\in[0,T]}, \mathbb{P})$, $W$ is a cylindrical Wiener process on $(\Omega, \mathscr{F}, \{\mathscr{F}_t\}_{t\in[0,T]}, \mathbb{P})$ taking values in a separable Hilbert space $U$, $b((\omega, s), X)$ is a $H$-valued random variable depending on $\omega \in \Omega$, on the time $s$, and on the path $X$, $\sigma((\omega, s), X)$ is a $L_2(U, H)$-valued random variable depending on $\omega \in \Omega$, on the time $s$, and on the path $X$, and $A$ is the generator of a $C_0$-semigroup $S$ on $H$. By using

This research has been partially supported by the ERC 321111 Rofirm.

M. Rosestolato (✉)
CMAP, École Polytechnique, Paris, France
e-mail: mauro.rosestolato@polytechnique.edu; mauro.rosestolato@gmail.com



S. N. Cohen et al. (eds.), *Frontiers in Stochastic Analysis - BSDEs, SPDEs and their Applications*, Springer Proceedings in Mathematics & Statistics 289,
https://doi.org/10.1007/978-3-030-22285-7_9

methods based on implicit functions associated with contractions in Banach spaces, we study continuity of the mild solution $X^{t,Y}$ of (1.1) with respect to $t, Y, A, b, \sigma$ under standard Lipschitz conditions on $b, \sigma$, Gâteaux differentiability of generic order $n \geq 1$ of $X^{t,Y}$ with respect to $Y$ under Gâteaux differentiability assumptions on $b, \sigma$, and continuity with respect to $t, Y, A, b, \sigma$ of the Gâteaux differentials $\partial_Y^n X^{t,Y}$.

Path-dependent SDEs in finite dimensional spaces are studied in [15]. The standard reference for SDEs in Hilbert spaces is [9]. More generally, in addition to SDEs in Hilbert spaces, also the case of path-dependent SDEs in Hilbert spaces is considered in [12, Chap. 3], but for the path-dependent case the study is there limited mainly to existence and uniqueness of mild solutions. Our framework generalize the latter one by weakening the Lipschitz conditions on the coefficients, by letting the starting process $Y$ belong to a generic space of paths contained in $B_b([0, T], H)$ ([1]) obeying few conditions, but not necessarily assumed to be $C([0, T], H)$, and by providing results on differentiability with respect to the initial datum and on continuity with respect to all the data.

In the literature on mild solutions to SDEs in Hilbert spaces, differentiability with respect to the initial datum is always proved only up to order $n = 2$, in the sense of Gâteaux (see e.g. [8, 9]) or Frečhet (see e.g. [12, 14]). In [8, Theorem 7.3.6] the case $n > 2$ is stated but not proved. There are no available results regarding differentiability with respect to the initial condition of mild solutions to SDEs of the type (1.1), with path-dependence. One of the contributions of the present work is to fill these gaps in the literature, by extending the results so far available to a generic order $n$ of differentiability and to the path-dependent case.

In case (1.1) is not path-dependent, the continuity of $X^{t,Y}$, $\partial_Y X^{t,Y}$, and $\partial_Y^2 X^{t,Y}$, separately with respect to $t$, $Y$ and $A, b, \sigma$, is considered and used in [8, Chap. 7]. We extend these previous results to the path-dependent case and to Gâteaux derivatives $\partial_Y^n X^{t,Y}$ of generic order $n$, providing joint continuity with respect to all the data $t, Y, A, b, \sigma$ of the system.

Similarly as in the cited literature, we obtain our results for mild solutions (differentiability and continuity with respect to the data) starting from analogous results for implicit functions associated with Banach space-valued contracting maps. Because of that, the first part of the paper is entirely devoted to study parametric contractions in Banach spaces and regularity of the associated implicit functions. In this respect, regarding Gâteaux differentiability of implicit functions associated with parametric contractions and continuity of the derivatives under perturbation of the data, we prove a general result, for a generic order $n$ of differentiability, extending the results in [1, 8, 9], that were limited to the case $n = 2$.

In a unified framework, our work provides a collection of results for mild solutions to path-dependent SDEs which are very general, within the standard case of Lipschitz-type assumptions on the coefficients, a useful toolbox for starting dealing with path-dependent stochastic analysis in Hilbert spaces. For example, the so called "vertical derivative" in the finite dimensional functional Itō calculus ([4, 10])

---

[1] $B_b([0, T], H)$ denotes the space of bounded Borel functions $[0, T] \to H$.

of functionals like $F(t, \mathbf{x}) = \mathbb{E}[\varphi(X^{t,\mathbf{x}})]$, where $\varphi$ is a functional on the space $\mathbb{D}$ of càdlàg functions and $\mathbf{x} \in \mathbb{D}$, is easily obtained starting from the partial derivative of $X^{t,\mathbf{x}}$ with respect to a step function, which can be treated in our setting by choosing $\mathbb{D}$ as space of paths (we refer to Remark 3.11 for further details). Another field in which the tools here provided can be employed is the study of stochastic representations of classical solutions to path dependent Kolmogorov equations, where second order derivatives are required. Furthermore, the continuity of the mild solution and of its derivatives with respect to all the data, including the coefficients, can be used e.g. when merely continuous Lipschitz coefficients need to be approximated by smoothed out coefficients, which is in general helpful when dealing with Kolmogorov equations in Hilbert spaces (path- or non-path-dependent) for which notions other than classical solutions are considered, as strong-viscosity solutions [5, 6] or strong solutions [1].

The contents of the paper are organized as follows. First, in Sect. 2, we recall some notions regarding strongly continuous Gâteaux differentiability and some basic results for contractions in Banach spaces. Then we provide the first main result (Theorem 2.9): the strongly continuous Gâteaux differentiability up to a generic order $n$ of fixed-point maps associated to parametric contractions which are differentiable only with respect to some subspaces. We conclude the section with a result regarding the continuity of the Gâteaux differentials of the implicit function with respect to the data (Proposition 2.11).

In Sect. 3 we consider path-dependent SDEs. After a standard existence and uniqueness result (Theorem 3.6), we move to study Gâteaux differentiability with respect to the initial datum up to order $n$ of mild solutions, in Theorem 3.9, which is the other main result and justifies the study made in Sect. 2. We conclude with Theorem 3.16, which concerns the continuity of the Gâteaux differentials with respect to all the data of the system (coefficients, initial time, initial condition).

## 2 Preliminaries

In this section we recall the notions and develop the tools that we will apply to study path-dependent SDEs in Sect. 3. We focus on strongly continuous Gâteaux differentiability of fixed-point maps associated with parametric contractions in Banach spaces.

### *2.1 Strongly Continuous Gâteaux Differentials*

We begin by recalling the basic definitions regarding Gâteaux differentials, mainly following [11]. Then we will define the space of strongly continuously Gâteaux differentiable functions, that will be the reference spaces in the following sections.

If $X, Y$ are topological vector spaces, $U \subset X$ is a set, $f: U \to Y$ is a function, $u \in U, x \in X$ is such that $[u - \epsilon x, u + \epsilon x] \subset U$ ([2]) for some $\epsilon > 0$, the directional derivative of $f$ at $u$ for the increment $x$ is the limit

$$\partial_x f(u) := \lim_{t \to 0} \frac{f(u+tx) - f(u)}{t}$$

whenever it exists. Also in the case in which the directional derivative $\partial_x f(u)$ is defined for all $x \in X$, it need not be linear.

Higher order directional derivatives are defined recursively. For $n \geq 1, u \in U$, the $n$th-order directional derivative $\partial^n_{x_1 \dots x_n} f(u)$ at $u$ for the increments $x_1, \dots, x_n \in X$ is the directional derivative of $\partial^{n-1}_{x_1 \dots x_{n-1}} f$ at $u$ for the increment $x_n$ (notice that this implies, by definition, the existence of $\partial^n_{x_1 \dots x_{n-1}} f(u')$ for $u'$ in some neighborhood of $u'$ in $U \bigcap (u + \mathbb{R}x_n)$)

If $Y$ is locally convex, we denote by $L_s(X, Y)$ the space $L(X, Y)$ endowed with the coarsest topology which makes continuous the linear functions of the form

$$L(X, Y) \to Y, \ \Lambda \mapsto \Lambda(x),$$

for all $x \in X$. Then $L_s(X, Y)$ is a locally convex space.

Let $X_0$ be a topological vector space continuously embedded into $X$. If $u \in U$, if $\partial_x f(u)$ exists for all $x \in X_0$ and $X_0 \to Y, \ x \mapsto \partial_x f(u)$, belongs to $L(X_0, Y)$, then $f$ is said to be Gâteaux differentiable at $u$ with respect to $X_0$ and the map $X_0 \to Y, \ x \mapsto \partial_x f(u)$, is the Gâteaux differential of $f$ at $u$ with respect to $X_0$. In this case, we denote the Gâteaux differential of $f$ at $u$ by $\partial_{X_0} f(u)$ and its evaluation $\partial_x f(u)$ by $\partial_{X_0} f(u).x$. If $\partial_{X_0} f(u)$ exists for all $u \in U$, then we say that $f$ is Gâteaux differentiable with respect to $X_0$, or, in case $X_0 = X$, we just say that $f$ is Gâteaux differentiable and we use the notation $\partial f(u)$ in place of $\partial_X f(u)$.

A function $f: U \to Y$ is said to be strongly continuously Gâteaux differentiable with respect to $X_0$ if it is Gâteaux differentiable with respect to $X_0$ and

$$U \to L_s(X_0, Y), \ u \mapsto \partial_{X_0} f(u)$$

is continuous. If $n > 1$, we say that $f$ is strongly continuously Gâteaux differentiable up to order $n$ with respect to $X_0$ if it is strongly continuously Gâteaux differentiable up to order $n-1$ with respect to $X_0$ and

$$\partial^{n-1}_{X_0} f : U \to \overbrace{L_s(X_0, L_s(X_0, \cdots L_s(X_0, Y)}^{n-1 \text{ times } L_s} \cdots))$$

exists and is strongly continuously Gâteaux differentiable with respect to $X_0$. In this case, we denote $\partial^n_{X_0} f := \partial_{X_0} \partial^{n-1}_{X_0} f$ and $\partial^n f := \partial \partial^{n-1} f$.

[2] If $x, x' \in X$, the segment $[x, x']$ is the set $\{\zeta x + (1 - \zeta)x' | \zeta \in [0, 1]\}$.

Let $X$, $X_0$ be topological vector spaces, with $X_0$ continuously embedded into $X$, let $U$ be an open subset of $X$, and let $Y$ be a locally convex space.

We denote by $\mathcal{G}^n(U, Y; X_0)$ the space of functions $f : U \to Y$ which are continuous and strongly continuously Gâteaux differentiable up to order $n$ with respect to $X_0$. In case $X_0 = X$, we use the notation $\mathcal{G}^n(U, Y)$ instead of $\mathcal{G}^n(U, Y; X)$.

Let $L_s^{(n)}(X_0^n, Y)$ be the vector space of $n$-linear functions from $X_0^n$ into $Y$ which are continuous with respect to each variable separately, endowed with the coarsest vector topology making continuous all the linear functions of the form

$$L_s^{(n)}(X_0^n, Y) \to Y, \ \Lambda \to \Lambda(x_1, \dots, x_n)$$

for $x_1, \dots, x_n \in X_0$. Then $L_s^n(X_0^n, Y)$ is a locally convex space. Trough the canonical identification (as topological vector spaces)

$$\overbrace{L_s(X_0, L_s(X_0, \cdots L_s(X_0, Y)}^{n \text{ times } L_s} \cdots)) \cong L_s^{(n)}(X_0^n, Y),$$

we can consider $\partial_{X_0}^n f$ as taking values in $L_s^{(n)}(X_0^n, Y)$, whenever $f \in \mathcal{G}^n(U, Y; X_0)$.

If $X_0$, $X$, $Y$ are normed spaces, $U$ is an open subset of $X$, $\partial_x f(u)$ exists for all $u \in U$, $x \in X_0$, $\partial_x f(u)$ is continuous with respect to $u$, for all $x \in X_0$, then $\partial_x f(u)$ is linear in $x$ (see [11, Lemma 4.1.5]).

The following proposition is a characterization for the continuity conditions on the directional derivatives of a function $f \in \mathcal{G}^n(U, Y; X_0)$, when $X_0$, $X$, $Y$ are normed spaces. The proof appears in the Appendix.

**Proposition 2.1** *Let $n \geq 1$, let $X_0$, $X$, $Y$ be normed spaces, with $X_0$ continuously embedded into $X$, and let $U$ be an open subset of $X$. Then $f \in \mathcal{G}^n(U, Y; X_0)$ if and only if $f$ is continuous, the directional derivatives $\partial_{x_1 \dots x_j}^j f(u)$ exist for all $u \in U$, $x_1, \dots, x_j \in X_0$, $j = 1, \dots, n$, and the functions*

$$U \times X_0^j \to Y, \ (u, x_1, \dots, x_j) \mapsto \partial_{x_1 \dots x_j}^j f(u) \tag{2.1}$$

*are separately continuous in each variable. In this case,*

$$\partial_{X_0}^j f(u).(x_1, \dots, x_j) = \partial_{x_1 \dots x_j}^j f(u) \qquad \forall u \in U, \ \forall x_1, \dots, x_j \in X_0, \ j = 1, \dots, n. \tag{2.2}$$

*Remark 2.2* If $X_0$ is Banach, $X$ is normed, $Y$ is locally convex, and $f \in \mathcal{G}^n(X, Y; X_0)$, then, by Proposition 2.1 and the Banach-Steinhaus theorem, if follows that the map

$$U \times X_0^n \to Y, \ (u, x_1, \dots, x_n) \mapsto \partial_{X_0}^n f(u).(x_1, \dots, x_n)$$

is continuous, jointly in $u, x_1, \dots, x_n$.

### 2.1.1 Chain Rule

In this subsection, we recall the classical Faà di Bruno formula and we prove a corresponding stability result, for derivatives of order $n \geq 1$ of compositions of strongly continuously Gâteaux differentiable functions. We will use this formula in order to prove the main results of Sect. 2.3 (Theorem 2.9 and Proposition 2.11).

Let $X_0, X_1$ be Banach spaces, with $X_0$ continuously embedded in $X_1$, and let $U$ be an open subset of $X_1$. Let $n \in \mathbb{N}$, $n \geq 1$, $\mathbf{x}_n := \{x_1, \ldots, x_n\} \subset X_0^n$, $j \in \{1, \ldots, n\}$. Then

- $P^j(\mathbf{x}_n)$ denotes the set of partitions of $\mathbf{x}_n$ in $j$ non-empty subsets.
- If $f \in \mathscr{G}^n(U, X_1; X_2)$ and $\mathbf{q} := \{y_1, \ldots, y_j\} \subset \mathbf{x}_n$, then $\partial^j_{\mathbf{q}} f(u)$ denotes the derivative $\partial^j_{y_1 \ldots y_j} f(u)$.
- $|\mathbf{q}|$ denotes the cardinality of $\mathbf{q}$.

**Proposition 2.3** [Faà di Bruno's formula] *Let $n \geq 1$. Let $X_0, X_1, X_2, X_3$ be Banach spaces, with $X_0$ continuously embedded in $X_1$, and let $U$ be an open subset of $X_1$. If $f \in \mathscr{G}^n(U, X_2; X_0)$ and $g \in \mathscr{G}^n(X_2, X_3)$, then $g \circ f \in \mathscr{G}^n(U, X_3; X_0)$. Moreover*

$$\partial^j_{\mathbf{x}_j} g \circ f(u) = \sum_{i=1}^{j} \sum_{\{\mathbf{p}^i_1, \ldots, \mathbf{p}^i_i\} \in P^i(\mathbf{x}_j)} \partial^j_{\partial^{|\mathbf{p}^i_1|}_{\mathbf{p}^i_1} f(u) \ldots \partial^{|\mathbf{p}^i_i|}_{\mathbf{p}^i_i} f(u)} g(f(u)). \tag{2.3}$$

*for all $u \in U$, $j = 1, \ldots, n$, $\mathbf{x}_j = \{x_1, \ldots, x_j\} \subset X_0^j$.*

**Proposition 2.4** *Let $n \geq 1$. Let $X_0, X_1, X_2, X_3$ be Banach spaces, with $X_0$ continuously embedded in $X_1$, and let $U$ be an open subset of $X_1$. Let*

$$\begin{cases} f \in \mathscr{G}^n(U, X_2; X_0) \\ f^{(k)} \in \mathscr{G}^n(U, X_2; X_0) \quad \forall k \in \mathbb{N} \\ g \in \mathscr{G}^n(X_2, X_3) \\ g^{(k)} \in \mathscr{G}^n(X_2, X_3) \qquad \forall k \in \mathbb{N}. \end{cases}$$

*Suppose that*

$$\begin{cases} \lim_{k \to \infty} f^{(k)}(u) = f(u) \\ \lim_{k \to \infty} \partial^j_{x_1 \ldots x_j} f^{(k)}(u) = \partial^j_{x_1 \ldots x_j} f(u) \quad \text{for } j = 1, \ldots, n, \end{cases}$$

*uniformly for $u$ on compact subsets of $U$ and $x_1, \ldots, x_j$ on compact subsets of $X_0$, and that*

$$\begin{cases} \lim_{k \to \infty} g^{(k)}(x) = g(x) \\ \lim_{k \to \infty} \partial^j_{x_1 \ldots x_j} g^{(k)}(x) = \partial^j_{x_1 \ldots x_j} f(x) \quad \text{for } j = 1, \ldots, n, \end{cases}$$

*uniformly for $x, x_1, \ldots, x_j$ on compact subsets of $X_2$. Then*

$$\begin{cases} \lim_{k\to\infty} g^{(k)} \circ f^{(k)}(u) = g \circ f(u) \\ \lim_{k\to\infty} \partial^j_{x_1\ldots x_j} g^{(k)} \circ f^{(k)}(u) = \partial^j_{x_1\ldots x_j} g \circ f(u) \quad \text{for } j = 1, \ldots, n, \end{cases}$$

*uniformly for $u$ on compact subsets of $U$ and $x_1, \ldots, x_j$ on compact subsets of $X_0$.*

*Proof.* Use recursively formula (2.3). ■

## 2.2 Contractions in Banach Spaces: Basic Results

In this section, we assume that $X$ and $Y$ are Banach spaces, and that $U$ is an open subset of $X$. We recall that, if $\alpha \in [0, 1)$ and $h\colon U \times Y \to Y$, then $h$ is said to be a *parametric $\alpha$-contraction* if

$$|h(u, y) - h(u, y')|_Y \le \alpha|y - y'| \qquad \forall u \in U,\ \forall y, y' \in Y.$$

By the Banach contraction principle, to any such $h$ we can associate a uniquely defined map $\varphi\colon U \to Y$ such that $h(u, \varphi(u)) = \varphi(u)$ for all $u \in U$. We refer to $\varphi$ as to *the fixed-point map associated with $h$*. For future reference, we summarize some basic continuity properties that $\varphi$ inherites from $h$.

The following lemma can be found in [13, p. 13].

**Lemma 2.5** *Let $\alpha \in [0, 1)$ and let $h(u, \cdot)\colon U \times Y \to Y$, $h_n(u, \cdot) : U \times Y \to Y$, for $n \in \mathbb{N}$, be parametric $\alpha$-contractions. Denote by $\varphi$ (resp. $\varphi_n$) the fixed-point map associated with $h$ (resp. $h_n$).*

*(i) If $h_n \to h$ pointwise on $U \times Y$, then $\varphi_n \to \varphi$ pointwise on $U$.*
*(ii) If $A \subset U$ is a set and if there exists an increasing concave function $w$ on $\mathbb{R}^+$ such that $w(0) = 0$ and*

$$|h(u, y) - h(u', y)|_Y \le w(|u - u'|_X) \qquad \forall u, u' \in A,\ \forall y \in Y, \tag{2.4}$$

*then*

$$|\varphi(u) - \varphi(u')|_Y \le \frac{1}{1-\alpha} w(|u - u'|_X) \qquad \forall u, u' \in A.$$

*(iii) If $h$ is continuous, then $\varphi$ is continuous.*

*Remark 2.6* If $h\colon U \times Y \to Y$ is a parametric $\alpha$-contraction ($\alpha \in [0, 1)$) belonging to $\mathscr{G}^1(U \times Y, Y; \{0\} \times Y)$, then

$$|\partial_Y h(u, y)|_{L(Y)} \le \alpha \qquad \forall u \in U,\ y \in Y, \tag{2.5}$$

where $|\cdot|_{L(Y)}$ denotes the operator norm on $L(Y)$. Hence $\partial_Y h(u, y)$ is invertible and the family $\{(I - \partial_Y h(u, y))^{-1}\}_{(u,y)\in U\times Y}$ is uniformly bounded in $L(Y)$. For what follows, it is important to notice that, for all $y \in Y$,

$$U \times Y \to Y,\ (u, y') \mapsto (I - \partial_Y h(u, y'))^{-1} y \tag{2.6}$$

is continuous, hence, because of the formula

$$(I - \partial_Y h(u, y'))^{-1} y = \sum_{n \in \mathbb{N}} \left(\partial_Y h(u, y')\right)^n y$$

and of Lebesgue's dominated convergence theorem (for series), $(I - \partial_Y h(u, y'))^{-1} y$ is jointly continuous in $u, y', y$.

The following proposition shows that the fixed-point map $\varphi$ associated with a parametric $\alpha$-contraction $h$ inherits from $h$ the strongly continuous Gâteaux differentiability.

**Proposition 2.7** *If $h \in \mathscr{G}^1(U \times Y, Y)$ is a parametric $\alpha$-contraction and if $\varphi$ is the fixed-point map associated with h, then $\varphi \in \mathscr{G}^1(U, Y)$ and*

$$\partial_x \varphi(u) = (I - \partial_Y h\,(u, \varphi(u)))^{-1}\,(\partial_x h\,(u, \varphi(u))) \qquad \forall u \in U, \forall x \in X. \tag{2.7}$$

*Proof.* For the proof, see [9, Lemma 2.9], or [1, Proposition C.0.3], taking into account also [1, Remark C.0.4], Lemma 2.5(iii), Remark 2.6. ∎

## 2.3 Gâteaux Differentiability of Order n of Fixed-Point Maps

In this section we provide a result for the Gâteux differentiability up to a generic order $n$ of a fixed-point map $\varphi$ associated with a parametric $\alpha$-contraction $h$, under the assumption that $h$ is Gâteaux differentiable only with respect to some invariant subspaces of the domain.

The main result of this section is Theorem 2.9, which is suitable to be applied to mild solutions of SDEs (Sect. 3.2). When $n = 1$, Theorem 2.9 reduces to Proposition 2.7. In the case $n = 2$, Theorem 2.9 is also well-known, and a proof can be found in [1, Proposition C.0.5]. On the other hand, when the order of differentiability $n$ is generic, the fact that the parametric $\alpha$-contraction is assumed to be differentiable only with respect to certain subspaces makes non-trivial the proof of the theorem. To our knowledge, a reference for the case $n \geq 3$ is not available in the literature. The main issue consists in providing a precise formulation of the statement, with its assumptions, that can be proved by induction.

For the sake of readability, we collect the assumptions of Theorem 2.9 in the following

**Assumption 2.8** (1) $n \geq 1$ and $\alpha \in [0, 1)$;
(2) $X$ is a Banach space and $U$ is an open subset of $X$.
(3) $Y_1 \supset Y_2 \supset \ldots \supset Y_n$ is a decreasing sequence of Banach spaces, with norms $|\cdot|_1, \ldots, |\cdot|_n$, respectively.

(4) For $k = 1, \dots, n$ and $j = 1, 2, \dots, k$, the canonical embedding of $Y_k$ into $Y_j$, denoted by $i_{k,j}\colon Y_k \to Y_j$, is continuous.
(5) $h_1\colon U \times Y_1 \to Y_1$ is a function such that $h_1(U \times Y_k) \subset Y_k$ for $k = 2, \dots, n$. For $k = 2, \dots, n$, we denote by $h_k$ the induced function

$$h_k\colon U \times Y_k \to Y_k,\ (u, y) \mapsto h_1(u, y). \tag{2.8}$$

(6) For $k = 1, \dots, n$, $h_k$ is continuous and satisfies

$$\left|h_k(u, y) - h_k(u, y')\right|_k \le \alpha|y - y'|_k \qquad \forall u \in U,\ \forall y, y' \in Y_k. \tag{2.9}$$

(7) For $k = 1, \dots, n$, $h_k \in \mathcal{G}^n(U \times Y_k, Y_k; X \times \{0\})$.
(8) For $k = 1, \dots, n-1$, $h_k \in \mathcal{G}^n(U \times Y_k, Y_k; X \times Y_{k+1})$
(9) For $k = 1, \dots, n$, $j = 1, \dots, n-1$, for all $u \in U$, $z_1, \dots, z_j \in X$, $y, z_{j+1} \in Y_k$, and for all permutations $\sigma$ of $\{1, \dots, j+1\}$, the directional derivative $\partial^{j+1}_{z_{\sigma(1)}\dots z_{\sigma(j+1)}} h_k(u, y)$ exists, and

$$U \times Y_k \times X^j \times Y_k \to Y_k,\ (u, y, z_1, \dots, z_j, z_{j+1}) \mapsto \partial^{j+1}_{z_{\sigma(1)}\dots z_{\sigma(j)} z_{\sigma(j+1)}} h_k(u, y) \tag{2.10}$$

is continuous.

The proof of the following theorem appears in the Appendix.

**Theorem 2.9** *Let Assumption 2.8 be satisfied and let $\varphi\colon U \to Y_1$ denote the fixed-point function associated with the parametric $\alpha$-contraction $h_1$. Then, for $j = 1, \dots, n$, we have $\varphi \in \mathcal{G}^j(U, Y_{n-j+1})$ and, for all $u \in U$, $x_1, \dots, x_j \in X$, $\partial^j_{x_1\dots x_j}\varphi(u)$ is given by the formula*

$$\begin{aligned}\partial^j_{x_1\dots x_j}\varphi(u) &= \left(I - \partial_{Y_1} h_1(u, \varphi(u))\right)^{-1} \partial^j_{x_1\dots x_j} h_1(u, \varphi(u)) \\ &+ \sum_{\substack{\mathbf{x} \in 2^{\{x_1,\dots,x_j\}} \\ \mathbf{x} \neq \emptyset}} \sum_{i=\max\{1,2-j+|\mathbf{x}|\}}^{|\mathbf{x}|} \sum_{\substack{\mathbf{p} \in P^i(\mathbf{x}) \\ \mathbf{p} = (\mathbf{p}_1,\dots,\mathbf{p}_i)}} \left(I - \partial_{Y_1} h_1(u, \varphi(u))\right)^{-1} \partial^j[\mathbf{x}^c, \mathbf{p}] h_1(u, \varphi(u))\end{aligned} \tag{2.11}$$

*where $2^{\{x_1,\dots,x_i\}}$ is the power set of $\{x_1, \dots, x_i\}$, $P^i(\mathbf{x})$ is the set of partitions of $\mathbf{x}$ in $i$ non-empty parts, $\mathbf{x}^c := \{x_1, \dots, x_j\} \setminus \mathbf{x}$, and $\partial^j[\mathbf{x}^c, \mathbf{p}] := \partial^{j-|\mathbf{x}|}_{\mathbf{x}^c} \partial^{|\mathbf{x}|}_{\partial^{|\mathbf{p}_1|}_{\mathbf{p}_1}\varphi(u), \dots, \partial^{|\mathbf{p}_i|}_{\mathbf{p}_i}\varphi(u)}$ [3].*

Theorem 2.9 says that $\varphi$ is $Y_n$-valued, continuous as a map from $U$ into $Y_n$, and, for $j = 1, \dots, n$, for all $u \in U$, $x_1, \dots, x_j \in X$, the directional derivative $\partial^j_{x_1\dots x_j}\varphi(u)$ exists, it belongs to $Y_{n-j+1}$, the map

$$U \times X^j \to Y_{n-j+1},\ (u, x_1, \dots, x_j) \mapsto \partial^j_{x_1\dots x_j}\varphi(u)$$

is continuous, and (2.11) holds true.

[3] Recall notation at p. 5.

Formula (2.11) can be useful e.g. when considering the boundedness of the derivatives of $\varphi$, or when studying convergences of derivatives under perturbations of $h$, as Corollary 2.10 and Proposition 2.11 show.

**Corollary 2.10** *Let Assumption 2.8 be satisfied. Suppose that there exists $M > 0$ such that*

$$\begin{cases} |\partial_y h_k(u, y')|_k \leq M|y|_k & \begin{cases} \forall u \in U, \\ \forall y, y' \in Y_k,\ k = 1, \dots, n \end{cases} \\ |\partial^j_{x_1 \dots x_j} h_k(u, y)|_k \leq M \prod_{l=1}^{j} |x_l|_X & \begin{cases} \forall u \in U,\ \forall x_1, \dots, x_j \in X, \\ \forall y \in Y_k,\ j, k = 1, \dots, n \end{cases} \\ |\partial^{j+i}_{x_1 \dots x_j y_1 \dots y_i} h_k(u, y)|_k \leq M \prod_{l=1}^{j} |x_l|_X \cdot \prod_{l=1}^{i} |y_l|_{k+1} & \begin{cases} \forall u \in U,\ \forall x_1, \dots, x_j \in X, \\ \forall y \in Y_k,\ \forall y_1, \dots, y_i \in Y_{k+1}, \\ k = 1, \dots, n-1, \\ j, i = 1, \dots, n-1,\ 1 \leq j + i \leq n-1. \end{cases} \end{cases} \tag{2.12}$$

*Then, for $k = 1, \dots, n$,*

$$\sup_{\substack{u \in U \\ x_1, \dots, x_k \in X \\ |x_1|_X = \dots = |x_k|_X = 1}} |\partial^k_{x_1 \dots x_k} \varphi(u)|_{n-k+1} \leq C(\alpha, M),$$

*where $C(\alpha, M) \in \mathbb{R}$ depends only on $\alpha$, $M$.*

*Proof.* Reason by induction taking into account (2.11) and (2.5). ∎

**Proposition 2.11** *Suppose that Assumption 2.8 holds true for a given $h_1$ and that $h_1^{(1)}, h_1^{(2)}, h_1^{(3)} \dots$ is a sequence of functions, each of which satisfies Assumption 2.8, uniformly with respect to the same $n$, $\alpha$. Let $h_k^{(m)}$ denote the map associated with $h_1^{(m)}$ according to (2.8) and let $\varphi^{(m)}$ denote the fixed-point map associated with the parametric $\alpha$-contraction $h_1^{(m)}$. Suppose that the following convergences occur.*

*(i) For $k = 1, \dots, n$, $y \in Y_k$,*

$$\lim_{m \to \infty} h_k^{(m)}(u, y) = h_k(u, y) \text{ in } Y_k \tag{2.13}$$

*uniformly for $u$ on compact subsets of $U$;*

*(ii) for $k = 1, \dots, n$,*

$$\begin{cases} \lim_{m \to \infty} \partial_x h_k^{(m)}(u, y) = \partial_x h_k(u, y) \text{ in } Y_k \\ \lim_{m \to \infty} \partial_y h_k^{(m)}(u, y') = \partial_y h_k(u, y') \text{ in } Y_k \end{cases} \tag{2.14}$$

*uniformly for $u$ on compact subsets of $U$, $x$ on compact subsets of $X$, and $y$, $y'$ on compact subsets of $Y_k$;*

*(iii) for all* $k = 1, \dots, n-1$, $u \in U$, $j, i = 0, \dots, n$, $1 \leq j + i \leq n$,

$$\lim_{m\to\infty} \partial^{j+i}_{x_1\dots x_j y_1\dots y_i} h_k^{(m)}(u, y) = \partial^{j+i}_{x_1\dots x_j y_1\dots y_i} h_k(u, y) \text{ in } Y_k \tag{2.15}$$

*uniformly for* $u$ *on compact subsets of* $U$, $x_1, \dots, x_j$ *on compact subsets of* $X$, $y$ *on compact subsets of* $Y_k$, $y_1, \dots, y_i$ *on compact subsets of* $Y_{k+1}$.

*Then* $\varphi^{(m)} \to \varphi$ *uniformly on compact subsets of* $Y_n$ *and, for all* $j = 1, \dots, n$

$$\lim_{m\to\infty} \partial^{j}_{x_1\dots x_j} \varphi^{(m)}(u) = \partial^{j}_{x_1\dots x_j} \varphi(u) \text{ in } Y_{n-j+1} \tag{2.16}$$

*uniformly for* $u$ *on compact subsets of* $U$ *and* $x_1, \dots, x_j$ *on compact subsets of* $X$.

*Proof.* Notice that (2.13) and the fact that each $h_k^{(m)}$ is a parametric $\alpha$-contraction (with the same $\alpha$) imply the uniform convergence $h_k^{(m)} \to h_k$ on compact subsets of $Y_k$. In particular, the sequence $h_k^{(1)}, h_k^{(2)}, h_k^{(3)}, \dots$ is uniformly equicontinuous on compact sets. Then, by Lemma 2.5(i), (ii), $\varphi^{(m)} \to \varphi$ in $Y_k$ uniformly on compact subsets of $Y_k$, for $k = 1, \dots, n$. Moreover, by (2.5), that holds for all $h_1^{(m)}$ uniformly in $m$, we have the boundedness of $(I - \partial_{Y_1} h_1^{(m)})^{-1}$, uniformly in $m$. Convergence (2.16) is then obtained by reasoning by induction on (2.11), taking into account the strong continuity of $(I - \partial_{Y_1} h_1)^{-1}$. ■

## 3 Path-Dependent SDEs in Hilbert Spaces

In this section we study mild solutions of path-dependent SDEs in Hilbert spaces. In particular, by applying the results of the previous section, we address differentiability with respect to the initial datum and stability of the derivatives.

Let $H$ and $U$ be real separable Hilbert spaces, with scalar product denoted by $\langle \cdot, \cdot \rangle_H$ and $\langle \cdot, \cdot \rangle_U$, respectively. Let $\mathfrak{e} := \{e_n\}_{n\in\mathscr{N}}$ be an orthonormal basis of $H$, where $\mathscr{N} = \{1, \dots, N\}$ if $H$ has dimension $N \in \mathbb{N} \setminus \{0\}$, or $\mathscr{N} = \mathbb{N}$ if $H$ has infinite dimension, and let $\mathfrak{e}' := \{e'_m\}_{m\in\mathscr{M}}$ be an orthonormal basis of $U$, where $\mathscr{M} = \{1, \dots, M\}$ if $U$ has dimension $M \in \mathbb{N} \setminus \{0\}$, or $\mathscr{M} = \mathbb{N}$ if $U$ has infinite dimension. If $\mathbf{x}\colon [0, T] \to \mathscr{S}$ is a function taking values in any set $\mathscr{S}$ and if $t \in [0, T]$, we denote by $\mathbf{x}_{t\wedge\cdot}$ the function defined by

$$\begin{cases} \mathbf{x}_{t\wedge\cdot}(s) := \mathbf{x}(s) & s \in [0, t] \\ \mathbf{x}_{t\wedge\cdot}(s) := \mathbf{x}(t) & s \in (t, T]. \end{cases}$$

For elements of stochastic analysis in infinite dimension used hereafter, we refer to [9, 12].

We begin by considering the SDE

$$\begin{cases} dX_s = (AX_s + b((\cdot, s), X))\,ds + \sigma((\cdot, s), X)\,dW_s & s \in (t, T] \\ X_s = Y_s & s \in [0, t], \end{cases} \tag{3.1}$$

where $t \in [0, T]$, $Y$ is a $H$-valued adapted process defined on a complete filtered probability space $(\Omega, \mathscr{F}, \mathbb{F} := \{\mathscr{F}_t\}_{t\in[0,T]}, \mathbb{P})$, $W$ is a $U$-valued cylindrical Wiener process defined on $(\Omega, \mathscr{F}, \mathbb{F}, \mathbb{P})$, $b((\omega, s), X)$ is a $H$-valued random variable depending on $\omega \in \Omega$, on the time $s$, and on the path $X$, $\sigma((\omega, s), X)$ is a $L_2(U, H)$-valued random variable depending on $\omega \in \Omega$, on the time $s$, and on the path $X$, and $A$ is the generator of a $C_0$-semigroup $S$ on $H$.

We introduce the following notation:

- $\mathbb{S}$ denotes a closed subspace of $B_b([0, T], H)$ ([4]) such that

$$\begin{cases} (a)\ C([0, T], H) \subset \mathbb{S} \\ (b)\ \mathbf{x}_{t\wedge\cdot} \in \mathbb{S},\ \forall \mathbf{x} \in \mathbb{S},\ \forall t \in [0, T] \\ (c)\ \text{for all } T \in L(H) \text{ and } \mathbf{x} \in \mathbb{S},\ \text{the map } [0, T] \to H,\ t \mapsto T\mathbf{x}_t,\ \text{belongs to } \mathbb{S}. \end{cases} \tag{3.2}$$

  Hereafter, unless otherwise specified, $\mathbb{S}$ will be always considered as a Banach space endowed with the norm $|\cdot|_\infty$. For example, $\mathbb{S}$ could be $C([0, T], H)$, the space of càdlàg functions $[0, T] \to H$, or $B_b([0, T], H)$ itself.
- $\Omega_T$ denotes the product space $\Omega \times [0, T]$ and $\mathscr{P}_T$ denotes the product measure $\mathbb{P} \otimes m$ on $(\Omega_T, \mathscr{F}_T \otimes \mathscr{B}_{[0,T]})$, where $m$ is the Lebesgue measure and $\mathscr{B}_{[0,1]}$ is the Borel $\sigma$-algebra on $[0, 1]$.
- $\mathscr{L}^0_{\mathscr{P}_T}(\mathbb{S})$ denotes the space of functions $X\colon \Omega_T \to H$ such that

$$\begin{cases} (a)\ \forall \omega \in \Omega,\ \text{the map } [0, T] \to H,\ t \mapsto X_t(\omega),\ \text{belongs to } \mathbb{S} \\ (b)\ (\Omega_T, \mathscr{P}_T) \to \mathbb{S},\ (\omega, t) \mapsto X_{t\wedge\cdot}(\omega) \text{ is measurable.} \end{cases} \tag{3.3}$$

  Two processes $X, X' \in \mathscr{L}^0_{\mathscr{P}_T}(\mathbb{S})$ are equal if and only if $\mathbb{P}(|X - X'|_\infty = 0) = 1$.
- For $p \in [1, \infty)$, $\mathscr{L}^p_{\mathscr{P}_T}(\mathbb{S})$ denotes the space of equivalence classes of functions $X \in \mathscr{L}^0_{\mathscr{P}_T}(\mathbb{S})$ such that $\Omega_T \to \mathbb{S},\ (\omega, t) \mapsto X_{t\wedge\cdot}(\omega)$ has separable range and

$$|X|_{\mathscr{L}^p_{\mathscr{P}_T}(\mathbb{S})} := \left(\mathbb{E}\left[|X|^p_\infty\right]\right)^{1/p} < \infty. \tag{3.4}$$

- For $p, q \in [1, \infty)$ and $\beta \in [0, 1)$, $\Lambda^{p,q,p}_{\mathscr{P}_T,S,\beta}(L(U, H))$ denotes the space of functions $\Phi\colon \Omega_T \to L(U, H)$ such that

$$\begin{cases} \Phi u\colon (\Omega_T, \mathscr{P}_T) \to H,\ (\omega, t) \mapsto \Phi_t(\omega)u,\ \text{is measurable, } \forall u \in U \\ |\Phi|_{p,q,S,\beta} := \left(\int_0^T \left(\int_0^t (t-s)^{-\beta q} \left(\mathbb{E}\left[|S_{t-s}\Phi_s|^p_{L_2(U,H)}\right]\right)^{q/p} ds\right)^{p/q} dt\right)^{1/p} < \infty. \end{cases}$$

[4] We recall that $B_b([0, T], H)$ is endowed with the norm $|\cdot|_\infty$.

The space $\Lambda^{p,q,p}_{\mathscr{P}_T,S,\beta}(L(U,H))$ is normed by $|\cdot|_{p,q,S,\beta}$ (see Remark 3.1 below).
- $\overline{\Lambda}^{p,q,p}_{\mathscr{P}_T,S,\beta}(L(U,H))$ denotes the completion of $\Lambda^{p,q,p}_{\mathscr{P}_T,S,\beta}(L(U,H))$. We keep the notation $|\cdot|_{p,q,S,\beta}$ for the extended norm.

It can be seen that $(\mathscr{L}^p_{\mathscr{P}_T}(\mathbb{S}), |\cdot|_{\mathscr{L}^p_{\mathscr{P}_T}(\mathbb{S})})$ is a Banach space ($\mathbb{F}$ is supposed to be complete).

*Remark 3.1* To see that $|\cdot|_{p,q,S,\beta}$ is a norm and not just a seminorm, suppose that $|\Phi|_{p,q,S,\beta} = 0$. In particular, for $u \in U$,

$$\int_{[0,T]^2} \mathbf{1}_{(0,T]}(t-s)(t-s)^{-\beta}\mathbb{E}\left[|S_{t-s}\Phi_s u|_H\right] ds \otimes dt = 0,$$

which entails, for $\mathbb{P} \otimes m$-a.e. $(\omega, s) \in \Omega_T$,

$$S_{t-s}\Phi_s(\omega)u = 0 \qquad m\text{-a.e. } t \in (s,T]. \tag{3.5}$$

Since $S$ is strongly continuous, (3.5) gives

$$\Phi_s(\omega)u = 0 \qquad \mathbb{P} \otimes m\text{-a.e. } (\omega, s) \in \Omega_T,$$

which provides $\Phi = 0$ $\mathbb{P} \otimes m$-a.e., since $U$ is supposed to be separable and $\Phi_s(\omega) \in L(U,H)$ for all $\omega, s$.

*Remark 3.2* The space $\overline{\Lambda}^{p,q,p}_{\mathscr{P}_T,S,\beta}(L(U,H))$ can be naturally identified with a closed subspace of the space of all those measurable functions

$$\zeta : (\Omega_T \times [0,T], \mathscr{P}_T \otimes \mathscr{B}_T) \to L_2(U,H)$$

such that

$$\begin{cases} \zeta((\omega,s),t) = 0, \ \forall((\omega,s),t) \in \Omega_T \times [0,T], \ s > t, \\ |\zeta|_{p,q,p} := \left(\int_0^T \left(\int_0^t \left(\mathbb{E}\left[|\zeta((\cdot,s),t)|^p_{L_2(U,H)}\right]\right)^{q/p} ds\right)^{p/q} dt\right)^{1/p} < \infty. \end{cases}$$

Indeed, if we denote by $L^{p,q,p}_{\mathscr{P}_T\otimes\mathscr{B}_T}(L_2(U,H))$ such a space, then $L^{p,q,p}_{\mathscr{P}_T\otimes\mathscr{B}_T}(L_2(U,H))$ endowed with $|\cdot|_{p,q,p}$ is a Banach space and the map

$$\iota : \Lambda^{p,q,p}_{\mathscr{P}_T,S,\beta}(L(U,H)) \to L^{p,q,p}_{\mathscr{P}_T\otimes\mathscr{B}_T}(L_2(U,H))$$

defined by

$$\iota(\Phi)(\omega,s,t) := \begin{cases} (t-s)^{-\beta}S_{t-s}\Phi_s(\omega) & \forall((\omega,s),t) \in \Omega_T \times [0,T], \ s \le t, \\ 0 & \text{otherwise.} \end{cases}$$

is an isometry.

The reason to introduce the space $\overline{\Lambda}^{p,q,p}_{\mathscr{P}_T,S,\beta}(L(U,H))$ is related to the existence of a continuous version of the stochastic convolution and to the factorization method used to construct such a version. Let $p > \max\{2, 1/\beta\}$, $t \in [0,T]$, and $\Phi \in \Lambda^{p,2,p}_{\mathscr{P}_T,S,\beta}(L(U,H))$. If we consider the two stochastic convolutions

$$Y_{t'} := \mathbf{1}_{[t,T]}(t') \int_t^{t'} S_{t'-s}\Phi_s dW_s, \quad Z_{t'} := \mathbf{1}_{[t,T]}(t') \int_t^{t'} (t'-s)^{-\beta} S_{t'-s}\Phi_s dW_s, \tag{3.6}$$

then $Y_{t'}$ is well-defined for all $t' \in [0,T]$, $Z_{t'}$ is well-defined for $m$-a.e. $t' \in [0,T]$, and $Y_{t'}$, $Z_{t'}$ belong to $L^p((\Omega, \mathscr{F}_{t'}, \mathbb{P}), H)$.

By using the stochastic Fubini theorem and the factorization method (see [9, 16]), we can find a predictable process $\widetilde{Z}$ such that:

(a) for $m$-a.e. $t \in [0,T]$, $\widetilde{Z}_t = Z_t$ $\mathbb{P}$-a.e.;
(b) for all $t' \in [0,T]$, the following formula holds

$$Y_{t'} = c_\beta \mathbf{1}_{[t,T]}(t') \int_t^{t'} (t'-s)^{\beta-1} \widetilde{Z}_s ds \qquad \mathbb{P}\text{-a.e.}, \tag{3.7}$$

where $c_\beta$ is a constant depending only on $\beta$.

By (3.6), (a), [7, Lemma 7.7], it follows that $\widetilde{Z}(\omega) \in L^p((0,T),H)$ for $\mathbb{P}$-a.e. $\omega \in \Omega$, hence, by [12, Lemma 3.2], the right-hand side of (3.7) is continuous in $t'$.

This classical argument shows that there exists a pathwise continuous process $S \overset{dW}{*_t} \Phi$ such that, for all $t' \in [0,T]$, $(S \overset{dW}{*_t} \Phi)_{t'} = Y_{t'}$ $\mathbb{P}$-a.e. In particular, $S \overset{dW}{*_t} \Phi \in \mathscr{L}^0_{\mathscr{P}_T}(C([0,T],H))$. By (3.6), (3.7), Hölder's inequality, and [7, Lemma 7.7], we also have

$$\mathbb{E}\left[|S \overset{dW}{*_t} \Phi|^p_\infty\right] \leq c^p_\beta \left(\int_0^T v^{\frac{(\beta-1)p}{p-1}} dv\right)^{p-1} \mathbb{E}\left[\int_0^T |\widetilde{Z}_s|^p_H ds\right] \leq c'_{\beta,T,p} |\Phi|^p_{p,2,S,\beta}, \tag{3.8}$$

where $c'_{\beta,T,p}$ is a constant depending only on $\beta$, $T$, $p$. This shows that the linear map $S \overset{dW}{*_t} \#$, defined as

$$\Lambda^{p,2,p}_{\mathscr{P}_T,S,\beta}(L(U,H)) \to \mathscr{L}^p_{\mathscr{P}_T}(C([0,T],H)), \ \Phi \mapsto S \overset{dW}{*_t} \Phi, \tag{3.9}$$

is well-defined and continuous. Then, we can uniquely extend (3.9) to a continuous linear map on $\overline{\Lambda}^{p,2,p}_{\mathscr{P}_T,S,\beta}(L(U,H))$, that we can see as $\mathscr{L}^p_{\mathscr{P}_T}(\mathbb{S})$-valued, since, by assumption, $C([0,T],H) \subset \mathbb{S}$. We end up with a continuous linear map, again denoted by $S \overset{dW}{*_t} \#$,

$$S \overset{dW}{*_t} \#\colon \overline{\Lambda}^{p,2,p}_{\mathscr{P}_T,S,\beta}(L(U,H)) \to \mathscr{L}^p_{\mathscr{P}_T}(\mathbb{S}). \tag{3.10}$$

Summarizing,

(1) the map $S \overset{dW}{*_t} \#$ is linear, continuous, $\mathscr{L}^p_{\mathscr{P}_T}(C([0,T],H))$-valued;
(2) the operator norm of $S \overset{dW}{*_t} \#$ depends only on $\beta, T, p$;
(3) if $\Phi \in \Lambda^{p,2,p}_{\mathscr{P}_T,S,\beta}(L_2(U,H))$, $S \overset{dW}{*_t} \Phi$ is a continuous version of the process $Y$ in (3.6).

Within the approach using the factorization method, the space $\overline{\Lambda}^{p,2,p}_{\mathscr{P}_T,S,\beta}(L(U,H))$ is then naturally introduced if we want to see the stochastic convolution as a continuous linear operator acting on a Banach space and providing pathwise continuous processes, and this perspective is useful when applying to SDEs the results based on parametric $\alpha$-contractions obtained in the first part of the paper.

We make some observations that will be useful later. Let $\hat{S}$ be another $C_0$-semigroup on $H$, and let $\Phi \in \Lambda^{p,2,p}_{\mathscr{P}_T,S,\beta}(L(U,H))$, $\hat{\Phi} \in \Lambda^{p,2,p}_{\mathscr{P}_T,\hat{S},\beta}(L(U,H))$. Then, by using the factorization formula (3.7) both with respect to the couples $(S,\Phi)$ and $(\hat{S},\hat{\Phi})$, and by an estimate analogous to (3.8), we obtain

$$\mathbb{E}\left[|S \overset{dW}{*_t} \Phi - \hat{S} \overset{dW}{*_t} \hat{\Phi}|^p_\infty\right] \leq$$
$$\leq c'_{\beta,T,p} \int_0^T \left( \int_0^t (t-s)^{-2\beta} \left( \mathbb{E}\left[ |S_{t-s}\Phi_s - \hat{S}_{t-s}\hat{\Phi}_s|^p_{L_2(U,H)} \right] \right)^{2/p} ds \right)^{p/2} dt. \tag{3.11}$$

For $0 \leq t_1 \leq t_2 \leq T$ and $\Phi \in \Lambda^{p,2,p}_{\mathscr{P}_T,S,\beta}(L(U,H))$, we also have

$$(S \overset{dW}{*_{t_1}} \Phi - S \overset{dW}{*_{t_2}} \Phi)_s = \mathbf{1}_{[t_1,t_2]}(s)(S \overset{dW}{*_{t_1}} \Phi)_s + \mathbf{1}_{(t_2,T]}(s) S_{s-t_2}(S \overset{dW}{*_{t_1}} \Phi)_{t_2} \quad \forall s \in [0,T]. \tag{3.12}$$

Since

$$\sup_{s\in[t_1,t_2]} |(S \overset{dW}{*_{t_1}} \Phi)_s|_H \leq |S \overset{dW}{*_{t_1}} (\mathbf{1}_{[t_1,t_2]}(\cdot)\Phi)|_\infty \qquad \mathbb{P}\text{-a.e.},$$

we obtain, by (3.8),

$$\lim_{t_2-t_1\to 0^+} \mathbb{E}\left[ \sup_{s\in[t_1,t_2]} |(S \overset{dW}{*_{t_1}} \Phi)_s|^p_H \right] \leq \lim_{t_2-t_1\to 0^+} c'_{\beta,T,p} |\mathbf{1}_{[t_1,t_2]}(\cdot)\Phi|^p_{p,2,S,\beta} = 0, \tag{3.13}$$

where the latter limit can be seen by applying Lebesgue's dominated convergence theorem three times, to the three integrals defining $|\cdot|_{p,2,S,\beta}$. Actually, since the linear map

$$\overline{\Lambda}^{p,2,p}_{\mathscr{P}_T,S,\beta}(L(U,H)) \to \overline{\Lambda}^{p,2,p}_{\mathscr{P}_T,S,\beta}(L(U,H)), \ \Phi \to \mathbf{1}_{[t_1,t_2]}(\cdot)\Phi$$

is bounded, uniformly in $t_1, t_2$, the limit (3.13) is uniform for $\Phi$ in compact subsets of $\overline{\Lambda}^{p,2,p}_{\mathscr{P}_T,S,\beta}(L(U,H))$ and $t_1, t_2 \in [0,T]$, $t_2 - t_1 \to 0^+$. Then, by (3.12) and (3.13), we finally obtain

$$\lim_{|t_2-t_1|\to 0} |S \overset{dW}{*}_{t_1} \Phi - S \overset{dW}{*}_{t_2} \Phi|_{\mathscr{L}^p_{\mathscr{P}_T}(\mathbb{S})} = 0 \tag{3.14}$$

uniformly for $\Phi$ in compact subsets of $\overline{\Lambda}^{p,2,p}_{\mathscr{P}_T,S,\beta}(L(U,H))$. In particular, thanks to the uniform boundedness of $\{S \overset{dW}{*}_t \#\}_{t\in[0,T]}$ (see (3.8)), the map

$$[0,T] \times \overline{\Lambda}^{p,2,p}_{\mathscr{P}_T,S,\beta}(L(U,H)) \to \mathscr{L}^p_{\mathscr{P}_T}(\mathbb{S}),\ (t,\Phi) \mapsto S \overset{dW}{*}_t \Phi \tag{3.15}$$

is continuous.

## 3.1 Existence and Uniqueness of Mild Solution

The following assumption will be standing for the remaining part of this manuscript. We recall that, if $E$ is a Banach space, then $\mathscr{B}_E$ denotes its Borel $\sigma$-algebra.

**Assumption 3.3**

(i) $b\colon (\Omega_T \times \mathbb{S}, \mathscr{P}_T \otimes \mathscr{B}_{\mathbb{S}}) \to (H, \mathscr{B}_H)$ is measurable;
(ii) $\sigma\colon (\Omega_T \times \mathbb{S}, \mathscr{P}_T \otimes \mathscr{B}_{\mathbb{S}}) \to L(U,H)$ is strongly measurable, that is $(\Omega_T \times \mathbb{S}, \mathscr{P}_T \otimes \mathscr{B}_{\mathbb{S}}) \to H$, $((\omega,t),\mathbf{x}) \mapsto \sigma((\omega,t),\mathbf{x})u$ is measurable, for all $u \in U$;
(iii) (non-anticipativity condition) for all $((\omega,t),\mathbf{x}) \in \Omega_T \times \mathbb{S}$, $b((\omega,t),\mathbf{x}) = b((\omega,t),\mathbf{x}_{t\wedge\cdot})$ and $\sigma((\omega,t),\mathbf{x}) = \sigma((\omega,t),\mathbf{x}_{t\wedge\cdot})$;
(iv) there exists $g \in L^1((0,T),\mathbb{R})$ such that

$$\begin{cases} |b((\omega,t),\mathbf{x})|_H \leq g(t)(1+|\mathbf{x}|_\infty) & \forall ((\omega,t),\mathbf{x}) \in \Omega_T \times \mathbb{S}, \\ |b((\omega,t),\mathbf{x}) - b((\omega,t),\mathbf{x}')|_H \leq g(t)|\mathbf{x}-\mathbf{x}'|_\infty & \forall (\omega,t) \in \Omega_T,\ \forall \mathbf{x},\mathbf{x}' \in \mathbb{S}; \end{cases}$$

(v) there exist $M > 0$, $\gamma \in [0,1/2)$ such that

$$\begin{cases} |S_t\sigma((\omega,s),\mathbf{x})|_{L_2(U,H)} \leq Mt^{-\gamma}(1+|\mathbf{x}|_\infty) & \forall ((\omega,s),\mathbf{x}) \in \Omega_T \times \mathbb{S},\ \forall t \in (0,T], \\ |S_t\sigma((\omega,s),\mathbf{x}) - S_t\sigma((\omega,s),\mathbf{x}')|_{L_2(U,H)} \leq Mt^{-\gamma}|\mathbf{x}-\mathbf{x}'|_\infty & \forall (\omega,s) \in \Omega_T,\ \forall t \in (0,T],\ \forall \mathbf{x},\mathbf{x}' \in \mathbb{S}. \end{cases}$$

*Remark 3.4* Assumption 3.3(iv) could be generalized to the form

$$\begin{cases} |S_t b((\omega,s),\mathbf{x})|_H \leq t^{-\gamma}g(s)(1+|\mathbf{x}|_\infty) & \forall ((\omega,s),\mathbf{x}) \in \Omega_T \times \mathbb{S},\ \forall t \in (0,T] \\ |S_t(b((\omega,s),\mathbf{x}) - b((\omega,s),\mathbf{x}'))|_H \leq t^{-\gamma}g(s)|\mathbf{x}-\mathbf{x}'|_\infty & \forall (\omega,s) \in \Omega_T,\ \forall t \in (0,T],\ \forall \mathbf{x},\mathbf{x}' \in \mathbb{S}, \end{cases}$$

with $g$ suitably integrable, and similarly for Assumption 3.3(v). The results obtained and the methods used hereafter can be adapted to cover these more general assumptions.

**Definition 3.5** (*Mild solution*) Let $Y \in \mathcal{L}^0_{\mathcal{P}_T}(\mathbb{S})$ and $t \in [0, T)$. A function $X \in \mathcal{L}^0_{\mathcal{P}_T}(\mathbb{S})$ is a *mild solution* to (3.1) if, for all $t' \in [t, T]$,

$$\mathbb{P}\left(\int_t^{t'} |S_{t-s}b(\cdot, s, X)|_H ds + \int_t^{t'} |S_{t-s}\sigma(\cdot, s, X)|^2_{L_2(U,H)} ds < \infty\right) = 1,$$

and

$$\begin{cases} \forall t' \in [0, t], & X_{t'} = Y_{t'} \ \mathbb{P}\text{-a.e.}, \\ \forall t' \in (t, T], & X_{t'} = S_{t'-t}Y_t + \int_t^{t'} S_{t'-s}b((\cdot, s), X)ds + \int_t^{t'} S_{t'-s}\sigma((\cdot, s), X)dW_s \ \mathbb{P}\text{-a.e.}. \end{cases}$$

Using a classical contraction argument, we are going to prove existence and uniqueness of mild solution in the space $\mathcal{L}^p_{\mathcal{P}_T}(\mathbb{S})$, when the initial datum $Y$ belongs to $\mathcal{L}^p_{\mathcal{P}_T}(\mathbb{S})$, for $p$ large enough. This will let us apply the theory developed in Sect. 2.

For $t \in [0, T]$ and

$$p > p^* := \frac{2}{1 - 2\gamma}, \quad \beta \in (1/p, 1/2 - \gamma),$$

we define the following maps:

$$\mathrm{id}_t^S : \mathcal{L}^p_{\mathcal{P}_T}(\mathbb{S}) \to \mathcal{L}^p_{\mathcal{P}_T}(\mathbb{S}), \ Y \mapsto \mathbf{1}_{[0,t]}(\cdot)Y + \mathbf{1}_{(t,T]}(\cdot)S_{\cdot-t}Y_t$$

$$F_b : \mathcal{L}^p_{\mathcal{P}_T}(\mathbb{S}) \to L^{p,1}_{\mathcal{P}_T}(H), \ X \mapsto b((\cdot, \cdot), X)$$

$$F_\sigma : \mathcal{L}^p_{\mathcal{P}_T}(\mathbb{S}) \to \overline{\Lambda}^{p,2,p}_{\mathcal{P}_T,S,\beta}(L(U, H)), \ X \mapsto \sigma((\cdot, \cdot), X)$$

$$S *_t \#: L^{p,1}_{\mathcal{P}_T}(H) \to \mathcal{L}^p_{\mathcal{P}_T}(\mathbb{S}), \ X \mapsto \mathbf{1}_{[t,T]}(\cdot) \int_t^{\cdot} S_{\cdot-s}X_s ds,$$

and we recall the map

$$S \overset{dW}{*_t} \#: \overline{\Lambda}^{p,2,p}_{\mathcal{P}_T,S,\beta}(L(U, H)) \to \mathcal{L}^p_{\mathcal{P}_T}(\mathbb{S}), \ \Phi \mapsto S \overset{dW}{*_t} \Phi.$$

Then $\mathrm{id}_t^S$ is well-defined, due to (*a*) and (*b*) in (3.2), because we can write

$$\mathrm{id}_t^S(Y) = Y_{t\wedge\cdot} + \mathbf{1}_{(t,T]}(\cdot)(S_{\cdot-t} - I)Y_t. \tag{3.17}$$

As regarding $F_b$, by Assumption 3.3(i), (iii), and by (*b*) in (3.2), the map

$$\Omega_T \to H, \ (\omega, t) \mapsto b((\omega, t), X(\omega)) = b((\omega, t), X_{t\wedge\cdot}(\omega))$$

is predictable. Moreover, by Assumption 3.3(ii), we have

$$\int_0^T \left(\mathbb{E}\left[|b(\cdot,t,X_{t\wedge\cdot})|^p\right]\right)^{1/p} dt \leq \int_0^T g(t)\left(\mathbb{E}\left[(1+|X|_\infty)^p\right]\right)^{1/p} dt \leq |g|_{L^1((0,T),\mathbb{R})}(1+|X|_{\mathcal{L}^p_{\mathcal{P}_T}(\mathbb{S})}),$$

which shows that $F_b(X) \in L^{p,1}_{\mathcal{P}_T}(H)$. By Assumption 3.3(iv), we also have that $F_b$ is Lipschitz, with Lipschitz constant dominated by $|g|_{L^1((0,1),\mathbb{R})}$. Similarly as done for $F_b$, by using Assumption 3.3(ii), one can see that, for $X \in \mathcal{L}^p_{\mathcal{P}_T}(\mathbb{S})$, the map

$$(\Omega_T, \mathcal{P}_T) \to L(U,H),\ (\omega,t) \mapsto \sigma((\omega,t), X_{t\wedge\cdot}(\omega))$$

is strongly measurable. Moreover, by Assumption 3.3(v), we have

$$\begin{aligned}|F_\sigma(X)|_{p,2,S,\beta} &= \left(\int_0^T \left(\int_0^t (t-s)^{-\beta 2}\left(\mathbb{E}\left[|S_{t-s}\sigma((\cdot,s),X_{s\wedge\cdot})|^p_{L_2(U,H)}\right]\right)^{2/p} ds\right)^{p/2} dt\right)^{1/p}\\ &\leq M\left(\int_0^T \left(\int_0^t v^{-(\beta+\gamma)2} dv\right)^{p/2} dt\right)^{1/p} (1+|X|_{\mathcal{L}^p_{\mathcal{P}_T}(\mathbb{S})})\end{aligned}$$

and the latter term is finite because $\beta < 1/2 - \gamma$ and $X \in \mathcal{L}^p_{\mathcal{P}_T}(\mathbb{S})$. Then $F_\sigma$ is well-defined. With similar computations, we have that $F_\sigma$ is Lipschitz, with Lipschitz constant depending only on $M, \beta, \gamma, p$. Regarding $S *_t \#$, if $X \in L^{p,1}_{\mathcal{P}_T}(H)$, then $X(\omega) \in L^1((0,T),H)$ for $\mathbb{P}$-a.e. $\omega \in \Omega$, hence it is easily checked that

$$[0,T] \to H,\ t' \mapsto \mathbf{1}_{[0,t]}(t')\int_t^{t'} S_{t'-s}X_s(\omega)ds$$

is continuous, and then it belongs to $\mathbb{S}$. Since $\mathbb{F}$ is complete, we can assume that $S *_t X(\omega)$ is continuous for all $\omega$, hence it is predictable, because it is $\mathbb{F}$-adapted. Since the trajectories are continuous, we also have the measurability of

$$(\Omega_T, \mathcal{P}_T) \to C([0,T],H) \subset \mathbb{S},\ (\omega,t') \mapsto (S *_t X)_{t'\wedge\cdot}(\omega).$$

Then, to show that $S *_t X \in \mathcal{L}^p_{\mathcal{P}_T}(\mathbb{S})$, it remains to verify the integrability condition. We have

$$|S *_t X|_{\mathcal{L}^p_{\mathcal{P}_T}(\mathbb{S})} \leq M'\left(\mathbb{E}\left[\left(\int_0^T |X_s|_H ds\right)^p\right]\right)^{1/p} \leq M'\int_0^T \left(\mathbb{E}\left[|X_s|^p_H\right]\right)^{1/p} ds = M'|X|_{p,1},$$

where

$$M' \text{ is any upper bound for } \sup_{t\in[0,T]} |S_t|_{L(H)}.$$

The good definition of $S \overset{dW}{*_t} \#$ was discussed above (observe that $p > \max\{2, 1/\beta\}$).

We can then build the map

$$\psi\colon \mathcal{L}^p_{\mathcal{P}_T}(\mathbb{S})\times\mathcal{L}^p_{\mathcal{P}_T}(\mathbb{S})\to\mathcal{L}^p_{\mathcal{P}_T}(\mathbb{S}),\ (Y,X)\mapsto \mathrm{id}^S_t(Y)+S*_t F_b(X)+S\overset{dW}{*_t}F_\sigma(X). \quad (3.18)$$

In what follows, whenever we need to make explicit the dependence of $\psi(Y,X)$ on the data $t,S,b,\sigma$, we write $\psi(Y,X;t,S,b,\sigma)$.

We first show that, for each $Y\in\mathcal{L}^p_{\mathcal{P}_T}(\mathbb{S})$, $\psi(Y,\cdot)$ has a unique fixed point $X$. Such a fixed point is a mild solution to (3.1).

The advantage of introducing the setting above is that it permits to see $\psi$ as a composition of maps that have different regularity and that can be considered individually when studying the regularity of the mild solution $X^{t,Y}$ with respect to $Y$ or the dependence of $X^{t,Y}$ with respect to a perturbation of the data $Y,t,S,b,\sigma$.

For $\lambda>0$, we consider the following norm on $\mathcal{L}^p_{\mathcal{P}_T}(\mathbb{S})$

$$|X|_{\mathcal{L}^p_{\mathcal{P}_T}(\mathbb{S}),\lambda}:=\left(\mathbb{E}\left[\sup_{t\in[0,T]}e^{-\lambda pt}|X_t|^p\right]\right)^{1/p}\qquad \forall X\in\mathcal{L}^p_{\mathcal{P}_T}(\mathbb{S}).$$

Then $|\cdot|_{\mathcal{L}^p_{\mathcal{P}_T}(\mathbb{S}),\lambda}$ is equivalent to $|\cdot|_{\mathcal{L}^p_{\mathcal{P}_T}(\mathbb{S})}$.

We proceed to show that there exists $\lambda>0$ such that $\psi$ is a parametric contraction.

For $X,X'\in\mathcal{L}^p_{\mathcal{P}_T}(\mathbb{S})$, $\lambda>0$, and $t'\in[0,T]$, we have

$$\begin{aligned}
e^{-\lambda pt'}|(S*_t F_b(X))_{t'}-(S*_t F_b(X'))_{t'}|^p_H &\le (M')^p\left(\int_0^{t'}e^{-\lambda t'}|b((\cdot,s),X)-b((\cdot,s),X')|_H ds\right)^p\\
&\le (M')^p\left(\int_0^{t'}e^{-\lambda(t'-s)}g(s)e^{-\lambda s}|X_{s\wedge\cdot}-X'_{s\wedge\cdot}|_\infty ds\right)^p\\
&\le C^p_{\lambda,g,M'}\sup_{s\in[0,T]}\left\{e^{-\lambda ps}|X_s-X'_s|^p_H\right\},
\end{aligned}$$

where $C_{\lambda,g,M'}:=M'\sup_{t'\in[0,T]}\int_0^{t'}e^{-\lambda v}g(t'-v)dv$. We then obtain

$$|S*_t F_b(X)-S*_t F_b(X')|_{\mathcal{L}^p_{\mathcal{P}_T}(\mathbb{S}),\lambda}\le C_{\lambda,g,M'}|X-X'|_{\mathcal{L}^p_{\mathcal{P}_T}(\mathbb{S}),\lambda}. \qquad (3.19)$$

It is not difficult to see that $C_{\lambda,g,M'}\to 0$ as $\lambda\to\infty$.

Now, if $\Phi\in\Lambda^{p,2,p}_{\mathcal{P}_T,S,\beta}(L(U,H))$, then $e^{-\lambda\cdot}\Phi\in\Lambda^{p,2,p}_{\mathcal{P}_T,e^{-\lambda\cdot}S,\beta}(L(U,H))$ for all $\lambda\ge 0$ and, for $\mathbb{P}$-a.e. $\omega\in\Omega$,

$$e^{-\lambda t'}(S\overset{dW}{*_t}\Phi)_{t'}=((e^{-\lambda\cdot}S)\overset{dW}{*_t}(e^{-\lambda\cdot}\Phi))_{t'}\qquad \forall t'\in[0,T]. \qquad (3.20)$$

For $X\in\mathcal{L}^p_{\mathcal{P}_T}(\mathbb{S})$, we have

$$\int_t^{t'} \mathbb{E}\left[|e^{-\lambda(t'-s)}S_{t'-s}(e^{-\lambda\cdot}F_\sigma(X))_s|^2_{L_2(U,H)}\right]ds < \infty \qquad \forall t' \in [t,T].$$

Then, for $X, X' \in \mathscr{L}^p_{\mathscr{P}_T}(\mathbb{S})$, $\lambda \geq 0$, and for all $t' \in [t,T]$, formula (3.7) provides

$$((e^{-\lambda\cdot}S) \overset{dW}{*_t} (e^{-\lambda\cdot}F_\sigma(X)))_{t'} - ((e^{-\lambda\cdot}S) \overset{dW}{*_t} (e^{-\lambda\cdot}F_\sigma(X')))_{t'} = c_\beta \int_t^{t'} (t'-s)^{\beta-1}\hat{Z}_s ds \qquad \mathbb{P}\text{-a.e.},$$

where $\hat{Z}$ is an $H$-valued predictable process such that, for a.e. $t' \in [t,T]$,

$$\hat{Z}_{t'} = \int_t^{t'} (t'-s)^{-\beta}e^{-\lambda(t'-s)}S_{t'-s}(e^{-\lambda\cdot}F_\sigma(X) - e^{-\lambda\cdot}F_\sigma(X'))_s dW_s \qquad \mathbb{P}\text{-a.e.}$$

By collecting the observations above, we can write, for $\lambda \geq 0$ and for all $t' \in [t,T]$,

$$e^{-\lambda pt'}|(S \overset{dW}{*_t} F_\sigma(X))_{t'} - (S \overset{dW}{*_t} F_\sigma(X'))_{t'}|^p_H \leq c^p_\beta \left(\int_0^T v^{\frac{(\beta-1)p}{p-1}}dv\right)^{p-1} \int_t^T |\hat{Z}_s|^p_H ds,$$

then, by applying [7, Lemma 7.7],

$$|S \overset{dW}{*_t} F_\sigma(X) - S \overset{dW}{*_t} F_\sigma(X')|^p_{\mathscr{L}^p_{\mathscr{P}_T}(\mathbb{S}),\lambda} \leq c'_{\beta,T,p}|e^{-\lambda\cdot}F_\sigma(X) - e^{-\lambda\cdot}F_\sigma(X')|^p_{p,2,e^{-\lambda\cdot}S,\beta}$$

where $c'_{\beta,T,p}$ is a constant depending only on $\beta, T, p$. Now, by using Assumption 3.3(v), we have

$$|e^{-\lambda\cdot}F_\sigma(X) - e^{-\lambda\cdot}F_\sigma(X')|^p_{p,2,e^{-\lambda\cdot}S,\beta} \leq M^p \left(\int_0^T \left(\int_0^t v^{-(\beta+\gamma)2}e^{-\lambda v}dv\right)^{p/2} dt\right)|X-X'|^p_{\mathscr{L}^p_{\mathscr{P}_T}(\mathbb{S}),\lambda}.$$

We finally obtain

$$|S \overset{dW}{*_t} F_\sigma(X) - S \overset{dW}{*_t} F_\sigma(X')|_{\mathscr{L}^p_{\mathscr{P}_T}(\mathbb{S}),\lambda} \leq c''_{\beta,\gamma,T,p,M,\lambda}|X-X'|_{\mathscr{L}^p_{\mathscr{P}_T}(\mathbb{S}),\lambda}, \tag{3.21}$$

where $c''_{\beta,\gamma,T,p,M,\lambda}$ is a constant depending only on $\beta, \gamma, T, p, M, \lambda$, and is such that

$$\lim_{\lambda\to\infty} c''_{\beta,\gamma,T,p,M,\lambda} = 0.$$

By (3.19) and (3.21), we have, for all $Y, X, Y', X'$,

$$|\psi(Y,X) - \psi(Y',X')|_{\mathscr{L}^p_{\mathscr{P}_T}(\mathbb{S}),\lambda} \leq M'|Y-Y'|_{\mathscr{L}^p_{\mathscr{P}_T}(\mathbb{S}),\lambda} + C'_{\lambda,g,\gamma,M',\beta,T,p,M}|X-X'|_{\mathscr{L}^p_{\mathscr{P}_T}(\mathbb{S}),\lambda}, \tag{3.22}$$

where $C'_{\lambda,g,\gamma,M',\beta,T,p,M}$ is a constant depending only on $\lambda$, $g$, $\gamma$, $M'$, $\beta$, $T$, $p$, $M$, such that

$$\lim_{\lambda\to\infty} C'_{\lambda,g,\gamma,M',\beta,T,p,M} = 0. \tag{3.23}$$

**Theorem 3.6** *Let Assumption 3.3 hold and let $t \in [0, T]$, $p > p^*$. Then there exists a unique mild solution $X^{t,Y} \in \mathcal{L}^p_{\mathcal{P}_T}(\mathbb{S})$ to SDE* (3.1)*. Moreover, there exists a constant $C$, depending only on $g$, $\gamma$, $M$, $M'$, $T$, $p$, such that,*

$$|X^{t,Y} - X^{t,Y'}|_{\mathcal{L}^p_{\mathcal{P}_T}(\mathbb{S})} \leq C|Y - Y'|_{\mathcal{L}^p_{\mathcal{P}_T}(\mathbb{S})} \qquad \forall Y, Y' \in \mathcal{L}^p_{\mathcal{P}_T}(\mathbb{S}).$$

*Proof.* Let us fix any $\beta \in (1/p, 1/2 - \gamma)$ and let $\psi$ be defined by (3.18). It is clear that any fixed point of $\psi(Y, \cdot)$ is a mild solution to SDE (3.1). Then, it is sufficient to apply Lemma 2.5 to $\psi$, taking into account (3.22) and (3.23), and recalling the equivalence of the norms $|\cdot|_{\mathcal{L}^p_{\mathcal{P}_T}(\mathbb{S})}$, $|\cdot|_{\mathcal{L}^p_{\mathcal{P}_T}(\mathbb{S}),\lambda}$. ■

*Remark 3.1* Since, for $p^* < p < q$, we have $\mathcal{L}^q_{\mathcal{P}_T}(\mathbb{S}) \subset \mathcal{L}^p_{\mathcal{P}_T}(\mathbb{S})$, then, if $Z \in \mathcal{L}^q_{\mathcal{P}_T}(\mathbb{S})$, the associated mild solution $X^{t,Z} \in \mathcal{L}^q_{\mathcal{P}_T}(\mathbb{S})$ is also a mild solution in $\mathcal{L}^p_{\mathcal{P}_T}(\mathbb{S})$ and, by uniqueness, it is *the* solution in that space. Hence the solution does not depend on the specific $p > p^*$ chosen.

### 3.2 Gâteaux Differentiability with Respect to the Initial Datum

We now study the differentiability of the mild solution $X^{t,Y}$ with respect to the initial datum $Y$.

**Assumption 3.7** Let $b, \sigma, g, \gamma$ be as in Assumption 3.3. Let $n \in \mathbb{N}$, $n \geq 1$.

(i) For all $(\omega, t) \in \Omega_T$ and $u \in U$, $b((\omega, t), \cdot) \in \mathcal{G}^n(\mathbb{S}, H)$, $\sigma((\omega, t), \cdot)u \in \mathcal{G}^n(\mathbb{S}, H)$.
(ii) There exists $M''$ and $c := \{c_m\}_{m\in\mathcal{M}} \in \ell^2(\mathcal{M})$ such that

$$\sup_{j=1,\dots,n} \sup_{\substack{\omega\in\Omega \\ \mathbf{x},\mathbf{y}_1,\dots,\mathbf{y}_j\in\mathbb{S} \\ |\mathbf{y}_1|_\infty=\dots=|\mathbf{y}_j|_\infty=1}} |\partial^j_{\mathbf{y}_1\dots\mathbf{y}_j} b((\omega, s), \mathbf{x})|_H \leq M'' g(s), \tag{3.24}$$

$$\sup_{j=1,\dots,n} \sup_{\substack{\omega\in\Omega \\ \mathbf{x},\mathbf{y}_1,\dots,\mathbf{y}_j\in\mathbb{S} \\ |\mathbf{y}_1|_\infty=\dots=|\mathbf{y}_j|_\infty=1}} |S_t \partial^j_{\mathbf{y}_1\dots\mathbf{y}_j} (\sigma((\omega, s), \mathbf{x})e'_m))|_H \leq M'' t^{-\gamma} c_m, \tag{3.25}$$

for all $s \in [0, T]$, $t \in (0, T]$, $m \in \mathcal{M}$.

In accordance with Assumption 3.7(i), by writing $\partial^j_{\mathbf{y}_1\dots\mathbf{y}_j}(\sigma((\omega,s),\mathbf{x})u)$, we mean the Gâteaux derivative of the map $\mathbf{x}\mapsto\sigma((\omega,s),\mathbf{x}).u$, for fixed $u\in U$.

**Lemma 3.8** *Suppose that Assumptions 3.3 and 3.7 are satisfied. Let $p>p^*$, $\beta\in(1/p,1/2-\gamma)$. Then, for $j=1,\dots,n$,*

$$F_b\in\mathscr{G}^j(\mathscr{L}^p_{\mathscr{P}_T}(\mathbb{S}),L^{p,1}_{\mathscr{P}_T}(H);\mathscr{L}^{jp}_{\mathscr{P}_T}(\mathbb{S})),$$

$$F_\sigma\in\mathscr{G}^j(\mathscr{L}^p_{\mathscr{P}_T}(\mathbb{S}),\overline{\Lambda}^{p,2,p}_{\mathscr{P}_T,S,\beta}(L(U,H));\mathscr{L}^{jp}_{\mathscr{P}_T}(\mathbb{S})).$$

*and, for $X\in\mathscr{L}^p_{\mathscr{P}_T}(\mathbb{S})$, $Y_1,\dots,Y_j\in\mathscr{L}^{jp}_{\mathscr{P}_T}(\mathbb{S})$, $u\in U$, $\mathbb{P}\otimes m$-a.e. $(\omega,t)\in\Omega_T$,*

$$\begin{cases}\partial^j_{Y_1\dots Y_j}F_b(X)(\omega,t)=\partial^j_{Y_1(\omega)\dots Y_j(\omega)}b((\omega,t),X(\omega))\\ \partial^j_{Y_1\dots Y_j}F_\sigma(X)(\omega,t)u=\partial^j_{Y_1(\omega)\dots Y_j(\omega)}(\sigma((\omega,t),X(\omega))u).\end{cases}\tag{3.26}$$

*Moreover,*

$$\sup_{j=1,\dots,n}\ \sup_{\substack{X\in\mathscr{L}^p_{\mathscr{P}_T}(\mathbb{S})\\ Y_1,\dots,Y_j\in\mathscr{L}^{jp}_{\mathscr{P}_T}(\mathbb{S})\\ |Y_1|_{\mathscr{L}^{jp}_{\mathscr{P}_T}(\mathbb{S})}=\dots=|Y_j|_{\mathscr{L}^{jp}_{\mathscr{P}_T}(\mathbb{S})}=1}}\left(|\partial^j_{Y_1\dots Y_j}F_b(X)|_{L^{p,1}_{\mathscr{P}_T}(H)}+|\partial^j_{Y_1\dots Y_j}F_\sigma(X)|_{p,2,S,\beta}\right)\le M''',$$

*where $M'''$ depends only on $T$, $p$, $\beta$, $\gamma$, $|g|_{L^1((0,T),\mathbb{R})}$, $M''$, $|c|_{\ell^2(\mathscr{M})}$.*

*Proof.* We prove the lemma by induction on $n$.
<u>Case $n=1$</u>. Let $X,Y\in\mathscr{L}^p_{\mathscr{P}_T}(\mathbb{S})$. First notice that the function

$$(\Omega_T,\mathscr{P}_T)\to H,\ (\omega,t)\mapsto\partial_{Y(\omega)}b((\omega,t),X(\omega))$$

is measurable. Let $\epsilon\in\mathbb{R}\setminus\{0\}$. Since $b((\omega,t),\cdot)\in\mathscr{G}^1(\mathbb{S},H)$ for all $(\omega,t)\in\Omega_T$, we can write

$$\begin{aligned}\Delta_{\epsilon Y}F_b(X)(\omega,t)&:=\epsilon^{-1}\left(F_b(X+\epsilon Y)(\omega,t)-F_b(X)(\omega,t)\right)\\&=\epsilon^{-1}\left(b((\omega,t),X(\omega)+\epsilon Y(\omega))-b((\omega,t),X(\omega))\right)\\&=\int_0^1\partial_{Y(\omega)}b((\omega,t),X(\omega)+\epsilon\theta Y(\omega))d\theta\qquad\mathbb{P}\otimes m\text{-a.e. }(\omega,t)\in\Omega_T.\end{aligned}\tag{3.27}$$

By (3.24), we also have

$$|\partial_{Y(\omega)}b((\omega,t),X(\omega)+\epsilon Y(\omega))|_H\le M''g(t)|Y(\omega)|_\infty\qquad\forall(\omega,t)\in\Omega_T,\ \forall\epsilon\in\mathbb{R}.\tag{3.28}$$

By (3.27) and (3.28), we can apply Lebesgue's dominated convergence theorem and obtain

$$\lim_{\epsilon\to 0}\int_0^T \left(\mathbb{E}\left[|\Delta_{\epsilon Y}F_b(X)(\cdot,t)-\partial_Y b((\cdot,t),X)|_H^p\right]\right)^{1/p}dt=0.$$

This proves that $F_b$ has directional derivative at $X$ for the increment $Y$ and that

$$\partial_Y F_b(X)(\omega,t)=\partial_{Y(\omega)}b((\omega,t),X(\omega)) \qquad \mathbb{P}\otimes m\text{-a.e. }(\omega,t)\in\Omega_T. \tag{3.29}$$

We now show that $\partial_Y F_b(X)$ is continuous in $(X,Y)\in\mathcal{L}^p_{\mathcal{P}_T}(\mathbb{S})$. Notice that, by (3.24), the linear map $\mathcal{L}^p_{\mathcal{P}_T}(\mathbb{S})\to L^{p,1}_{\mathcal{P}_T}(H)$, $Y\mapsto\partial_Y F_b(X)$, is bounded, uniformly in $X$. Then it is sufficient to verify the continuity of $\partial_Y F_b(X)$ in $X$, for fixed $Y$. Let $X_k\to X$ in $\mathcal{L}^p_{\mathcal{P}_T}(\mathbb{S})$. By (3.24), (3.29), and Lebesgue's dominated convergence theorem, we have

$$\lim_{k\to\infty}\partial_Y F_b(X_k)=\partial_Y F_b(X)\ \text{in}\ L^{p,1}_{\mathcal{P}_T}(H).$$

This concludes the proof that $F_b\in\mathcal{G}^1(\mathcal{L}^p_{\mathcal{P}_T}(\mathbb{S}),L^{p,1}_{\mathcal{P}_T(H)})$ and that the differential is uniformly bounded.

Similarly, as regarding $F_\sigma$, we have that, for all $u\in U$, the function

$$(\Omega_T,\mathcal{P}_T)\to H,\ (\omega,t)\mapsto\partial_{Y(\omega)}(\sigma(t,X(\omega))u)$$

is measurable, and

$$\begin{aligned}\Delta_{\epsilon Y}(F_\sigma(X)u)(\omega,t)&:=\epsilon^{-1}\left((F_\sigma(X+\epsilon Y)u)(\omega,t)-(F_\sigma(X)u)(\omega,t)\right)\\&=\epsilon^{-1}\left(\sigma((\omega,t),X(\omega)+\epsilon Y(\omega))u-\sigma((\omega,t),X(\omega))u\right)\\&=\int_0^1\partial_{Y(\omega)}(\sigma((\omega,t),X(\omega)+\epsilon\theta Y(\omega))u)d\theta \qquad \mathbb{P}\otimes m\text{-a.e. }(\omega,t)\in\Omega_T.\end{aligned} \tag{3.30}$$

By (3.25), for all $0\le s<t\le T$, $\omega\in\Omega$, $\epsilon\in\mathbb{R}$, $m\in\mathcal{M}$,

$$|S_{t-s}\partial_{Y(\omega)}(\sigma((\omega,s),X(\omega)+\epsilon Y(\omega))e'_m)|_H\le M''(t-s)^{-\gamma}c_m|Y(\omega)|_\infty. \tag{3.31}$$

By repeatedly applying Lebesgue's dominated convergence theorem, we have that

$$\int_0^T\left(\int_0^t(t-s)^{-2\beta}\left(\mathbb{E}\left[\left(\sum_{m\in\mathcal{M}}|S_{t-s}\left(\Delta_{\epsilon Y}F_\sigma(X)(\cdot,s).e'_m-\partial_Y(\sigma((\cdot,s),X).e'_m)\right)|_H^2\right)^{p/2}\right]\right)^{2/p}ds\right)^{p/2}dt$$

goes to 0 as $\epsilon\to 0$. This proves that $F_\sigma$ has directional derivative at $X$ for the increment $Y$ and, taking into account the separability of $U$, that

$$\partial_Y F_\sigma(X)(\omega,t)=\partial_{Y(\omega)}(\sigma((\omega,t),X(\omega))\#) \qquad \mathbb{P}\otimes m\text{-a.e. }(\omega,t)\in\Omega_T. \tag{3.32}$$

By (3.31) and arguing similarly as done for $\partial_Y F_b(X)$, in order to show the continuity of $\partial_Y F_\sigma(X)$ in $(X,Y) \in \mathcal{L}^p_{\mathcal{P}_T}(\mathbb{S})$, it is sufficient to verify the continuity of $\partial_Y F_\sigma(X)$ in $X$, for fixed $Y$. Let $X_k \to X$ in $\mathcal{L}^p_{\mathcal{P}_T}(\mathbb{S})$. By (3.25), (3.32), and Lebesgue's dominated convergence theorem, we have

$$\lim_{k\to\infty} \partial_Y F_\sigma(X_k) = \partial_Y F_\sigma(X) \text{ in } \overline{\Lambda}^{p,2,p}_{\mathcal{P}_T,S,\beta}(L(U,H)).$$

This shows that $F_\sigma \in \mathcal{G}^1(\mathcal{L}^p_{\mathcal{P}_T}(\mathbb{S}), \overline{\Lambda}^{p,2,p}_{\mathcal{P}_T,S,\beta}(L(U,H)))$ and that the differential is uniformly bounded.

<u>Case $n>1$</u>. Let $X \in \mathcal{L}^p_{\mathcal{P}_T}(\mathbb{S})$ and $Y_1,\ldots,Y_n \in \mathcal{L}^{np}_{\mathcal{P}_T}(\mathbb{S})$. By inductive hypothesis, we can assume that $\partial^{n-1}_{Y_1\ldots Y_{n-1}} F_b(X) \in L^{p,1}_{\mathcal{P}_T}(H)$ exists, jointly continuous in $X \in \mathcal{L}^p_{\mathcal{P}_T}(\mathbb{S})$ and $Y_1,\ldots,Y_{n-1} \in \mathcal{L}^{(n-1)p}_{\mathcal{P}_T}(H)$, and that

$$\partial^{n-1}_{Y_1\ldots Y_{n-1}} F_b(X)(\omega,t) = \partial^{n-1}_{Y_1(\omega)\ldots Y_{n-1}(\omega)} b((\omega,t),X(\omega)) \qquad \mathbb{P}\otimes m\text{-a.e. } (\omega,t)\in\Omega_T.$$

The argument goes like the case $n=1$. Let $\epsilon \in \mathbb{R}\setminus\{0\}$. Since $b((\omega,t),\cdot) \in \mathcal{G}^n(\mathbb{S},H)$ for $(\omega,t)\in\Omega_T$, we can write, for $\mathbb{P}\otimes m$-a.e. $(\omega,t)\in\Omega_T$,

$$\begin{aligned}
\Delta_{\epsilon Y_n}\partial^{n-1}_{Y_1\ldots Y_{n-1}} F_b(X)(\omega,t) &:= \epsilon^{-1}\left(\partial^{n-1}_{Y_1\ldots Y_{n-1}} F_b(X+\epsilon Y_n)(\omega,t) - \partial^{n-1}_{Y_1\ldots Y_{n-1}} F_b(X)(\omega,t)\right)\\
&= \epsilon^{-1}\left(\partial^{n-1}_{Y_1(\omega)\ldots Y_{n-1}(\omega)} b((\omega,t),X(\omega)+\epsilon Y_n(\omega)) - \partial^{n-1}_{Y_1(\omega)\ldots Y_{n-1}(\omega)} b((\omega,t),X(\omega))\right)\\
&= \int_0^1 \partial^n_{Y_1(\omega)\ldots Y_{n-1}(\omega)Y_n(\omega)} b((\omega,t),X(\omega)+\epsilon\theta Y_n(\omega))d\theta.
\end{aligned}$$

By (3.24) we have

$$|\partial^n_{Y_1(\omega)\ldots Y_n(\omega)} b((\omega,t),X(\omega)+\epsilon Y_n(\omega))|_H \le M'' g(t) \prod_{j=1}^n |Y_j(\omega)|_\infty \qquad \forall(\omega,t)\in\Omega_T,\ \forall\epsilon\in\mathbb{R}.$$

Since $Y_j \in \mathcal{L}^{np}_{\mathcal{P}_T}(H)$, by the generalized Hölder inequality $\prod_{j=1}^n |Y_j|_\infty \in L^p((\Omega,\mathcal{F}_T,\mathbb{P}),\mathbb{R})$. Then we can apply Lebesgue's dominated convergence theorem twice to obtain

$$\lim_{\epsilon\to 0}\int_0^T \left(\mathbb{E}\left[|\Delta_{\epsilon Y_n}\partial^{n-1}_{Y_1\ldots Y_{n-1}} F_b(X)(\cdot,t) - \partial^n_{Y_1\ldots Y_n} b((\cdot,t),X)|^p_H\right]\right)^{1/p} dt = 0.$$

This proves that $\partial^{n-1}_{Y_1\ldots Y_{n-1}} F_b$ has directional derivative at $X$ for the increment $Y_n$ and that

$$\partial^n_{Y_1\ldots Y_{n-1}Y_n} F_b(X)(\omega,t) = \partial^n_{Y_1(\omega)\ldots Y_n(\omega)} b((\omega,t),X(\omega)) \qquad \mathbb{P}\otimes m\text{-a.e. } (\omega,t)\in\Omega_T. \tag{3.33}$$

The continuity of $\partial^n_{Y_1\ldots Y_{n-1}Y_n} F_b(X)$ in $X \in \mathcal{L}^p_{\mathcal{P}_T}(\mathbb{S})$, $Y_1,\ldots,Y_n \in \mathcal{L}^{np}_{\mathcal{P}_T}(H)$, is proved similarly as for the case $n=1$, again by invoking the generalized Hölder inequality.

This concludes the proof that $F_b \in \mathscr{G}^n(\mathscr{L}^p_{\mathscr{P}_T}(\mathbb{S}), L^{p,1}_{\mathscr{P}_T(H)}; \mathscr{L}^{np}_{\mathscr{P}_T}(H))$. The uniform boundedness of the differentials is obtained by (3.24), (3.33), and the generalized Hölder inequality.

Finally, as regarding $F_\sigma$, let again $X \in \mathscr{L}^p_{\mathscr{P}_T}(\mathbb{S})$ and $Y_1, \ldots, Y_n \in \mathscr{L}^{np}_{\mathscr{P}_T}(\mathbb{S})$. By inductive hypothesis, we can assume that $\partial^{n-1}_{Y_1 \ldots Y_{n-1}} F_\sigma(X) \in \overline{\Lambda}^{p,2,p}_{\mathscr{P}_T,S,\beta}(L(U,H))$ exists, that it is continuous in $X \in \mathscr{L}^p_{\mathscr{P}_T}(\mathbb{S})$, $Y_1, \ldots, Y_{n-1} \in \mathscr{L}^{(n-1)p}_{\mathscr{P}_T}(\mathbb{S})$, and that, for all $u \in U$,

$$\partial^{n-1}_{Y_1 \ldots Y_{n-1}} F_\sigma(X)(\omega,t)u = \partial^{n-1}_{Y_1(\omega)\ldots Y_{n-1}(\omega)}(\sigma((\omega,t),X(\omega))u) \qquad \mathbb{P}\otimes m\text{-a.e. } (\omega,t) \in \Omega_T.$$

For $\epsilon \in \mathbb{R} \setminus \{0\}$, by strongly continuous Gâteaux differentiability of

$$x \mapsto \partial^{n-1}_{Y_1(\omega)\ldots Y_{n-1}(\omega)}(\sigma(t,x)u),$$

we can write,

$$\begin{aligned}
\Delta_{\epsilon Y_n} \partial^{n-1}_{Y_1 \ldots Y_{n-1}} F_\sigma(X)(\omega,t)u &:= \epsilon^{-1}\left(\partial^{n-1}_{Y_1 \ldots Y_{n-1}} F_\sigma(X+\epsilon Y_n)(\omega,t)u - \partial^{n-1}_{Y_1 \ldots Y_{n-1}} F_\sigma(X)(\omega,t)u\right)\\
&= \epsilon^{-1}\left(\partial^{n-1}_{Y_1(\omega)\ldots Y_{n-1}(\omega)}(\sigma((\omega,t),X(\omega)+\epsilon Y_n(\omega))u) - \partial^{n-1}_{Y_1(\omega)\ldots Y_{n-1}(\omega)}(\sigma((\omega,t),X(\omega))u)\right)\\
&= \int_0^1 \partial^n_{Y_1(\omega)\ldots Y_n(\omega)}(\sigma((\omega,t),X(\omega)+\epsilon\theta Y_n(\omega))u)d\theta.
\end{aligned}$$

By (3.25) we have, for all $\omega \in \Omega$, $\epsilon \in \mathbb{R}$, $0 \leq s < t \leq T$, $m \in \mathscr{M}$,

$$|S_{t-s}\partial^n_{Y_1(\omega)\ldots Y_n(\omega)}(\sigma((\omega,s),X(\omega)+\epsilon Y_n(\omega))e'_m)|_H \leq M''(t-s)^{-\gamma} c_m \prod_{j=1}^n |Y_j(\omega)|_\infty.$$

By the generalized Hölder inequality and by Lebesgue's dominated convergence theorem, we conclude

$$\begin{aligned}
\lim_{\epsilon\to 0} \int_0^T \Bigg( \int_0^t (t-s)^{-2\beta} \Bigg( \mathbb{E}\Bigg[\Bigg( \sum_{m\in\mathscr{M}} \Bigg| S_{t-s}\Big(\Delta_{\epsilon Y_n}\partial^{n-1}_{Y_1\ldots Y_{n-1}} F_\sigma(X)(\omega,s)e'_m \\
-\partial^n_{Y_1(\omega)\ldots Y_n(\omega)}(\sigma((\cdot,s),X)e'_m)\Big)\Bigg|^2_H\Bigg)^{p/2}\Bigg]\Bigg)^{2/p} ds\Bigg)^{p/2} dt = 0.
\end{aligned} \tag{3.34}$$

Then $\partial^{n-1}_{Y_1 \ldots Y_{n-1}} F_\sigma$ has directional derivative at $X$ for the increment $Y_n$, given by, for all $u \in U$,

$$\partial_{Y_n}\partial^{n-1}_{Y_1 \ldots Y_{n-1}} F_\sigma(X)(\omega,t)u = \partial^n_{Y_1(\omega)\ldots Y_n(\omega)}(\sigma((\omega,t),X(\omega))u) \qquad \mathbb{P}\otimes m\text{-a.e. } (\omega,t) \in \Omega_T.$$

The continuity of $\partial_{Y_n}\partial^{n-1}_{Y_1 \ldots Y_{n-1}} F_\sigma(X)$ with respect to $X \in \mathscr{L}^p_{\mathscr{P}_T}(\mathbb{S})$, $Y_1, \ldots, Y_n \in \mathscr{L}^{np}_{\mathscr{P}_T}(H)$, is proved as for the case $n = 1$. Then $F_\sigma \in \mathscr{G}^n(\mathscr{L}^p_{\mathscr{P}_T}(\mathbb{S}), \overline{\Lambda}^{p,2,p}_{\mathscr{P}_T,S,\beta}$

$(L(U,H)); \mathscr{L}^{np}_{\mathscr{P}_T}(H))$. The uniform boundedness of the differentials is obtained by (3.25), (3.34), and the generalized Hölder inequality. ∎

Due to the fact that $X^{t,Y}$ is the fixed point of $\psi(Y,\cdot)$ and due to the structure of $\psi$, the previous lemma permits to easily obtain the following

**Theorem 3.9** *Suppose that Assumption 3.7 is satisfied. Let $t \in [0,T]$, $p > p^*$, $p \geq n$. Then the map*

$$\mathscr{L}^{p^n}_{\mathscr{P}_T}(\mathbb{S}) \to \mathscr{L}^{p}_{\mathscr{P}_T}(\mathbb{S}),\ Y \mapsto X^{t,Y} \tag{3.35}$$

*belongs to $\mathscr{G}^n(\mathscr{L}^{p^n}_{\mathscr{P}_T}(\mathbb{S}), \mathscr{L}^{p}_{\mathscr{P}_T}(\mathbb{S}))$ and the Gâteaux differentials up to order $n$ are uniformly bounded by a constant depending only on $T, p, \gamma, g, M, M', M'', |c|_{\ell^2(\mathscr{M})}$.*

*Proof.* Let $\beta \in (1/p, 1/2-\gamma)$. We have $p^k > p^*$ and $\beta \in (1/p^k, 1/2-\gamma)$ for all $k = 1,\dots,n$. Then, for $k = 1,\dots,n$, the map

$$\psi_k \colon \mathscr{L}^{p^k}_{\mathscr{P}_T}(\mathbb{S}) \times \mathscr{L}^{p^k}_{\mathscr{P}_T}(\mathbb{S}) \to \mathscr{L}^{p^k}_{\mathscr{P}_T}(\mathbb{S}),\ (Y,X) \mapsto \mathrm{id}^S_t(Y) + S *_t F_b(X) + S \overset{dW}{*_t} F_\sigma(X)$$

is well-defined, where we have implicitly chosen the space $L^{p^k,1}_{\mathscr{P}_T}(H)$ as codomain of $F_b$ and $\overline{\Lambda}^{p^k,2,p^k}_{\mathscr{P}_T,S,\beta}(L(U,H))$ as codomain of $F_\sigma$. Since the functions

$$\begin{gathered}\mathscr{L}^{p^k}_{\mathscr{P}_T}(\mathbb{S}) \to \mathscr{L}^{p^k}_{\mathscr{P}_T}(\mathbb{S})\\ S *_t \#\colon L^{p^k,1}_{\mathscr{P}_T}(H) \to \mathscr{L}^{p^k}_{\mathscr{P}_T}(\mathbb{S})\\ S \overset{dW}{*_t} \#\colon \overline{\Lambda}^{p^k,2,p^k}_{\mathscr{P}_T,S,\beta}(L(U,H)) \to \mathscr{L}^{p^k}_{\mathscr{P}_T}(\mathbb{S})\end{gathered}$$

are linear and continuous, with an upper bound for the operator norms depending only on $\beta, M', T, p$, we have, by applying Lemma 3.8, for $k, j = 1,\dots,n$,

$$\psi_k \in \mathscr{G}^j(\mathscr{L}^{p^k}_{\mathscr{P}_T}(\mathbb{S}) \times \mathscr{L}^{p^k}_{\mathscr{P}_T}(\mathbb{S}), \mathscr{L}^{p^k}_{\mathscr{P}_T}(\mathbb{S}); \mathscr{L}^{p^k}_{\mathscr{P}_T}(\mathbb{S}) \times \mathscr{L}^{jp^k}_{\mathscr{P}_T}(\mathbb{S})),$$

with differentials bounded by a constant depending only on g, $\gamma, M, M', M''$, $|c|_{\ell^2(\mathscr{M})}$, $T$, on $p^k$ (hence on $p$), and on $\beta$, which depends on $p, \gamma$. In particular, since $np^k \leq p^{k+1}$, we have, for the rescritions $\psi_{k|\mathscr{L}^{p^n}_{\mathscr{P}_T}(\mathbb{S}) \times \mathscr{L}^{p^k}_{\mathscr{P}_T}(\mathbb{S})}$ of $\psi_k$ to $\mathscr{L}^{p^n}_{\mathscr{P}_T}(\mathbb{S}) \times \mathscr{L}^{p^k}_{\mathscr{P}_T}(\mathbb{S})$,

$$\begin{cases}\psi_{k|\mathscr{L}^{p^n}_{\mathscr{P}_T}(\mathbb{S}) \times \mathscr{L}^{p^k}_{\mathscr{P}_T}(\mathbb{S})} \in \mathscr{G}^1(\mathscr{L}^{p^n}_{\mathscr{P}_T}(\mathbb{S}) \times \mathscr{L}^{p^k}_{\mathscr{P}_T}(\mathbb{S}), \mathscr{L}^{p^k}_{\mathscr{P}_T}(\mathbb{S}))\\ \psi_{k|\mathscr{L}^{p^n}_{\mathscr{P}_T}(\mathbb{S}) \times \mathscr{L}^{p^k}_{\mathscr{P}_T}(\mathbb{S})} \in \mathscr{G}^n(\mathscr{L}^{p^n}_{\mathscr{P}_T}(\mathbb{S}) \times \mathscr{L}^{p^k}_{\mathscr{P}_T}(\mathbb{S}), \mathscr{L}^{p^k}_{\mathscr{P}_T}(\mathbb{S}); \mathscr{L}^{p^n}_{\mathscr{P}_T}(\mathbb{S}) \times \mathscr{L}^{p^{k+1}}_{\mathscr{P}_T}(\mathbb{S}))\end{cases}$$

for $k = 1,\dots,n$, with the Gâteaux differentials that are uniformly bounded by a constant depending only on $g$,$\gamma$, $M$,$M'$,$M''$,$|c|_{\ell^2(\mathscr{M})}$, $T$, on $\beta$ (hence on $p,\gamma$), and on $p^n$, $p^k$, $p^{k+1}$ (hence on $p$).

By (3.22) and (3.23) (where $p$ should be replaced by $p^k$), there exists $\lambda > 0$, depending only on $g, \gamma, M, M', \beta, T$, and on $p^k$ (hence on $p$), such that $\psi_k$ is a parametric 1/2-contraction with respect to the second variable, uniformly in the first one, when the space $\mathscr{L}^{p^k}_{\mathscr{P}_T}(\mathbb{S})$ is endowed with the equivalent norm $|\cdot|_{\mathscr{L}^{p^k}_{\mathscr{P}_T}(\mathbb{S}),\lambda}$. Then we can assume that the uniform bound of the Gâteaux differentials of $\psi_k$, for $k = 1, \ldots, n$, holds with respect to the equivalent norms $|\cdot|_{\mathscr{L}^{p^k}_{\mathscr{P}_T}(\mathbb{S}),\lambda}$, and is again depending only on $g, \gamma, M, M', M'', |c|_{\ell^2(\mathscr{M})}, T, p$.

Now consider Assumption 2.8, after setting:

- $\alpha := 1/2$;
- $U := X := (\mathscr{L}^{p^n}_{\mathscr{P}_T}(\mathbb{S}), |\cdot|_{\mathscr{L}^{p^n}_{\mathscr{P}_T}(\mathbb{S}),\lambda})$;
- $Y_1 := (\mathscr{L}^{p}_{\mathscr{P}_T}(\mathbb{S}), |\cdot|_{\mathscr{L}^{p}_{\mathscr{P}_T}(\mathbb{S}),\lambda}), \ldots, Y_k := (\mathscr{L}^{p^k}_{\mathscr{P}_T}, |\cdot|_{\mathscr{L}^{p^k}_{\mathscr{P}_T}(\mathbb{S}),\lambda}), \ldots, Y_n := (\mathscr{L}^{p^n}_{\mathscr{P}_T}, |\cdot|_{\mathscr{L}^{p^n}_{\mathscr{P}_T}(\mathbb{S}),\lambda})$;
- $h_1 := \psi_{1|\mathscr{L}^{p^n}_{\mathscr{P}_T}(\mathbb{S})\times\mathscr{L}^{p}_{\mathscr{P}_T}(\mathbb{S})}, \ldots, h_k := \psi_{k|\mathscr{L}^{p^n}_{\mathscr{P}_T}(\mathbb{S})\times\mathscr{L}^{p^k}_{\mathscr{P}_T}(\mathbb{S})}, \ldots, h_n := \psi_{n|\mathscr{L}^{p^n}_{\mathscr{P}_T}(\mathbb{S})\times\mathscr{L}^{p^n}_{\mathscr{P}_T}(\mathbb{S})}$.

The discussion above, together with the smooth dependence of $h_k$ on the first variable, shows that Assumption 2.8 is verified. We can then apply Theorem 2.9, which provides

$$((3.35) =)\ \mathscr{L}^{p^n}_{\mathscr{P}_T}(\mathbb{S}) \to \mathscr{L}^{p}_{\mathscr{P}_T}(\mathbb{S}),\ Y \mapsto X^{t,Y}, \in \mathscr{G}^n(\mathscr{L}^{p^n}_{\mathscr{P}_T}(\mathbb{S}), \mathscr{L}^{p}_{\mathscr{P}_T}(\mathbb{S})).$$

Finally, by applying Corollary 2.10, we obtain the uniform boundedness of the Gâteaux differentials up to order $n$ of (3.35), with a bound that depends only on $T, \gamma, g, M, M', M'', |c|_{\ell^2(\mathscr{M})}, p$. ■

*Remark 3.10* As we have said in the introduction, we obtain the Gâteaux differentiability of $x \mapsto X^{t,x}$ by studying the parametric contraction providing $X^{t,x}$ as its unique fixed point, similarly as done in [8] for the non-path-dependent case. A different approach consists in studying directly the variations $\lim_{h\to 0} \frac{X^{t,x+hv}-X^{t,x}}{h}$, showing that the limit exists (under suitable smooth assumptions on the coefficients) and is continuous with respect to $v$, for fixed $t, x$. This would provide the existence of the Gâteaux differential $\partial X^{t,x}$. Usually, in this way one shows also that $\partial X^{t,x}.v$ solves an SDE. By using this SDE, one could go further and prove that the second order derivative $\partial^2 X^{t,x}.(v, w)$ exists, and that it is continuous in $v, w$, for fixed $t, x$. This would provide the second order Gâteaux differentiability of $x \mapsto X^{t,x}$. In this way, it is possible also to study the continuity of the Gâteaux differentials, by considering the SDEs solved by the directional derivatives, and to obtain Fréchet differentiability (under suitable assumptions on the coefficients, e.g. uniformly continuous Fréchet differentiability). By doing so, first- and second-order Fréchet differentiability are proved in [14]. But if one wants to use these methods to obtain derivatives of a generic order $n \geq 3$, then a recursive formula providing the SDE solved by the $(n-1)$th-order derivatives is needed, hence we fall back to a statement like Theorem 2.9.

Let $n = 2$ and let $h_1$ be as in the proof of Theorem 3.9. By continuity and linearity of $\mathrm{id}_t^S$, $S *_t \#$, $S \overset{dW}{*_t} \#$, and by recalling Lemma 3.8, we have, for $Y, Y_1, Y_2 \in \mathscr{L}^{p^2}_{\mathscr{P}_T}(\mathbb{S})$ (the space of the first variable of $h_1$), $X, X_1, X_2 \in \mathscr{L}^p_{\mathscr{P}_T}(\mathbb{S})$ (the space of the second variable of $h_1$),

$$\begin{cases} \partial_{Y_1} h_1(Y, X) = \mathrm{id}_t^S(Y_1) \\ \partial_{X_1} h_1(Y, X) = S *_t \partial_{X_1} F_b(X) + S \overset{dW}{*_t} \partial_{X_1} F_\sigma(X) \\ \partial^2_{Y_1 Y_2} h_1(Y, X) = \partial^2_{Y_1 X_1} h_1(Y, X) = 0 \\ \partial^2_{X_1 X_2} h_1(Y, X) = S *_t \partial^2_{X_1 X_2} F_b(X) + S \overset{dW}{*_t} \partial^2_{X_1 X_2} F_\sigma(X). \end{cases}$$

Then, by Theorem 2.9, we have

$$\partial_{Y_1} X^{t,Y} = \mathrm{id}_t^S(Y_1) + S *_t \partial_{\partial_{Y_1} X^{t,Y}} F_b(X^{t,Y}) + S \overset{dW}{*_t} \partial_{\partial_{Y_1} X^{t,Y}} F_\sigma(X^{t,Y}) \tag{3.36a}$$

$$\begin{aligned} \partial^2_{Y_1 Y_2} X^{t,Y} =& S *_t \partial_{\partial^2_{Y_1 Y_2} X^{t,Y}} F_b(X) + S \overset{dW}{*_t} \partial_{\partial^2_{Y_1 Y_2} X^{t,Y}} F_\sigma(X) \\ &+ S *_t \partial^2_{\partial_{Y_1} X^{t,Y} \partial_{Y_2} X^{t,Y}} F_b(X) + S \overset{dW}{*_t} \partial^2_{\partial_{Y_1} X^{t,Y} \partial_{Y_2} X^{t,Y}} F_\sigma(X) \end{aligned} \tag{3.36b}$$

where the equality (3.36a) holds in the space $\mathscr{L}^{p^2}_{\mathscr{P}_T}(\mathbb{S})$ and the equality (3.36b) holds in the space $\mathscr{L}^p_{\mathscr{P}_T}(\mathbb{S})$. Formulae (3.36a) and (3.36b) generalize to the present setting the well-known SDEs for the first- and second-order derivatives with respect to the initial datum of mild solutions of non-path-dependent SDEs ([9, Theorems 9.8 and 9.9]).

*Remark 3.11* Suppose that $\mathbb{S} = \mathbb{D}$, where $\mathbb{D}$ is the space of right-continuous left-limited functions $[0, T] \to H$. Notice that $\mathbb{D}$ satisfies all the properties required at p. 11. Then our setting applies and (3.36a)–(3.36b) provide equations for the first- and second-order directional derivatives of $X^{t,Y}$ with respect to vectors belonging to $\mathscr{L}^{p^2}_{\mathscr{P}_T}(\mathbb{D})$. In particular, if $\varphi\colon \mathbb{D} \to \mathbb{R}$ is a suitably regular functional, then the so-called "vertical derivatives" in the sense of Dupire of $F(t, \mathbf{x}) := \mathbb{E}[\varphi(X^{t,\mathbf{x}})]$, used in the finite dimensional Itô calculus developed by [2–4, 10] to show that $F$ solves a path-dependent Kolmogorov equation associated with $X$, can be classically obtained by the chain rule starting from the Gâteaux derivatives $\partial_{Y_1} X^{t,Y}$, $\partial^2_{Y_1 Y_2} X^{t,Y}$, where $y_1, y_1 \in H$ and $Y_1 := \mathbf{1}_{[t,T]}(\cdot) y_1$, $Y_2 := \mathbf{1}_{[t,T]}(\cdot) y_2$.

### 3.3 Perturbation of Path-Dependent SDEs

In this section we study the stability of the mild solution $X^{t,Y}$ and of its Gâteaux derivatives with respect to perturbations of the data $t, Y, S, b, \sigma$.

Let us fix sequences $\mathbf{t} := \{t_j\}_{j \in \mathbb{N}} \subset [0, T]$, $\{S_j\}_{j \in \mathbb{N}} \subset L(H)$, $\{b_j\}_{j \in \mathbb{N}}$, $\{\sigma_j\}_{j \in \mathbb{N}}$, satisfying the following assumption.

**Assumption 3.12** Let $b$, $\sigma$, $g$, $\gamma$, $M$, be as in Assumption 3.3. Assume that

(i) $\{t_j\}_{j\in\mathbb{N}}$ is a sequence converging to $\hat{t}$ in $[0, T]$;
(ii) for all $j \in \mathbb{N}$, $b_j : (\Omega_T \times \mathbb{S}, \mathscr{P}_T \otimes \mathscr{B}_{\mathbb{S}}) \to (H, \mathscr{B}_H)$ is measurable;
(iii) for all $j \in \mathbb{N}$, $\sigma_j : (\Omega_T \times \mathbb{S}, \mathscr{P}_T \otimes \mathscr{B}_{\mathbb{S}}) \to L(U, H)$ is strongly measurable;
(iv) for all $j \in \mathbb{N}$ and all $((\omega, t), \mathbf{x}) \in \Omega_T \times \mathbb{S}$, $b_j((\omega, t), \mathbf{x}) = b_j((\omega, t), \mathbf{x}_{t\wedge\cdot})$ and $\sigma_j((\omega, t), \mathbf{x}) = \sigma_j((\omega, t), \mathbf{x}_{t\wedge\cdot})$;
(v) for all $j \in \mathbb{N}$,

$$\begin{cases} |b_j((\omega, t), \mathbf{x})|_H \le g(t)(1 + |\mathbf{x}|_\infty) & \forall ((\omega, t), \mathbf{x}) \in \Omega_T \times \mathbb{S}, \\ |b_j((\omega, t), \mathbf{x}) - b_j((\omega, t), \mathbf{x}')|_H \le g(t)|\mathbf{x} - \mathbf{x}'|_\infty & \forall (\omega, t) \in \Omega_T,\ \forall \mathbf{x}, \mathbf{x}' \in \mathbb{S}; \end{cases}$$

(vi) for all $j \in \mathbb{N}$,

$$\begin{cases} |(S_j)_t\sigma_j((\omega, s), \mathbf{x})|_{L_2(U,H)} \le Mt^{-\gamma}(1 + |\mathbf{x}|_\infty) & \forall ((\omega, s), \mathbf{x}) \in \Omega_T \times \mathbb{S},\ \forall t \in (0, T], \\ |(S_j)_t\sigma_j((\omega, s), \mathbf{x}) - (S_j)_t\sigma_j((\omega, s), \mathbf{x}')|_{L_2(U,H)} \le Mt^{-\gamma}|\mathbf{x} - \mathbf{x}'|_\infty \forall (\omega, s) \in \Omega_T, \forall \mathbf{x}, \mathbf{x}' \in \mathbb{S}, \forall t \in (0, T]; \end{cases}$$

(vii) for all $t \in [0, T]$, $\{(S_j)_t\}_{j\in\mathbb{N}}$ converges strongly to $S_t$, that is

$$\lim_{j\to\infty} (S_j)_t x = S_t x \qquad \forall x \in H;$$

(viii) the following convergences hold true:

$$\begin{cases} \lim_{j\to\infty} |b((\omega, t), \mathbf{x}) - b_j((\omega, t), \mathbf{x})|_H = 0 & \forall (\omega, t) \in \Omega_T,\ \forall \mathbf{x} \in \mathbb{S} \\ \lim_{j\to\infty} |S_t\sigma((\omega, s), \mathbf{x}) - (S_j)_t\sigma_j((\omega, s), \mathbf{x})|_{L_2(U,H)} = 0 & \forall (\omega, s) \in \Omega_T,\ \forall t \in (0, T],\ \forall \mathbf{x} \in \mathbb{S}. \end{cases}$$

Under Assumption 3.12, for $p > p^*$ and $\beta \in (1/p, 1/2 - \gamma)$, we define $\mathrm{id}_{t_j}^{S_j}$, $F_{b_j}$, $F_{\sigma_j}$, $S_j *_{t_j} \#$, $S_j \overset{dW}{*_{t_j}} \#$, $\psi_j$, similarly as done for $\mathrm{id}_t^S$, $F_b$, $F_\sigma$, $S *_t \#$, $S \overset{dW}{*_t} \#$, $\psi$, that is

$$\mathrm{id}_{t_j}^{S_j} : \mathscr{L}^p_{\mathscr{P}_T}(\mathbb{S}) \to \mathscr{L}^p_{\mathscr{P}_T}(\mathbb{S}),\ Y \mapsto \mathbf{1}_{[0,t_j]}(\cdot)Y + \mathbf{1}_{(t_j,T]}(\cdot)(S_j)_{\cdot-t_j}Y_{t_j}$$

$$F_{b_j} : \mathscr{L}^p_{\mathscr{P}_T}(\mathbb{S}) \to L^{p,1}_{\mathscr{P}_T}(H),\ X \mapsto b_j((\cdot, \cdot), X)$$

$$F_{\sigma_j} : \mathscr{L}^p_{\mathscr{P}_T}(\mathbb{S}) \to \overline{\Lambda}^{p,2,p}_{\mathscr{P}_T,S_j,\beta}(L(U, H)),\ X \mapsto \sigma_j((\cdot, \cdot), X)$$

$$S_j *_{t_j} \#: L^{p,1}_{\mathscr{P}_T}(H) \to \mathscr{L}^p_{\mathscr{P}_T}(\mathbb{S}),\ X \mapsto \mathbf{1}_{[t_j,T]}(\cdot)\int_{t_j}^{\cdot} (S_j)_{\cdot-s}X_s ds$$

$$S_j \overset{dW}{*_{t_j}} \#: \overline{\Lambda}^{p,2,p}_{\mathscr{P}_T,S_j,\beta}(L(U, H)) \to \mathscr{L}^p_{\mathscr{P}_T}(\mathbb{S}),\ \Phi \mapsto (S_j) \overset{dW}{*_{t_j}} \Phi.$$

$$\psi^{(j)}: \mathscr{L}^p_{\mathscr{P}_T}(\mathbb{S}) \times \mathscr{L}^p_{\mathscr{P}_T}(\mathbb{S}) \to \mathscr{L}^p_{\mathscr{P}_T}(\mathbb{S}),\ (Y, X) \mapsto \mathrm{id}_{t_j}^{S_j}(Y) + S_j *_{t_j} F_{b_j}(X) + S_j \overset{dW}{*_{t_j}} F_{\sigma_j}(X).$$

In a similar way as done for $\psi$, we can obtain (3.22) for each $\psi^{(j)}$, with a constant $C'_{\lambda,g,\gamma,M',\beta,T,p,M}$ independent of $j$. In particular, there exists $\lambda_0$ large enough such that, for all $\lambda > \lambda_0$ and all $Y, X \in \mathcal{L}^p_{\mathcal{P}_T}(\mathbb{S})$,

$$\begin{aligned} &|\psi^{(j)}(Y,X) - \psi^{(j)}(Y',X')|_{\mathcal{L}^p_{\mathcal{P}_T}(\mathbb{S}),\lambda} \leq \\ &\leq M'|Y-Y'|_{\mathcal{L}^p_{\mathcal{P}_T}(\mathbb{S}),\lambda} + \tfrac{1}{2}|X-X'|_{\mathcal{L}^p_{\mathcal{P}_T}(\mathbb{S}),\lambda}, \qquad \forall j \in \mathbb{N}, \end{aligned} \tag{3.38}$$

where

$$M' \text{ is any upper bound for } \sup_{\substack{t\in[0,T]\\ j\in\mathbb{N}}} |(S_j)_t|_{L(H)}.$$

Let $A_j$ denotes the infinitesimal generator of $S_j$. By arguing as done in the proof of Theorem 3.6, we have that, for each $j \in \mathbb{N}$, there exists a unique mild solution $X_j^{t,Y}$ in $\mathcal{L}^p_{\mathcal{P}_T}(\mathbb{S})$ to

$$\begin{cases} d(X_j)_s = \left(A_j(X_j)_s + b_j\left((\cdot,s),X_j\right)\right)ds + \sigma_j\left((\cdot,s),X_j\right)dW_s & s\in(t_j,T] \\ (X_j)_s = Y_s & s\in[0,t_j], \end{cases} \tag{3.39}$$

and that, due to the equivalence of the norms $|\cdot|_{\mathcal{L}^p_{\mathcal{P}_T}(\mathbb{S}),\lambda}$, the map $\mathcal{L}^p_{\mathcal{P}_T}(\mathbb{S}) \to \mathcal{L}^p_{\mathcal{P}_T}(\mathbb{S})$, $Y \mapsto X_j^{t_j,Y}$ is Lipschitz, with Lipschitz constant bounded by some $C''_{g,\gamma,M,M',T,p}$ depending only on $g,\gamma,M,M',T,p$ and independent of $j$.

For a given set $B \subset [0,T]$, let us denote

$$\mathbb{S}_B := \{\mathbf{x}\in\mathbb{S} \colon \forall t\in B,\ \mathbf{x} \text{ is continuous in } t\}.$$

Then $\mathbb{S}_B$ is a closed subspace of $\mathbb{S}$ and it satisfies all the three conditions required for $\mathbb{S}$ at p. 11. Moreover, if $t\in[0,T]$ and $Y\in\mathcal{L}^p_{\mathcal{P}_T}(\mathbb{S}_B)$, then $X^{t,Y}\in\mathcal{L}^p_{\mathcal{P}_T}(\mathbb{S}_B)$, because $X^{t,Y}$ is continuous on $[t,T]$ (recall that $S *_t \#$ and $S \overset{dW}{*_t} \#$ are $\mathcal{L}^p_{\mathcal{P}_T}(C([0,T],H))$-valued) and coincides with $Y$ on $[0,t]$.

**Proposition 3.13** *Suppose that Assumptions 3.3 and 3.12 are satisfied and let $p > p^*$. Then*

$$\lim_{j\to\infty} X_j^{t_j,Y} = X^{\hat{t},Y} \tag{3.40}$$

*in $\mathcal{L}^p_{\mathcal{P}_T}(\mathbb{S}_{\{\hat{t}\}})$, uniformly for $Y$ on compact subsets of $\mathcal{L}^p_{\mathcal{P}_T}(\mathbb{S}_{\{\hat{t}\}})$.*

*Proof.* Let $\psi^{(j)}$ be defined as above (p. 26). It is clear that, if $Y\in\mathcal{L}^p_{\mathcal{P}_T}(\mathbb{S}_{\{\hat{t}\}})$ and $X\in\mathcal{L}^p_{\mathcal{P}_T}(\mathbb{S})$, then $\psi(Y,X)\in\mathcal{L}^p_{\mathcal{P}_T}(\mathbb{S}_{\{\hat{t}\}})$, because it is continuous on $[\hat{t},T]$ and coincides with $Y$ on $[0,\hat{t}]$. Similarly, $\psi^{(j)}(Y,X)$ is continuous on $[t_j,T]$ and coincides with $Y$ on $[0,t_j]$, than also $\psi^{(j)}(Y,X)\in\mathcal{L}^p_{\mathcal{P}_T}(\mathbb{S}_{\{\hat{t}\}})$. Then, if the claimed convergence occurs, it does in $\mathcal{L}^p_{\mathcal{P}_T}(\mathbb{S}_{\{\hat{t}\}})$.

In order to prove the convergence, we consider the restrictions

$$\begin{cases} \hat{\psi}^{(j)} := \psi^{(j)}_{|\mathcal{L}^p_{\mathcal{P}_T}(\mathbb{S}_{\{\hat{t}\}})\times\mathcal{L}^p_{\mathcal{P}_T}(\mathbb{S})} & \forall j \in \mathbb{N} \\ \hat{\psi} := \psi_{|\mathcal{L}^p_{\mathcal{P}_T}(\mathbb{S}_{\{\hat{t}\}})\times\mathcal{L}^p_{\mathcal{P}_T}(\mathbb{S})}, \end{cases}$$

which are $\mathcal{L}^p_{\mathcal{P}_T}(\mathbb{S}_{\{\hat{t}\}})$-valued, as noticed above. Clearly (3.38) still holds true with $\hat{\psi}^{(j)}$, $\hat{\psi}$ in place of $\psi^{(j)}$, $\psi$, respectively, and then

$$\mathcal{L}^p_{\mathcal{P}_T}(\mathbb{S}_{\{\hat{t}\}}) \to \mathcal{L}^p_{\mathcal{P}_T}(\mathbb{S}_{\{\hat{t}\}}),\ Y \mapsto X_j^{t_j,Y}$$

is Lipschitz in $Y$, uniformly in $j$. We then need only to prove the convergence

$$X_j^{t_j,Y} \to X^{\hat{t},Y} \text{ in } \mathcal{L}^p_{\mathcal{P}_T}(\mathbb{S}_{\{\hat{t}\}}), \forall Y \in \mathcal{L}^p_{\mathcal{P}_T}(\mathbb{S}_{\{\hat{t}\}}).$$

Thanks to Lemma 2.5(i), the latter convergence reduces to the pointwise convergence

$$\hat{\psi}^{(j)} \to \hat{\psi}.$$

Let $Y \in \mathcal{L}^p_{\mathcal{P}_T}(\mathbb{S}(\{\hat{t}\}))$. Due to the continuity of $Y(\omega)$ in $\hat{t}$ for $\mathbb{P}$-a.e. $\omega \in \Omega$, the strong continuity of $S_j$ and $S$, and the strong convergence $S_j \to S$, we have $\mathrm{id}_{t_j}^{S_j}(Y) \to \mathrm{id}_{\hat{t}}^{S}(Y)$ in $\mathcal{L}^p_{\mathcal{P}_T}(\mathbb{S}_{\{\hat{t}\}})$ for all $Y \in \mathcal{L}^p_{\mathcal{P}_T}(\mathbb{S})$ (this can be seen by (3.17)).

We show that $S_j \overset{dW}{*_{t_j}} F_{\sigma_j}(X) \to S \overset{dW}{*_{\hat{t}}} F_\sigma(X)$, for all $X \in \mathcal{L}^p_{\mathcal{P}_T}(\mathbb{S})$. Write

$$S_j \overset{dW}{*_{t_j}} F_{\sigma_j} - S \overset{dW}{*_{\hat{t}}} F_\sigma = (S_j \overset{dW}{*_{t_j}} F_{\sigma_j} - S \overset{dW}{*_{t_j}} F_\sigma) + (S \overset{dW}{*_{t_j}} F_\sigma - S \overset{dW}{*_{\hat{t}}} F_\sigma).$$

By Lebesgue's dominated convergence theorem and by Assumption 3.12, we have, for $\beta \in (1/p, 1/2 - \gamma)$,

$$\lim_{j\to\infty} \int_0^T \left( \int_0^t (t-s)^{-2\beta} \left( \mathbb{E}\left[ |(S_j)_{t-s}\sigma_j((\cdot,s),X)) - S_{t-s}\sigma((\cdot,s),X))|^p_{L_2(U,H)} \right]\right)^{2/p} ds \right)^{p/2} dt = 0$$

Then, by (3.11) (which holds uniformly in $t$),

$$S_j \overset{dW}{*_{t_j}} F_{\sigma_j}(X) - S \overset{dW}{*_{t_j}} F_\sigma(X) \to 0 \text{ in } \mathcal{L}^p_{\mathcal{P}_T}(\mathbb{S}).$$

By (3.14), we also have

$$S \overset{dW}{*_{t_j}} F_\sigma(X) - S \overset{dW}{*_{\hat{t}}} F_\sigma(X) \to 0 \text{ in } \mathcal{L}^p_{\mathcal{P}_T}(\mathbb{S}).$$

Then, we conclude

$$S_j \overset{dW}{*_{t_j}} F_{\sigma_j} - S \overset{dW}{*_{\hat{t}}} F_\sigma \to 0 \text{ in } \mathcal{L}^p_{\mathcal{P}_T}(\mathbb{S}).$$

By arguing in a very similar way as done for $S_j \overset{dW}{*_{t_j}} F_{\sigma_j} - S \overset{dW}{*_{\hat{t}}} F_\sigma$, one can prove that

$$\forall X \in \mathcal{L}^p_{\mathcal{P}_T}(\mathbb{S}),\ S_j *_{t_j} F_{b_j}(X) - S *_{\hat{t}} F_b(X) \to 0 \text{ in } \mathcal{L}^p_{\mathcal{P}_T}(\mathbb{S}).$$

Then $\hat{\psi}^{(j)} \to \hat{\psi}$ pointwise and the proof is complete. ■

The following result provides continuity of the mild solution with respect to perturbations of all the data of the system.

**Theorem 3.14** *Suppose that Assumptions 3.3 and 3.12 are satisfied, let $p > p^*$, $Y \in \mathcal{L}^p_{\mathcal{P}_T}(\mathbb{S}_{\{\hat{t}\}})$, and let $\{Y_j\}_{j\in\mathbb{N}} \subset \mathcal{L}^p_{\mathcal{P}_T}(\mathbb{S})$ be a sequence converging to $Y$ in $\mathcal{L}^p_{\mathcal{P}_T}(\mathbb{S})$. Then*

$$\lim_{j\to\infty} X_j^{t_j,Y_j} = X^{\hat{t},Y} \text{ in } \mathcal{L}^p_{\mathcal{P}_T}(\mathbb{S}).$$

*Proof.* Write

$$X^{\hat{t},Y} - X_j^{t_j,Y_j} = (X^{\hat{t},Y} - X_j^{t_j,Y}) + (X_j^{t_j,Y} - X_j^{t_j,Y_j}), \tag{3.41}$$

The term $X^{\hat{t},Y} - X_j^{t_j,Y}$ tends to 0 by Proposition 3.13, whereas the term $X_j^{t_j,Y} - X_j^{t_j,Y_j}$ tends to 0 by uniform equicontinuity of the family

$$\left\{ \mathcal{L}^p_{\mathcal{P}_T}(\mathbb{S}) \to \mathcal{L}^p_{\mathcal{P}_T}(\mathbb{S}),\ Y \mapsto X_j^{t_j,Y} \right\}_{j\in\mathbb{N}}.$$

■

We end this chapter with a result regarding stability of Gâteaux differentials of mild solutions.

**Assumption 3.15** Let $b, \sigma, g, \gamma, n, c, M''$ be as in Assumption 3.7, and let $\{b_j\}_{j\in\mathbb{N}}$, $\{\sigma\}_{j\in\mathbb{N}}$, $\{S_j\}_{j\in\mathbb{N}}$, be as in Assumption 3.12. Assume that

(i) for all $j \in \mathbb{N}$, $(\omega, t) \in \Omega_T$, and $u \in U$, $b_j((\omega, t), \cdot) \in \mathcal{G}^n(\mathbb{S}, H)$ and $\sigma_j((\omega, t), \cdot)u \in \mathcal{G}^n(\mathbb{S}, H)$;
(ii) for all $s \in [0, T]$,

$$\sup_{\substack{i=1,\ldots,n \\ j\in\mathbb{N}}} \sup_{\substack{\omega\in\Omega \\ \mathbf{x},\mathbf{y}_1,\ldots,\mathbf{y}_j\in\mathbb{S} \\ |\mathbf{y}_1|_\infty=\ldots=|\mathbf{y}_i|_\infty=1}} |\partial^i_{\mathbf{y}_1\ldots\mathbf{y}_i} b_j((\omega, s), \mathbf{x})|_H \le M'' g(s), \tag{3.42}$$

and, for all $s \in [0, T]$, $t \in (0, T]$, and all $m \in \mathcal{M}$,

$$\sup_{\substack{i=1,\ldots,n \\ j\in\mathbb{N}}} \sup_{\substack{\omega\in\Omega \\ \mathbf{x},\mathbf{y}_1,\ldots,\mathbf{y}_i\in\mathbb{S} \\ |\mathbf{y}_1|_\infty=\ldots=|\mathbf{y}_i|_\infty=1}} |(S_j)_t \partial^i_{\mathbf{y}_1\ldots\mathbf{y}_i}(\sigma_j((\omega, s), \mathbf{x})e'_m))|_H \le M'' t^{-\gamma} c_m; \tag{3.43}$$

(iii) for all $X \in \mathbb{S}$,

$$\begin{cases} \lim_{j\to\infty} |\partial^i_{\mathbf{y}_1\dots\mathbf{y}_i} b((\omega,t),\mathbf{x}) - \partial^i_{\mathbf{y}_1\dots\mathbf{y}_i} b_j((\omega,t),\mathbf{x})|_H = 0 & \forall(\omega,t)\in\Omega_T \\ \lim_{j\to\infty} |S_t \partial^i_{\mathbf{y}_1\dots\mathbf{y}_i}(\sigma((\omega,s),\mathbf{x})e'_m) - (S_j)_t \partial^i_{\mathbf{y}_1\dots\mathbf{y}_i}(\sigma_j((\omega,s),\mathbf{x})e'_m)|_H = 0 & \begin{cases} \forall \omega\in\Omega, \\ \forall s\in[0,T], \forall t\in(0,T], \\ \forall m \in \mathscr{M}. \end{cases} \end{cases}$$

**Theorem 3.16** *Suppose that Assumptions 3.3 and 3.12 are satisfied, and that, for some $n \in \mathbb{N}$, $n \geq 1$, Assumptions 3.7 and 3.15 are satisfied. Let $p > p^*$, $p \geq n$. Then, for $i = 1, \dots, n$,*

$$\partial^i_{Y_1\dots Y_i} X_j^{t_j,Y} \to \partial^i_{Y_1\dots Y_i} X^{\hat{t},Y} \ in \ \mathscr{L}^p_{\mathscr{P}_T}(\mathbb{S}_{\{\hat{t}\}}), \tag{3.44}$$

*uniformly for $Y, Y_1, \dots, Y_i$ in compact subsets of $\mathscr{L}^{p^n}_{\mathscr{P}_T}(\mathbb{S}_{\{\hat{t}\}})$.*

*Proof.* By Theorem 3.9, $\mathscr{L}^{p^n}_{\mathscr{P}_T}(\mathbb{S}) \to \mathscr{L}^p_{\mathscr{P}_T}(\mathbb{S})$, $Y \mapsto X_j^{t_j,Y}$ belongs to $\mathscr{G}^n(\mathscr{L}^{p^n}_{\mathscr{P}_T}(\mathbb{S}), \mathscr{L}^p_{\mathscr{P}_T}(\mathbb{S}))$. Then, since $X_j^{t_j,Y} \in \mathscr{L}^p_{\mathscr{P}_T}(\mathbb{S}_{\{\hat{t}\}})$ if $Y \in \mathscr{L}^p_{\mathscr{P}_T}(\mathbb{S}_{\{\hat{t}\}})$, the map $\mathscr{L}^{p^n}_{\mathscr{P}_T}(\mathbb{S}_{\{\hat{t}\}}) \to \mathscr{L}^p_{\mathscr{P}_T}(\mathbb{S}_{\{\hat{t}\}})$, $Y \mapsto X_j^{t_j,Y}$ belongs to $\mathscr{G}^n(\mathscr{L}^{p^n}_{\mathscr{P}_T}(\mathbb{S}_{\{\hat{t}\}}), \mathscr{L}^p_{\mathscr{P}_T}(\mathbb{S}_{\{\hat{t}\}}))$.

To prove (3.44), we wish to apply Proposition 2.11. In the proof of Theorem 3.9, we associated the map $\psi$ and the spaces $\mathscr{L}^{p^k}_{\mathscr{P}_T}(\mathbb{S})$ with Assumption 2.8. In the same way, here, we associate the restrictions

$$\psi^{(1)}_{|\mathscr{L}^{p^n}_{\mathscr{P}_T}(\mathbb{S}_{\{\hat{t}\}}) \times \mathscr{L}^p_{\mathscr{P}_T}(\mathbb{S}_{\{\hat{t}\}})}, \psi^{(2)}_{|\mathscr{L}^{p^n}_{\mathscr{P}_T}(\mathbb{S}_{\{\hat{t}\}}) \times \mathscr{L}^p_{\mathscr{P}_T}(\mathbb{S}_{\{\hat{t}\}})}, \psi^{(3)}_{|\mathscr{L}^{p^n}_{\mathscr{P}_T}(\mathbb{S}_{\{\hat{t}\}}) \times \mathscr{L}^p_{\mathscr{P}_T}(\mathbb{S}_{\{\hat{t}\}})}, \dots,$$

respectively to the functions $h_1^{(1)}, h_1^{(2)}, h_1^{(3)}, \dots$ appearing in the assumption of Proposition 2.11, and, to each $h_1^{(m)}$, we associate the functions $h_k^{(m)}$, for $k = 1, \dots, n$, defined by $h_k^{(m)} := \psi_{k|\mathscr{L}^{p^n}_{\mathscr{P}_T}(\mathbb{S}_{\{\hat{t}\}}) \times \mathscr{L}^{p^k}_{\mathscr{P}_T}(\mathbb{S}_{\{\hat{t}\}})}$ and considered as $\mathscr{L}^{p^k}_{\mathscr{P}_T}(\mathbb{S})$-valued functions.

As argued several times above, we can choose $\lambda > 0$ such that, for $m = 1, 2, \dots$ and $k = 1, \dots, n$, each function $h_k^{(m)}$ is a parametric 1/2-contractions with respect to the norm $|\cdot|_{\mathscr{L}^{p^k}_{\mathscr{P}_T}(\mathbb{S}),\lambda}$. With respect to this equivalent norm, for each $h_1^{(m)}$, Assumption 2.8 can be verified in exactly the same way as it was verified for the function $h_1$ appearing in the proof of Theorem 3.9. Then, in order to apply Proposition 2.11, it remains to verify hypotheses (i), (ii), (iii) appearing in the statement of that proposition. Since the norms $|\cdot|_{\mathscr{L}^{p^k}_{\mathscr{P}_T}(\mathbb{S}),\lambda}$, $\lambda \geq 0$, are equivalent, the three hypotheses reduce to the following convergences:

(i) for all $k = 1, \dots, n$, $X \in \mathscr{L}^{p^k}_{\mathscr{P}_T}(\mathbb{S}_{\{\hat{t}\}})$,

$$\psi^{(j)}(Y, X) \to \psi(Y, X) \text{ in } (\mathscr{L}^{p^k}_{\mathscr{P}_T}(\mathbb{S}_{\{\hat{t}\}}), |\cdot|_{\mathscr{L}^{p^k}_{\mathscr{P}_T}(\mathbb{S})}) \tag{3.45}$$

uniformly for $Y$ on compact subsets of $\mathscr{L}^{p^n}_{\mathscr{P}_T}(\mathbb{S}_{\{\hat{t}\}})$;

(ii) for $k = 1, \ldots, n$

$$\begin{cases} \lim_{j\to\infty} \partial_{Y'}\psi^{(j)}(Y,X) = \partial_{Y'}\psi(Y,X) & \text{in } (\mathcal{L}^{p^k}_{\mathcal{P}_T}(\mathbb{S}_{\{\hat{t}\}}), |\cdot|_{\mathcal{L}^{p^k}_{\mathcal{P}_T}(\mathbb{S})}) \\ \lim_{j\to\infty} \partial_{X'}\psi^{(j)}(Y,X) = \partial_{X'}\psi(Y,X) & \text{in } (\mathcal{L}^{p^k}_{\mathcal{P}_T}(\mathbb{S}_{\{\hat{t}\}}), |\cdot|_{\mathcal{L}^{p^k}_{\mathcal{P}_T}(\mathbb{S})}) \end{cases} \tag{3.46}$$

uniformly for $Y, Y'$ on compact subsets of $\mathcal{L}^{p^n}_{\mathcal{P}_T}(\mathbb{S}_{\{\hat{t}\}})$ and $X, X'$ on compact subsets of $\mathcal{L}^{p^k}_{\mathcal{P}_T}(\mathbb{S}_{\{\hat{t}\}})$;

(iii) for all $k = 1, \ldots, n-1$, $Y \in \mathcal{L}^{p^n}_{\mathcal{P}_T}(\mathbb{S}_{\{\hat{t}\}})$, $l, i = 0, \ldots, n$, $1 \le l + i \le n$,

$$\lim_{j\to\infty} \partial^{l+i}_{Y_1\ldots Y_l X_1\ldots X_i}\psi^{(j)}(Y,X) = \partial^{l+i}_{Y_1\ldots Y_l X_1\ldots X_i}\psi(Y,X) \text{ in } (\mathcal{L}^{p^k}_{\mathcal{P}_T}(\mathbb{S}_{\{\hat{t}\}}), |\cdot|_{\mathcal{L}^{p^k}_{\mathcal{P}_T}(\mathbb{S})}) \tag{3.47}$$

uniformly for $Y, Y_1, \ldots, Y_l$ on compact subsets of $\mathcal{L}^{p^n}_{\mathcal{P}_T}(\mathbb{S}_{\{\hat{t}\}})$, $X$ on compact subsets of $\mathcal{L}^{p^k}_{\mathcal{P}_T}(\mathbb{S}_{\{\hat{t}\}})$, $X_1, \ldots, X_i$ on compact subsets of $\mathcal{L}^{p^{k+1}}_{\mathcal{P}_T}(\mathbb{S}_{\{\hat{t}\}})$.

Taking into account the equicontinuity of the family $\{\psi^{(j)}\}_{j\in\mathbb{N}}$ with respect to the second variable, (i) is contained in the proof Proposition 3.13. As regarding (ii) and (iii), since the linear term $\mathrm{id}^{S_j}_{t_j}$ is easily treated in $\mathcal{L}^{p}_{\mathcal{P}_T}(\mathbb{S}_{\{\hat{t}\}})$ (as shown in the proof of Proposition 3.13), the only comments to make are about the convergences of the derivatives

$$\begin{cases} \partial_{Y'}(S_j *_{t_j} F_{b_j})(X) \\ \partial_{X'}(S_j *_{t_j} F_{b_j})(X) \\ \partial_{Y'}(S_j \overset{dW}{*_{t_j}} F_{\sigma_j})(X) \\ \partial_{X'}(S_j \overset{dW}{*_{t_j}} F_{\sigma_j})(X) \end{cases} \quad \text{and} \quad \begin{cases} \partial^{l+i}_{Y_1\ldots Y_l X_1\ldots X_i}(S_j *_{t_j} F_{b_j})(X) \\ \partial^{l+i}_{Y_1\ldots Y_l X_1\ldots X_i}(S_j *_{t_j} F_{\sigma_j})(X). \end{cases}$$

Due to linearity and continuity of the convolution operators, to the independence of the first variable of $F_b$ and $F_\sigma$, and to Lemma 3.8, the above derivatives are respectively equal to

$$\begin{cases} 0 \\ S_j *_{t_j} (\partial_{X'}F_{b_j})(X) \\ 0 \\ S_j \overset{dW}{*_{t_j}} (\partial_{X'}F_{\sigma_j})(X) \end{cases} \quad \text{and} \quad \begin{cases} \begin{cases} S_j *_{t_j} (\partial^i_{X_1\ldots X_i}F_{b_j})(X) & \text{if } l = 0 \\ 0 & \text{otherwise} \end{cases} \\ \begin{cases} S_j *_{t_j} (\partial^i_{X_1\ldots X_i}F_{\sigma_j})(X) & \text{if } l = 0 \\ 0 & \text{otherwise.} \end{cases} \end{cases} \tag{3.48}$$

Let us consider, for example, the difference

$$S_j *_{t_j} (\partial^i_{X_1\ldots X_i}F_{\sigma_j})(X_j) - S *_{\hat{t}} (\partial^i_{X_1\ldots X_i}F_{\sigma})(X) \tag{3.49}$$

for some sequence $\{X_j\}_{j\in\mathbb{N}}$ converging to $X$ in $\mathcal{L}^{p^k}_{\mathcal{P}_T}(\mathbb{S})$. We can decompose the above difference as done in (3.41), and then use the same arguments, together with expressions (3.26), the bounds (3.42) and (3.42), the generalized Hölder inequality, the pointwise convergences in Assumption 3.15(iii), and Lebesgue's dominated convergence theorem, to conclude

$$S_j *_{t_j} (\partial^i_{X_1\dots X_i} F_{\sigma_j})(X_j) - S *_{\hat{t}} (\partial^i_{X_1\dots X_i} F_\sigma)(X) \to 0$$

in $\mathscr{L}^{p^k}_{\mathscr{P}_T}(\mathbb{S}_{\{\hat{t}\}})$, for all $X_1, \dots, X_i \in \mathscr{L}^{p^{k+1}}_{\mathscr{P}_T}(\mathbb{S}_{\{\hat{t}\}})$. By recalling the continuity of $X \mapsto \partial^i_{X_1\dots X_i} F_\sigma(X)$ (Lemma 3.8), this shows the convergence

$$S_j *_{t_j} (\partial^i_{X_1\dots X_i} F_{\sigma_j})(X) - S *_{\hat{t}} (\partial^i_{X_1\dots X_i} F_\sigma)(X) \to 0, \tag{3.50}$$

uniformly for $X$ on compact sets of $\mathscr{L}^{p^k}_{\mathscr{P}_T}(\mathbb{S}_{\{\hat{t}\}})$, for fixed $X_1, \dots, X_i \in \mathscr{L}^{p^{k+1}}_{\mathscr{P}_T}(\mathbb{S}_{\{\hat{t}\}})$. But, since by Lemma 3.8 the derivatives (3.48) are jointly continuous in $X, X'$, $X_1, \dots, X_i$, and uniformly bounded, the convergence (3.50) occurs uniformly for $X$ on compact sets of $\mathscr{L}^{p^k}_{\mathscr{P}_T}(\mathbb{S}_{\{\hat{t}\}})$ and $X_1, \dots, X_i$ on compact sets of $\mathscr{L}^{p^{k+1}}_{\mathscr{P}_T}(\mathbb{S}_{\{\hat{t}\}})$. The arguments for the other derivatives are similar. This shows that we can apply Proposition 2.11, which provides (3.44). ■

# 4 Appendix

**Proof of Proposition** 2.1 Suppose that the derivatives $\partial^j_{x_1\dots x_j} f(u)$ exists for all $u \in U$, $x_1, \dots, x_j \in X_0$, $j = 1, \dots, n$, separately continuous in $u, x_1, \dots, x_j$. We want to show that $f \in \mathscr{G}^n(U, Y; X_0)$.

We proceed by induction on $n$. Let $n = 1$. Since $\partial_x f(u)$ is continuous in $u$, for all $x \in X_0$, we have that $X_0 \to Y$, $x \mapsto \partial_x f(u)$ is linear ([11, Lemma 4.1.5]). By assumption, it is also continuous. Hence $x \mapsto \partial_x f(u) \in L(X_0, Y)$ for all $u \in U$. This shows the existence of $\partial_{X_0} f$. The continuity of $U \to L_s(X_0, Y)$, $u \mapsto \partial_{X_0} f(u)$, comes from the separate continuity of (2.1) and from the definition of the locally convex topology on $L_s(X_0, Y)$. This shows that $f \in \mathscr{G}^1(U, Y; X_0)$.

Let now $n > 1$. By inductive hypothesis, we may assume that $f \in \mathscr{G}^{n-1}(U, Y; X_0)$ and

$$\partial^j_{X_0} f(u).(x_1, \dots, x_j) = \partial^j_{x_1\dots x_j} f(u) \qquad \forall u \in U,\ \forall j = 1, \dots, n-1,\ \forall (x_1, \dots, x_j) \in X_0^j.$$

Let $x_n \in X_0$. The limit

$$\lim_{t\to 0} \frac{\partial^{n-1}_{X_0} f(u + t x_n) - \partial^{n-1}_{X_0} f(u)}{t} = \Lambda \tag{4.1}$$

exists in $L_s^{(n-1)}(X_0^{n-1}, Y)$ if and only if $\Lambda \in L_s^{(n-1)}(X_0^{n-1}, Y)$ and, for all $x_1, \ldots, x_{n-1} \in X_0$, the limit

$$\lim_{t \to 0} \frac{\partial^{n-1}_{x_1 \ldots x_{n-1}} f(u + t x_n) - \partial^{n-1}_{x_1 \ldots x_{n-1}} f(u)}{t} = \Lambda(x_1, \ldots, x_{n-1}) \tag{4.2}$$

holds in $Y$. By assumption, the limit (4.2) is equal to $\partial^n_{x_1 \ldots x_{n-1} x_n} f(u)$, for all $x_1, \ldots, x_{n-1}$. Since, by assumption, $\partial^n_{x_1 \ldots x_{n-1} x_n} f(u)$ is separately continuous in $u, x_1, \ldots, x_{n-1}, x_n$, we have that the limit (4.1) exists in $L_s^{(n-1)}(X_0^{n-1}, Y)$ and is given by

$$\partial_{x_n} \partial^{n-1}_{X_0} f(u).(x_1, \ldots, x_{n-1}) = \Lambda(x_1, \ldots, x_{n-1}) = \partial^n_{x_1 \ldots x_{n-1} x_n} f(u) \qquad \forall x_1, \ldots, x_{n-1} \in X_0.$$

Since $u$ and $x_n$ were arbitrary, we have proved that $\partial_{x_n} \partial^{n-1}_{X_0} f(u)$ exists for all $u$, $x_n$. Moreover, for all $x_1, \ldots, x_n \in X_0$, the function

$$U \to Y, \ u \mapsto \partial_{x_n} \partial^{n-1}_{X_0} f(u).(x_1, \ldots, x_{n-1}) = \partial_{x_n} \partial^n_{x_1 \ldots x_{n-1}} f(u)$$

is continuous, by separate continuity of (2.1). Then $\partial^n_{x_1 \ldots x_{n-1} x_n} f(u)$ is linear in $x_n$. The continuity of

$$X_0 \to L_s^{(n-1)}(X_0^{n-1}, Y), \ x \mapsto \partial_x \partial^{n-1}_{X_0} f(u) \tag{4.3}$$

comes from the continuity of $\partial^n_{x_1 \ldots x_{n-1} x} f(u)$ in each variable, separately. Hence (4.3) belongs to $L_s(X_0, L_s^{n-1}(X_0^{n-1}, Y))$ for all $u \in U$. This shows that $\partial^{n-1}_{X_0} f$ is Gâteaux differentiable with respect to $X_0$ and that

$$\partial^n_{X_0} f(u).(x_1, \ldots, x_n) = \partial^n_{x_1 \ldots x_n} f(u) \qquad \forall u \in U, \ \forall x_1, \ldots, x_n \in X_0,$$

and shows also the continuity of

$$U \to L_s^{(n)}(X_0^n, Y), \ u \mapsto \partial^n_{X_0} f(u),$$

due to the continuity of the derivatives of $f$, separately in each direction. Then we have proved that $f \in \mathcal{G}^n(U, Y; X_0)$ and that (2.2) holds.

Now suppose that $f \in \mathcal{G}^n(U, Y; X_0)$. By the very definition of $\partial_{X_0} f$, $\partial_x f(u)$ exists for all $x \in X_0$ and $u \in U$, it is separately continuous in $u, x$, and coincides with $\partial_{X_0} f(u).x$. By induction, assume that $\partial^{n-1}_{x_1 \ldots x_{n-1}} f(u)$ exists and that

$$\partial^{n-1}_{X_0} f(u).(x_1, \ldots, x_{n-1}) = \partial^{n-1}_{x_1 \ldots x_{n-1}} f(u) \qquad \forall u \in U, \ \forall x_1, \ldots, x_{n-1} \in X_0. \tag{4.4}$$

Since $\partial^{n-1}_{X_0} f(u)$ is Gâteaux differentiable, the directional derivative $\partial_{x_n} \partial^{n-1}_{X_0} f(u)$ exists. Hence, by (4.4), the derivative $\partial^n_{x_1 \ldots x_{n-1} x_n} f(u)$ exists for all $x_1, \ldots, x_{n-1}, x_n \in X_0$. The continuity of $\partial^n_{x_1 \ldots x_{n-1} x_n} f(u)$ with respect to $u$ comes from the continuity of

$\partial^n_{X_0} f$. The continuity of $\partial^n_{x_1\ldots x_j\ldots x_n} f(u)$ with respect to $x_j$ comes from the fact that, for all $x_{j+1}, \ldots, x_n \in X_0$, $u \in U$,

$$X_0^j \to Y,\ (x_1', \ldots, x_j') \mapsto \partial^n_{X_0} f(u).(x_1', \ldots, x_j', x_{j+1}, \ldots, x_n)$$

belongs to $L_s^{(j)}(X_0^j, Y)$. ■

**Proof of Theorem** 2.9 The proof is by induction on $n$. The case $n = 1$ is provided by Proposition 2.7.

Let $n \geq 2$. Clearly, it is sufficient to prove that $\varphi \in \mathscr{G}^n(U, Y_n)$ and that (2.11) holds true for $j = n$. Since we are assuming that the theorem holds true for $n-1$, we can apply it with the data

$$\widetilde{h}_1 \colon U \times \widetilde{Y}_2 \to \widetilde{Y}_2,\ \ldots, \widetilde{h}_{n-1} \colon U \times \widetilde{Y}_n \to \widetilde{Y}_n,$$

where $\widetilde{h}_k := h_{k+1}$, $\widetilde{Y}_k := Y_{k+1}$, for $k = 1, \ldots, n-1$. According to the claim, the fixed-point function $\widetilde{\varphi}$ of $\widetilde{h}_1$ belongs to $\mathscr{G}^j(U, \widetilde{Y}_{(n-1)-j+1})$, for $j = 1, \ldots, n-1$, and formula (2.11) holds true for $\widetilde{\varphi}$ and $j = 1, \ldots, n-1$. Since $\varphi(u) = (i_{2,1} \circ \widetilde{\varphi})(u)$, for $u \in U$, we have $\varphi \in \mathscr{G}^j(U, \widetilde{Y}_{n-j}) = \mathscr{G}^j(U, Y_{n-j+1})$, for $j = 1, \ldots, n-1$, and

$$\partial^j_{x_1\ldots x_j}\varphi(u) = \partial^j_{x_1\ldots x_j}\widetilde{\varphi}(u) \in \widetilde{Y}_{n-j} = Y_{n-j+1}, \qquad \forall u \in U,\ \forall x_1, \ldots, x_j \in X.$$

Then (2.11) holds true for $\varphi$ up to order $j = n-1$. In particular $\varphi \in \mathscr{G}^{n-1}(U, Y_2)$, hence, for $x_1, \ldots, x_n \in X$, $\epsilon > 0$, we can write

$$\begin{aligned}
&\partial^{n-1}_{x_1\ldots x_{n-1}}\varphi(u+\varepsilon x_n) - \partial^{n-1}_{x_1\ldots x_{n-1}}\varphi(u) \\
&= \Big(\partial_{Y_1} h_1(u+\varepsilon x_n, \varphi(u+\varepsilon x_n)).\partial^{n-1}_{x_1\ldots x_{n-1}}\varphi(u+\varepsilon x_n) - \partial_{Y_1} h_1(u, \varphi(u)).\partial^{n-1}_{x_1\ldots x_{n-1}}\varphi(u)\Big) \\
&\quad + (\mathscr{S}(u+\varepsilon x_n) - \mathscr{S}(u)) \\
&=: \mathbf{I} + \mathbf{II},
\end{aligned} \tag{4.5}$$

where $\mathscr{S}(\cdot)$ denotes the sum

$$\mathscr{S}(v) := \partial^{n-1}_{x_1\ldots x_{n-1}} h_1(v, \varphi(v)) + \sum_{\substack{\mathbf{x} \in 2^{\{x_1,\ldots,x_{n-1}\}} \\ \mathbf{x} \neq \emptyset}} \sum_{i=\max\{1, 2-(n-1)+|\mathbf{x}|\}}^{|\mathbf{x}|} \sum_{\substack{\mathbf{p} \in P^i(\mathbf{x}) \\ \mathbf{p} = (\mathbf{p}_1, \ldots, \mathbf{p}_i)}} \partial^{n-1}[\mathbf{x}^c, \mathbf{p}] h_1(v, \varphi(v)),$$

for $v \in U$. By recalling that $\varphi \in \mathscr{G}^j(U, Y_{n-j+1})$, $j = 1, \ldots, n-1$, hence by taking into account with respect to which space the derivatives of $\varphi$ are continuous, we write

$$\begin{aligned}
\mathbf{I} &= \partial_{\partial^{n-1}_{x_1\ldots x_{n-1}}\varphi(u+\varepsilon x_n)} h_1(u+\varepsilon x_n, \varphi(u+\varepsilon x_n)) - \partial_{\partial^{n-1}_{x_1\ldots x_{n-1}}\varphi(u)} h_1(u, \varphi(u)) \\
&= \int_0^1 \partial_{x_n} \partial_{\partial^{n-1}_{x_1\ldots x_{n-1}}\varphi(u+\varepsilon x_n)} h_1(u+\theta\varepsilon x_n, \varphi(u+\varepsilon x_n))\varepsilon d\theta
\end{aligned}$$

$$
\begin{aligned}
&+\int_0^1 \partial_{\frac{\varphi(u+\varepsilon x_n)-\varphi(u)}{\varepsilon}} \partial_{\partial^{n-1}_{x_1\dots x_{n-1}}\varphi(u+\varepsilon x_n)} h_1(u,\varphi(u)+\theta(\varphi(u+\varepsilon x_n)-\varphi(u)))\varepsilon d\theta\\
&+\partial_{\partial^{n-1}_{x_1\dots x_{n-1}}\varphi(u+\varepsilon x_n)-\partial^{n-1}_{x_1\dots x_{n-1}}\varphi(u)} h_1(u,\varphi(u))\\
=&\mathbf{I_1}+\mathbf{I_2}+\partial_{Y_1}h_1(u,\varphi(u)).\left(\partial^{n-1}_{x_1\dots x_{n-1}}\varphi(u+\varepsilon x_n)-\partial^{n-1}_{x_1\dots x_{n-1}}\varphi(u)\right),
\end{aligned}
\tag{4.6}
$$

with ([5])

$$
\lim_{\varepsilon\to 0}\frac{\mathbf{I_1}}{\varepsilon}=\partial_{x_n}\partial_{\partial^{n-1}_{x_1\dots x_{n-1}}\varphi(u)}h_1(u,\varphi(u)) \qquad \text{and} \qquad \lim_{\varepsilon\to 0}\frac{\mathbf{I_2}}{\varepsilon}=\partial_{\partial_{x_n}\varphi(u)}\partial_{\partial^{n-1}_{x_1\dots x_{n-1}}\varphi(u)}h_1(u,\varphi(u)).
$$

In a similar way,

$$
\begin{aligned}
\lim_{\varepsilon\to 0}\frac{\mathbf{II}}{\varepsilon}=&\ \partial_{x_n}\partial^{n-1}_{x_1\dots x_{n-1}}h_1(u,\varphi(u))+\partial_{\partial_{x_n}\varphi(u)}\partial^{n-1}_{x_1\dots x_{n-1}}h_1(u,\varphi(u))\\
&+\sum_{\substack{\mathbf{x}\in 2^{\{x_1,\dots,x_{n-1}\}}\\ \mathbf{x}\neq\emptyset}}\ \sum_{i=\max\{1,2-(n-1)+|\mathbf{x}|\}}^{|\mathbf{x}|}\ \sum_{\substack{\mathbf{p}\in P^i(\mathbf{x})\\ \mathbf{p}=(\mathbf{p}_1,\dots,\mathbf{p}_i)}}\partial_{x_n}\partial^{n-1}[\mathbf{x}^c,\mathbf{p}]h_1(u,\varphi(u))\\
&+\sum_{\substack{\mathbf{x}\in 2^{\{x_1,\dots,x_{n-1}\}}\\ \mathbf{x}\neq\emptyset}}\ \sum_{i=\max\{1,2-(n-1)+|\mathbf{x}|\}}^{|\mathbf{x}|}\ \sum_{\substack{\mathbf{p}\in P^i(\mathbf{p})\\ \mathbf{p}=(\mathbf{p}_1,\dots,\mathbf{p}_i)}}\Bigg(\partial_{\partial_{x_n}\varphi(u)}\partial^{n-1}[\mathbf{x}^c,\mathbf{p}]h_1(u,\varphi(u))\\
&+\sum_{j=1}^{i}\partial^{|\mathbf{x}^c|}_{\mathbf{x}^c}\partial_{\partial^{|\mathbf{p}_1|}_{\mathbf{p}_1}\varphi(u)}\cdots\partial_{\partial^{|\mathbf{p}_{j-1}|}_{\mathbf{p}_{j-1}}\varphi(u)}\partial_{\frac{\partial}{\partial x_n}\partial^{|\mathbf{p}_j|}_{\mathbf{p}_j}\varphi(u)}\partial_{\partial^{|\mathbf{p}_{j+1}|}_{\mathbf{p}_{j+1}}\varphi(u)}\cdots\partial_{\partial^{|\mathbf{p}_i|}_{\mathbf{p}_i}\varphi(u)}h_1(u,\varphi(u))\Bigg).
\end{aligned}
\tag{4.7}
$$

Notice that

$$
\begin{aligned}
&\sum_{\mathbf{x}\in 2^{\{x_1,\dots,x_{n-1}\}}\,\mathbf{x}\neq\emptyset}\ \sum_{i=\max\{1,2-(n-1)+|\mathbf{x}|\}}^{|\mathbf{x}|}\ \sum_{\substack{p_\pi\in P^i(\mathbf{x})\\ \mathbf{p}=(\mathbf{p}_1,\dots,\mathbf{p}_i)}}\partial_{x_n}\partial^{n-1}[\mathbf{x}^c,\mathbf{p}]h_1(u,\varphi(u))\\
=&\sum_{\substack{\mathbf{x}\in 2^{\{x_1,\dots,x_n\}}\\ \mathbf{x}\neq\emptyset\\ x_n\notin\mathbf{x}}}\ \sum_{i=\max\{1,2-n+|\mathbf{x}|\}}^{|\mathbf{x}|}\ \sum_{\substack{\mathbf{p}\in P^i(\mathbf{x})\\ \mathbf{p}=(\mathbf{p}_1,\dots,\mathbf{p}_i)}}\partial^n[\mathbf{x}^c,\mathbf{p}]h_1(u,\varphi(u))-\partial_{x_n}\partial_{\partial^{n-1}_{x_1\dots x_{n-1}}\varphi(u)}h_1(u,\varphi(u))
\end{aligned}
\tag{4.8}
$$

[5] The limits should be understood in the suitable spaces $Y_k$. For instance, when computing $\lim_{\varepsilon\to 0}\frac{\mathbf{I_1}}{\varepsilon}$, the object $\partial^{n-1}_{x_1\dots x_{n-1}}\varphi(u+\varepsilon x_n)$ should be considered in the space $Y_2$, which can be done thanks to the inductive hypothesis.

and

$$\sum_{\substack{\mathbf{x}\in 2^{\{x_1,\dots,x_{n-1}\}}\\ \mathbf{x}\neq\emptyset}} \sum_{i=\max\{1,2-(n-1)+|\mathbf{x}|\}}^{|\mathbf{x}|} \sum_{\substack{\mathbf{p}\in P^i(\mathbf{x})\\ \mathbf{p}=(\mathbf{p}_1,\dots,\mathbf{p}_i)}} \partial_{\partial_{x_n}\varphi(u)}\partial^{n-1}[\mathbf{x}^c,\mathbf{p}]h_1(u,\varphi(u))$$

$$= \sum_{\substack{\mathbf{x}\in 2^{\{x_1,\dots,x_n\}}\\ x_n\in\mathbf{x}\\ \mathbf{x}\neq\{x_n\}}} \sum_{i=\max\{1,2-n+|\mathbf{x}|\}}^{|\mathbf{x}|} \sum_{\substack{\mathbf{p}\in P^i(\mathbf{x})\\ \mathbf{p}=(\mathbf{p}_1,\dots,\mathbf{p}_i)\\ \{x_n\}\in\mathbf{p}}} \partial^n[\mathbf{x}^c,\mathbf{p}]h_1(u,\varphi(u)) \tag{4.9}$$

$$- \partial_{\partial_{x_n}\varphi(u)}\partial_{\partial^{n-1}_{x_1\dots x_{n-1}}\varphi(u)}h_1(u,\varphi(u))$$

and

$$\sum_{\substack{\mathbf{x}\in 2^{\{x_1,\dots,x_{n-1}\}}\\ \mathbf{x}\neq\emptyset}} \sum_{i=\max\{1,2-(n-1)+|\mathbf{x}|\}}^{|\mathbf{x}|} \sum_{\substack{\mathbf{p}\in P^i(\mathbf{x})\\ \mathbf{p}=(\mathbf{p}_1,\dots,\mathbf{p}_i)}} \sum_{j=1}^{i} L(\mathbf{p},j;u)$$

$$= \sum_{\substack{\mathbf{x}\in 2^{\{x_1,\dots,x_n\}}\\ x_n\in\mathbf{x}\\ \mathbf{x}\neq\{x_n\}}} \sum_{i=\max\{1,2-n+|\mathbf{x}|\}}^{|\mathbf{x}|} \sum_{\substack{\mathbf{p}\in P^i(\mathbf{x})\\ \mathbf{p}=(\mathbf{p}_1,\dots,\mathbf{p}_i)\\ \{x_n\}\notin\mathbf{p}}} \partial^n[\mathbf{x}^c,\mathbf{p}]h_1(u,\varphi(u)) \tag{4.10}$$

where

$$L(\mathbf{p},j;u) := \partial^{|\mathbf{x}^c|}_{\mathbf{x}^c}\partial^{|\mathbf{x}|}_{\partial^{|\mathbf{p}_1|}_{\mathbf{p}_1}\varphi(u)\dots\partial^{|\mathbf{p}_{j-1}|}_{\mathbf{p}_{j-1}\varphi(u)}\partial_{x_n}\partial^{|\mathbf{p}_j|}_{\mathbf{p}_j}\varphi(u)\partial^{|\mathbf{p}_{j+1}|}_{\mathbf{p}_{j+1}}\varphi(u)\dots\partial^{|\mathbf{p}_i|}_{\mathbf{p}_i}\varphi(u)}h_1(u,\varphi(u)).$$

By collecting (4.7), (4.8), (4.9), (4.10), we obtain

$$\lim_{\varepsilon\to 0}\frac{\mathbf{II}}{\varepsilon} = \partial_{\partial_{x_n}\varphi(u)}\partial^{n-1}_{x_1\dots x_{n-1}}h_1(u,\varphi(u)) + \partial^n_{x_1\dots x_n}h_1(u,\varphi(u)) - \partial_{\partial_{x_n}\varphi(u)}\partial_{\partial^{n-1}_{x_1\dots x_{n-1}}\varphi(u)}h_1(u,\varphi(u))$$
$$+ \sum_{\substack{\mathbf{x}\in 2^{\{x_1\dots x_n\}}\\ \mathbf{x}\neq\emptyset\\ \mathbf{x}\neq\{x_n\}}} \sum_{i=\max\{1,2-n+|\mathbf{x}|\}}^{|\mathbf{x}|} \sum_{\substack{\mathbf{p}\in P^i(\mathbf{x})\\ \mathbf{p}=(\mathbf{p}_1,\dots,\mathbf{p}_i)}} \partial^n[\mathbf{x}^c,\mathbf{p}]h_1(u,\varphi(u)) - \partial_{x_n}\partial_{\partial^{n-1}_{x_1\dots x_{n-1}}\varphi(u)}h_1(u,\varphi(u))$$
$$= \partial^n_{x_1\dots x_n}h_1(u,\varphi(u)) - \partial_{\partial_{x_n}\varphi(u)}\partial_{\partial^{n-1}_{x_1\dots x_{n-1}}\varphi(u)}h_1(u,\varphi(u))$$
$$+ \sum_{\substack{\mathbf{x}\in 2^{\{x_1,\dots,x_n\}}\\ \mathbf{x}\neq\emptyset}} \sum_{i=\max\{1,2-n+|\mathbf{x}|\}}^{|\mathbf{x}|} \sum_{\substack{\mathbf{p}\in P^i(\mathbf{x})\\ \mathbf{p}=(\mathbf{p}_1,\dots,\mathbf{p}_i)}} \partial^n[\mathbf{x}^c,\mathbf{p}]h_1(u,\varphi(u)) - \partial_{x_n}\partial_{\partial^{n-1}_{x_1\dots x_{n-1}}\varphi(u)}h_1(u,\varphi(u)).$$

Hence

$$\lim_{\varepsilon\to 0}\left(\frac{\mathbf{I_1}}{\varepsilon}+\frac{\mathbf{I_2}}{\varepsilon}+\frac{\mathbf{II}}{\varepsilon}\right) = \sum_{\substack{\mathbf{x}\in 2^{\{x_1,\dots,x_n\}}\\ \mathbf{x}\neq\emptyset}} \sum_{i=\max\{1,2-n+|\mathbf{x}|\}}^{|\mathbf{x}|} \sum_{\substack{\mathbf{p}\in P^i(\mathbf{x})\\ \mathbf{p}=(\mathbf{p}_1,\dots,\mathbf{p}_i)}} \partial^n[\mathbf{x}^c,\mathbf{p}]h_1(u,\varphi(u)) + \partial^n_{x_1\dots x_n}h_1(u,\varphi(u)),$$

and, by recalling (4.5), (4.6), we obtain

$$\lim_{\varepsilon\to 0}\left(I-\partial_{Y_1}h_1(u,\varphi(u))\right).\frac{\partial^{n-1}_{x_1\dots x_{n-1}}\varphi(u+\varepsilon x_n)-\partial^{n-1}_{x_1\dots x_{n-1}}\varphi(u)}{\varepsilon}$$
$$=\sum_{\substack{\mathbf{x}\in 2^{\{x_1,\dots,x_n\}}\\ \mathbf{x}\neq\emptyset}}\ \sum_{i=\max\{1,2-n+|\mathbf{x}|\}}^{|\mathbf{x}|}\ \sum_{\substack{\mathbf{p}\in P^i(\mathbf{x})\\ \mathbf{p}=(\mathbf{p}_1,\dots,\mathbf{p}_i)}}\partial^n[\mathbf{x}^c,\mathbf{p}]h_1(u,\varphi(u))+\partial^n_{x_1\dots x_n}h_1(u,\varphi(u)).$$

Finally, we can conclude the proof by recalling that $I-\partial_{Y_1}h_1(u,\varphi(u))$ is invertible with strongly continuous inverse. ■

## References


1. Cerrai, S.: Second-Order PDE's in Finite and Infinite Dimension. Springer (2001)
2. Cont, R., Fournié, D.-A.: Change of variable formulas for non-anticipative functionals on path space. J. Funct. Anal. **259**, 1043–1072 (2010)
3. Cont, R., Fournié, D.-A.: A functional extension of the Itô formula. C. R. Math. Acad. Sci. Paris, Ser. I **348**, 57–61 (2010)
4. Cont, R., Fournié, D.-A.: Functional Itô calculus and stochastic integral representation of martingales. Ann. Probab. **41**, 109–133 (2013)
5. Cosso, A., Di Girolami, C., Russo, F.: Calculus via regularizations in Banach spaces and Kolmogorov-type path-dependent equations. In: Probability on Algebraic and Geometric Structures, pp. 43–65. American Mathematical Society (2016)
6. Cosso, A., Russo, F.: A regularization approach to functional Itô calculus and strong-viscosity solutions to path-dependent PDEs. Preprint arXiv:1401.5034 (2014)
7. Da Prato, G., Zabczyck, J.: Stochastic Equations in Infinite Dimensions. Cambridge University Press (1992)
8. Da Prato, G., Zabczyck, J.: Second Order Partial Differential Equations in Hilbert Spaces. Cambridge University Press (2002)
9. Da Prato, G., Zabczyck, J.: Stochastic Equations in Infinite Dimensions, 2 edn. Cambridge University Press (2014)
10. Dupire, B.: Functional Itô Calculus. Bloomberg Portfolio Research Paper (2009)
11. Flett, T.M.: Differential Analysis. Cambridge University Press (1980)
12. Gawarecki, L., Mandrekar, V.: Stochastic Differential Equations in Infinite Dimensions. Springer (2011)
13. Granas, A., Dugundji, J.: Fixed Point Theory. Springer (2003)
14. Knoche, C., Frieler, K.: Solutions of stochastic differential equations in infinite dimensional Hilbert spaces and their dependence on initial data. Diplomarbeit. Universität Bielefeld, Fakultät für Mathematik (2001)
15. Mohammed, S.-E.A.: Stochastic Functional Differential Equations. Pitman (1984)
16. Rosestolato, M.: A note on stochastic Fubini's theorem and stochastic convolution. arXiv:1606.06340 (2016)

CPI Antony Rowe
Eastbourne, UK
September 03, 2019